BIOTECHNOLOGY IN AGRICULTURE SERIES

General Editor: Gabrielle J. Persley, Science Adviser, The Agricultural Research Group, Environmental and Sustainable Development, The World Bank, Washington DC, USA.

For a number of years, biotechnology has held out the prospect for major advances in agricultural production, but only recently have the results of this new revolution started to reach application in the field. The potential for further rapid developments is, however, immense.

The aim of this book series is to review advances and current knowledge in key areas of biotechnology as applied to crop and animal production, forestry and food science. Some titles focus on individual crop species, others on specific goals such as plant protection or animal health, with yet others addressing particular methodologies such as tissue culture, transformation or immunoassay. In some cases, relevant molecular and cell biology and genetics are also covered. issues of relevance to both industrialized and developing countries are addressed and social, economic and legal implications are also considered. Most titles are written for research workers in the biological sciences and agriculture, but some are also useful as textbooks for senior-level students in these disciplines.

BIOTECHNOLOGY IN AGRICULTURE SERIES

Titles Available:

1: Beyond Mendel's Garden: Biotechnology in the Service of World Agriculture *G.J. Persley*
2: Agricultural Biotechnology: Opportunities for International Development *Edited by G.J. Persley*
3: The Molecular and Cellular Biology of the Potato *Edited by M.E. Vayda and W.D. Park*
4: Advanced Methods in Plant Breeding and Biotechnology *Edited by D.R. Murray*
5: Barley: Genetics, Biochemistry, Molecular Biology and Biotechnology *Edited by P.R. Shewry*
6: Rice Biotechnology *Edited by G.S. Khush and G.H. Toenniessen*
7: Plant Genetic Manipulation for Crop Protection *Edited by A. Gatehouse, V. Hilder and D. Boulter*
8: Biotechnology of Perennial Fruit Crops *Edited by F.A. Hammerschlag and R.E. Litz*
9: Bioconversion of Forest and Agricultural Plant Residues *Edited by J.N. Saddler*
10: Peas: Genetics, Molecular Biology and Biotechnology *Edited by R. Casey and D.R. Davies*
11: Laboratory Production of Cattle Embryos *I. Gordon*
12: The Molecular and Cellular Biology of the Potato, 2nd edn *Edited by W.R. Belknap, M.E. Vayda and W.D. Park*
13: New Diagnostics in Crop Sciences *Edited by J.H. Skerritt and R. Appels*
14: Soybean: Genetics, Molecular Biology and Biotechnology *Edited by D.P.S. Verma and R.C. Shoemaker*
15: Biotechnology and Integrated Pest Management *Edited by G.J. Persley*
16: Biotechnology of Ornamental Plants *Edited by R.L. Geneve, J.E. Preece and S.A. Merkle*
17: Biotechnology and the Improvement of Forage Legumes *Edited by B.D. McKersie and D.C.W. Brown*

Biotechnology and Integrated Pest Management

Edited by

Gabrielle J. Persley

The World Bank
Washington DC
USA

CAB INTERNATIONAL

CAB INTERNATIONAL
Wallingford
Oxon OX10 8DE
UK

Tel: +44 (0)1491 832111
Fax: +44 (0)1491 833508
E-mail: cabi@cabi.org
Telex: 847964 (COMAGG G)

A catalogue record for this book is available from the British Library.

ISBN 0 85198 930 6

Typeset in 10/12 pt Plantin by York House Typographic Ltd, London
Printed and bound at the University Press, Cambridge

Contents

Section Two: Case Studies of IPM Implementation and of Using Biocontrol Agents in IPM Systems

List of Contributors

Roger Beachy, Department of Molecular Biology, Research Institute of Scripps Clinic, 10666 North Torrey Pines Road, La Jolla, CA 95616, USA

Ivan W. Buddenhagen, Department of Agronomy and Range Management, University of California, Davis, CA 95616, USA

Christopher F. Curtis, Department of Medical Parasitology, London School of Hygiene and Tropical Medicine, University of London, London WC1E 7HT, UK

David Dall, Division of Entomology, CSIRO, GPO Box 1700, Canberra, ACT 2601, Australia

Peter Dart, Department of Agriculture, University of Queensland, St Lucia, Qld 4072, Australia

Elizabeth Evans, Agricultural Sciences, The Rockefeller Foundation, 1133 Avenue of the Americas, New York, NY 10036, USA

David A. Fischhoff, Monsanto Company, 700 Chesterfield Parkway North, St Louis, MO 63198, USA

G.P. Fitt, Division of Entomology, CSIRO, PO Box 59, Narrabri, NSW 2390, Australia

Fred Gould, Department of Entomology, North Carolina State University, Raleigh, NC 27695, USA

Hans R. Herren, International Centre of Insect Physiology and Ecology, PO Box 30772, Nairobi, Kenya

Marjorie A. Hoy, Department of Entomology and Nematology, University of Florida, PO Box 11620, Gainsville, FL 32611-0620, USA

Michael E. Irwin, Department of Agricultural Entomology, University of

Illinois, 172 Natural Resources Building, MC-652, Urbana, IL 61801, USA

Richard A. Jefferson, CAMBIA, c/o Division of Entomology, CSIRO, GPO Box 1700, Canberra, ACT 2601, Australia

Nandini V. Katre, 6107 Jordan Ave, El Cerrito, CA 94530, USA

Peter E. Kenmore, Food and Agriculture Organization of the United Nations, PO Box 1864, Manila, The Philippines

Lim Guan Soon, CAB International, Asian Regional Office, PO Box 11872, 50760 Kuala Lumpur, Malaysia (Present address: Food and Agriculture Organization of the United Nations, PO Box 1864, Manila, The Philippines)

D.J. Llewellyn, Division of Plant Industry, CSIRO, PO Box 1600, Canberra, ACT 2601, Australia

Pamela Marrone, Entotech, Inc., 1497 Drew Avenue, Davis, CA 95616-4880, USA (Present address: AgraQuest, Inc., 1105 Kennedy Place, Suite 4, Davis, CA 95616, USA)

Ben J. Miflin, CIBA-Geigy Ltd, Rosenthal AG 5-1, CH-4002 Basle, Switzerland (Present address: Institute of Arable Crops Research, Rothamsted, Harpenden, Herts AL5 2JQ, UK)

Flavio Moscardi, EMBRAPA/CNPSo, Caixa Postal 1061, 86001 Londrina, Parana, Brazil

Lowell R. Nault, Ohio Agricultural Research and Development Center, Ohio State University, Woodster, OH 44691, USA

Rebecca J. Nelson, Division of Plant Pathology, International Rice Research Institute, PO Box 933, 1099 Manila, Philippines

W. James Peacock, Division of Plant Industry, CSIRO, GPO Box 1600, Canberra, ACT 2601, Australia

Gabrielle J. Persley, ESDAR, The World Bank, 1818 H Street NW, Washington, DC 20433, USA

Veronica Rodrigues, Molecular Biology Unit, Tata Institute of Fundamental Research, Hami Bhabha Road, Colaba, Bombay 400 005, India

Richard T. Roush, Department of Entomology, Cornell University, 2119 Comstock Hall, Ithaca, NY 14853, USA

D.R. Sosa-Gómez, EMBRAPA/CNPSo, Caixa Postal 1061, 86001 Londrina, Parana, Brazil

K. VijayRaghavan, Molecular Biology Unit, Tata Institute of Fundamental Research, Hami Bhabha Road, Colaba, Bombay 400 005, India

Gary Toenniessen, Agricultural Sciences Division, The Rockefeller Foundation, 1133 Ave of the Americas, New York, NY 10036, USA

Jeff Waage, International Institute of Biological Control, Silwood Park, Buckhurst Road, Ascot, Berkshire SL5 7TA, UK

Mark E. Whalon, Department of Entomology, Pesticide Research Center, Michigan State University, East Lansing, MI 48824-1115, USA
Max J. Whitten, Division of Entomology, CSIRO, GPO Box 1700, Canberra ACT 2601, Australia

Foreword

Over the next thirty years food production in developing countries must more than double, with most of the increase coming from land that is already in production. Reducing crop losses caused by pests and pathogens can and should make an important contribution toward the necessary increases in yield and productivity. Heavy reliance on chemical pesticides, however, is not a viable strategy. Pesticides, at best, provide ephemeral benefits, often with adverse side effects, and in some cases actually worsen farmers' overall pest problems.

New biologically based pest management strategies are needed which, in addition to being ecologically sound, are reliable, economical, and practical. Such integrated pest management (IPM) is based on knowledge of the natural ecosystem and of farmers' overall production systems, with crop losses reduced through careful integration of a number of pest monitoring and control techniques, which maximize use of biological and cultural components.

Advances in biotechnology can help to make IPM strategies more robust and more sustainable. They can help generate a more in-depth understanding of local and regional ecosystems, improve the sensitivity and reliability of pest monitoring, and strengthen biologically based control measures including host plant resistance and biocontrol agents. Unfortunately, however, many within the IPM community seem reluctant to embrace the idea that biotechnology has much to offer, and many within the biotechnology community seem unaware of, or uninterested in, the needs of IPM programmes. Greater dialogue, constructive interaction and collaborative research between these two groups needs to be encouraged and facilitated. This book presents the results of one such effort organized by The World Bank and The Rockefeller Foundation, with additional financial support from the United Nations Development Programme (UNDP) and the Australian Centre for International Agricultural Research (ACIAR).

This publication is significant in that it is the first major volume on the role biotechnology can play in IPM. The chapters have been contributed by leading scientists worldwide and the result is a valuable compendium on this difficult and challenging topic – increasing worldwide food production while reducing the use of harmful agricultural chemicals.

Robert W. Herdt
Director, Agricultural Sciences
Rockefeller Foundation

Preface

Excessive pesticide use is a threat to human health and the environment. Proponents of integrated pest management (IPM) advocate the use of IPM systems to reduce or eliminate the use of chemical pesticides in agriculture. Proponents of biotechnology believe that the use of novel products, such as transgenic plants with insect resistance, will reduce the need for chemical pesticides. However, the use of such novel products within IPM systems is not welcomed by all advocates of IPM, who see potential risks in their deployment.

This dichotomy of views led the Rockefeller Foundation to host a conference on Biotechnology and Integrated Pest Management, at its Bellagio Center in Italy, from 4 to 8 October 1993. Twenty-five scientists and policy-makers represented a spectrum of views on the benefits and risks in the use of biotechnology in IPM systems. Participants included IPM practitioners from developing countries, international development agencies and non-government organizations, as well as scientists from public-sector research institutes, biotechnology companies and multinational companies. They represented a wide range of views as to the opportunities and threats offered by the applications of new biotechnologies in pest and disease control. Thus lively discussion was ensured. Their purpose was to assess the likelihood of novel products developed by the applications of new biotechnologies being usefully incorporated into IPM systems. The specific types of new biotechnologies which were responsive to farmers' perceived needs and which would be most useful to facilitate the wider use of IPM strategies were also discussed.

Integrated pest management is a system which utilizes all suitable methods, in a compatible manner, to maintain pest populations below levels causing economic injury. Integrated pest management programmes involve farmers and pest-control technicians in continuing and careful assessments of the populations of pests and their natural enemies, in order to decide if and when to intervene with control measures that range from the use of biological control agents to resistant varieties and the carefully targeted use of chemical pesticides. The range of IPM options is defined by agroecological, socioeconomic and institutional factors. The key to successful implementation of IPM is the development by farmers and pest-control technicians of a practical understanding of the ecology of their crops, their pests and their natural enemies, and the translation of this knowledge into decision tools and practical control tactics to solve particular pest problems.

This volume discusses the potential benefits of and constraints to the applications of biotechnology in IPM systems, especially in developing countries, and the related policy issues confronting decision-makers in national agricultural research systems and international development agencies. It contains the revised text of 22 papers commissioned for the study. These cover the needs and opportunities for linking biotechnology and IPM in developing countries; case-studies on the applications of IPM on rice in Asia, soybeans in Brazil and cassava in Africa; an assessment of the use of modern biotechnology to develop new biocontrol agents; an analysis of the linkages between biotechnology and plant breeding, in relation to pest and disease resistance; case-studies on the use of transgenic plants in IPM systems; alternative strategies as to how these transgenes may be best deployed in relation to insect resistance; the development of new diagnostic systems; new trends in virus and vector control, including the control of vectors of human and animal diseases; industry views as to the probable commercial applications of biotechnology; and the policy issues affecting the investments by international development agencies in biotechnology and IPM.

The Bellagio conference led on to an interagency study by three United Nations agencies and the World Bank, on the future needs of IPM in developing countries. This has resulted in a decision in 1995 to establish an international IPM initiative cosponsored by the United Nations Food and Agriculture Organization (FAO), the United Nations Development Programme (UNDP), the United Nations Environment Programme (UNEP) and the World Bank. The purpose of this IPM facility is to foster the wider implementation of farmer-led, participatory approaches to IPM, in a wide range of commodities and countries. In fostering effective implementation of IPM, the facility will also encourage the assessment of new applications of biotechnology in IPM systems, for some of the novel products described in this volume.

In 1970, Price-Jones described IPM as a 'mixture of idealism, evangelism, pursuit of fashion, fund-raising and even empire-building. The movement has indeed acquired the impetus and character of a religious revival'. Some proponents of IPM would say the same of biotechnology today.

The purpose of the Bellagio Conference and this volume is to establish a dialogue amongst concerned parties, with disparate views as to likely threats and opportunities emerging in the use of the tools of modern biotechnology, to contribute to more sustainable pest management and greater food security for many people in developing countries.

Gabrielle J. Persley

Acknowledgements

The valuable contribution of all authors and participants in the Bellagio conference is gratefully acknowledged. The participants in Bellagio provided a spirited discussion of issues, and I have attempted to reflect their views in this volume.

Gary Toenniessen was the Rockefeller Foundation's representative collaborating on the study, and his many valuable contributions and wise advice are gratefully acknowledged. The assistance of Elizabeth Evans and of, Warren Weaver, a Fellow of the Rockefeller Foundation, in the organization of the Bellagio conference is also gratefully acknowledged. Peter Dart, Jack Doyle, Peter Kenmore and Jeff Waage reviewed the drafts of various chapters and their many helpful suggestions are much appreciated.

Skilful administrative and secretarial assistance was provided by Pamela George of the World Bank. Reg MacIntyre undertook the technical editing of the manuscripts and his contributions are warmly acknowledged. The continued encouragement and patience of Tim Hardwick, Pippa Smart and Emma Critchley at CAB INTERNATIONAL are much appreciated.

Several members of the sponsoring organizations provided encouragement for the study: Mr Michel Petit and Mr Douglas Forno, the World Bank; Dr Bob Herdt, Rockefeller Foundation; Drs Al App and Nyle Brady, United Nations Development Programme (UNDP); and Dr Paul Ferrar, Australian Centre for International Agricultural Research (ACIAR). Their support and guidance are warmly acknowledged.

Gabrielle J. Persley

Needs and Opportunities 1

Max J. Whitten, Richard A. Jefferson and
David Dall

Introduction

This overview of the needs and opportunities for linking integrated pest management (IPM) and biotechnology has been developed with the concerns of the past and the opportunities of the future in mind. We first cover the period of pesticide dependency, followed by the period when IPM became popular, to the present when we are grappling to implement some of the promising new technologies. The review turns frequently to the interface between science and society, and the need for scientists to understand how the technologies they develop relate to the world-view of the users. We need to let the user's world-view shape the technologies. We argue the necessity for a paradigm shift in the way we identify the role of science and scientists before we can deliver the benefits we seem so confidently to believe will inexorably flow from our endeavours. We argue that intelligent and informed management and system changes are the most effective tools in sustainable crop protection, and that molecular biology and biotechnology should be focused here.

Dependency on Synthetic Pesticides

The availability of broad-spectrum and persistent pesticides since the Second World War has been a major contributor to global increases in food and fibre production. Their availability has also been a key element in controlling disease where both invertebrate and vertebrate vectors have enabled trans-

 Biotechnology and Integrated Pest Management (ed. G.J. Persley)

mission of bacterial, viral or protozoan pathogens. During the 1950s the view began to emerge that synthetic pesticides could represent a satisfactory and permanent solution for disease and pest control.

Advances in plant physiology and biochemistry in the 1950s allowed the development of a large range of chemicals that served as broad-spectrum or specific herbicides (National Research Council, 1989). By the 1970s, herbicides had eclipsed insecticides and fungicides around the globe, both in terms of numbers of chemicals available and tonnage applied to crops and pastures (see Fig. 1.1 for patterns in the USA). The range of herbicides available, together with new classes of insecticide cyclodienes, organophosphates, carbamates and pyrethroids, which supplanted dichlorodiphenyltrichloroethane (DDT), suggested to farmers and the agrochemical industry that the synthetic pesticide paradigm would enjoy secure tenure.

In the case of bacterial, fungal and protozoan infections of humans and their livestock, another suite of man made chemicals, the antibiotics, became central to disease control (Fig. 1.2). This dependency on, agrochem-

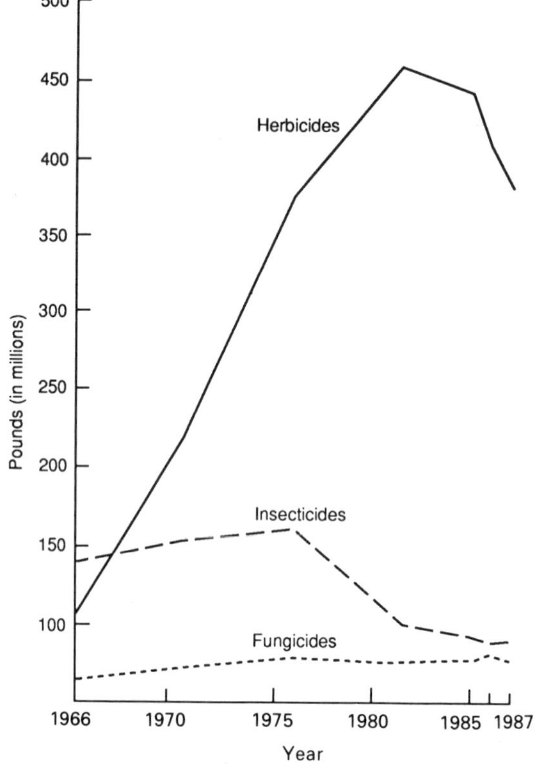

Fig. 1.1. Herbicide, insecticide and fungicide use estimates in the USA (after National Research Council, 1989).

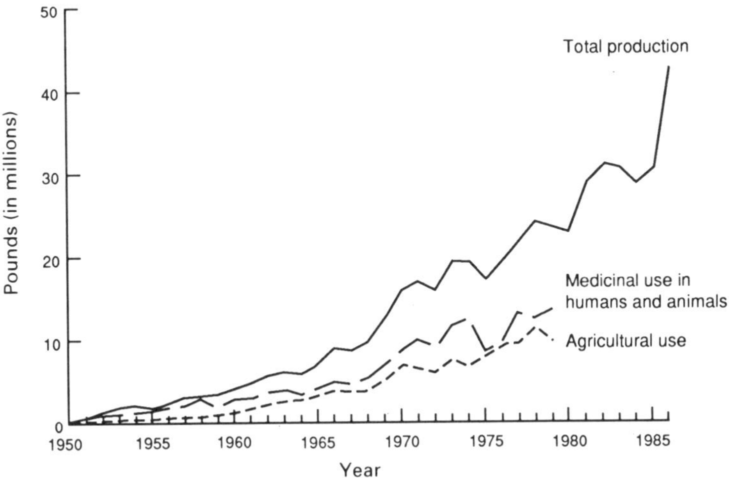

Fig. 1.2. US antibiotic production and use in animal feed (after National Research Council, 1989).

icals, which typified the period 1940–80, has been termed 'chemism' by Zadoks (1993). According to Zadoks (1993), 'The change to the chemotherapeutical paradigm is dated 1940, though in reality it took over a century. However, penicillin did it! The change affected all of us so deeply that we are hardly aware of it.' The chemotherapeutical philosophy pervades marketing strategies worldwide. In Asia, pesticides are described by the agrochemical industry as plant medicines and they sport such macho trade names as matador, ambush, rip-cord, drag-net, kill-top, round-up and target.

The chemism philosophy proved to be highly compatible with another agricultural development of the 1950s, the green revolution. Indeed, the high-input/high-output agricultural practices of the green revolution, featuring dwarf varieties of rice and wheat, saw pesticides take their place alongside fertilizers as essential elements of modern agriculture in both industrial and developing countries.

This relationship is well illustrated in Indonesia, where the impressive increase in rice production depended on intensive use of pesticides to accompany the dwarf rice varieties released in the 1970s (Fig. 1.3). In 1973, the Indonesian government implemented a fertilizer subsidy policy to encourage the widespread use of these new varieties. However, a subsidy provision for pesticide production was a largely unnoticed and hidden element of the fertilizer subsidy. In 1973 the annual value of the pesticide subsidy was around $US7 million. It peaked at $US160 million in 1981 and plateaued around $US140 million by 1985 (Fig. 1.4) (FAO, 1991). It had reached the stage in Indonesia where 'funds to purchase seeds and fertiliser could not be borrowed without assurance that pesticides would be purchased and applied' (Useem *et al.*, 1992).

What amounted to a pesticide economy had come to characterize the green revolution for rice production in Indonesia until 1985. A similar phenomenon existed in other Asian countries, including the Philippines, Thailand and Japan, although the details varied (FAO, 1991). In Africa, crop production systems built around high-yielding varieties are also found, such as cotton production in Sudan, where, in some years, a crop with a market value of $US95 million was treated with chemicals worth $US90 million on the international market.

In Western Europe (Zadoks, 1993) and in the USA (National Research Council, 1989) pesticide dependency established itself as a hallmark of agricultural production between 1950 and 1990. In other countries, such as China and the former USSR, a pesticide economy emerged, where enormous amounts of pesticides were applied to field crops, driven either by a command economy or because a treadmill between pest and pesticide had

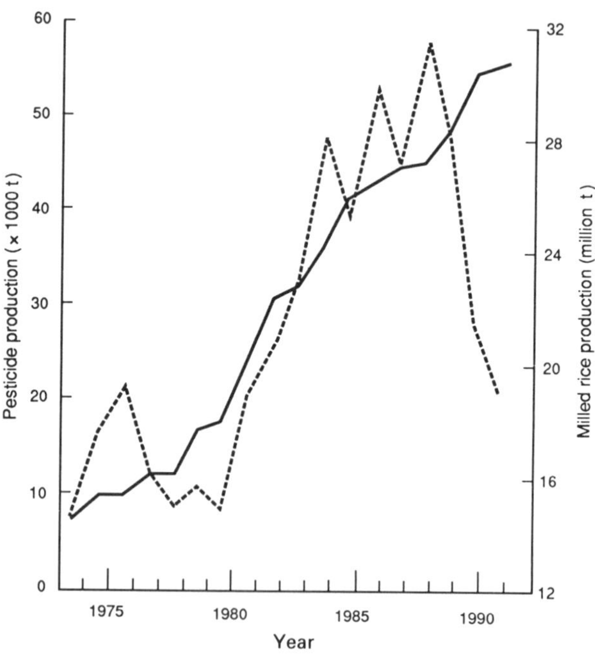

Fig. 1.3. Pesticide usage and rice production, Indonesia (after FAO, 1991). __, Rice; ----, pesticides.

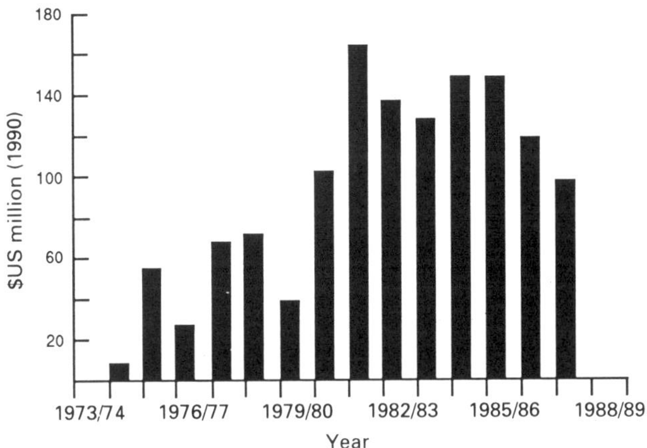

Fig. 1.4. Pesticide subsidies for rice in Indonesia (after FAO, 1991).

commenced. The 'chemism' paradigm was entrenched globally.

Press reports on agricultural technology in former Eastern bloc countries, such as Poland and East Germany, have revealed a situation possibly far worse than in the European Union (EU) or other industrial regions of the world. There, pollution and lack of sustainable production systems are not even partly offset by temporarily enhanced efficiency in production, transport or storage.

More reliable information was provided on 'socialist' agriculture during the Food and Agriculture Organization (FAO)/United Nations Development Programme (UNDP)/USSR workshop in Kishinev, Moldavia, USSR, on IPM in June 1990. This workshop, attended by members of the FAO panel of experts on IPM, compared experiences with Soviet scientists from a range of research institutions, hosted by the then All-Union Institute of Biological Methods of Plant Protection (AUIBMPP) in Kishinev. According to Filippov (1990), during 1989 some 250,000–300,000 t of more than 300 pesticides were applied over some 170 million ha in the USSR. This accounted for 5.3% of pesticide usage worldwide. Some 2 billion roubles were spent on pesticide production and application during 1989, adding about 8 billion roubles in value to crops, including grain (17–20 million t), potatoes (8 million t), sugar beet (19 million t), raw cotton (1.8 million t) and fruit and vegetables (6.5 million t). The 4 : 1 benefit-to-cost ratio is remarkably similar to estimates in industrial countries (Morley *et al.*, 1989). However, Filippov (1990) also acknowledged that the cost/benefit analysis ignored the serious negative effects: environmental pollution and the destruction of natural ecosystems, pollinators, natural enemies, wild animals and birds.

According to Novozhilov (1990), 40 species of arthropods and fungi were registered as resistant to one or more pesticides and, in 1989, 5.7% of foodstuffs were contaminated with pesticides, reportedly a threefold decrease since 1985.

The record of control of human and livestock disease vectors on the world scene was similarly characterized during these 40 years by heavy dependence on broad-spectrum and persistent pesticides. Malaria, filariasis, yellow fever, dengue and river blindness were just some of the vector-borne diseases targeted by the World Health Organization (WHO) for global eradication or control programmes in the 1950s. Broad-spectrum pesticides were expected to be the principal weapon.

The Emergence of Environmentalism and IPM

Widespread resistance to synthetic pesticides, inability to find replacement chemicals at the same rate and unacceptable negative consequences, especially environmental pollution and damage to non-target organisms, have since necessitated a re-evaluation of chemical pesticides as the final and comprehensive solution to pest and disease management (Zadoks, 1993). Zadoks (1993) arbitrarily identifies 1990 as the date for a shift from the paradigm of chemism to that of environmentalism.

As with most paradigms, it is easier to characterize expired or alien paradigms than ones in which we are currently immersed. Consequently, we can readily recognize the diagnostic elements of chemism – a period where chemical inputs were increasingly cheap and 'liberally applied as a "sleeping pill," to buy a good night's rest' (Zadoks, 1989), supported by agricultural scientists and sometimes as a part of government policy. It is sometimes described as the 'magic-bullet' syndrome.

Environmentalism, in contrast, is a philosophy that recognizes the integration of the production system with its resource base, including soil, water, air, soil biota and biodiversity. It addresses the renewability, durability or persistence of these resources, as well as a concern over the productivity of these systems for generations ahead. It recognizes our generation as being just one in a succession, and that the resources we inherited are also those we should pass on to later generations. The term intergenerational equity is sometimes used to tag this concept. Environmentalism can, in some forms, recognize the valuable role of chemical inputs, such as fertilizers and pesticides, but it also recognizes the ecological, social and economic consequences of chemical pesticide usage as well. It goes beyond the question of supply and price of chemicals. It recognizes that technical solutions must be embedded in the social and political fabric of communities if they are to serve the short- and long-term interests of those communities (Waibel, 1990). It is not surprising, then, to see a rise in prominence of resource economists and environmental and social scientists as the era of environmentalism unfolds.

It appears that the People's Republic of China was the first major country to address the problem of pesticide dependency at a central government level (FAO, 1991). China was sufficiently concerned about its so-called '3R problem' – resistance, resurgence and residues – for it to develop and adopt as central government policy, by 1975, the philosophy of IPM. The concept and the term were in fact developed and promoted elsewhere, especially in the USA and internationally by FAO. It was widely elaborated and implemented by entomologists worldwide in the 1970s (Frisbie *et al.*, 1988).

In 1992, FAO's panel of experts on IPM in agriculture reconfirmed its initial 1967 definition of IPM, which states:

> For the purposes of this Panel, Integrated Control is defined as a pest management system that, in the context of the associated environment and the population dynamics of the pest species, utilises all suitable techniques and methods in as compatible a manner as possible and maintains the pest populations at levels below those causing economic injury. In its restricted sense it refers to the management of single pest species on specific crops or in particular places. In a more general sense, it applies to the co-ordinated management of all pest populations in the agricultural or forest environment. It is not simply the juxtaposition or superimposition of two control techniques (such as chemical and biological controls) but the integration of all suitable management techniques with the natural regulating and limiting elements of the environment.

This definition falls short of encompassing a systems awareness in which the 'pest' is often a created entity and in which achieving balance is the key to management. Before we outline the salient features of IPM, we indicate some of the factors which undermined the pesticide paradigm and created the conditions for the emergence of the IPM philosophy. Firstly, resistance necessitated the pursuit of new chemicals 'immune' to the widespread resistance genes. New classes of chemicals have been increasingly difficult to identify and develop and more costly to register and commercialize. The prospect of a shorter commercial life in the interval between registration and expiration of patent term and the requirement for increased specificity have together reduced the investment attractiveness to chemical companies. We might recognize the possibilities emerging with modern pesticides (Moffat, 1993), but their commercial viability will always prove a challenge because of their specificity. Secondly, the more comprehensive economic analysis of high-input/high-output agriculture, with its negative consequences (especially pollution of soils, atmosphere and groundwater), also with its covert or indirect subsidies and the storage and dumping of subsidized commodities on the international market, has made many countries realize that chemical pest control is less attractive economically than originally envisaged (Waibel, 1990). By the same arguments, biorationals (biocontrol agents), although more costly to produce and apply, are probably cheaper in the long run if their fewer negative consequences are recognized. Policy changes by governments may be necessary if biorationals are to be given a fair run.

Dutch farmers, for example, are now moving towards low-input/low-output agriculture with less dependence on pesticides and greater reliance on IPM (Zadoks, 1993). The Netherlands experience and its current policy to reduce pesticide dependency by the year 2000 have counterparts in other countries. In the USA, use of pesticides in cotton production has dropped over 50% in the past 10 years (Fig. 1.5). Again, environmental and economic reasons were major elements in the shift to IPM (Hearn and Fitt, 1992).

There can be no doubt that the high-input/high-output agriculture of the 1980s increased net yield, but it also brought with it increased variation in yield, both spatially and temporally. This uncertainty factor, the direct result of periodic pest and disease outbreaks, required contingency planning and the stockpiling of commodity reserves, since demand through population increases and growing per capita consumption was rising in step with production increases. For example, in assessing the economic losses due to misuse of pesticides for control of brown planthopper in rice production in Indonesia, the 'uncertainty of supply' was a significant factor (FAO, 1991).

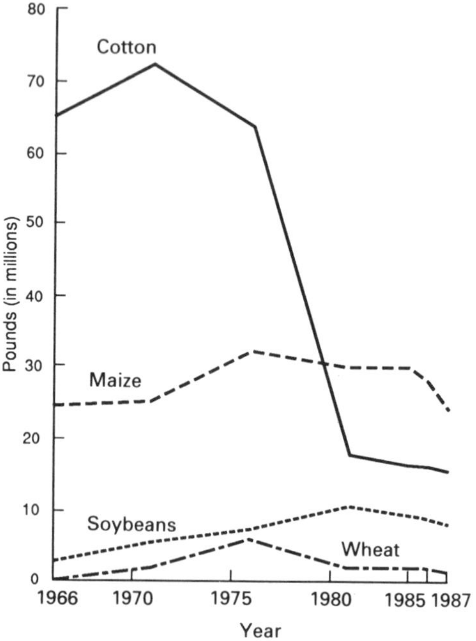

Fig. 1.5. Insecticide use estimates on maize, cotton, soybeans and wheat in the USA (after National Research Council, 1989).

Yield variation is only a symptom of a far more serious problem – a probable major collapse of the production system because non-renewable resources are depleted or because environmental pollution and damage to health simply become intolerable. Rice production was heading in that direction in China by 1970 and in Indonesia by 1986 (FAO, 1991). Similarly, agricultural production in general was taking the same course in The Netherlands (Zadoks, 1993). When pest resistance, supply of replacement chemicals and the consequences (social, economic and environmental) are all accepted as real and permanent problems of agriculture in the late 20th century, it becomes obvious that we have adopted production systems that are not sustainable, ecologically, socially or economically, in the long run.

This has been the crisis that has spawned an obvious, but novel, term, which had not been necessary in the vocabulary of agricultural production before the late 20th century. The term is 'sustainability'. It was not used explicitly by President Soeharto of Indonesia when he issued presidential decree INPRES3/86 in 1986, which banned 57 of 61 rice pesticides, adopted IPM as national policy and declared a phasing out of the pesticide

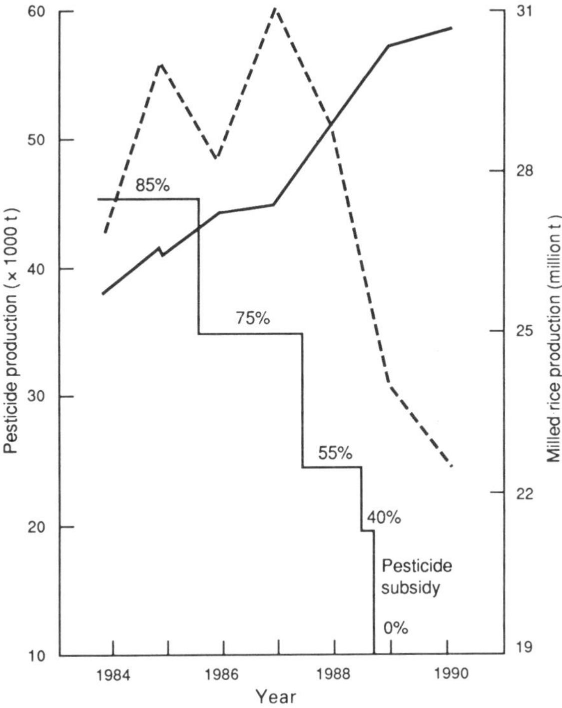

Fig. 1.6. Rice pesticides in Indonesia (after FAO, 1991). __, Rice; ----, pesticides.

subsidy (Figs 1.4 and 1.6). However, he recognized that the green revolution, which had increased rice yields threefold between 1973 and 1983 (Fig. 1.3), was no longer economically or environmentally sustainable. The soundness of the presidential decree is illustrated by the fact that rice production in Indonesia has continued to rise despite a 60% reduction in the amount of pesticide applied by farmers (Figs 1.3 and 1.6). Policy and its implementation had succeeded in uncoupling the green revolution and pesticide dependency.

The Chinese government had reached a similar conclusion in 1975 with its national system of rice production. The Netherlands government also decided in 1989 that it was time for action. In accepting the Multi-Year Crop Protection Plan, the Dutch set the target of reducing pesticide usage by 50% before the year 2000 (Langeweg, 1989). The Netherlands was moving socially and politically into an era that Zadoks (1993) has termed 'environmentism'. In accepting the feasibility of these targets, despite opposition from its national agrochemical industry, the Dutch government was cognizant of what had been achieved in rice production in Indonesia and cotton production in the USA (Peter Kenmore, personal communication).

The Rabbinge (1991) report outlines some remarkable goals for agricultural production in Europe. Its conclusions, as summarized by Zadoks (1993), are that

> The European Community (EC) today exploits some 130 Mha but it could do with only 40 to 50 Mha, a reduction of over 60%. The EC uses over 400 Mkg of pesticide, but 40 to 60 Mkg would suffice – a reduction of over 80%. The EC employs a labour force of 10 million person years but it could get by with 2 to 5, a reduction of over 50%. Even so, the EC would be self-supporting in food, and it would no longer disturb the international markets . . . and keep farmers in developing countries bound in poverty (De Hoogh, 1987a, b).

Intensive agriculture, dependent on high inputs of agrochemicals applied to high-yielding cultivars, has often scored low on yield stability and lower still on sustainability as measured in decades, let alone on any longer time-scale. Although the details might differ for different cropping systems, under various political regimes from socialism to capitalism, and in developing or industrial countries, current agricultural practices are inadequate to feed and clothe the world's rapidly growing population. Without a fundamental change in our approach, the outlook is bleak. Similarly, despite the promise of the early 1950s that pesticides would provide the final solution to the major vector-borne diseases of the world, many of these programmes have also been abandoned, and diseases such as malaria are on the increase (Desowitz, 1993).

Role of IPM in Plant and Animal Protection

As we emerge from the chemical dependency era, we have an immense problem to feed and clothe humans, to feed livestock and to keep humans and livestock healthy. Pesticides have contributed to a temporary increase in the carrying capacity of the globe, and a sceptic might observe that, in absolute terms, we have more hungry mouths and diseased bodies to contend with now than existed in the 1940s, and that the remainder who are better fed and healthier face an uncertain future because of the major deficiency of late 20th-century agriculture – non-sustainability of production systems. We also face a future with unprecedented losses of biodiversity.

What role can IPM play in delaying or reversing the process of non-sustainability? Does IPM offer a solution to the problems of food shortage, uncertain annual yield and the likelihood of declining supplies in the longer term? Does biotechnology significantly add to the options for higher and more stable yields, and with reduced negative consequences? What are the opportunities for IPM and biotechnology to provide technical options to social problems, namely adequate food and health for an expanding human population, with an acceptable impact on biodiversity and the environment?

Although many of the elements and philosophy of IPM have their genesis in times prior to 'chemism', these were largely lost in the misguided euphoria of the chemical era. Pests, pathogens and weeds obviously predated the pesticide era but they were dealt with by a variety of practices, such as plant and animal breeding, biological control, physical containment and an array of cultural and husbandry practices. Pesticides certainly diminished the need to retain some of these options and they tended to recede in importance. The resurgence of primary pests, such as the brown planthopper, diamondback moth, codling moth and *Helicoverpa armigera*, to their former pest status, despite continued widespread use of pesticides, caused a major rethink of pest-management philosophy. It was especially the emergence of secondary pests, such as phytophagous mites, aphids and mirids, which had previously been economically unimportant that drew attention to the biotic causes of pest outbreaks. Previously unrecognized faunas, such as the large number of species of predatory mite, had gone largely unnoticed until the emergence of secondary pests like spider mites (McMurtry, 1983).

It was the desire to combine the perceived benefits of continued pesticide usage and yet revitalize the benefits of biological control, whether inoculative releases of exotic natural enemies or the resurgence of local, depressed natural enemies, that stimulated the development of IPM as a specific philosophy. Integrated pest management owes its genesis largely to the need to resolve the conflict between two specific approaches to pest management – pesticides and biological control. A colourful illustration of the polarization between ecologists and the agrochemical movement is captured in the emotive text of Robert Van den Bosch's (1978) *The Pesticide Conspiracy*.

Its seems likely that some of the reductions in pesticide usage attributed to early use of IPM resulted, not so much from integrating a suite of approaches, but rather from recognizing situations where there was an excessively high and counterproductive usage. This seems to be the case for cotton production in Texas, where pest problems had never been major and where pesticide usage was more the fashion than the necessity. Pesticide usage for cotton pests elsewhere in the USA remains relatively high because the pests are more complex and severe and pesticides have represented a useful option for growers (Hearn and Fitt, 1992).

Integrated pest management has subsequently broadened into a concept which recognizes the costly limitations of the promotion of specific technologies, such as pesticides, plant breeding or biological control, to the exclusion of other options. Rather, IPM philosophy now acknowledges that all methods of control, including physical, chemical, biological and cultural strategies and plant and animal breeding, should be recognized, exploited and integrated if they can conceivably have a role in cost-effective and lasting pest management. The IPM approach cannot be faulted if it claims no more than recognizing the value and place of all control measures and their sensible integration into economically and ecologically sustainable pest management.

We have avoided referring to IPM as a technology. Peter Kenmore (personal communication) at a Global IPM Meeting in Bangkok, September 1993, described IPM in the following terms:

> IPM is not a technology. It is a problem-solving process where farmers learn from their own experiences and carry out their own research to create solutions to their crop management problems. These problems are perceived and defined by farmers within their local ecosystems, so research and extension put farmers first. The fundamental role of farmers as the owners and implementors of IPM in this process knits farmers, extension agents and research workers into IPM implementing networks.

It will be ironical if farmers in the developing world become experts at managing their crops through committed adoption of IPM, while in the industrialized world the linear model of technology transfer remains the norm.

With the recognition of two or more options for pest management, the rules for choosing between these, in terms of economics, efficacy, consequences and durability, need to be established. Cultural practices, then biological control and finally plant or animal breeding are often preferred in field situations, while physical containment or controlled atmospheres usually emerge as preferred options in commodity storage.

Prominent non-government organizations (NGOs), such as Pesticide Action Network and the International Organization of Consumer Unions, would like FAO to adopt the definition of IPM as articulated in its *Field Programme Circular* No. 8/92 of December 1992, which states:

> The presence of pests does not automatically require control measures, as damage may be insignificant. When plant protection measures are deemed necessary, a system of non-chemical pest methodologies should be considered before a decision is taken to use pesticides. Suitable pest control methods should be used in an integrated manner and pesticides should be used on an as-needed basis only, and as a last resort component of an IPM strategy. In such a strategy, the effects of pesticides on human health, the environment, sustainability of the agricultural system and the economy should be carefully considered.

The NGOs go further and argue that the least toxic materials should be used if pesticides are deemed necessary. Stacking the options in this way should only stem from ecological and economic considerations. However, it seems important not to assume that biological control is always available, or indeed is intrinsically superior to chemical methods. This amounts to an ideology of naturalism and leads to unproductive tension between biological-control supporters and the pesticide industry.

We have laboured this point about the development and philosophy of IPM for an important reason. In practice, its validity is obvious, and consequently its essence is frequently misunderstood. It should not be viewed as simply another alternative in the long list of control measures. Sometimes we can observe individuals who pursue a particular ideology, even biological control, under the guise of IPM; what is really being pursued in this latter case, and sometimes with missionary zeal, is a form of naturalism with an obsessive, if understandable, distrust of synthetic pesticides. Equally unhelpful is the technology push, with pesticides as the principal and superior weapon. In these cases, it is occasionally difficult to identify exactly what elements are being integrated. Overstating the merits of any one avenue of attack can be an unnecessary cause of conflict between commercial or research interests promoting biological control, pesticide usage, plant breeding and, more recently, biotechnology options, especially transgenic plants or livestock and genetically engineered natural enemies and pathogens.

IPM on rice in Asia

The FAO rice IPM project in Asia demonstrates the massive distortions that can arise when the technology alienates, rather than empowers, the user of the technology (FAO, 1991; Kenmore, 1991). In Indonesia, the $US160 million annual subsidy of production of rice pesticides served to enrich bureaucrats and international chemical companies. The army was used to ensure that farmers applied, on a calendar basis, pesticides supplied at the beginning of each season. In Vietnam, pesticide sales for rice pests and diseases provided the operating funds for the Ministry of Agriculture. Farmers were locked into a production regime driven by external social, political and commercial forces which alienated them from personal control of their farms. They had simply lost the appropriate level of autonomy over their own production systems, a power they had exercised for thousands of years. Insect control solely with pesticides had disempowered them from controlling their own farm operations. The Indonesian and Philippines governments recognized these aberrations during the 1980s and changed central government policy and farmer education practices with remarkable success. Farmers are beginning to climb back into the driver's seat. Bangladesh, India, Malaysia, Vietnam and Thailand have now all introduced central policies consistent with IPM thinking; some have also placed restrictions on pesticide usage.

A significant event in the development of IPM policy occurred at the United Nations Conference on Environment and Development (UNCED) in Rio de Janeiro in 1992. An entire section of the Agenda 21 plan of action adopted at UNCED is devoted to IPM, calling on all signatory governments to promote IPM at farm level through networks of farmers, extension workers and researchers.

Despite the Rio outcome, we should note the additional pressures that will come to bear, especially on Asian countries, to increase dependency on pesticides, once the Japanese market is open to international trade in rice. Japan, which produces less than 2% of the world's rice, consumes 58% (by value) of the world's 'rice' pesticides in its production (Kurodo, 1993). This distorted pesticide consumption stems from sheltered Japanese rice-growers receiving six to ten times the world price for rice. To maintain revenue levels, pesticide manufacturers will need to expand production and sell the additional supplies off-shore, especially in Asian countries. Preliminary evidence of this trend was detected in Vietnam during the study of rice production in 1990 (FAO, 1991). Japan exported 50,000 t of pesticide worldwide in 1991. About 8% was shipped as aid; the bulk of the remaining exports was sold through government subsidy programmes. For example, 90% of Japanese pesticide exports to Africa (1957.5 t) in 1991 came as aid under the heading 'aid for increased food productions'. This programme, unique to Japan, emerged from the tariff reduction negotiations at the 1967 Kennedy General Agreement on Tariffs and Trade (GATT) round. Apparently, Cambodia, with no prior history of large-scale pesticide usage, is destined shortly to receive 30 t of insecticides as part of Japan's aid package. The Japanese government supports this outcome despite the conclusion by FAO and International Rice Research Institute (IRRI) scientists that Cambodia's efforts to intensify food production should be based on sustainable approaches, including IPM. It represents a further global threat to farmers controlling their own production systems. It illustrates the distortions that can arise when local factors are not paramount in local decision-making.

Undoubtedly the most significant achievement of the rice IPM project so far is the demonstration that peasant farmers, through non-formal field-orientated education experiences, can become experts at managing their production systems (FAO, 1991). This remarkable revolution was reported by Professor Michael Useem and Louis Setti in the 11 December 1990 issue of the *Boston Globe*, which stated:

> Given that millions of Third World rice farmers lack even the rudiments of literacy and numeracy, let alone an understanding of entomology and hydrology, the task was daunting. But since the ecology of each rice field was different, there was no other way. Farmers had to become their own experts. Integrated pest management distributed information to the grass roots, decentralised power to use the information, and promoted local farming networks to ensure effective application. Combined with tangible and immediate gains, you have an organisational fuel that has proven irresistible to even the most sceptical . . . American organisations would also do well to watch the new art of Javanese management.

Useem *et al.* (1992), in a formal analysis of the success of the Indonesian IPM programme, identify two key elements for mobilizing and aligning action in large organizations – informed decision-making and appropriate contingent incentives. Both of these are met in the Indonesian IPM experience.

Biotechnology and Its Role in Pest and Disease Management

If the promise of biotechnology coupled with the traditional technologies is to be delivered both to developing and industrial countries, two principles must be satisfied.

1. Integrated pest management and each of its component technologies must empower the users, including both the extension worker and the farmer.
2. The proponents of each technology must recognize their proper role in the integration of scientific practice in order to ensure that their technology is a coherent element in IPM practice.

Traditional technology options for IPM

There is a need for the continued development of 'traditional' options for integration into pest-management packages and they should not be ignored simply because they are 'low-tech' or 'traditional' in conception. Several important options are listed below, including some opportunities with conventional biotechnologies.

Biological control

Classical biological control

This has enjoyed something of a renaissance since the 1960s. Some conspicuous examples of major recent successes include control of: skeleton weed (*Chondrilla junicea*) with the rust *Puccinia chondrillina*; *Salvinia molesta* with *Cyrtobagous salvinia*; cassava mealy bug with the parasitic wasp *Epidinocarsis lopezi*; and *Sirex noctilio* with the nematode *Deladenus siricidicola* (for a review of these projects see Whitten and Hoy, 1994).

Several observations about classical biological control seem pertinent. There are a significant number of weeds, invertebrates, vertebrates and pathogens that have colonized new territories, often through unwitting human intervention, and have flourished without the suite of organisms that either competed with them or predated or parasitized them in their traditional environs. Introducing one or several of these 'natural enemies' can create a novel biological situation leading to a dramatic reduction in abundance of the organism which had assumed pest status. The diamondback moth, water hyacinth and *Phytophthora cinnamoni* are classic examples of exotic organisms that assumed pest status in many countries because of a biotic imbalance. Moving to a 'natural balance' would seem an obvious and first step at dealing with such pests. Sometimes the level of suppression is sufficient for further intervention to be unwarranted.

However, there are cases where biological control has itself been viewed as a narrow technology fix. Instead, it should be integrated into a broader approach. For example, some exotic plants have assumed weed status due to pressures within the production system, e.g. overgrazing or too few plant species in the pasture. In such instances, biological control is more effectively viewed as but one element in an integrated pasture (or crop) management programme. Again, it is not helpful to regard some species as intrinsically weeds or pests; rather, they should be viewed as a symptom of a management regime that requires adjustment. For example, biological control of weeds in Australia, especially in heavily grazed pasture, is currently moving from the narrow technological approach to being viewed as one, albeit important, cog in a systems management approach for pastures.

Inundative release of natural enemies

Inoculative, or classical, biological control relies heavily on effective initial release of a suitable natural enemy, although it does recognize the considerable scope for manipulation of subsequent conditions for optimal and lasting impact of that organism. However, there are many cases where the natural enemy cannot survive over time and, in some instances, this may not even be desirable. Genetically engineered pathogens could be one such example. This possibility has led to a whole field of research and development (R & D) around the selection of suitable agents, establishing conditions for mass rearing, packaging, marketing and release so that an economic impact is realized with little or no adverse effects on the environment.

Inundative releases cover a wide range of arthropods, bacteria, nematodes, fungi and viruses, and their targets include virtually all classes of organisms. In the former USSR alone, in 1990 some 1700 factories supplied mass-produced natural enemies for horticulture, field crops and forestry. Public enterprises are frequently concerned with the supply of arthropod natural enemies, whereas private enterprise seems more comfortable with pathogenic agents or biorationals where capital-intensive facilities such as fermenters are required, and where the products can be handled and marketed in much the same way as chemical pesticides. Examples of biorationals include *Bacillus thuringiensis* formulations, baculoviruses, fungi and entomopathogenic nematodes.

In a significant breakthrough, spores of the insect-pathogenic fungus *Metarhizium flavoviride* can be successfully formulated in oil and applied by ultralow-volume (ULV) sprayers for control of grasshoppers and locusts under hot, dry conditions. Thus two important constraints, low virulence and the need for high humidity, on the development of a mycoinsecticide for biological control of these pests have been removed. Recent studies in Australia, in collaboration with the Australian Plague Locust Commission and the International Institute of Biological Control (Milner and Prior, 1994; Milner *et al.*, 1994) have shown that one isolate (originally from an Australian acridid) is highly pathogenic for the plague locust and the wingless grasshopper and can be formulated in oil for ULV application. Thus, for the first time ever, there is the real prospect of developing an environmentally benign biological method for control of these pests.

Much of what has been said about the strengths and limitations of synthetic pesticides is pertinent to biorationals. For example, resistance remains an issue, although the prospect of exploiting genetic diversity in the natural enemy to counter the genetic response in the target organism becomes a distinct possibility. Negative side-effects are likely to be much lower with biorationals. Indeed, biorationals would be far more viable commercially if the price of chemical pesticides were forced to include the environmental and social costs. These costs are normally hidden or indirect, but are nevertheless real and borne by others (Waibel, 1990; Holmes, 1993).

Biological control using entomopathogenic nematodes
Nematodes have enormous potential for inoculative and inundative release and control of a wide range of insect pests, in both the industrial and the developing world, and can be used as a single technology or as part of an IPM system. A brief summary of the state of this technology follows.

The first nematode to be used on a large scale for insect control, *Deladenus siricidicola*, is an example of a classical biological-control agent. It completely sterilizes female wasps, *Sirex noctilio*, which none the less disperse the nematode throughout infested pine forests during pseudo-oviposition. Because of the nematode's density dependence and extraordinary life cycle, populations of the pest crash rapidly. The parasitic cycle is followed by a series of free-living cycles which enable the nematode to breed in vast numbers while feeding on the symbiotic fungus of *Sirex* as it spreads throughout the tree (Bedding, 1972, 1984a; Haugen *et al.*, 1990).

Mermithids are certainly the most obvious nematode parasites of insects, occasionally reaching 0.5 m or more in length in large insects, and eventually filling the entire body cavity of their host, which dies soon after the nematodes emerge as preadults. A major problem with manipulating them for insect control, apart from their usually very long life cycles, is that, like the obligate parasitic tylenchids, it has not been possible to culture them satisfactorily outside an insect host, in spite of numerous attempts. With *Filipjevimermis leipsandra*, which unlike many mermithids has a relatively short life cycle, considerable progress has been made towards *in vitro* culture (Fassuliotis and Creighton, 1982) and by 2000 this nematode could conceivably be used against the very important maize root worm, *Diabrotica undecimpunctata*. Various species of mermithids are known to heavily parasitize populations of short-winged grasshoppers on the tablelands of south-east Australia (Baker, 1986).

It is the entomopathogenic *Heterorhabditis* and *Steinernema* species that are currently providing the most exciting possibilities for controlling a wide range of insect pests. During the next decade they should be established among the most important insect pathogens used for this purpose. There are at least 30 species of heterorhabditid and steinernematid nematodes, and several hundred isolates have been collected from around the world (Bedding, 1990). Infective juveniles of these nematodes seek out insects, penetrate natural orifices (Poinar, 1979) or interskeletal membranes (Bedding and Molyneux, 1982) and enter the haemocele. The nematodes release specific symbiotic bacteria of the genus *Xenorhabdus* into the insect's haemolymph; this causes septicaemia and death of the insect within a day or so (Poinar and Thomas, 1966). The bacteria are highly pathogenic to insects, provide essential nutrients for the nematodes and produce antibiotics that preserve the insect cadaver while the nematodes multiply (Akhurst, 1982). Within about 2 weeks, infective juveniles (about 1 million per gram of insect) migrate from the host cadaver and may survive many months or infect new hosts immediately.

At the first sight it might seem that, because nematodes require a moisture film in which to migrate, their use to control insects could be severely limited despite their broad host range. However, over 90% of insects spend at least part of their life in soil and there are many plant-boring insects; in these situations nematodes have a distinct advantage over other microbials and insecticides in that they can seek out their hosts. There are also possibilities for microenvironmental manipulation which could allow their use in other situations (Begley, 1990). On foliage there is considerable potential for the use of water-thickeners, which not only allow excess water to be retained on the plant (Webster and Bronskil, 1968) but also provide the necessary viscosity for nematode migration and penetration; exsheathment and chemical activation of nematodes will also speed up the infection process, which, together with artificial selection for more rapidly acting strains and improved application technology, may enable high levels of infection before the nematodes desiccate (Bedding *et al.*, 1993).

Baiting is another means of avoiding the problem of nematode desiccation. Already it is proving important in controlling the banana weevil, *Cosmopolites sordidus*, in Australia (Treverrow and Bedding, 1990, 1993) and this will doubtless extend to other countries. It also has possibilities for controlling the termite *Mastotermes darwiniensis* (Miller and R.A. Bedding, personal communication) and cockroaches (Deseo, personal communication).

It is only recently that methods for the mass production (Bedding, 1984b, 1990; Bedding *et al.*, 1991) and large-scale storage (Bedding, 1988, 1990) of entomopathogenic nematodes have made their use possible for commercial control of some insect pests. To date, commercial utilization has been on a relatively small scale and on high-value crops.

Use of entomopathogenic nematodes to control insect pests is ideally suited to developing countries. Nematode-based control can be relatively inexpensive, and production can be adapted as a local cottage industry (Bedding, 1990). Small-scale control has been achieved against the live-wood termite *Glyptotermes dilatatustus* in Sri Lanka (Dantharayana and Vitarana, 1987). Injections of *Steinernema carpocapsae* are now the standard means of eliminating carpenter worm, *Holcocerus insularis*, from shade trees in some of China's major northern cities (Qin *et al.*, 1988; Yang, H. *et al.*, 1993) where over 1 million trees have already been treated. Of even greater significance is the realization that China's major apple pest, the peach borer *Carposina nipponensis*, could be effectively and economically controlled using soil applications of *Steinernema carpocapsae* instead of insecticides (Wang and Li, 1987; Doeleman, 1990; Wang, 1990, 1993). For this to be accomplished, nearly three-quarters of a million hectares would require treatment, using some 300 t of nematodes (6×10^{14} worms) over 6–8 weeks each year. Currently CSIRO (Commonwealth Scientific and Industrial Research

Organization), Australia, the Biological Control Institute, Beijing, and the Guangdong Entomological Institute, Guangzhou, are cooperating, under the auspices of the Australian Centre for International Agricultural Research (ACIAR), to make this possible.

For developing countries, mass-rearing using monoxenic solid culture, in self-aerating trays (Bedding, 1988; Bedding *et al.*, 1993) may be most economical because of low labour cost. However, commercial interests in Australia, the United States of America and the United Kingdom have been investigating the potential of liquid culture to produce huge batches of entomopathogenic nematodes with limited labour input (Pace *et al.*, 1986; Friedman *et al.*, 1989) and significantly better economies of scale (Friedman, 1990). Whether such methods will soon be suitable for developing countries, or flexible enough to cope with a range of different species, remains to be seen.

Biological control within the IPM context

In the case of primary pests, such as codling moth, brown planthopper, cotton ear worm, cotton boll weevil, and diamondback moth, more judicious use of pesticides has often enabled the local natural enemies to re-establish a regulating influence on the primary pest.

A recent example of progress in this field has been the integrated use of pheromones and insect growth regulators to reduce or replace usage of azinphosmethyl in codling moth control (R.A. Vickers, personal communication). The challenge for pest managers here is to monitor pest status of some ten other phytophagous arthropods in the pome orchard system. With a better appreciation of the role played by pesticides in the emergence of secondary pests, there has been valuable research done on integrating pesticide usage for primary pests, genetic improvement of natural enemies, especially in relation to pesticide resistance, and better exploitation of natural enemies of potential secondary pests. Perhaps the best example of this approach to IPM has been with the control of phytophagous mites in various pome and almond orchards (Hoy, 1990a.).

Improved pesticide application technology

Traditional or discarded chemicals can sometimes be given a new lease of life with improved application techniques; for example, cylinder supply or on-site gas generation (carbon dioxide (CO_2), nitrogen (N_2)) or flow-through fumigants such as phosphine (PH_3), using processes such as SIRO-FLO® (Champ *et al.*, 1990). Felton and colleagues (Felton, 1991) have designed spray boom equipment, attaching a 'green' sensor with a solenoid control valve to each nozzle, which allows a reduction of 90% of herbicide usage on fallow weed control.

The abundance of disease vectors, especially in Africa can be effectively reduced by a range of community-based control procedures. Curtis (1991) describes an array of traps, as physical exclusion, and environmental modifications which succeed because of their simplicity and community participation.

Attention to pesticide concentrations and chemical rotation has assisted in the development of a pesticide resistance strategy for delaying the spread of pesticide-resistance alleles in *Helicoverpa armigera* (Roush and McKenzie, 1987). These are just some examples where continued research on conventional technology represents a sound investment of R & D money.

Physical methods of pest management

Quite remarkable advances have been made in this technology and its application, with sealed storage, controlled atmospheres using inert gases, and amorphous silica dusts, bunker storage, grain drying and heat and cold disinfestation, both on the farm and in bulk storage facilities. Such technologies are especially advanced in Australia (Ripp *et al.*, 1984; Sutherland, 1986, 1988; Sutherland *et al.*, 1987) but they also have application in developing countries. Sealed packages (Mitsuda and Yamamoto, 1980) and sealed-bag stack storage using CO_2 and N_2 have been demonstrated under temperate and tropical conditions (Annis, 1990), and the use of controlled flow-through systems for fumigants, such as the SIROFLO® system with PH_3 (Winks, 1989, 1992) has demonstrated the practical value of a new method of application in a wide diversity of storage situations. The stored-commodity environment is sufficiently understood and controllable for it to lend itself to the application of decision-support and expert systems. These have only recently been introduced into Australia (Longstaff, 1993; Longstaff and Cornish, 1994), yet the knowledge and technology have been in hand for some 10 years. Technology does not seem to be the rate-limiting process. Again the explanation for this tardiness must lie in the social or political domains. Indeed, it could be asserted that there is no technological basis for the significant deterioration and loss of many commodities that currently occurs during storage and transport in modern agriculture.

New pesticides

The recent demonstration that the simple molecule carbonyl sulphide ($O = C = S$) has insecticidal and nematocidal properties and represents a candidate to replace methyl bromide as a fumigant for quarantine, stored commodities and structures indicates that opportunities for discovering new pesticides still exist. It is interesting to wonder why such a simple substance as carbonyl sulphide, the most common form of sulphur in the atmosphere (being emitted as a by-product of natural breakdown of organic matter), has been overlooked and not patented previously, in the comprehensive search for new pesticides. Some large companies, such as Bayer, screen over 14,000 compounds annually for insecticidal properties. And yet this naturally occurring and ubiquitous substance, if it replaces methyl bromide, will do more to replenish the ozone layer in the short term (Watson *et al.*, 1992). Its discovery resulted from some simple and neat lateral thinking.

Despite reservations expressed earlier about designer insecticides, molecular biology and sophisticated screening of biologically active molecules do enhance the prospects of finding new classes of pesticide (Moffat, 1993).

Pheromones in pest management

The potential of sex pheromones as a tool in insect pest management was recognized some 50 years ago by Gotz (1940), who wrote that 'the use of attractants to create artificial sources of stimulation, unknown to nature in the same strength, could fundamentally change pest control, which today is based on toxic chemicals'. This assessment is even more valid today.

Gotz (1940) envisaged mass trapping as the means by which control could be achieved, and, although this particular technique has since met with limited success, the role of pheromones in pest management is now well established. In a survey of the uses to which commercially available pheromones were being put, Inscoe *et al.* (1990) revealed that monitoring was the most common application, followed by lure-and-kill traps and mating disruption (Table 1.1).

Monitoring

Pheromone traps are generally species-specific. They are biologically sensitive, require little maintenance and can be operated by non-entomologists, all of which make them ideally suited for monitoring purposes. Their main application is to determine the seasonal abundance of pest species, making possible the judicious application of control measures by careful interpretation of trap catch. Because of the many variables that influence trap catch, such interpretation is often difficult and, in industrial countries, where damage thresholds may be as low as 1%, there is often so little margin for error that monitoring traps have limited value. In developing countries, however, the threshold is sometimes higher and monitoring traps play a valuable role in the timing of insecticide applications. Examples here include the bollworm *Heliothis armigera* in India, the potato moth *Pthorimaea operculella* in South America and the African army worm *Spodoptera exempta* in eastern and central Africa (Campion and Nesbitt, 1981).

Monitoring traps also serve a useful purpose as quarantine tools, where they are used to indicate whether or not a particular species is present. Examples include their use to detect infestations of fruit flies (*Bactrocera* (= *Dacus*) and *Ceratatis* species (Bateman, 1982), gypsy moth (*Lymantria*

Table 1.1. Uses listed by suppliers in 1988 for commercially available pheromones (after Inscoe *et al.*, 1990).

	No. arthropod species		
Order	Monitoring	Mating disruption	Lure and kill
Blattodea	3		1
Coleoptera	27		4
Diptera	7		5
Homoptera	4		
Hymenoptera			1*
Lepidoptera	189	18	15
Total	230	18	26

* Swarm trapping.

dispar) and the Japanese beetle (*Popillia japonica*) in the USA (Kennedy, 1981) and codling moth (*Cydia pomonella*) in Australia.

Pheromone traps have been used recently to monitor the incidence of insecticide resistance. Small quantities of the insecticide are incorporated in the sticky surface of the trap and mortality of the trapped males is compared with those caught at insecticide-free traps. Alternatively, topical applications of insecticide are given to males caught in pheromone traps. The incidence of resistance to azinphos methyl has been determined for the tufted apple bud moth (*Platynota idaeusalis*) (Knight and Hull, 1989) and for the light brown apple moth (*Epiphyas postvittana*) (Suckling *et al.*, 1987) using these methods.

In New Zealand, pheromone traps have been used since 1985 to guide insecticide use at trial sites in Nelson and Hawkes Bay. These traps monitored the population density of three major leafroller pests of apples (*Epiphyas postvittana*, *Ctenopseustis obliquana* and *Planotortrix excessana*) and allow insecticide sprays to be withheld during periods of low moth activity. This has enabled two or three sprays to be omitted each season, which represents a reduction of up to 38% while achieving the same level of control as the recommended programme. This IPM strategy is being adopted by orchardists in New Zealand (Walker *et al.*, 1989): again, local problems and local solutions but benefiting from modern technologies.

Mating disruption

Mating disruption is a technique that involves release into the atmosphere of large amounts (relative to those produced by the pest population) of synthetic pheromone. Communication between the sexes is disrupted, mating prevented or at least severely curtailed and the population suppressed. For some species the method has worked successfully as a control method in its own right, e.g. the oriental fruit moth *Cydia molesta* (Vickers *et al.*, 1985) and the pink bollworm *Pectinophora gossypiella* (Critchley *et al.*, 1985). For other pests, such as the bollworms *Earias insulana* and *E. vitella*, reductions in the number of insecticide applications were made possible by releasing the main component of *Earias* pheromone as a mating disruptant to assist in reducing population levels (McVeigh *et al.*, 1990).

Attract and kill

This method relies upon the removal of sufficient adults to limit mating to such an extent that the population can no longer maintain itself. Removal may be achieved through trapping ('mass trapping') or by combining insecticide with the pheromone ('attracticide') so that adults contacting the lure are either killed by the insecticide or are so debilitated by it that they are unable to mate.

Removal by trapping is more likely to succeed in species where females or both sexes are attracted to the lure rather than males alone, which, at least in most lepidopteran pest species, are capable of mating several times. Campion and McVeigh (1984) considered that, for species where only males are trapped, over 95% would have to be removed to have a significant impact. Even at low population densities this may require the deployment and maintenance of so many traps that labour costs alone make the method uneconomic unless the crop is of particularly high value, as is the case with the citrus flower moth *Prays citri* in Israel (Sternlicht *et al.*, 1990).

Where the attractant is extremely potent, mass trapping can be economical and effective, even if only males are removed. Methyl eugenol, an empirical fruit-fly attractant, is more potent than any known natural source and has been used, for example, to eliminate occasional introductions of the Medfly *Ceratitis capitata* and other Tephritidae from Florida and California. While the future of methyl eugenol is in some doubt because of its possible carcinogenic properties, several non-carcinogenic alternatives have now been identified (Lanier, 1990).

The mass trapping of both sexes has been employed mainly as an aid in the control of coleopteran pests. There is circumstantial evidence to suggest that an outbreak of the spruce bark beetle *Ips typographus* in Norway and Sweden was curtailed in a massive trapping programme that removed an estimated 7.4 billion beetles from Norway alone (Bakke, 1982). In another example, tree mortality caused by the western pine beetle *Dendroctonus brevicomus* in central California fell from 227 per year to less than 73 per year in plots of 2.6 sq km, following a mass trapping programme (Bedard and Wood, 1974).

Prospects for pheromones

By 1982 compounds acting as sex attractants had been identified for over 900 species, the majority of them being lepidopteran (Klassen *et al.*, 1982). The total now is almost certainly well over 1000, and yet from Table 1.1 it can be seen that pheromones are commercially available for the management of fewer than 280 species.

One reason for there being such a discrepancy between the number of identifications and commercially available pheromones is that many of the former relate to species of little or no economic importance. However, a major factor mitigating against the more widespread use of pheromones is the effectiveness, relatively cheap price and ease of use of insecticides.

While pheromones have an obvious role to play in pest management, their full potential is still to be realized. Improvements in the techniques used to isolate, identify and synthesize pheromone components have made what was once considered one of the most difficult steps in the processes leading to pheromone use in pest management into one of the easiest.

Similar advances are now required in our understanding of insect behaviour, particularly in the context of mating disruption. Despite several often-repeated hypotheses about probable mating-disruption mechanisms, we still cannot point to any particular one as being operative against species for which the method is known to work. A sound understanding of the way in which behaviour is manipulated by the release of synthetic pheromones is likely to lead to further successes with other species. This provides a further example of the value of sound basic research underpinning novel methods of pest control.

In the context of mating disruption, further improvements are required in the technology used to dispense pheromones. Most current dispensing systems are at least a decade old and, while they may well be adequate for the few species for which they have been successfully employed, they are often unsuitable for releasing pheromone components such as aldehydes and dienes that are particularly susceptible to oxidation and ultra-violet degradation. The development of a dispenser that provides this sort of protection will facilitate mating disruption of a wider range of pest species.

There are a number of other essentially technical problems which, if overcome, will greatly improve the prospects of pheromone application in pest management. However, there are also obstacles of a non-technical nature which need to be confronted. Standardizing and simplifying the requirements to register pheromones for commercial use will encourage manufacturers to bring to market new products by reducing the costs associated with their registration. A willingness on the part of governments to promote alternative pest-control strategies will also encourage their use. In the case of pheromones, this may require education programmes to make growers aware of the benefits to be gained by using less pesticides, the provision of expertise to assist them in establishing and maintaining such strategies and encouragement in the form of subsidies to make adoption of the new strategies financially viable, bearing in mind that it is not just the grower that will benefit, but the whole community.

The potential of pheromones to make a significant contribution to pest management has still to be fully realized. Their use should not be considered in isolation from other control measures but, where appropriate, as part of the arsenal in IPM programmes. If only some of the problems outlined here can be overcome by the turn of the century, we shall have gone a long way towards realizing that potential.

New technology options

While traditional technology options can continue to contribute to management of pest and disease problems, under appropriate circumstances, biotechnology offers a powerful resource which complements and extends those capabilities. In some cases, biotechnology options represent extensions of techniques already in use; in others, they are genuinely novel strategies which can assist in solving important problems (Whitten, 1989, 1991; Whitten and Oakeshott, 1990). Some important applications of biotechnology in integrated pest management systems are outlined below.

Identifying pests and diseases

Accurate identification of pests is crucial to implementation of successful control programmes; interactive programmes and expert systems are only as useful as user identification of pests allows them to be. When the appearance of a pest, or of the symptoms that it induces, is sufficiently distinctive to allow accurate visual identification, there is no need for further intervention. When, however, the organism is microscopic, polymorphic or part of a sibling species complex, or induces symptoms that cannot be easily interpreted, other methods must be employed. In such cases, biotechnology can be used to find molecular characteristics that identify the relevant taxon, and can use these to provide a basis for user-friendly, field-level identification.

Regardless of ideological preferences, finding and assessing characters of this type are a process currently best done in a high-tech laboratory where an organism can be studied in many ways. Characteristics of the organism's genome and the proteins it encodes provide a wealth of information which can assist identification. Empirically chosen pre-existing genetic markers can sometimes be used; digestion of genomic deoxyribonucleic acid (DNA) with a restriction enzyme uses the distribution of the recognition sequence as a marker, and the resulting fragment length pattern may be directly or indirectly used as an identifier. The technique of restriction fragment length polymorphism (RFLP) mapping is based on this approach. In other circumstances, reproducible and identifiable fragment patterns can be synthesized from an organism's DNA, as with use of the polymerase chain reaction (PCR) in random amplified polymorphic DNA (RAPD) screening protocols. Each of these techniques has been used successfully for laboratory-based identification of pests and diseases which do not present easily utilized morphological characters. A new and extremely powerful synthesis of RFLP and PCR, called 'A' fragment length polymorphisms (AFLPs), developed in The Netherlands, may increase the utility and power of these techniques by orders of magnitude.

Ultimately, nucleotide sequence itself can be determined and, whereas this depth of analysis is rarely necessary for purposes of identification, it will be increasingly useful for establishing higher-order relationships between groups of organisms, and will be the basis for a sound understanding of 'biotypes'. It has also been a key factor in the development of many of the genetic engineering options discussed below.

One or more of the proteins of an organism may also be distinctive in mobility, activity or antigenicity, and this too can provide a specific identifier. Many assays and techniques have been used to determine and identify species differences based on these characters. The sampling density of these technologies has been low.

Once a taxon can be identified in the laboratory, a further challenge for biotechnology is to use relevant molecular characters as the basis for reliable field identification. Antigenic characteristics of proteins have been widely exploited for this purpose, and antisera raised against proteins purified from positively identified specimens have provided the bases of many diagnostic tests. Enzyme-linked immunosorbent assay (ELISA)-based kits that can identify a range of plant viruses are already commercially available at a cost of a few pence per test. Kits of this type can be prepared in ways that make them simple to use, and relatively robust with respect to variations in protocol. In such a form they should be readily adaptable to conditions of field use in the developing world.

As well as providing the practical and conceptual bases for these identification methods, developments in biotechnology continue to enhance their scope of application and level of sophistication. Thus, techniques for the production of antibodies have moved from reliance on captive immunized vertebrates and the polyclonal sera they produce, to generation of monoclonal antibodies in large-scale cultures of hybridoma cells and, most recently, to use of recombinant bacterial viruses. The extraordinary specificity of monoclonal antibodies allows differentiation between pest species which cannot be separated by morphological criteria, for example, the important heliothines *Helicoverpa armigera* and *H. punctigera* (Trowell *et al.*, 1993). Similarly, developments in production technology continue to reduce the real cost of producing and purifying reagents necessary for these kits; assuming that savings are passed to end-users, it seems that biotechnology will be able to contribute to this aspect of IPM. The rapid field-based separation of *Helicoverpa* spp. in field crops such as cotton allows for local and timely information to be made available to the pest manager. These decision-makers are thus empowered to make informed decisions. The example demonstrates that the concepts of empowerment and contingent incentives can also apply to high-tech solutions implemented in the industrial world.

Incorporating resistance to diseases and pests

Biotechnology also presents options for making humans, animals and plants more resistant to the attacks of diseases and pests. Perhaps the most immediately familiar of these is vaccination, a technique that is only useful for the protection of vertebrates. Biotechnology can facilitate both production and administration aspects of the vaccination process.

Production of the quantities of material required for major immunization programmes of human or animal populations, in forms that are effective at inducing a useful immune response, is a major undertaking. Large-scale industrial production of antigenic materials through biotechnological means can be both cheap and efficient, and considerable work now centres on the use of recombinant insect viruses for this purpose. Studies on a range of antigens produced by recombinant baculoviruses have shown that they induce protection against challenge with infectious agents, and it is likely that they will also prove cheaper and safer than inactivated or attenuated whole-organism vaccines (Luckow and Summers, 1988). In an exciting extension of this methodology, nuclear polyhedrosis virus (NPV)-based production of self-assembled multiprotein antigens for use against bluetongue, an insect-transmitted disease of livestock, has recently been reported (Loudon and Roy, 1991).

Although traditional vaccines require inoculation of each susceptible individual, use of biotechnology to create and produce live transmissible vaccines offers the prospect of immunizing wild animal populations. This could then be simply achieved by capture, infection and release of a few individuals from any such population. While induction of protective immunity is the most common use of vaccination, antigens which induce reactions against an individual's own reproductive system have potential to suppress breeding of vertebrate pest populations; a programme based on this approach is being developed for control of the feral rabbit population of Australia, and is potentially applicable to vertebrate pests elsewhere (Morell, 1993). Use of live transmissible vaccines may also be useful in situations where it is difficult to assemble herds and flocks for treatment.

Although plants cannot be immunized in the traditional sense, resistance to pests and diseases can be achieved by incorporation and intracellular expression of selected foreign genes. Transformation of many plant species has now been achieved and, in theory at least, this presents great scope for improving production and empowering the grower. Advances that might be anticipated from this technology include protection against pests and diseases, flow of information to growers about plant stress levels or nutritional requirements, enhanced ripening or nutritional characteristics, and reduction in requirements for external chemical inputs.

Large investments are currently being directed to production of transgenic plants, and in particular to plants containing *Bacillus thuringiensis* toxin genes. While such plants have improved resistance to insect attack, it seems that questions about the major beneficiaries of the work remain to be answered.

Transformation of plants to express traits such as resistance to viruses, tolerance to herbicides and production of proteins from non-vegetative sources has been demonstrated, at least at the laboratory level; field trials of many of these are now in progress, and commercial availability of some is likely.

We should like to see production of transgenic plants which are both useful to and empowering for the end-user in the developing world. For example, plants transformed in a manner that allows production of a clear signal when a predetermined stress type (e.g. nitrogen deficiency) occurs could be planted in small numbers and act as 'sentinels' for the rest of the crop. Although it sounds fanciful now, it is not beyond expectation that these plants could show a clear 'non-vegetative' pink or blue colour that would serve to inform the grower of stress-related problems in his/her crop.

Transformed animals are also the subject of large amounts of research development, and similar concerns apply. Where vertebrates are involved, much of that effort relates to production of pharmaceuticals for human and veterinary use and, as such, lies outside the scope of this review. Transgenic insects could, however, be expected to be of value in many different IPM settings (Whitten and Hoy, 1994). At present, the technology for successful insect transformation is limited to drosophilids, but there is now intensive effort to extend the scope of this work to other insect groups.

Other dipterans are particularly attractive targets, given their evolutionary proximity to drosophilids and their pre-eminent status as pests and vectors. Genes which would be a priority target for mobilization encode the factor(s) which make some strains of malaria vectors refractory for transmission of the disease (Miller, 1992). Another valuable outcome of engineered dipterans would be increased efficiency of production of mutated strains for use in sterile release programmes. It is possible to envision flies being engineered to actually carry in their germ line and to secrete and disseminate virus that is pathogenic to themselves and other members of the same species when contacted through the oral route. These transformed insects would be expected to be highly effective as vectors, since they would go to the places that flies go and shed the virus at places where flies feed. Thus, we have the prospect of both vertical and horizontal transfer of virus particles in pest insects.

Another hypothetically useful facet of engineering for insects would be to increase the fitness of beneficial species, particularly with respect to the conditions encountered in agricultural settings. We suggest incorporating tolerance to commonly used pesticides into beneficials such as honey-bees, parasitoids and predatory species of lacewings and mites. This would assist in maintaining population levels of these species even in circumstances where chemical intervention is necessary, and could prevent subsequent pest resurgence. In practice, however, it raises the same worries about colonization of unforeseen niches by engineered organisms that are currently being examined for transformed plants (Crawley *et al.*, 1993; Karieva, 1993).

Nevertheless, and despite the fact that transposable elements have been found in non-drosophilid insects (Kidwell, 1993; Robertson, 1993; Atkinson, personal communication), it is likely that it will be some time before widespread engineering of insects becomes reality, so we have a period in which to discuss the merits of these possibilities.

Controlling pests and diseases

Biotechnology now presents options for control of organisms whose abundance or location does not suit our specific purposes. These options are primarily related to genetic engineering and/or large-scale production of microbial agents or nematodes for control of weeds and insect pests.

Insect viruses have long been considered as desirable agents for control of pests, primarily because of the specificity which could be achieved through their use. These agents are always present in the field, and it is now suspected that they keep outbreak species naturally in check most of the time. Such viruses can differ greatly in their rate of kill or time taken to cessation of feeding.

Despite many characteristics that suit them for use in an IPM context, the relatively slow rate of kill of most viruses prevents their use as knockdown insecticides. However, exceptions also exist, e.g. a small ribonucleic acid (RNA) virus that causes rapid cessation of feeding in *Heliothis* spp. (Hanzlik *et al.*, 1993). Many viruses can, however, be used in circumstances where speed of action is not a crucial factor, or when outbreak timing or damage windows can be closely predicted. A wild-type NPV of the velvetbean caterpillar has been used to protect more than 1 million ha of soybeans in Brazil over four growing seasons (see Moscardi and Sosa-Gómez, Chapter 5, this volume). Micro-scale production of other NPV isolates is used elsewhere. These unmodified viruses can be readily produced in larvae on a local scale for application as crude extracts to crops. Such an activity both empowers local farmers and agricultural workers and makes use of their knowledge. One challenge for biotechnology is to help develop appropriate quality-control technologies for use in such locally controlled rearing facilities.

Techniques of molecular biology allow NPVs (and other potentially useful groups of insect viruses, such as entomopoxviruses) to be engineered in ways that significantly increase the rate at which they kill their hosts. This involves inserting a gene that codes for an insect-specific toxin or other metabolic disruptant into the genome of the virus. In this state the gene is inert until a host is infected, and the toxin is then delivered directly to its site of action inside the insect. Products of this type can be used as knock-down insecticides, but harm only insects that are susceptible to infection. Non-target mortality and the associated problem of pest resurgence should thus be essentially eliminated. Further, insects infected with these viruses die so rapidly that very little progeny virus is synthesized; only very small amounts of engineered virus would thus be expected to persist in the environment. Production technology to produce these viruses on the scale required for commercial use is now becoming available.

Wild-type forms of insect-pathogenic bacteria, principally of *Bacillus thuringiensis*, are already commercially available for agricultural use, and there is intensive effort to engineer these to vary characters such as host range and persistence. This work continues despite the fact that there is already evidence of development of resistance to *B. thuringiensis* in pests in the field (Gibbons, 1991; McGaughey and Whalon, 1992; Gould, Chapter 16 this volume) and that the situation is further complicated by the question of *B. thuringiensis* toxins in transgenic plants. Nevertheless, the use of *B. thuringiensis* strains for control of non-lepidopteran pests remains an attractive prospect; *B. thuringiensis* var. *israelensis*, for example, is highly larvicidal for mosquitoes, contains a toxin which is only distantly related to that which kills caterpillars, and is likely to be a benefit to IPM in public-health settings throughout the developing world.

Other groups of insect pathogens, such as nematodes and fungi, can also be effective agents for pest control, and there is a voluminous literature describing characteristics such as their growth, host range, pathogenic effects and so on. It is likely that use of these agents will benefit more from advances in production aspects of biotechnology than from those of genetic engineering. Although selected species from each have been successfully transformed, as for most other groups of animals and plants, the technology cannot yet be extended to species at will. None the less, improvements to practice and theory of large-scale culture should allow useful amounts of the agents to be produced, formulated and distributed in an active form; once again, the limiting factor to their use in developing countries is probably socio-political, relating in particular to their cost for farmers.

Reverse Genetics: Power and Limitations

Traditional plant and animal breeding identifies desirable phenotypic goals; it marshals heritable variation relevant to the desired traits and selects as parents a portion of the population with phenotypic values leading towards the selected goal. From ancient times until the green revolution, the formal discipline of genetics was not central to the success of the vast majority of plant breeding, and even less so for livestock improvement. Often, little to nothing was known about the genes, their number or their mode of action that contributed to the rapid phenotypic response on which many valuable modern crop cultivars are based. For example, the genetic basis for the commercial cultivars of cotton bred in Australia for pest resistance is not well characterized, nor need it be so.

Crop and livestock improvement based on transgenic technologies, in contrast, necessarily starts with very well-characterized genes or specific modifications of nucleic acid sequences. Usually the gene product of a target gene is also well characterized, as is its biochemical function. The genetic engineer would like to do either of two things: target a specific gene and replace it with some variant that might come from another species (of plant, animal or microorganism); or simply add a novel gene to the genome. Unfortunately, targeted gene replacement is not yet available to the plant genetic engineer, although it is available in animals. The introduction of herbicide and virus resistance genes or parts of genes and the insertion of bacterial genes coding for insect toxins are the most widespread examples of plant genetic engineering.

Disrupting the function of specific genes to alter specific pathways, such as the ripening process in fruit, demonstrates that there are important examples of genomic modifications which will generate desirable phenotypic shifts. These modifications are predicated on a thoroughly researched characterization of the target, for instance the biochemistry of ethylene production or the structure of plant cell walls. For gene addition and disruption via suppression technology to be of real value, it is necessary to have a clear idea of the scope of phenotypic shifts inducible by the potentially infinite variety of mutations that could be generated for any given locus. We can contemplate altering the timing of gene action, influencing translation or transcription efficiency or altering the biochemical or structural function of the novel gene product.

Thus ‘reverse genetics’, i.e. manipulating the genotype to induce specific phenotypic shifts, represents a powerful technology at the DNA level. However, its great weakness emerges when we talk about what specific phenotypic shifts are induced by what DNA modifications, in terms of both what we want to induce and the undesirable pleiotropic side-effects. Herbicide and viral resistance and postharvest properties in plants probably represent an exception to this difficulty. These are obviously preferred early candidates for plant transgenesis. However, other phenotypic shifts, especially for insect resistance and agronomic traits, are likely to encounter difficulties for a variety of reasons, principally because of the extraordinary complexities and vagaries of the environment and the plant's response to it.

The use of *B. thuringiensis* toxin genes is so widely publicized as the model for genetically engineered insect resistance that it serves as a good basis for discussion. In the formative years of plant transformation technology, it was difficult to find molecular biology laboratories which were not contemplating introducing these toxin genes into some crop plant where some lepidopteran pest could be found. With the prospect of *B. thuringiensis* strains specific for other insect groups, especially coleopterans and dipterans, it was näively assumed that a final solution to insect pest problems was again within our grasp. There is a distressing similarity with the early euphoria of the chemism era. This disease of plant molecular biologists is so distinctive it invites the label '*B. thuringiensis*-ism' (Bt-ism). The excesses manifested by Bt-ism represent poor science; worse still, they represent a significant opportunity cost to the pursuit of intelligent and creative science.

Given the serious limitations with transgenic plants, especially perennials, for pest and disease control we should be exploring novel options beyond transgenesis. For example, endophytic and epiphytic bacteria, nematodes and fungi should be isolated and developed for predicted pests, and then held in reserve to cope with the occasional outbreak. The pathogen option would seem more flexible and more easily targeted to the problem as and when it arises. We have to acknowledge that these options would not be as attractive to commercial interests, especially those with valuable intellectual property in the form of patented genes and associated constructs. Furthermore, personal prejudice and institutional prestige will also exert their influence on priority-setting and direction of research. It would therefore be näive to assume that we shall not witness certain technologies being promoted contrary to the public interest. Thus, biotechnology has the same capacity as chemism to set its own agenda, to disempower the end-user and to result in non-sustainable production systems.

More broadly, and more importantly, there is an urgent need for increased strategic research on disease/insect/plant interactions so that a much wider set of opportunities for IPM can be identified and developed. We cite one example. The cotton boll weevil's traditional host plant is not cotton but kapok. Some of the traditional natural enemies of the boll weevil recognize kapok as the host plant by certain chemical cues. Consequently, cotton plants are not recognized as likely sites by some natural enemies for locating the boll weevil. Further knowledge is required to determine whether it is feasible to transform cotton so that the suite of the boll weevil's natural enemies could be enriched. Research directed at such goals would be more appropriate to strategic research agencies than research in competition with commercial interests to transform plants with *B. thuringiensis* when the latter will surely accomplish such goals in reasonable time frames. There are many more examples where innovative research on insect/plant interactions will ultimately provide a valuable base for biotechnologists to draw upon in transforming plants.

Future Directions

When the throughput of research is low and the cost high, then the luxury of first taking a measured and balanced approach to understanding a system and only secondly proposing enhancements is sometimes not available. The vigorous promotion of particular 'solutions' in crop protection violates this approach, which has been so valuable in the development of modern science. The approach can be briefly summarized as developing a hypothesis and then trying as hard as humanly possible to falsify that hypothesis. The alternative approach is to prove the efficacy of a particular 'solution'. By judicious choice of conditions and systems – a narrow subset of the actual conditions of practice – it often proves to be a successful 'solution'.

So what can we do to harness the enormously powerful engine of molecular biology to a socially balanced and environmentally sound IPM approach? That is the key to linking IPM and biotechnology, or, more properly, sustainable crop protection and molecular biology.

First, we must examine whether the methodology that we currently have can stimulate local involvement – even to the point of local rejection of imposed options. We should develop methods that will ensure, for instance, that farmers themselves are allowed to make judgements and to develop, appreciate and use their knowledge of local conditions and needs to control the protection of their crops. For instance, if we wish to place a transgene in a crop that will afford protection from a pest or even a group of potential pests, would it not be both socially and ecologically desirable to have that gene silent until it was needed to prevent catastrophic yield loss? To develop systems for farmer control of gene activity? This would eliminate the chronic selection pressure that current strategies flirt with and, more importantly, it would place the farmer in the role of an active participant in the process. By assessing the constraints of the local farming communities and letting these predicate the conditions for the innovations, this could be done.

For instance, there are genes in bacteria whose activity is induced thousands-fold by application of compounds found in animal urine. The compounds that achieve this induction are not normally found in plants, nor are the enzymes that hydrolyse them. The sequences of the repressors and operators and the mechanisms are now known. The proof of principle, of having such control sequences function in plants, has been achieved. Could these control sequences be used as effective vehicles to restrict activity of transgenes until the farmer makes a judgement that it is time to protect, for instance, from an insect predator? Using locally available material, such as animal urine, which, while cheap, is not of unlimited availability, would provide a fail-safe mechanism.

Secondly, can we develop methods that increase the heuristic values to the farmer – that is, can we develop situations, perhaps using molecular methods, to develop robust field diagnostics – where the farmer has more and better information in a form that is immediately relevant and useful, and where the extension and the research community that is responsible for assisting in that farmer's livelihood is further empowered with this information? Can we, for instance, develop the concept of sentinel plants, or bioindicators, which will provide simple colour changes that are proportional to particular biotic stresses that are otherwise difficult to score, or where early prophylaxis or systems adjustments, before the onset of symptoms, could be ameliorative? Or can we use the craft of molecular biology to develop diagnostic systems that can be fully utilized in impoverished rural areas? Can the components of such diagnostics – for instance, for virus-free planting material – be made in villages and the craft of using them passed on as indigenous village knowledge? In the longer term, we believe these goals are achievable. The question is how and by what mechanism we can establish this reversal of roles. Once determined, the sophisticated molecular-research community will not serve as final providers of solutions, but will be in the service of local decision-makers.

Thirdly, we can pay particular attention to the cost and throughput of experimentation. If we are being driven to assertive imposition of central problem-solving by the difficulty of cost recovery, can we reduce costs and increase diversity by methodological innovations? Can we devise strategies where transgenic plants of numerous species and varieties can be trivially and cheaply made, to stimulate truly balanced experimentation? There may be an outcry that this is a trivial statement and exactly what is currently desired, but is this true? While there is considerable investment in using available methods – be they tissue culture, molecular selections or whatever – there is little investment in developing new methods, especially now that highly capitalized laboratories have ready access to these technologies.

Information Technology

This review has acknowledged that, despite technological advances in many areas of biological science, there remain major challenges in the production of adequate global supplies of food and fibre and the prevention of disease in the growing human population. In many instances, the knowledge is available but is not accessible in a timely manner to those who require it – the farmers, downstream handling authorities, manufacturers and their technical advisers. Even in industrial societies, it is often not effectively used by central authorities charged with bulk storage and transport of commodities. For instance, spoilage and contamination of bulk grains are unacceptably high in the USA, South Africa and other industrial countries; in developing countries they can be even more substantial.

One of the tragedies of modern science is to witness commodity losses during production or storage simply because critical information is unavailable or incorrect information forms the basis for decision-making. Sometimes a pest is incorrectly identified, its resistance status is unknown or the list (sometimes extensive) of corrective measures and how to apply them is not readily available. Powerful software in the form of decision support and expert systems either exists or could readily be adapted for the pest manager. Furthermore, with recent hardware developments and compact disc (CD) technologies, such as compact disc read-only memory (CD-ROM) and CD.I, knowledge should be much more readily available to consultants and technical advisers. Given that a domestic television set and a CD reader are the minimal hardware required for the operation of this software (e.g. CD.I), far more effort should be devoted to the development of relevant and accessible interactive packages appropriate to both the industrial and the developing world.

An excellent example of the application of information technology is the current development of *The Electronic Compendium for Crop Protection* by CAB INTERNATIONAL. For the *Electronic Compendium* to achieve its objectives fully and expeditiously, it recognizes the need to work closely with information technologists to develop packages that give user-friendly and ready access to the vast store of relevant and accurate information which currently exists. When we are looking at the needs and opportunities for linking biotechnology and IPM in developing countries, it is highly likely that far more effort should be devoted to information technology applications.

Integrated Pest Management and Biotechnology: An Analysis of Their Potential for Integration

2

JEFF WAAGE

Introduction

Biotechnology has considerable potential to increase agricultural production, through improving the yield and nutritive value of crops, extending crop production into inhospitable habitats and improving the protection of crops against pests. This latter possibility, biotechnology for crop protection, is receiving considerable attention today. Many studies are now under way to improve, through genetic engineering, the value of microbial pesticides, while studies to date on transgenic plants have been dominated by efforts to achieve herbicide resistance and expression of insect and virus resistance genes. Where the potential for biotechnology in developing countries is discussed, crop protection always emerges as a key area for research (Persley, 1990; Swaminathan, 1991; Ananthakrishnan, 1992; Kajiwara, 1992; Thottappilly *et al.*, 1992).

Biotechnology enters the history of crop protection at a turbulent time, when the paradigm of chemical control is giving way to that of integrated pest management (IPM) and when interpretations of what IPM means are diverse and contrasting. At first glance, the 'greenness' of biotechnology, as an alternative to chemical pesticides, makes it seem appropriate to IPM. On the other hand, its sophistication appears to distance it from the 'farmer-first' concept with which IPM is increasingly associated.

The potential relationship between biotechnology and IPM is of particular interest to the development assistance community, which is currently investing in both. Clearly, it would be convenient and satisfying if it could be demonstrated that an investment in biotechnology for pest management will benefit the development and adoption of IPM methods in developing countries. The aim of this chapter is to determine whether this is likely. More specifically, I shall ask whether current efforts in biotechnology for pest management are complementary or antagonistic to the progress of IPM. To do this, I shall first review the ecological basis of pest management and the development of IPM, with an emphasis on the developing-country experience. Then, I shall survey current research into biotechnologies for pest management and analyse its strategy and direction. Finally, I shall compare the needs of IPM with what biotechnology has to offer, and draw conclusions.

I shall not discuss in this chapter the potential risks of biotechnology, for instance, the potential of transgenic plants to become weeds by virtue of their new properties or of ecologically undesirable genes to enter new species through outcrossing of engineered species (Williamson, 1988; Tiedje *et al.*, 1989; Altman, 1992; Doyle and Persley, 1996). The risk that biotechnology might create, as well as solve, pest problems warrants greater discussion and research.

Pest Management from an Ecological Perspective

The emotive and topical image of a locust swarm epitomizes the popular view that pest outbreaks are inevitable, natural phenomena. However, most of the pest problems we see today are not entirely 'natural', but the result, at least in part, of changes we have made to the crop environment, making it more favourable to pest population growth. Furthermore, these changes are particularly associated with programmes of agricultural intensification, and hence with recent agricultural trends in developing countries. From an ecological perspective, we can see these changes affecting the spatial and temporal dimensions of pest dynamics.

The concentration of a single plant species in ever larger and more extensive monoculture increases its apparency to pests and the number of pest species that colonize it (Strong *et al.*, 1984). As crop land is increasingly put down to genetically uniform, high-yielding crop varieties, conditions for pest colonization, spread and rapid population growth improve.

On this same, spatial dimension, agricultural intensification frequently leads to a reduction in that natural vegetation which conserves the natural enemies so important to the suppression of pest populations. These changes mean that natural enemies from increasingly smaller and more distant non-crop reservoirs enter crops too late or in too small numbers to prevent pest outbreaks.

On a temporal dimension, synchrony of planting of large areas, where it provides a local abundance of food for an immigrating or emerging pest population, can enable pest populations to outstrip the density-dependent action of natural enemies and diseases. Intensification has also frequently resulted in the reduction of intervals between the planting of the same crop or in the overlapping of crops, which gives more continuous resources for pests. Even where a range of crops are grown in an intensification scheme, this can be a problem, because some of the worst pest species are polyphagous (Moran, 1983). Thus, the well-known pest control crisis in the Gezira cotton scheme in Sudan was fuelled by the introduction of vegetable crops in the intercrop period between cotton, which allowed the polyphagous cotton bollworm, *Helicoverpa armigera*, to maintain numbers between cotton crops (Kiss and Meerman, 1991).

Associated with the spatial and temporal intensification of agriculture, the search for better crop varieties has accelerated the movement of plant material around the developing world, and with it the movement of new pests. Plant breeders, commercial importers, distributors of food aid and general commerce have all been responsible for inadvertent introduction of an increasing number of exotic pests into developing countries. These invertebrates, vertebrates, weeds and plant pathogens, removed from their own, natural controlling factors, can cause even more serious problems than indigenous pest species.

Recent damaging introductions of exotic pests into developing countries include the golden snail, *Pomacea canaliculata*, on rice (from Latin America to South-East Asia), the larger grain borer, *Prostephanus truncatus*, on maize (from Central America to Africa), the mango mealy bug, *Rastrococcus invades* (from South Asia to West Africa), the coffee berry borer, *Hypothenemus hampei* (from Africa to Central America and India) and itchgrass, *Rottboellia cochinchinensis*, a weed of cereals and sugarcane (from South Asia to Central America), to name but a few.

The distribution of crops into new tropical regions creates the potential not only for exotic pests, but for the adaptation of local species to become new pest problems. Thus, a number of serious tropical crop diseases have arisen through the ecological or evolutionary shift of a pathogen from a native plant to an introduced crop (Hokkanen, 1985).

Last but not least, there is one quite specific human intervention which has contributed dramatically to increasing problems with pests. This is, paradoxically, our own efforts at pest control and our over-reliance on chemical pesticides.

For almost half a century, chemical pesticides have provided a simple and frequently cost-effective means of checking the growth of insect, disease and weed populations in crops. They have proved particularly appealing to intensification programmes where capital is relatively easily available and labour often limiting.

In recent decades, problems with reliance on pesticides have come to light with the control of insect pests. The intensive and extensive use of insecticides has led to the expression of insecticide resistance by a wide range of insect pests (Georghiou, 1990). Many insecticides have also been shown to remove an important form of natural pest control, that provided by natural enemies, many of which are also insects.

In varying degrees, resistance of insect pests to insecticides and the quite separate effect of insecticides on eliminating natural enemies combine to create a phenomenon called the 'pesticide treadmill', which goes something like this. Pest resurgence following pesticide applications poses farmers with a dilemma. Having little or no experience with non-chemical approaches to pest control, they have little option but to respond to new or greater pest numbers with the application of more pesticides. This further aggravates the problem, by making pests more resistant to chemicals and killing even more of their natural enemies. At the same time, new, secondary pests emerge, freed from their control by the elimination of their natural enemies. Insecticide use increases further, driving up the cost of production, while income continues to decline, as uncontrollable pests reduce yield. Ultimately, the cropping system may become unsustainable and be abandoned.

Is the problem just with insects?

Plant pathologists, weed scientists and nematologists often despair at the way in which the problems associated with pesticides in insect pest management are extrapolated into demands for dramatic changes in pest management as a whole, without thought as to the very different situation and experience of pesticide use against weeds, nematodes and plant diseases.

However, the fundamental elements associated with over-reliance on pesticides are to be found in these other pest systems. Herbicide, nematocide and fungicide resistance exists and is increasing (Davies, 1992). Natural enemies are limiting factors in population growth and spread of weeds, nematodes and plant diseases, although we still know relatively little about their population ecology and regulatory nature. Further, for weeds and plant diseases, natural enemies *sensu stricto* (e.g. parasites, predators) is too narrow a concept: competition and antagonism from other plants and micro-organisms may be more important in regulating population. The wrong fungicide might be as prone to eliminate the fungal antagonists of a plant disease as the wrong insecticide to remove the predators of an insect pest.

Further, we know already that insect pest problems can be caused not only by insecticides, but by herbicides and fungicides as well, through a range of subtle food-chains and community processes (Waage, 1989). Why should this not be true for other kinds of pests and pesticides? Acknowledging the substantial gaps in our knowledge, and trusting our grasp of general ecological and evolutionary processes, it seems sensible to anticipate the need to reduce dependence on pesticides for all pest problems, not just insects.

IPM – the new technology of pest control

Integrated pest management was born of the problems in insect pest control caused by over-reliance on chemical insecticides. It set out to reduce reliance on insecticides and increase use of other, less environmentally damaging methods, including biological control, cultural methods, behavioural methods (e.g. pheromones, sterile insect release) and host-plant resistance.

Integrated pest management has had a complex evolution, and has spun off many interpretations in the process. Today, most groups involved in pest management have embraced IPM, but often with quite different perspectives on what it means. As we shall see, this has particular implications for how we interpret the benefits of biotechnology to IPM in developing countries.

While some veterans of pest management in tropical plantation crops will rightly note that they applied IPM concepts long before its current popularity, history will probably come to associate the birth of IPM with events in the USA in the 1970s. Here, the new concept of IPM was based on restricting pesticide use through the imposition of economic thresholds for spraying, and replacing insecticides themselves with alternative, more biological products or methods, including biopesticides, pheromones, intercropping, etc. In this way, the action of natural enemies could be protected and the risks of pesticide side-effects reduced.

Perkins (1982) has noted that those entomologists who designed IPM did so in the knowledge that their clients were 'capital-intensive, technologically sophisticated farmers facing a mass-exodus of labour from agricultural production'. As a result, IPM was first designed to the specifications of the same high-tech chemical pest control tradition which it endeavoured to replace. A seminal volume reporting a grand, national initiative in IPM at that time summed up this perspective in its title – *New Technologies of Pest Control* (Huffaker, 1979).

This perspective of IPM as a technology has engendered a rich working vocabulary of IPM tools, IPM component technologies, IPM packages, etc. In this form, it has attracted the interest of the crop-protection industry which services agriculture in industrial countries, admittedly after some initial scepticism. Increasingly, agrochemical companies are accepting the concept of economic thresholds, as a means of ensuring the life of their products ('pesticide resistance management') and, of course, a continuing place for pesticides in the IPM concept. At the same time, companies have invested in research and development (R & D) for non-chemical pest-control products to replace older, chemical ones, particularly biopesticides, and to a lesser extent pheromones and mass-produced insect natural enemies. Biotechnologies, including transgenic plants and engineered microbials, are, conceptually, the latest in this R & D portfolio of potential IPM products. As such, they elaborate an approach that I call 'technological IPM'.

While this technological vision of IPM has widespread appeal, practical experience with the implementation of IPM in recent years has taken the concept down a different path. Much of this experience has been gained in developing countries, for some very good reasons.

IPM as an ecological approach

The problems created by excessive use of insecticides in the largely agricultural economies of developing countries have frequently been severe. There have been, literally, economic crises in crop production. In the past few decades, declines or failures in production resulting from insecticide treadmills in plantation fruit, oil and beverage crops, vegetables, cotton and rice have been recurrent phenomena, appearing predictably in different geographical regions (Ooi *et al.*, 1992; Anon., 1994).

In these crisis situations, insecticide use has frequently been shown to be so inappropriate and excessive that it can often be completely eliminated without significantly reducing crop production (Ooi *et al.*, 1992; Waage, 1993). In some such systems, scientists and farmers have seen, after elimination of pesticide use, the recovery of a satisfactory, self-renewing, natural level of pest control by local natural enemies.

Where excessive use of insecticides has led to problems in crop production on a range of tropical and subtropical crops, the dramatic nature of both the problem and the solution has had a substantial impact on the thinking of policymakers, scientists and farmers.

Firstly, it has identified the key role that local indigenous natural enemies can play in the suppression of pests, a role which is still poorly understood, particularly in tropical systems. For the resource-poor farmer, this control can be free and, if it is understood and appreciated, it can be conserved and manipulated.

Secondly, these experiences have enhanced suspicion of the value of imported interventions, like pesticides, and the top–down process by which their use has been extended from the research station to the farm. In the crises that have occurred, this process has been shown to deliver inappropriate pest-management messages for the local situation.

All of this has led to a recognition of the need to understand better the local ecology of pests, to develop and implement IPM with reference to local conditions and resources and to place greater emphasis on local processes, such as natural pest control. This involves accessing indigenous knowledge and ensuring the participation of farmers in the development of IPM methods.

Where IPM is developed from an understanding of the local situation, relying particularly on the self-renewing action of natural enemies and local cultural practices, the concept of IPM becomes more ecological than technological. Furthermore, greater emphasis is placed on people, because IPM development involves more than scientists on research stations. Farmers, scientists and extensionists become involved in a collaboration to develop solutions to local pest problems. For convenience, I shall refer to this concept of IPM as 'ecological IPM', on the understanding that farmers are a key part of the local ecology of agriculture, and indeed shape that ecology to both create and solve pest problems, as described earlier.

This concept of 'ecological IPM' differs in some important, if subtle, ways from that of 'technological IPM'. Both base IPM on underlying natural processes of pest regulation, particularly the action of natural enemies. For the technological approach, it is these processes which make thresholds for intervention possible, but they remain something of a black box and the emphasis is on intervention. The more ecological concept of IPM makes these processes the central emphasis of IPM, seeks to understand and use them, and does not recognize the need for pest-control interventions. Gallagher (1992) presents a particularly clear comparison between the technological and ecological, farmer-participatory approach to IPM for control of insect pests of rice in Indonesia.

To say that IPM initiatives can all be characterized as ecological or technological and that one is more associated with developing countries than the other is clearly an oversimplification. These are hypothetical extremes, but they have parallels in other aspects of development today. Chambers (1991), for instance, identifies two similar and contrasting paradigms for agricultural development as a whole, one a technological, science-driven 'blueprint' approach, practised by professionals (e.g. biologists, economists, engineers), and the other a people-driven, participatory 'learning-process' approach, centred on farmers. He notes that both have strengths and weaknesses, but observes that the people-centred approach may be more appropriate to the solution of complex, local problems faced by small-scale, risk-prone farmers in developing countries.

Table 2.1. Emerging paradigms in the evolution of IPM (adapted from Chambers, 1991).

	'Blueprint' technological IPM	'Process' ecological IPM
Goals	Predetermined	Evolving
Keyword	Planning	Participation
Methods	Package	Basket
Clients are	Motivated	Empowered
Outputs	Infrastructure	Competence

In Table 2.1, I present Chambers' concept and adapt it to the concepts of ecological and technological IPM. In the technological paradigm of IPM, the goals are predetermined, we want to plan a package for pest management which we shall motivate farmers to adopt, and the output will be a crop-protection infrastructure for delivering this package – products, marketing and extension systems. In the ecological paradigm, the goals are evolving and location-specific. We seek to encourage participation of scientists and farmers in the development of a basket of methods from which farmers, empowered and competent in pest management, can choose.

Current developments in IPM in developing countries

As a result of national and donor interest in IPM, a series of scientific consultancies were held in the 1980s (NRI, 1992; Teng, 1992). In September 1991, crop-protection leaders and senior decision-makers from 21 developing Asian countries met to create an action plan for the regional development and implementation of IPM (Ooi *et al.*, 1992). A similar exercise, involving senior representatives from 17 African countries, occurred in April 1993 (Anon., 1994).

Both meetings drew similar conclusions regarding IPM development. Progress in IPM implementation was seen to require an elimination of structures that perpetuate unnecessary pesticide use (e.g. subsidies). A greater awareness of IPM is needed, associated with training at the research, extension and farm level. Research on IPM must be designed to fit the needs of farmers, and hence must involve farmers and be undertaken as far as possible on-farm.

In the light of these developments, and particularly of the strong support lent to IPM by the United Nations Conference on Environment and Development (UNCED) and its Agenda 21, major international development-assistance institutions are revising their perspective on IPM. Stimulated in the 1980s by environmental concerns about pesticide provision in development assistance, IPM guidelines of donor agencies focused on pesticide reduction, safe use and, generally, a technological approach to IPM. Now, as guidelines are reconsidered and rewritten in the light of IPM experience, the future emphasis may be more on IPM as a process in human resource development, and future guidelines are likely to emphasize the need to develop IPM locally with farmer involvement.

Current Trends in Biotechnology for Pest Management

Although biotechnology is a term applicable to a range of activities, including conventional but important activities like fermentation, it would seem that the overwhelming tendency today is to interpret biotechnology as involving the creation and use of genetically modified organisms (GMOs). Later in this chapter, I shall suggest that this interpretation is too limiting, but for the moment I shall concentrate on where the money is being concentrated, the engineering of organisms to improve plant protection.

Biotechnology for plant protection can currently be characterized by three approaches to genetic engineering: (i) the engineering of natural enemies of pests to be more effective agents of biological control; (ii) the engineering of plants or their microbial associates with genes that protect them from pests; and (iii) the engineering of natural enemies or plants with genes for pesticide resistance, which increases options for pesticide use.

Engineering natural enemies

Most efforts to improve natural enemies through biotechnology have concentrated on entomopathogenic microorganisms, but not all taxa have received the same attention. Emphasis has been placed on bacteria and viruses, in contrast to fungi, protozoa and nematodes. This is, in part, because bacterial and viral genomes are better understood and more easily manipulated than those of other natural enemy groups.

Bacillus thuringiensis

Research on bacteria has concentrated on *Bacillus thuringiensis*, which is used to control lepidopteran and coleopteran pests of crops and storage and dipteran disease vectors. At present, *B. thuringiensis* is the most widely used biopesticide: its sale, in a range of strains and formulations, comprises over 90% of all biopesticide sales (see Marrone, Chapter 8, this volume).

Efforts to improve *B. thuringiensis* through engineering have focused primarily on increasing host range, by combining the different genetic forms of the delta endotoxin, which have different virulence within and between insect orders. This can be done by conventional means, through plasmid fusion, but engineering offers a more versatile approach. Efforts to improve host range through engineering have been most promising, while efforts to improve virulence and time to kill through mixing of different delta endotoxin genes have not (van Frankenhuyzen, 1993).

Baculoviruses

Insect baculoviruses are important natural regulating factors for insect populations, particularly Lepidoptera. Their commercial development has been limited by the need to produce them *in vivo* and by their generally high specificity, which, while desirable from an environmental point of view, limits markets.

A few commercial formulations exist, largely against forest Lepidoptera, but there are also effective viruses for some of the key lepidopteran pests of crops, such as *Helicoverpa* spp. and *Spodoptera* spp., and their wider commercialization is likely.

Research on engineered baculoviruses has been in the vanguard of environmental studies on released GMOs, and much work to date has focused therefore on genetic markers and other manipulations (e.g. removal of polyhedrin gene, which produces the protective viral protein coat) to investigate persistence in the field (Corey, 1991).

Specific work to improve baculoviruses as control agents has focused on incorporating genes that produce proteins which will accelerate the death of infected hosts (Bonning and Hammock, 1992). Baculoviruses can take several days to kill lepidopteran hosts in the field, and are generally less virulent against older larvae, which will be the most damaging. Thus, improving speed of kill could make a viral product more attractive to farmers.

To date, successfully incorporated genetic material includes that which codes for the delta endotoxin of *B. thuringiensis*, insect juvenile hormone (which reduces feeding and arrests moulting) and a component of spider venom. Virtually all work has been done on the *Autographa californica* nuclear polyhedrosis virus, which has an unusual broad host range. The results of these manipulations has a range from nil to about 50% reduction in time to death (J. Corey, personal communication).

Other biological agents of pest control

Next to *B. thuringiensis*, entomophilic nematodes are the most commercialized of biological control agents today, largely owing to the fact that, as animals, they have effectively escaped some expensive regulatory constraints in their development. Genetic research on nematode species closely related to those used in insect pest control suggests that manipulation of the nematode genome is possible, but greater short-term promise may lie in the engineering of the bacteria *Xenorhabdus* spp., which are vectored by the nematodes into insects and which are, in fact, the agents that kill the pest. Engineering of these bacteria, which are closely related to *E. coli*, may improve problems found with their loss of virulence in production (Ehler and Peters, 1993).

Fungi are particularly promising biological control agents, in so far as virulent forms are known from virtually all insect orders (unlike bacteria and viruses) and they frequently cause natural epizootics in pest populations. Further, fungal species are the primary focus of attention in the development of biological pesticides for weeds and plant diseases. They have received little attention with respect to genetic improvement, partly due, perhaps, to the limited development of fungal-based biopesticides, but also because the fungal genome is comparatively complex and the genetic basis of virulence and other key properties are poorly understood.

Arthropods are better known than microorganisms as important biological control agents of insect pests in crops, and also of weeds in some situations (e.g. with exotic weed species). The relative complexity of their genome poses a constraint to their genetic manipulation, although recent success in engineering microbial genes into *Drosophila* is encouraging (Hoy, 1992b). More importantly, their insect- (or plant-) killing behaviour is complex, and not likely to be governed by simple, easily manipulated, genetic systems.

Transgenic plants and plant-associated microbes

Engineering of plants for crop protection has concentrated on incorporating resistance against insects and diseases into the plant genome or into the genome of plant-associated microorganisms.

One of the most extensively studied manipulations has been the incorporation of genes which produce the *B. thuringiensis* delta endotoxin into crops, primarily to confer resistance against Lepidoptera and Coleoptera. Work to date has focused on tobacco, tomato, cotton and potato, where good expression of the toxin has been obtained, leading to effective control in laboratory and field trials. Now, development assistance agencies are supporting similar studies in the major tropical food crops.

More limited work has been done on incorporating genes for plant-defensive chemicals into new plant species. A gene for an enzyme that inhibits protein digestion, cowpea trypsin inhibitase, has been transferred from cowpea into other plants (Boulter, 1992). It affects insects by reducing their capacity to assimilate plant proteins, which in turn reduces feeding and leads to starvation. Similar problems would exist for vertebrate herbivores, but the selection of enzymes with activity only in the alkaline guts of some insect pests confers a degree of specificity. Cowpea trypsin inhibitase has been engineered into tobacco and shown to substantially reduce damage by Lepidoptera (Hilder *et al.*, 1993).

Another group of enzymes with potential for plant protection are chitinases (Broglie *et al.*, 1993). High levels of production of certain chitinases at the plant surface could protect plants from invasion by fungi and bacteria, and this makes them a promising target for engineering.

One of the most exciting prospects for transgenic plant protection is the creation of virus-resistant plants by incorporation of viral deoxyribonucleic acid (DNA) into crops. This has been achieved so far in a range of vegetable crops by the incorporation of genes for viral coat proteins into crop genomes (Hull, 1990; Beachy, Chapter 14, this volume).

Engineering plant-associated microorganisms

A genetically modified strain of the bacterium, *Agrobacterium radiobacter*, was the first genetically manipulated organism to be released for pest management. Inoculation with this organism can protect a number of horticultural crops from infection by the crown gall bacterium, *Agrobacterium tumefaciens*. While a natural, avirulent strain of *A. radiobacter* is effective for this purpose, the engineered strain eliminates the possibility that genes from the natural strain could transfer to virulent *A. tumefaciens*, making it even more aggressive (Jones and Kerr, 1989; Ryder and Jones, 1990).

Some attention is currently being paid to microbes that suppress plant diseases in the soil, notably the fluorescent pseudomonads which colonize plant roots. Exploitation of natural variation from different soils to find the most disease-suppressive forms has potential, as does the improvement of natural forms by engineering. Such engineering, for instance, could be directed at greater expression of antibiotics or siderophores that suppress pathogenic fungi (Dowling *et al.*, 1993).

An interesting approach to using plant-colonizing bacteria for biological control is to be found in a recently marketed *B. thuringiensis* product, CellCap. CellCap consists of a common bacterium of plant surfaces, *Pseudomonas fluorescens*, engineered to express the *B. thuringiensis* delta endotoxin gene, for control of leaf-feeding Lepidoptera. While the bacterium is killed before formulation, the better persistence of this product on plant surfaces, compared with *B. thuringiensis* itself, may be associated with physical properties of the epiphyte, *P. fluorescens* (Soares and Quick, 1992).

Endophytic bacteria offer another site for the incorporation of plant-protective genes. Because they are themselves bacteria, manipulation can be less problematic. The *B. thuringiensis* delta endotoxin gene, for instance, has been engineered into *Clavibacter xyli*, a xylem-inhabiting bacterium of maize and other Graminae, and can confer protection against maize stem-boring moths (*Ostrinia* spp.) (van Frankenhuyzen, 1993).

Engineering for pesticide resistance

Incorporating into crops genes that confer herbicide resistance is an innovative means of using biotechnology to address weed control (Hinchee *et al.*, 1993). This allows crops to be safely treated with chemicals that would otherwise be damaging, and permits a shift in weed control from pre- to post-emergence treatment.

Hoy (1992b) has discussed the potential for engineering pesticide resistance into insect natural enemies to facilitate their use in systems where pesticides are used against other pests (see also Hoy, Chapter 9, this volume). Substantial success of this kind has already been achieved by conventional selection methods.

Recent success in incorporating benomyl resistance into the entomopathogenic fungus, *Metarhizium anisopliae*, could make this agent more useful against insects on crops where fungicides are used (Goettel *et al.*, 1990).

Analysis

What does this brief survey tell us about the broad strategy and objectives of current biotechnology for pest management? Firstly, it reveals a striking technological consistency: for all their diverse manipulations, virtually all current initiatives involve the manipulation of fragments of bacterial or viral genome into other viruses and bacteria or into plants.

This pattern reflects the substantial extent to which current developments have been based on accessible technologies. For the companies and institutions involved, the opportunity to realize short-term, economic returns on substantial investments in biotechnology has inclined research towards simple, well-known microbial systems. Leaders of these programmes have noted that the development of markets for these technologies has proceeded somewhat more slowly than the technologies themselves (Wochok, 1991). To this extent, it might be fair to say that current initiatives have been as much technology- as demand-driven.

A second characteristic common to current initiatives is the conservatism of their targets or potential markets. Many address problems for which there are already alternative products or processes, often natural forms of the same organisms. Thus, engineered viruses and bacteria are variants of existing biopesticidal or antagonistic formulations of the same species, while some transgenic plants or plant-associated bacteria seek to improve on what can already be achieved by topical application of a particular microorganism. In the case of engineering pesticide resistance into crops or natural enemies, the manipulation has no inherent value but reinforces conventional technologies of chemical control.

If we look more closely at the particular technical objectives of these manipulations, we see that many are directed at improving the performance of an engineered product relative to its 'wild-type' competitor by:

- broadening the target spectrum of the product (e.g. the insect host range of *B. thuringiensis*, or the expression of cowpea trypsin inhibitase in a range of crops);
- increasing the speed of action of the product (e.g. engineering insect venoms into baculoviruses); and
- enhancing the delivery of the product to the pest (e.g. by distributing *B. thuringiensis* endotoxin or plant virus coat proteins throughout plant tissues).

These objectives have much in common with the objectives of chemical pesticide development, and with the perception that what the farmer wants from a pest-control product is dependability, easy handling, quick kill and a broad-spectrum action. The implication is that markets for new products of biotechnology will be won through direct competition with existing alternatives.

Overall, a brief analysis suggests that biotechnology for pest management has been technology-driven and conservative in its tendency to keep new technologies close to existing, pesticide-like models.

Biotechnology and IPM for Developing Countries

Comparing recent developments in IPM and in biotechnology for pest management leads to the following impression: current biotechnological approaches to IPM are born of a 'technological IPM' paradigm and destined for the 'IPM packages' of that approach. Many have been modelled on existing products, as competitors and replacements for them. Current IPM in developing countries, on the other hand, is leaning towards a different paradigm, that of 'ecological IPM', which looks for a more local and participatory approach to the selection and development of pest-control methods and an emphasis on methods that are appropriate and self-sustaining.

On the surface, the sometimes technology-driven nature of new biotechnologies may limit their value in IPM systems that are increasingly identifying their needs locally and designing the appropriate inputs required. But can we be more specific about where the mismatching of biotechnology and IPM might occur?

The fact that some novel plant-protection biotechnologies have been developed as close analogues of existing pest-control methods may be useful. It allows us to examine the future of these existing methods in an IPM context, and then to consider how related biotechnological methods will do in comparison.

I shall attempt this analysis with two model systems: biopesticides/antagonists, with which genetically engineered *Bacillus*, *Pseudomonas*, *Agrobacterium*, nematodes and baculoviruses may be compared, and pest-resistant plants, with which plants made transgenic for insect or disease resistance may be compared.

The future of biopesticides

Biopesticides have been slow to find a use in pest management. Only with the recent demand for pesticide replacements has there been the incentive in public and private institutions to make the improvements necessary to make them into dependable, viable products. Most of these improvements have focused on constraints in production, quality control, storage and formulation for good coverage and persistence in the field. Note that these are not the same processes as those presently targeted from improvement of biopesticides through genetic engineering.

Current registered biopesticide products include bacteria (104 products on the market, mostly *B. thuringiensis*), nematodes (44 products), fungi (12 products), viruses (eight products), protozoa (six products) and arthropod natural enemies (107 products) (Lisansky, 1993). Arthropod natural enemies, such as predatory mites, ladybirds and parasitic wasps, are often considered as biopesticides when they are mass-reared and released against particular pest populations.

The current commercial development of biopesticides is largely undertaken by multinational agrochemical or pharmaceutical companies in industrialized countries. Some of these products will find their way, like their chemical counterparts, into developing-country markets. The potential for biopesticides in developing countries, however, goes beyond what might be imported from abroad.

Biopesticides, unlike the chemical pesticides that they can replace, can be produced with technologies that are well within the reach of most developing countries, and on an appropriate scale. This could make possible the development of local biopesticide products that have the advantage of targeting local pests (Prior, 1989). Such pests may not create sufficient markets to attract the interest of multinationals. Further, locally produced biopesticides could generate employment and business, reduce environmental and health risks relative to chemical alternatives and reduce the need to buy in expensive foreign pesticides, whose supply and appropriateness are often uncertain.

Like any living control agent, the effectiveness of a biopesticide depends on two factors: its capacity to kill pests and its capacity to reproduce on pests and thereby continue and compound its killing action – in ecological terms, its functional and numerical responses.

By and large, the crop-protection industry, in developing commercial biopesticides, has ignored the numerical response of natural enemies and concentrated only on their direct killing action. Not only has the numerical response of pathogens been ignored in biopesticide development, but the species selected are often incapable of such a response because they are alien to the crop environment and not adapted ecologically to it.

Thus, *B. thuringiensis*, the most widely used biopesticide in field crops, is naturally associated with soils (Meadows, 1993), possibly tree surfaces (Smith and Couche, 1991), and rarely exhibits epizootics in pest populations resident in crop vegetation, where it is usually applied. The toxin-based nature of its action means that its actual reproduction is relatively unimportant to its controlling action. A similar situation exists for entomophilic nematodes, the second most commercialized biopesticides. Heterorhabditid nematodes are also soil-dwelling organisms that survive poorly in crops and only rarely cause epizootics in insects (Kaya and Gaugler, 1993) – indeed they seem particularly closely associated in nature with sandy seaside beaches (Hara *et al.*, 1991; Amarasinghe, 1993; Griffin *et al.*, 1993).

B. thuringiensis and nematodes exemplify the traditional focus of commercial biopesticide development, in that they are products selected for their virulence, not for their efficiency in suppressing pest populations. They are truly 'pesticidal' biopesticides. Over the next decade, products like *B. thuringiensis* and nematodes will play a key role as pesticide replacements in IPM systems where intervention is regularly required and where their safety to natural enemies is desirable. They will contribute to the recovery of crop systems where pesticide use has been excessive and to the re-establishment of natural control. If introduced on an at-need basis, they may become long-lasting components of IPM systems. However, as 'pesticidal biopesticide' they will encounter increasingly many of the problems of chemical insecticides, including the risk of resistance development. Their use in a prophylactic manner will greatly increase this risk and reduce their value to IPM, as has been evidenced by the experience of *B. thuringiensis* use on diamondback moth in South-East Asia (Waage, 1992).

In the longer term, IPM will encourage the self-renewing properties of natural enemies, that is, their numerical responses. Biopesticides of greatest value to IPM systems will be ones that exploit an organism's capacity to persist and increase once released. Rather than being frequently released pesticidal products, they will be infrequently released inoculative products, subsequently managed for their continuing impact.

This trend is already seen across a range of present and potential biopesticidal systems, often as a result of 'accidents of experience' where a biopesticide proves to be more persistent than expected. Thus, the fungus *Beauveria brongniartii*, applied as a biopesticide for control of grass-feeding scarabs in the soil, has proved in some situations to persist and cycle in the environment, suppressing populations for years (Keller, 1992). Similarly, fungi developed as mycoherbicides for weed control have sometimes proved so persistent as to greatly reduce the need for repeated applications (Charudattan, 1991).

Discoveries like this encourage the concept of inoculating pest populations with pathogens for long-lasting effects. One clever method for doing this, currently under commercial development for insect pests, is a pheromone- or attractant-baited trap that lures in pests, doses them with a fungus or virus and then releases them to spread the disease through the population.

Similarly, in the commercial use of predators and parasitoids for insect control, there is a distinct move away from mass production and mass release to strategic release of small numbers early in a season, which build up to pest-controlling numbers later on. For instance, small early season inoculations of parasitoids, such as *Trichogramma* spp. or *Encarsia* spp. against moth pests and whiteflies, respectively, are capable of season-long suppression and are less expensive than traditional, calendar-based inundative releases of far greater numbers.

Finally, for the control of soil-borne plant diseases, such as take-all disease of wheat, the strategy of mass application of particular bacterial or fungal competitors or antagonists is seen increasingly as less desirable than the strategy of encouraging and managing the local flora of disease-suppressive soils, perhaps with occasional inoculations of particular species (Cook, 1990).

Implications for genetically engineered biopesticides

What are the implications of these predicted trends for biotechnology and its application to improve biopesticides? My first conclusion is that *B. thuringiensis* may not be the ideal choice for a biopesticide to meet the longer-term needs of IPM. While engineered *B. thuringiensis* products may be widely used as pesticide replacements, their pesticidal nature may lead them ultimately to be less favoured in IPM systems, however much engineering improves their spectrum or killing action.

Secondly, the process of engineering itself may reduce the valuable persistence of a biopesticide relative to its natural counterpart. Biotechnologists, confronted with the potential risks of introducing engineered organisms into the environment, argue that engineering inevitably reduces fitness relative to wild types, because of the cost of carriage and expression of recombinant DNA. While this may be only partly true from a population genetic perspective (Lenski and Nguyen, 1988), to the extent that it occurs it will weaken the capacity of a biopesticide to have a continuing effect. Indeed, the perceived risks associated with persistence may lead biotechnologists deliberately to engineer non-persistence into new biopesticides. An example, albeit one done for experimental rather than commercial purposes, has been the removal of the viral coat protein gene from insect baculoviruses, which reduces their survival on plant surfaces. Natural viruses have a considerable potential to cause epizootics in pest populations; these engineered viruses may not.

Finally, the current objective of biotechnology to make biopesticides more virulent and faster-acting has fundamental, ecological implications for their ability to reproduce and spread in pest populations. Population models of insect-pathogen systems reveal a distinct dynamical trade-off between virulence and transmission (Anderson and May, 1981): where long-term suppression of pest densities is the goal, pathogens of intermediate virulence may be more effective.

The future for pest-resistant plants

Breeding for plant resistance to pests is a major activity of the well-resourced international agriculture research centres, as well as national programmes, and has resulted over recent decades in the production of substantial resistance in crops to plant diseases and some resistance to insects. Buddenhagen (1991) has argued that far more could be done in this area, and that the popularity of insecticides, and even insect-based IPM, has reduced efforts in resistance breeding (see also Buddenhagen, Chapter 11, this volume).

Most resistance breeding to date has focused on methods that result in qualitative, vertical resistance. This approach is convenient because high levels of resistance can be achieved and because the method is compatible with breeding schemes used for enhancing crop performance through control of major genes. However, its gene-for-gene nature can sometimes lead to its breakdown through the evolution of resistance-breaking pest genotypes. Classic examples of such resistance breakdown in insect pests include the hessian fly, *Mayetiola destructor*, and greenbug, *Schizaphis graminum*, on wheat, and the brown planthopper, *Nilaparvata lugens*, and green leafhopper, *Nephotettix virescens*, on rice.

While plant-resistance breeding is often identified as a component of IPM, its actual integration with other pest-control methods in IPM systems has been limited. In most cases, breeding programmes have sought single-technology solutions to pest problems (i.e. complete resistance), much like chemical pesticides. This is often hard to achieve for some pests, e.g. stem-boring moths in cereals, and, while partial resistance has been possible, it has generally not been used. Partially resistant varieties, however, may have a value when viewed in an IPM context, where the contribution of natural enemies and other factors can complement their effect (Thomas and Waage, 1993).

An IPM approach can also contribute to the durability of plant resistance to pests. The action of mortality factors, such as natural enemies, can reduce pest populations and selection pressure to overcome resistance factors. The role of natural enemies in maintaining durability in this manner has been shown theoretically and experimentally (Gould *et al.*, 1991a; Gould, Chapter 16, this volume).

Another approach to more durable resistance is to select for horizontal rather than vertical resistance. Horizontal resistance is quantitative in nature, and involves mass selection techniques in existing susceptible populations. Where successful, this leads to a durable, polygenic form of resistance, resulting from many different mechanisms (Robinson, 1991). Horizontal resistance techniques have been developed for plant–pathogen systems, but can be applied to insect populations as well (Dent, 1991; Bosque-Perez and Buddenhagen, 1992). The partial nature of horizontal resistance, at present seen as a disadvantage by plant breeders, may in fact be an advantage in IPM, where it can support the continuity and contribution of natural enemies and their important density-dependent effects. The greater emphasis which horizontal resistance breeding might place on selecting with local material for local conditions is compatible with current ecological approaches to IPM.

Implications for transgenic plants

The potential contributions of biotechnology to the creation of pest-resistant plants is enormous. This is particularly true where biotechnology offers possibilities of resistance that are not easily achievable by conventional means, as perhaps with virus resistance in some crops and insect resistance for particular pests, such as cereal stem-borers.

Current work with transgenic plants, particularly the incorporation of the *B. thuringiensis* delta endotoxin into crops for control of insects, appears to be proceeding on a vertical resistance model, based on complete resistance conferred by one or a few genes. These varieties, like those produced through conventional vertical resistance breeding, will be susceptible to the development of pest resistance. Further, they may undervalue the benefits of an IPM approach suggested above.

Many will now agree that the risk of resistance development to *B. thuringiensis* engineered transgenic plants is considerable, and probably greater than that for a *B. thuringiensis* formulation topically applied to a crop on a need-based IPM basis, simply because of its more continuous and extensive expression. Indeed, some entomologists have come together to express their concern that use of transgenic plants may waste a valuable resource by encouraging resistance development (Anon., 1992). The pest-control industry itself has established a *B. thuringiensis* Management Working Group, which funds research into resistance management (Anon., 1993b; see Gould, Chapter 16, Roush, Chapter 15, and Peacock *et al.*, Chapter 13, this volume).

Suggested solutions to resistance problems involve more complex strategies of gene deployment, for instance, mixed or intercropped populations of resistant and susceptible plants, or genetic methods to restrict expression of genes to certain parts of plants or certain times. Resistance management is therefore a strong possibility, but the track record of resistance management for chemical pesticides is not encouraging.

Overall, the present strategy of protecting plants with engineered, single-gene defences may offer some real opportunities for improving pest management in the short term, particularly on crops where conventional breeding cannot achieve resistance. However, as with the strategy of conventional breeding for vertical resistance, it runs the risk of breakdown and its contribution to sustainable IPM systems over the longer term may be limited by this. In its emphasis on complete resistance, it may not only reduce its durability, but also undervalue the contribution of other components of IPM systems.

Alternative Applications of Biotechnology for Pest Management

The first generation of biotechnologies for pest management has considerable potential but also some problems with respect to their integration into IPM systems for developing countries. Are there other possible applications of biotechnology to IPM which might also be applicable to, and possibly more compatible with, trends in IPM? I would like to suggest three: (i) the use of biotechnology in the characterization of pests and natural enemies; (ii) the culture and mass production of natural enemies; and (iii) the evaluation of natural enemies in IPM systems. These applications, like those discussed above, may draw upon genetic engineering, but they may also make good use of more conventional biotechnologies, such as fermentation of microbials, enzyme-linked immunosorbent assay (ELISA) techniques and amino acid analysis, particularly in the developing world, where their application is still limited (Ananthakrishnan, 1992).

Biotechnology and biosystematics

The characterization of organisms, biosystematics, is essential to pest management. As IPM steers pest management towards greater reliance on sustainable biological processes, such as plant resistance, natural enemies and manipulating pest reproduction and behaviour, accurate characterization of the players (plants, pests and natural enemies) becomes even more important. Improper characterization may, for instance, lead to the introduction of a disease-resistant plant that is not resistant to the local form of the disease or the use of the wrong natural enemy for a particular pest problem.

Communicating the local experience of IPM between countries also requires the common language of taxonomy. In the absence of a detailed understanding of local ecology, taxonomic names often reveal a great deal about ecological niches. For instance, a parasitic wasp of the genus *Cotesia* can usually be counted on to confer significant larval mortality wherever it is found attacking caterpillars on crops, whereas a parasitoid of the similarly distributed genus *Pristomerus* cannot.

The characterization of natural enemies is particularly challenging in tropical developing countries, where the biodiversity is enormous, probably very important and extremely poorly known (Hawksworth, 1991; Waage, 1991). Pest management generally has little biosystematic support in developing countries, which are still over-reliant on the great European, North American and Australasian repositories of centuries of tropical collection. Recent initiatives, particularly BioNET, have set out to redress this (Jones, 1994).

Today, biosystematics is challenged by the need to formalize important biological variability below the species level, where conventional morphological means of taxonomy often cease to be dependable. Thus, some forms of plant pathogens, such as *Fusarium oxysporum* and *Colletotrichum gloeosporioides*, are morphologically identical but attack different crops. Some forms even attack weeds in crops and therefore have potential as bioherbicides.

Biochemical means have proved useful in distinguishing between similar pest and non-pest forms, but biotechnology now offers even more powerful tools. Taxonomic techniques based on biotechnological innovation, including monoclonal antibodies, random amplified polymorphin DNA (RAPD), polymerase chain reaction (PCR) and other methods, offer considerable promise for improving characterization across a range of pest taxa (Hawksworth, 1994). The development of easily used probes for pest identification, based on these techniques, is well under way for various pest taxa, including the plant-pathogenic fungi (Brown, 1993).

For natural enemies, genetic methods for characterization have accelerated biopesticide development. Thus, scientists spent decades seeking and characterizing the diverse and useful forms of *B. thuringiensis* which we use today. Now, biotechnological methods have helped to identify at least ten different genetic forms, producing different toxins with predictably different specificity to insect groups. As a result, probes developed for these can quickly screen new isolates for their host range without the need for biological characterization (van Frankenhuyzen, 1993).

Even where pests and natural enemies can be identified by conventional morphological or biological characters, biotechnological probes offer a rapid and efficient alternative. This is of particular value in quarantine, where the presence or spread of exotic pests needs to be quickly and extensively assessed, usually by non-experts.

Biotechnology for improved natural enemy culture and production

Biological control is the neglected discipline of IPM, partly because we still know too little about the identity and biology of natural enemies to conserve and use them to maximal advantage in pest management. Beyond the taxonomic problems, we also face the difficulty of rearing natural enemies with complex life cycles, or producing them in numbers for release against pests. Biotechnology has the potential to solve many of these problems, through increasing our understanding of the genetics and physiology of natural enemy reproduction and by enabling us to manipulate it.

For instance, insect viruses, some entomopathogenic fungi (key genera of the Entomophthorales) and some nematodes (e.g. Mermithidae) remain constrained in their development as biopesticides by the need to produce them *in vivo*. The development of economical insect tissue culture techniques to which biotechnology can contribute might remove the present constraint of mass-producing the host insect in order to produce the natural enemy.

For some parasitic natural enemies, rearing and economical mass production is possible on alternate, 'factitious' hosts, which are easier to rear than the target host (Waage *et al.*, 1985). However, coevolutionary immunological interactions between host and parasite often make rearing on a non-host impossible. Biotechnology could be applied to a broad natural enemy host range by manipulating these immunological systems. With insect parasitoids, for instance, biotechnology is helping us to understand and manipulate the mutualistic polydnaviruses that appear to determine host range in some parasitoid–caterpillar systems.

Other natural enemies, by virtue of their saprophytic potential, can be cheaply produced on simple nutrient substrates. These include *B. thuringiensis* and some plant- and insect-pathogenic fungi. Even here, however, there is the potential, through the biotechnology of fermentation and a better genetic understanding of physiological requirements, to greatly improve production efficiency. For instance, the deuteromycete fungi *Beauveria* and *Metarhizium* spp. are frequently mass-produced for insect control in a two-stage process of mycelial production in liquid and spore production in air. Modifications to the fungi, or to their substrates, which could permit efficient single-stage production of spores would greatly reduce production costs and increase productivity.

Biotechnology may enable us, eventually, to substitute wholly synthetic, 'meridic' diets for even more complex natural enemies, such as insect predators and parasitoids (where *in vitro* rearing is already possible), through techniques that enable us to better understand and manipulate nutritional physiology.

Biotechnology for improved evaluation of natural enemies

Wherever IPM is developed, the action of natural enemies emerges as both important and subtle. Measuring the precise contribution of natural enemies is often difficult: many predators kill pests and leave little trace; key bacteria or fungi antagonistic to soil-borne plant pathogens are often hard to determine in the rich flora of the soil; some viruses lie latent in the genomes of caterpillar hosts, released to kill by still poorly understood mechanisms.

Biotechnology has the potential to contribute to all of these problems of evaluation. For instance, crops contain a great diversity of predators of insect pests. But which ones actually feed on the pest, and which are the most important? These questions can now be answered in some crops by the use of monoclonal antibodies to assay predator gut contents for pest-specific proteins. Similar techniques can also be used to detect diseased or parasitized pests from a population sample, without the need to culture pest individuals until the parasites develop.

The fate – and hence the value – of a released natural enemy is often difficult to trace when natural populations exist as well. In such circumstances, genetic markers can be valuable. In our current programme for development of a mycoinsecticide against the desert locust, for instance, the strain of *Metarhizium flavoviride* used has been characterized genetically and biochemically, so that the chance of distinguishing it from local strains is maximized. Thus, recovering this genotype from dead pests will confirm its impact on pest populations, while recovering it from non-target species will determine its environmental impact on non-target species relative to indigenous genotypes. These methods are also useful, of course, to determine the fate of engineered microbials in the environment.

Conclusion

Perceptions of IPM are at present diverse, and I have suggested two particular approaches to IPM, which might be thought of as ecological and technological, both interpreting differently a simple ground truth, the need to create a more diverse and sustainable basis for pest management. The ecological approach to IPM has particular popularity in developing-country agriculture, partly because of its emphasis on farmer empowerment, participatory research and local, sustainable and self-renewing pest control methods, and partly because of negative experience with external technologies, such as pesticides.

In contrast, current biotechnology for pest management is born of the scientific culture that has produced a technological approach to IPM. It has followed a conservative blueprint designed by the pest-control industry to produce substitutes for existing pesticides and other products. While these products will find a place in their intended markets in the highly technological agriculture of industrial countries, they may prove less appropriate to an ecological IPM approach adopted by some developing countries. Given that this biotechnology is inevitable, and that developing countries deserve to benefit from it where possible, there emerges an urgent need to examine the value of these technologies in the local IPM context of developing countries, through a more participatory approach. Where they are deployed, this should be done as far as possible in an ecological IPM context, to improve their cost-effectiveness and durability.

The deliberate development of pest-management biotechnologies for developing countries (e.g. transgenic forms of tropical subsistence crops) must be viewed in the context of a broader IPM approach. There may be a considerable opportunity cost to steering research into particular developing-country pest problems down a biotechnological track.

Should we, for instance, be investing heavily into engineering *B. thuringiensis* to improve pest control in developing countries, when those countries still lack the potential to characterize, develop and commercialize their own indigenous, natural *B. thuringiensis* and other pathogens as conventional pest-control products? Should we be investing in the (possibly temporary) benefits of transgenic cotton for protection against bollworms in developing countries, when most of these countries still lack the knowledge to implement known IPM technologies which would get them off the pesticide treadmill forever? Would it not be wiser to first help to build the IPM capability that enables farmers, extensionists and researchers to make decisions about whether inputs like biotechnology are useful in their crop systems?

This is not to say that engineering Bt into tropical crops, for instance, should not be done, it simply should not be done as a technological alternative to establishing good local IPM systems for those target pests, but be done as a development activity in pest management. There are other reasons for undertaking research into such biotechnology, not the least of which is to improve national capability in biotechnology, but then let us get more excited about the process, and less about the product.

For the next generation of biotechnologies for crop protection, I suggest that we should be looking first at methods which would help us with the characterization, use and evaluation of pests, resistant crops and natural enemies.

Then we should look further, and consider possibilities beyond our current preconceptions. For instance, if sustainable pest management is the ultimate target, we should perhaps be looking at ways in which biotechnology could increase the persistence and numerical response of biological control agents in the field, to make their effects more long-lasting. To do this, we would need to challenge the current position that good engineered organisms should not persist, multiply or work themselves into ecosystems. If biotechnology is going to offer us something truly sustainable in pest management, I suggest that we grasp this nettle, and begin to do the ecological research which will help us determine whether such engineering would in fact be useful, and safe.

Most importantly, and most immediately, we should be investing in IPM training and research so as to create an informed customer base for future opportunities in IPM. Farmers expert in IPM, working with researchers, are the best, if not the only, way to ensure a future demand for useful biotechnological initiatives.

Integrated Pest Management in Developing Countries 3

LIM GUAN SOON

Introduction

Agricultural production in developing countries is seriously affected by the action of pests, including all agents that cause crop losses, such as insects, weeds, fungi, vertebrates and other invertebrates.

Pest problems in developing countries today are largely associated with the accidental introduction of exotic pests and with efforts to intensify agricultural productivity in order to meet a growing demand for food and export income (Waage, 1993). Contributing factors in increasing productivity include expanding areas under monoculture, shortened fallow periods and sometimes overlapping of crops, the planting of genetically uniform, high-yielding but susceptible varieties and the use of agrochemicals.

The use of pesticides dominates crop-protection efforts in many developing countries, and reliance on them is growing. The main reasons are their ability to control short-term pest problems (e.g. against certain pests that suddenly become an epidemic) and their potential for replacing labour-intensive work in some situations (e.g. weeding). The strong marketing campaign of the pesticide industry, coupled with weak government extension, is also a major contributing factor.

However, the heavy reliance on pesticides has resulted in many problems and is causing much concern, particularly in the areas of human health and social aspects, the natural ecology and the environment. These are examined later, including the need for integrated pest management (IPM), as well as the present scope and shortcomings in the practice of IPM.

Pest Control Practices with Pesticides

Although a wide range of control tactics is available, few farmers consciously attempt to integrate them into a system for managing pests. Usually there is a tendency to use only one or two methods of control, and often only one, depending on the crops and the pest involved. For reasons already stated, many farmers still rely on chemical control, particularly for cash crops.

A common practice for vegetables is frequent pesticide spraying on a calendar basis. At least 50% of farmers in Malaysia spray two to three times per week against the diamondback moth (*Plutella xylostella*) (Ooi and Sudderuddin, 1978). In Honduras, an average of six to ten applications are made per crop season, and Costa Rican farmers applied chemicals 16 times (Andrews *et al.*, 1992). Dosages in many cases have also been higher than those normally recommended.

Another prevalent practice is mixing pesticides, also commonly known as 'cocktails'. This practice is particularly common in South-East Asia (Ooi and Sudderuddin, 1978; Magallona *et al.*, 1982; Iman *et al.*, 1986; Rushtapakornchai and Vattanatangum, 1986).

The application of pesticides on crops right up to harvest is also widespread, largely because farmers want to produce clean and unblemished vegetables in response to demands by both consumers and exporters.

In general, farmers have little information on or understanding of the hazards of pesticides. For example, most are unaware of the danger of skin absorption and do not take adequate precautions during spray operations. Pesticide poisonings are the result.

In the absence of pesticide regulatory schemes, and with poor enforcement where regulation exists, coupled with strong marketing compaigns by the pesticide industry, farmers quickly grow to rely on pesticides. Because of overuse, many problems now exist.

Problems Arising from Pesticide Overuse

Pesticides causing pest outbreaks

As pesticide use increases, with frequent overuse and misuse, problems with insect pests, diseases and weeds also increase. For example, heavy insecticide applications often precede the outbreaks of rice brown planthopper (Kenmore, 1991). Today, many insecticides used commonly in the rice field will result in a resurgence of the brown planthopper, resulting in 'hopper burn' crop damage. Some examples include sprays of carbofuran, deltamethrin, diazinon and methyl parathion (Heinrichs, 1979).

The use of broad-spectrum insecticides on plantation crops to control

Table 3.1. Number of insecticides to which diamondback moth has developed resistance in some South-East Asian countries.

Insecticide group	Indonesia	Malaysia	Philippines	Thailand	Vietnam
Organochlorine	3	4	1	–	4
Organophosphate	15	9	9	12	10
Carbamate	3	4	1	3	1
Pyrethroid	5	4	2	8	–
Insect growth regulator	1	2	0	5	–
Others	1	?	1	?	–

– No recorded cases; ?, resistance suspected in *Bacillus thuringiensis*.

outbreaks of cocoa bark-boring caterpillars (*Endoclita hosei* and *Zeuzera* spp.) and oil-palm defoliators (mainly *Mahasena corbetti*) have also caused outbreaks of many other pests (Wood, 1971).

Examples of pesticide-induced pest outbreaks now exist for many pests on different crops, such as cotton in Peru (Boza-Barducci, 1972), banana in Costa Rica (Hansen, 1987), mango in Pakistan (Waage, 1993) and vegetables in Asia (Talekar, 1992). In fact, pesticide-induced pests are a feature of non-sustainable agriculture practised in many parts of Asia today, and this phenomenon poses one of the greatest problems for the future (Whitten, 1992). The insecticides usually destroy more of the natural enemies than the pests themselves (Waage, 1989). This was especially evident in the case of the brown planthopper of rice (Kenmore, 1980; Ooi, 1988).

Pesticide-resistance development

Over 500 species of insects and mites have now become resistant to one or more pesticides (Georghiou, 1990). Some are resistant to all major pesticide groups (e.g. diamondback moth; Table 3.1). The problem is especially acute in South-East Asia, where applied dosages have to be increased and the products changed frequently (Lim, 1990). There has been a shift from intensive use of organochlorines to organophosphates, then to carbamates, pyrethroids, insect growth regulators, and then to *Bacillus thuringiensis*. Resistance to the latter has been confirmed in Hawaii (Tabashnik *et al.*, 1992a) and early signs have already been detected in Thailand (Miyata *et al.*, 1988) and Cameron Highlands, Malaysia (Syed Raman, 1993, personal communication).

Resistance to herbicides and fungicides is less prevalent but growing (Davies, 1992). Examples include a benomyl-resistant strain of *Colletotrichum capsici* from chilli (Meon, 1986) and *Echinochloa colona* weed developing resistance to propanil (Garro *et al.*, 1991).

The trend in resistance development suggests that many new cases will be found as researchers explore a wider range of pests. Pesticide-resistance development is a serious problem and, when combined with pesticides killing off natural enemies, it creates the 'pesticide treadmill' phenomenon of repeated pest resurgences and increasing use of pesticides until crop production becomes uneconomical and collapses (Waage, 1993).

Human pesticide poisonings

Pesticides harm human health. For example, increasing mortality rates among males aged 15–54 years in rice areas of central Luzon in the Philippines was found to coincide with increases in insecticide use (Loevinsohn, 1987).

Excessive use of pesticides without protective measures or regard for safety has been the main cause of pesticide poisonings. In Thailand, for example, there were 4046 poisoning cases with 289 deaths from 55 provinces in 1985 (Kritalugsana, 1988). In Sri Lanka, 17.1% of the annual average of 13,000 cases from 1975 to 1980 were due to occupational exposure (Foo, 1989). In the plantation sector, all workers applying paraquat herbicide suffer from skin rash (Arumugam, 1992).

Pesticide contamination of produce

Unacceptable levels of pesticide residues in agricultural produce are causing increasing concern, particularly to urban consumers. For example, tomatoes from several markets in Malaysia had residues of dithiocarbamate fungicides ranging from 0.21 to 15.8 mg CS_2/kg (Lim *et al.*, 1983). In Indonesia, residues of endosulfan were found in carrot in 1983, diazinon in onion in 1984, benomyl in potato in 1985 and dithiocarbamates in tomato in 1985 (Soekardi, 1988). Cabbage in Honduras had 5.5 and 4.3 ppm of cypermethrin and chlordane, respectively, during the 1989 dry season (Andrews *et al.*, 1992). In 1981, 56 of 58 samples of food products in Cape Verde ware found to contain high levels of dichlorodiphenyltrichloroethane (DDT) or lindane and were considered not suitable for consumption (Abou-Jawdeh, 1992).

Excessive residues on market vegetables not only cause concern about general consumer health and the envirorment, but also have significant trade implications. For example, in 1987, contaminated vegetables exported to Singapore from Malaysia were rejected (Lim, 1990). Also, the Guatemalan broccoli export industry suffered a major setback in 1990 when the USA rejected large quantities of frozen broccoli found to be tainted with residues of phenthoate (Andrews *et al.*, 1992). Like the development of resistance to pesticides, increasing cases of unacceptable residues will be found as more tests on residues are made in the future.

Environmental impact and hazards to non-target organisms

Pesticides can adversely affect the environment in various ways. For example, rice-field fish declined in areas where certain insecticides (e.g. endosulfan, endrin, dieldrin) were introduced. In the Philippines, all insecticides used in the rice-field are toxic to fish (Cagauan, 1990). There is also evidence of a negative impact on many other aquatic species (Pingali, 1992). Pesticide accumulations have been detected in snails, fish and frogs (Ocampo *et al.*, 1991) and also in nearby well water (Medina *et al.*, 1991).

Hazards of pesticides to natural enemies can greatly influence success in the control of a pest. For example, by switching from hazardous chemical insecticides to the use of the relatively safer *B. thuringiensis*, diamondback moth's parasitoids (particularly *Diadegma semiclausum*) are able to increase in numbers to exert effective suppression of the moth. Consequently, diamondback moth in the highlands of Malaysia and Taiwan has declined and is no longer a serious problem in many cultivated areas there (Talekar, 1992).

The natural enemies of insect pests are often more susceptible to pesticides than are the pests themselves. Thus, a pesticide may remove a degree of future control and thereby create a serious insect pest situation (Waage, 1989).

Rise in production costs

Increased production costs are largely due to the rapid development of resistance by pests to chemicals, leading to a need for higher dosages and more frequent applications. In some countries, such as the Philippines, the supply problem is further compounded by foreign exchange restrictions (Magallona, 1986).

The Need for IPM

Pesticide use in crop production in many developing countries has created a number of problems, some of them acute. Continued use, therefore, is increasingly challenged for socioeconomic, health, ecological and environmental reasons. An unexpected consequence of the use of pesticides is their capacity to create rather than control some pest problems.

More national and international organizations, including farmer groups, are re-evaluating the use and need for pesticides. They conclude that IPM is the best option to reduce the many problems arising from reliance on pesticides. Considerable evidence now exists to show that routine chemical treatment is both unnecessary and undesirable. Past experiences indicate that successful pest control programmes are usually those combining methods in a single coordinated management system. The IPM strategy maximizes the self-renewing effects of the natural enemies of pests (otherwise known as biological control) and cultural components, including host-plant resistance. Chemicals are used as a supplementary measure and only when necessary.

Successful IPM programmes yield many benefits (Ooi *et al.*, 1992). These include lower costs of IPM and significant savings to farm households (Oka, 1990). At the national level, many governments annually save millions of dollars spent on pesticide subsidy schemes (Kenmore, 1991). Because nearly all pesticides in developing countries are imported, huge savings in foreign exchange are realized. The stability of agricultural production which IPM provides has important benefits for political stability, particularly where agriculture is the dominant economic sector. Moreover, the potential that IPM offers for reducing pesticide use is environmentally significant in improving water quality, reducing farmer and consumer risks from pesticide poisoning and related hazards, and in contributing to ecological sustainability through conserving natural resources.

Integrated pest management is not a technology package, but rather a process that engages farmers in experiential learning and dynamic local research that continuously reshapes solutions to their pest problems. As such, IPM plays an important role in the process of developing local (village) self-reliance in rural development. Through the discovery of new skills, IPM farmers are stimulated to seek greater management control of other production activities such as irrigation, soil conservation, credit and marketing. In this way, they become empowered through their own training and the process of human resource development. IPM therefore has broad and long-lasting socioeconomic benefits far beyond plant protection activities.

It is now well established that the IPM model has many advantages. The cost of not adopting IPM is high, particularly for developing countries, where pesticide regulations and their enforcement are mostly lacking and reliance on pesticides is growing rapidly. Integrated pest management also offers a unique opportunity for economic development within the context of the United Nations Conference on Environment and Development (UNCED) Agenda 21, which affirms the global recognition of IPM as a requirement in any agricultural development project (UNCED, 1992).

IPM Examples and Experiences

Many attempts at IPM have been made in Asia and elsewhere (Wiebers, 1991). However, most involved experimental studies that still require large-scale verification at farm level. How IPM can be developed, and, in particular, the practice at farm level, is best illustrated by examining programmes already in operation. Such examples can provide lessons on how pest-control methods are assembled into an IPM strategy, including anticipated constraints to be avoided, as well as attributes that need to be incorporated, and other requirements. Common features of the underlying causes for success or failure may also be identified.

The selected cases described below include a geographic mix, covering a diversity of approaches using different IPM elements. Both cash crops and plantation crops are considered. Two outstanding examples that specifically involve the close participation of farmers – rice in Asia and soybean in Brazil – are not included because they will be described elsewhere in this book.

Insect pests of oil-palm and cocoa

In the late 1950s some west Malaysian oil-palm estates suffered severe outbreaks of the bagworm *Mahasena corbetti* and other leaf-eating caterpillars. Intensive spraying with long-residual, contact insecticides (e.g. DDT, dieldrin, endrin) led to outbreaks of several other pest species. As a result, from the end of 1962 it was recommended that spraying broad-spectrum contact insecticides be stopped. Since then, high populations of natural enemies of the outbreak pests have developed. Major outbreaks rapidly declined and complete natural control has been re-established in the affected area. The occasional and localized resurgence was readily controlled by selective spraying.

Similar situations occurred for some bark-boring caterpillars of cocoa in Sabah, Malaysia. Indiscriminate spraying with benzene hexachloride (BHC), DDT, dieldrin and endrin caused outbreaks of several new pests, such as aphids, mealy bugs and planthoppers. Discontinuation of spraying eventually allowed the re-establishment of natural enemies, bringing the new pests completely under control. Damage from the bark-borers also rapidly declined when their secondary host tree, *Trema cannabina*, was eventually removed from the vicinity.

Diamondback moth on vegetables in the highlands of Asia

The diamondback moth has become the most important pest of crucifers wherever its effective natural enemies are absent. The problem is exacerbated by excessive use of insecticides, largely because of the rapid development of resistance. It has now acquired resistance to almost all insecticides (Lim, 1990; Table 3.1), underscoring the danger of relying solely on chemical control.

Bacillus thuringiensis, a biological insecticide, is now being used. This has allowed an increase in crucifer crops, because they are not harmed by the biological insecticide. In areas where farmers have stopped using chemical insecticides, the parasitoids continue to keep the moth in check.

Rhinoceros beetle on coconut in Asia and the Pacific

When the rhinoceros beetle was accidentally introduced into the Pacific, it was free from the natural enemies that normally keep it under control. The beetle was then able to breed unhindered into large populations, causing severe damage to coconut palms.

An important biological control agent of the beetle is a baculovirus discovered in Malaysia in the late 1960s. This agent, together with good sanitation in removing dead trunks as breeding sites, constitutes the main IPM strategy against the beetle.

Cassava mealy bug in Africa

First reported in 1973 in the Congo, the cassava mealy bug (*Phenacoccus manihoti*) spread rapidly throughout 31 African countries within a period of 15 years. It inflicted up to 80% yield loss at an estimated annual cost of $US2 billion. Insecticides were used initially, particularly in large cassava plantations, but were largely ineffective.

Backed by strong research efforts and funding support, the International Institute of Tropical Agriculture (IITA), together with the International Institute of Biological Control (IIBC), embarked on a search for indigenous mealy bugs and their natural enemies in their suspected South American homeland. After several years, the same mealy bug was found in Paraguay, together with two of its enemies, a ladybird beetle (*Diomus* sp.) and a parasitic wasp (*Epidinocarsis lopezi*). Methods for mass-rearing were developed and the wasp was released at many sites in more than 15 African countries. It spread rapidly and has become so well established that the cassava mealy bug has now been brought under control.

Mangoes in Pakistan

Mangoes are grown in a number of regions of Pakistan, largely for local consumption. Farmers apply insecticides about five times per year against four major insect pests, but still the problems persist.

Females of the mealy bug *Drosicha stebbingi* lay eggs in the soil and young larvae move up the leaves in spring. Encouraging farmers to hoe around the bases of trees to expose and kill eggs in winter helps to reduce the mealy bugs, which are also heavily preyed upon by the ladybird *Sumnius renardi*. Also, incorporating artificial winter shelters through simple sack banding around mango trunks encourages earlier ladybird activity, because they need not migrate from overwintering sites outside the orchard.

As an alternative to chemicals, attractant traps made from cheap local materials and baited with methyl eugenol proved highly effective in reducing infestations of fruit flies (*Bactrocera dorsalis*) from 35 to 3%.

Scale insects (primarily *Aspidiotus destructor*) were mainly secondary pests brought about by spraying against fruit flies and mango hoppers (*Amritodus* and *Idioscopus* spp.), which eliminated their effective natural enemies. With the use of traps for fruit-fly control, the scale insect problem diminished.

Mango hoppers also have several effective natural enemies. Because of the hopper distribution on plants, supplemental spraying need only be done on portions below 5 m. This reduces the amount of chemical applied, and also the risk of upsetting biological control of mealy bugs and scales.

Overall, the IPM programme reduces the annual sprays from five to one, with a 14-fold cost reduction. Currently growers of about 25% of 13,000 ha of mango in the Punjab are practising IPM.

The case-study demonstrates the importance of on-farm research involving farmers, the need for understanding their pest-control practices and the local ecology of pests and natural enemies, and the use of simple but appropriate control methods.

Cotton in the Canete Valley of Peru

One of the earliest warnings of the dangers of over-reliance on pesticides occurred in the mid-1950s in Canate Valley, Peru, where cotton production for export dominated. Synthetic insecticides were introduced in 1949: first DDT, BHC and toxaphene, then aldrin, dieldrin and endrin, and finally parathion. By 1955, many cotton pests had developed resistance, requiring heavier dosages and more frequent applications. Six new species, all secondary pests, had appeared, raising the number of serious pests from seven to 13. Cotton yields were dropping sharply; by 1956, the average yield was the lowest in over a decade (Boza-Barducci, 1972).

In response to the crisis, the Peruvian government issued an IPM plan in July 1956. This included banning synthetic pesticides and reintroducing beneficial insects and adopting certain cultural practices, such as planting early-maturing varieties, planting by established deadlines and destroying crop residues. With the introduction of this programme, pest problems declined dramatically and pest-control costs were reduced (Boza-Barducci, 1972). The secondary pests quickly dropped to their former innocuous levels, primary pest outbreaks decreased in intensity and cotton yields reached an all-time high.

Bananas in Costa Rica

During the 1940s and 1950s, large areas of virgin lowland rain forest in south-west Costa Rica were cleared and planted with banana as monoculture plantations. Before the mid-1950s, only two pests, the banana rust thrip (*Chaetanaphotrips orchidii*) and the banana weevil (*Cosmopolites sordidus*), were considered economically important.

Mass application of dieldrin to control them began in 1954. Within 5 years, six species of lepidopteran pests became more severe while insecticide use increased rapidly. Studies revealed that many natural enemies of these pests were present but their numbers were constantly reduced by pesticides. They were particularly abundant in unsprayed zones between the forest and the plantations.

Finally, by 1973, entomologists persuaded the plantation owners to stop all insecticide sprays. Within 2 years, most of the pest species had almost disappeared as their natural enemies re-established themselves (Hansen, 1987). Thrips remained a pest, but damage was prevented by covering the fruits with plastic bags. Since 1973 no insecticides have been sprayed on these plantations. There is, however, excessive fungicide use at present in an attempt to control black sigatoka disease.

Discussion

Despite their serious limitations, the use of pesticides is rising rapidly in developing countries, accounting for about 20% of global pesticide sales (Waage, 1993). For example, for rice alone, Asia, with 90% of the world rice land and nearly 92% of the world production, uses more than 90% of the global end-user market value of $US2100 million for pesticides (Woodburn, 1990). IPM, in spite of its many advantages, is not yet widely practised.

Several important constraints are responsible for the limited application of IPM. Some have been discussed earlier. They provide valuable lessons on key issues in IPM development and implementation (Hansen, 1987; Teng and Heong, 1988; Wiebers, 1991; Ooi *et al.*, 1992). These include constraints that should be avoided and the requirements needed for success. The key features are summarized in Table 3.2, and the critical issues are discussed below.

Perception and raising awareness of IPM

It is important to shed the myth that IPM is too complicated. Experience has shown that IPM can work well in developing countries, and it is within the reach of virtually all growers if they are given appropriate help and training in terms they can understand. Not only does IPM work in developing countries but it has been demonstrated to be economically more efficient than a pesticide-only approach. Pesticide use can be reduced substantially (often more than 50%, sometimes even completely) while yields stay the same, or even increase.

However, IPM still remains poorly understood outside the scientific community (Waage, 1993). It is important that IPM successes be made better known to others, such as policy-makers, extension workers, farmers and the general public. Raising the awareness of IPM to those who are involved in or will be responsible for intensifying farming systems is especially crucial. In many developing countries, intensification is still anticipated in a range of crops. Kiss and Meerman (1991) indicate that farmers in Africa not yet used to chemical control may be more responsive to alternatives. The challenge, therefore, is to persuade governments and donors to invest in an IPM approach for an intensification programme in advance of perceived pest problems (Waage, 1993).

Policy issues

A positive policy environment for IPM, one that discourages the use of pesticides, is important. In some instances this may even be a prerequisite in initiating IPM programmes.

Government economic policies can sometimes undermine a scientifically well-designed IPM programme. For example, many agricultural credit programmes in developing countries focus on the use of technology 'packages' that include substantial use of pesticides. Furthermore, governments also commonly distribute free or subsidized pesticides to small growers, and spray their fields for them when there are signs of pest buildup or a pest outbreak (Repetto, 1985). These practices encourage overuse of pesticides and reduce the economic incentive to use alternatives.

Of even greater importance is a lack of firm support for IPM at the policy level. Through a commitment to IPM, policy-makers and government officials can better prepare to structure their research and extension system in their organizations to facilitate IPM development and implementation, and to provide the training and resources to support this effort (Waage, 1993). Interministerial linkages may also be established to guide IPM implementation, and particularly, to coordinate research and extension activities, which are still poorly linked in many developing countries.

The policy impact on IPM is especially important in developing countries because of the more direct effect on large rural populations, for example, in terms of providing the credit and supoort services for agricultural production. In this regard, IPM on a national scale seems possible only in countries where the government adopts IPM as a central component of agricultural projects.

Farmer and extension worker involvement in IPM research

Researchers must involve extension workers and farmers when developing IPM technical content. This is important to establish elements likely to be useful in IPM practice. Since the farmer is the final participant, he/she will

Table 3.2. Key features of some successful IPM programmes in developing countries.

Feature	B	C	CA	CN	CO	CR	M	OP	R	SB
Pesticide subsidies contributed to reliance on chemical control					×	×		×	×	
Reliance on chemical control led to more pest outbreaks because natural enemies eliminated	×	×			×	×	×	×	×	×
Reliance on chemical control resulted in serious problems of resistance development		×				×			×	
Failure of chemical control led to IPM adoption	×	×			×	×	×	×	×	×
IPM strategy emphasized reduction and/or use of selective pesticides	×	×			×	×	×	×	×	×
Reducing/stopping use of pesticides restored effective natural control	×	×			×	×	×	×	×	×
Pest problems declined after reducing/stopping use of pesticides	×	×			×	×	×	×	×	×
Research focused on field problems	×	×	×	×	×	×	×	×	×	×
Close cooperation of researchers, extension workers and growers		×			×	×	×	×	×	×
Exotic pest resulted in chronic outbreaks			×	×		×				
Introduction of natural enemies from the native area provided effective control of exotic pests			×	×		×				
IPM programme has strong biological control component	×	×	×	×	×	×	×	×	×	×
Biopesticides have provided substantial contribution				×		×				×
IPM programme also included other non-chemical methods (mainly cultural methods)		×		×	×		×		×	

Table 3.2 *continued*

Feature	B	C	CA	CN	CO	CR	M	OP	R	SB
IPM programme also included limited use of pesticides		×			×	×	×	×	×	×
Training of growers given importance in IPM implementation						×		×	×	×
Farmers learned about IPM through field experiments									×	
Strong government/management support exists		×	×	×	×	×		×	×	×

B, Banana (rust thrip and weevil in Costa Rica); C, cotton (insect pests in Canete Valley); CA, cassava (mealy bug in Africa); CN, coconut (rhinoceros beetle in Asia/Pacific); CO, Cocoa (bark-boring caterpillars in Malaysia); CR, Crucifers (diamondback moth in Asia); M, mango (range of mango pests in Pakistan); OP, oil-palm (defoliators, mainly bagworm in Malaysia); R, rice (mainly brown planthopper in tropical Asia); SB, soybean (several pests, mainly velvetbean caterpillar); ×, the feature was present to a significant extent.

decide on whether a developed IPM technology is relevant to his/her needs.

On-farm research not only compels the researcher to consider all pests and their local ecology together, but also makes possible the involvement of farmers, and the inclusion of their knowledge, perception and practical economic constraints in the design of IPM methods. Once trained to recognize pests and natural enemies, farmers have a unique capability to be research partners and innovators, because of their knowledge of the crop environment. The involvement of extension specialists and farmers in IPM development at an early stage will greatly facilitate subsequent adoption of IPM practices. Failing to recognize this is the reason why many top-down, science-driven IPM efforts have failed.

Priority of biological control

The self-renewing impact of biological control agents is crucial for sustainability of insect IPM. Its absence or disruption when present is usually the cause of continuous insect problems. This has featured repeatedly in most IPM programmes.

In the developing tropics, special consideration should be given to the naturally occurring and rich biological control agents that usually exist there.

However, where effective species are absent in an area, particularly for exotic pests, their introduction should be an important first step in developing an IPM programme. Past experiences have shown that biological control pays back at a rate of $US100–200 or more for every dollar invested (Greathead and Waage, 1983).

Integration of people

Most successful IPM programmes not only give focus to growers and adopt a problem-solving approach, but also emphasize and foster strong integration of different constituencies concerned with IPM. The diverse groups of people – farmers, village workers, crop protection, economic and non-formal education experts, policy-makers, supportive donor community (local, sectoral and international) and environmentally conscious citizens – become an integral team, with the growers as recipients of the IPM technology. This is a critical requirement. Kogan (1988) aptly pointed out that, although IPM is usually conceived at three levels of integration (tactics, multiple pest stresses and systems integration), successful IPM actually hinges on the capacity to reach yet a fourth level of integration – the integration of people.

Human resource development

Even when a sound IPM programme can be formulated, it cannot be effectively implemented on a large scale unless there are adequately trained people to execute the programme. Building the human resource base would involve training trainers, developing appropriate training curricula and training materials, providing backup in IPM problem-solving areas, data handling and the development of relevant management systems and skills. Ensuring that the appropriate training approach is used is also important, such as the non-formal education method. To sustain IPM, farmers need to be encouraged to discover the role of natural enemies, to learn by doing and to ultimately become IPM experts (Gallagher, 1992).

Financial support

The question is not whether financial support is needed but on what area the support should focus. Among many possible areas, the following should receive priority if IPM is to become more widely practised by farmers:

- Development of human resources. This is critical since IPM is not simply a technical intervention in a crop production programme, but an exercise in human resource development intimately linked with other related activities in the agricultural sector.

- Farmer demand-driven IPM research (mainly on-farm). By also encouraging the participation of local governments, community organizations and other non-government organizations (NGOs), such research activities can ensure that the research addressed by national and international research institutions will be less top-down and technology-driven.
- Supporting IPM implementation by farmers as a broad educational programme that also involves other related groups. This will promote working together by farmers, government, scientists and others to effect a positive change in agricultural development, and prevent IPM being viewed only as a technical component for expert consultation.

Role of biotechnology in IPM

The need for IPM research to be demand-driven and farmer-focused has often been highlighted. Researchers in biotechnology, especially, should pay particular attention to this when developing IPM technical content, because their work is frequently more confined to laboratory activities and isolated from the field.

The contribution of biotechnological science to pest management is only just beginning. Early developments (such as transgenic plants with built-in protection against diseases) have generated strong enthusiasm. However, caution is necessary. We should remember the early achievements of synthetic pesticides, which were exploited with few questions being asked about their long-term contributions and possible side-effects.

The impact of biotechnology in IPM in the future is unlikely to be limited by shortcomings in technological capabilities. A far more serious impediment to the successful application of these methodologies will be the recurrent problems of product development into forms acceptable to growers, consumers and the community at large, and product availability at prices which growers can afford (Whitten, 1992). This can prove formidable because public institutions and large enterprises usually take the initiative in researching such frontier sciences and developments in technology. Such a situation is one reason why developments in science and technology may be subtly taken out of the hands of individual scientists and technologists, with a consequent increase in the influence of politicians and technocrats (Hyong, 1987). This condition is worrying because of the potential danger that the neutrality of science and technology will be affected by employers, sponsors, administrators and others. Researchers in biotechnology, therefore, must always be wary of this possibility as they seek creative solutions, in partnership with others interested in the wider application of IPM.

Integrated Pest Management in Rice 4

PETER E. KENMORE

> Diversity is a basic characteristic of all agricultural enterprise, and we have ignored it to our cost. In the formulation of research objectives, in plant-breeding, in soil fertility, in water conservation and management, in the planning and execution of economic policy, and in the amelioration and reform of social practice and custom, the fundamental importance of local circumstances is brought out . . .
>
> This is not . . . reassuring . . . It destroys the illusion that agricultural problems can be solved by massive centrally planned research, and directs the investigator to the village and the field as the places where understanding must be gained if progress is to be made. But surely we know that we were deluding ourselves, and that in the long run we must go back to the field.
>
> Sir Joseph Hutchinson (1977)

This case-study will review the development and implementation of integrated pest management (IPM) in Asian tropical and subtropical rice, covering research, policy change and farmers' training. It will then consider some areas where biotechnology might strengthen IPM practice by farmers in village communities. Finally, recent developments where the Asian rice IPM model is being introduced to other regions through national and international initiatives will be described.

Rice Production Stability

Rice supply is certainly not the only factor in Asian economic and political development, but without that supply governments dissolve, markets collapse, the environment is plundered and development degenerates to survival. Total rice production in an Asian market locality is not a sufficient guarantee of supply to the people living there (Sen, 1981; Dreze and Sen, 1989) but it is a necessary precondition. Asian self-sufficiency in rice had been achieved with total production from the mid-1960s through the early 1980s staying ahead of population growth by an adequate margin.

For tropical and subtropical Asia much of that margin came from production advances in irrigated rice systems pioneered in the Philippines at the International Rice Research Institute (IRRI). Work by IRRI economists (Pingali *et al.*, 1990) has shown that:

- rice production has been tapering off, with growth rates falling from 4% per year to 2.24% per year (dangerously close to population growth and estimated 1990s growth rates of demand of 2.1–2.6% per year) in intensified rice-growing Asia;
- yields on research stations not only have not exceeded ceilings achieved more than 20 years ago, but the highest yields obtainable on these stations have been falling steadily;
- farmers' yields had increased but yield growth decelerated by 1990, so that food production increases will depend more on local improvements in knowledge-intensive technology that narrow yield gaps among farmers, and
- farmers are the key actors in making local yield-enhancing improvements. Training is the surest way they will acquire scientific skills for problem-solving that cannot be packaged in simple messages. Integrated pest management training is more than pest control; it is a gateway to scientific farming.

IPM Research Paradigm Shifts from Technology Components to Ecosystems

Triggered by insect pest upsurges in the late 1960s IPM research for rice in Asia began in the early 1970s at IRRI and national research systems, by combining single components of insect pest-management systems that had shown the best promise: host-plant resistance and selective insecticide use based on economic thresholds for single pest species (Pathak and Dyck, 1974; Kiritani, 1979). Initial successes were led by host-plant resistance for rice brown planthopper (BPH) and green leafhopper vectors of rice tungro virus. A 'boom-and-bust' cycle followed, with pest outbreaks, varietal replacement and new outbreaks in the Philippines, India, Sri Lanka, Indonesia and Thailand (Aquino and Heinrichs, 1979; Cariño *et al.*, 1982; Claridge *et al.*, 1982).

These cycles led researchers, by the end of the 1970s, to investigate natural population dynamics and evolution of virulence on host-plant varieties in pests like BPH (Kenmore *et al.*, 1984; Cook and Perfect, 1985, 1989a, b; Gallagher *et al.*, 1994; Rombäch and Gallagher, 1994) and the role of a large and diverse community of natural enemies (Triwidodo *et al.*, 1992; Way and Heong, 1994) in regulating insect population densities. This emphasis on ecological dynamics bore fruit, and a change in attitude gradually spread through the institutions of IPM research through the early 1990s. Field, laboratory and simulation studies on natural enemies, community structure and the influence of habitat variability became more fashionable and common (Denno and Roderick, 1990; Hidaka, 1993; Ooi and Shepard, 1994; Ooi and Waage, 1994). The importance of crop physiology in insect feeding–damage relations was another new front for researchers, often combining crop simulation models with elegant field and laboratory studies (Rubia *et al.*, 1989). Researchers' technological recommendations reflected a deeper understanding of evolution, population regulation and host-plant response in intensified rice systems.

Chemically based strategies were seen to be increasingly unsustainable and not cost-effective due to insecticide-induced outbreaks of insect pests, rising costs and the evolution of resistance to insecticides (Heinrichs *et al.*, 1982; Heinrichs and Mochida, 1984; Triwidodo *et al.*, 1992). The negative effects of insecticide use in rice on human health, for many years ignored or dismissed as unquantifiable, have recently been investigated in detail, and quantitative estimates of rates of illness and economic costs have now been estimated for both acute and chronic occupational poisoning (Rola and Pingali, 1993; Kishi *et al.*, 1995). The cost in health care and days of work lost in normal occupational exposure to insecticides are higher than the cash prices of those insecticides.

Policy Reform to Support IPM

Policy-makers came to understand that the foundation of IPM for intensified rice production should be the conservation of natural enemies, which are natural resources, in every rice field. Agricultural policy-makers took steps to remove the earlier policy incentive to higher insecticide use. These distorting policy instruments, based on untested, unsubstantiated, but common beliefs that insecticides were essential to higher food security, included:

- direct government purchases to reduce prices paid by farmers for insecticides;
- distribution systems through government supported or endorsed 'marketing' outlets;

- government-sponsored marketing and promotion efforts through agricultural extension systems, packages of production technologies or small-farmer credit packages;
- preferential tariffs to importers of insecticides versus importers of other inputs; and
- pervasive policy messages that greater insecticide consumption led to higher rice production.

Beginning in 1986 in Indonesia, the Philippines, India and, most recently, Vietnam, direct price subsidies of up to 85% for insecticides were reduced or eliminated, contributing to higher end-user prices and lower field use, without reducing the rates of growth of rice production. In Indonesia, an annual pesticide subsidy that reached more than $US140 million in 1987 was eliminated by 1989. In the Philippines, the national pesticide subsidy for rice reached $US12 million per year in 1986; it was eliminated by 1988. India eliminated its central government subsidy for insecticides of about $US35 million per year by 1993. Restrictions, based on reviews of ecologically disruptive effects on rice production, on the range of insecticides legally permitted on rice – as in Indonesia, or in the country – as in the Philippines and Vietnam, followed reductions in subsidies. Vietnam eliminated all national government direct subsidies for rice pesticides by 1993.

In early 1994 the Finance Ministry of India, under 'the principle of making polluters pay', imposed a 10% excise tax on all pesticides. This bold policy step generated instantaneous – and ferocious – public criticism, which it countered with strong technical support from the Ministry of Agriculture. The decision added about $US60 million annually to government revenue. In the tax's first year, 1994, India's annual total food grain production reached the highest level in its history.

IPM in Practice

A common IPM implementation model has emerged over the past decade in rice-growing Asia. The model has two special characteristics: farmers' field schools (FFSs) and national IPM training programmes. In FFSs and their associated follow-up activities, farmers' participation intensifies community scientific and organizational capabilities. National IPM training programmes include changes in national and local government policies to support, protect and extend participatory IPM. The Asian model influenced the United Nations Conference on Environment and Development (UNCED) and its Agenda 21, which called (in Chapter 14, section I) for networks of farmers, extensionists and researchers to implement participatory IPM as the best, most sustainable crop protection option for the future. The model also received strong endorsement from national IPM planners and administrators from Africa, Latin America and the Near East in a global IPM field study tour and meeting held in August–September 1993 (FAO, 1993, 1994).

Farmers' field schools for IPM of rice

What is a farmers' field school?

Small groups of farmers meet every week through an entire rice crop season with trainers to carry out field observation, analyse data, draw conclusions and debate these conclusions using agroecosystem analyses; these groups are the FFSs and they maintain common field areas where they carry out season-long experiments.

The FFS model presumes that:

- farmers are rational, sceptical and curious and do experiments (Richards, 1995);
- farmers will make important field decisions;
- farmers should be encouraged to act upon their decisions;
- local ecology is significant in those decisions;
- farmers should have scientific field skills and concepts;
- farmers who have concepts, theories and skills will make scientific cases by public debate on their local observations;
- these cases will consider alternative explanations, analysis, predictions and evaluations and come to decisions upon which to act.

How farmers' field schools work

An FFS group can be set up from scratch or built upon an existing community group. The members are familiar with each other, and able to speak without unnecessary formality. The FFS is rooted in fields and fieldwork. The usual arrangement is a shared study field where ecosystem progress and development can be observed and discussed. This can even take the form of shifting the location of meeting from week to week, in order to cover different members' field problems. The important quality that defines each session of the FFS is shared field time. The field allows and encourages participants to ask questions, to propose explanations and to offer alternatives without threat of being silenced. Local names for plants, symptoms, animals, soil conditions and water conditions are worked out and agreed by the team in the field. A school of about 25 farmers is divided into teams of five for agroecosystem observations. Members measure plants, recognize, count and record animals, note water, soil and plant conditions and take samples for reference. In the field the team alternates observation and discussion of each plant or point sampled, recording data after each discussion. The previous experiences of each group member are brought out to interpret the biological or physical material at hand.

Once out of the field, each team summarizes its fieldwork and then draws up the agroecosystem chart for that week. A plant is drawn at the correct growth stage, and each type of insect drawn, herbivores on one side of the plant, carnivores on the opposite side, decomposers below. Measurements of plant height and, tiller number and densities of animals are recorded, as are physiological symptoms and the day's weather.

Each team presents its summary to the entire school. Discussions again alternate with reported observations. The experiences of the school's members are brought to bear on each summary, so that a larger pool of expertise is applied. After each presentation and again at the end of all presentations, a more wide-ranging discussion takes place. Participants are encouraged to explain the natural history of their observations, to use concepts of energy flow, of predation, of life cycles and of crop physiology to anchor major observations from the field to ecological concepts and theories. Speculative questions and comments are encouraged. Active comparisons with earlier seasons or other villages' experiences are made, so that discussions are not limited to the data from the day. These comparisons may be put into historical context – first from the perspective of the local community, then from the larger frame of national agricultural development through intensification of production. These are concrete, hard-headed sessions firmly linked to that day's or that season's experiences. Finally, questions on future action are raised: 'What do we need to do to conserve natural enemies? How should we present these conclusions to other farmers in the community?'

At some point during each 2–3-hour presentation and discussion period, special topics are often presented. These focus on aspects of the agroecosystem being explored through structured experiments in the FFS: crop compensatory growth after defoliation, feeding rates by predators in cages, varietal or crop nutrient comparisons in the field. They may involve games and simulations of genetics, of population growth (e.g. of rodents) or of economic decision-making with data generated through group discussions.

Underpinning the FFS process is a special relationship between farmer participants and trainer-facilitators. The FFS presumes that farmers are concerned with practical performance of technology rather than its hypothetical potential. It presumes that farmers conduct science in much the same cognitive fashion as researchers: by making informed guesses (using current theories) and testing those guesses, often by experimental manipulation, systematic observations and public debate. The FFS respects the previous experiences of its members not as the unchallengeable voice of authority, but to interpret fresh observations, just as researchers do. The FFS does not seek to supplant institution-based researchers, but to assist its farmer members to be better able to hold public dialogues with researchers.

Trainer-facilitators in FFSs should have special skills. They must be good listeners and very good questioners. They should not lecture. They influence the flow of each session by reminding the group of agreements made at the beginning of the season to do fieldwork, to encourage everyone to participate and to respect each others' opinions and expressions. They treat every question by farmers as an opportunity to encourage discovery learning. 'What is this insect?' is one of the most frequent questions posed by farmers to trainers. It takes self-discipline and confidence for the trainer not to give a name, but to ask in return, 'Looks like a bug . . . Where did you find it? Were there others? What were they doing? How many were there? What did they seem to be eating? How did they move?' and so on. A good trainer can keep this up for 10 minutes or more in the field. After this interaction, the farmer has, from seeking answers to simple, common-language questions that flow naturally from one to another, built a strong ecological picture of the insect in question. The insect is remembered functionally, not nominally, and the whole agroecosystem is understood more strongly by the linkages of that one insect.

A similar questioning process works in the presentation discussions. Facilitators ask questions both to bring out observations or experiences and to introduce larger issues into discussions. These questions can be 'What was your experience with this variety last season? Why do you think different sections of the study fields had different insect densities? What happened when you used insecticides? When was this virus disease a major outbreak problem in this village? How can we conserve natural enemies? What can we show other farmers about our FFS fields?'

Facilitators must be able to introduce new concepts or data from outside the farmers' experience. As Bentley (1994) correctly put it: 'What farmers don't know can't help them.' A rice virus disease causal agent is impossible to see with village equipment. It is possible, however, to satisfy Koch's postulates and produce symptoms of tungro or grassy stunt disease only in presence of vector insects from symptomatic plants, with simple field cages. As FFSs are using these cages – called 'insect zoos' – for predator–prey studies and to verify the trophic level (herbivore or carnivore) of insects caught in the field, modifying them for virus work is possible. A good trainer should provide information that generates new questions for testing in the field. Also useful are new concepts (and language) to obtain more information from extension, researchers, agricultural teachers or inputs sales agents. Information that illustrates or extends the application of concepts is good. Information that is publicly shared is better, so that more farmers' experiences and analyses can be tapped.

Fakih (1993) has evaluated the FFS process as a non-formal educator. His conclusions included the following:

- facilitators did not dominate, and freely let others facilitate discussions;
- the majority of farmers participated in leadership roles at one time or another in each session;
- discussions were frank, open and without intimidation;
- attendance was remarkably good, because farmers enjoyed the sessions and felt the learning was important;
- educational goals were set by participants, not from an outsider's agenda, with active involvement by most participants;
- FFS experiences were structured, not free-form, but learner-led and characterized by independent discovery;
- trainers did not control knowledge or the learning process, but they catalysed both; and
- materials for learning were not produced by others and brought into the FFSs; instead, materials, mostly drawn from the field, were created by the learners.

Participatory evaluation exercises showed that farmers most approved of the following aspect of FFSs:

- the learning environment is a welcome break from routine;
- the process is enjoyable itself: farmers would be unhappy to miss sessions;
- the relationship of facilitators and farmers is natural, and makes real connections among people, not empty formalities;
- the curricula are all connected to farmers' lives, they fit farmers' needs and they move towards achievable goals; and
- the explicit long-term objectives are to change the fate of farmers in desirable ways, not simply to change behaviour.

Farmers gain self-confidence, both within their communities and for dealing with outside agencies. Farmers are motivated to tell others about their experiences and knowledge generated in the FFSs. They feel they own the learning process and the results of their discovery efforts. They gain skills in conducting discussions, in making and interrogating a scientific case, in analysing field observations and outside information and in keeping a local scientific society intact for weeks and months.

Technical curriculum of FFSs

Integrated pest management field skills grow from a core of entomology because insects are big enough to see, touch, taste, smell, hear and manipulate. Insects are fascinating, attractive, worrisome, stimulating, available, fast-responding and ever-present. Perhaps from millennia of evolution, Asian rice ecosystems have hundreds of species present in every hectare of insecticide-free rice (Way and Heong, 1994).

In the entomological core of IPM practice, farmers apply theories to natural entities in/through:

- recognition;
- classification;
- observation (behavioural, habitat, life cycle, etc.);
- contextualization (e.g. 'What are the other species that interact with this one?');
- manipulation;
- making new theories;
- prediction + intervention = experiment;
- hypothesis testing, hypothesis rejection, hypothesis modification;
- analyses and conclusions;
- publication to the FFS – scientific presentation; and
- recommendation for action.

Moving concentrically outwards from entomology, the FFSs cover most of the following during the first rice crop season:

- crop physiology and damages;
- plant nutrition, water, soil;
- diseases, rodents, weeds; and
- varieties, selection, evolution.

The material is confronted through field exercises and steadily deeper observation. The exercises always begin with questions and the expressed interest of the farmers in answering the questions. For example, farmers carry out field experiments on the yield effect of removing rice leaves to simulate defoliating damage. In all cases the degree of defoliation that was tolerated by the crop without yield loss was much higher than supposed.

Team building in farmers' field schools: group dynamics

Rice farmers learn powerful field and decision-making skills by working in the field with skilled trainers from government agencies or non-governmental organizations (NGOs). These skills are concrete and re-inforced each time a farmer observes the field to improve a decision. Training processes must support and develop these decision-making skills. Non-formal education uses experiential learning processes to involve participants in the learning process such that discussion, analysis, decision-making and presentation are incorporated into all activities. These learning processes transform trainers and trainees through direct discovery and understanding of basic ecology. Trained farmers are able to use field-based methods to train other farmers because the process focuses on actively discovering what is already present and not on lecture presentation of facts.

Non-formal education processes in IPM training provide an opening for farmers to change their own behaviour and their own institutions.

Impact on farmers' practice

A summary of over 400 location-seasons across seven Asian countries showed that, after IPM rice training, farmers' yields increased by about 9% (FAO, 1994). Studies from individual countries (FAO, 1993) showed yield increases from 4% to 13%, with concurrent reductions in insecticide application frequency and cost from 35 to 100%. In Vietnam and Korea fungicide frequency and cost was reduced by 30–40% after the FFS training. Herbicide use is not consistently affected by the first season's FFS. In all cases, farm-level profits increased, both from cost reduction and from yield increases.

IPM and farmers' organizations: making communities stronger

How IPM builds on farmers' organizations

The FFSs are often formed through village-level farmers' organizations that wish to strengthen the technical/scientific skills of their members (Fakih, 1993). These community organizations, defined pragmatically as groups with membership larger than single families, are found in every village. Examples include: religious institution-based groups, political or issue-centred groups, sports groups, arts (music, drama) groups, credit groups, age/school-year cohorts and miscellaneous clubs. Integrated pest management enhances ecological awareness, decision-making and other business skills and farmer confidence in local community organizations. These groups are the best vehicle for institutionalizing IPM within communities.

IPM strengthens farmers' organizations

Farmers' field schools often take root in pre-existing farmers' organizations, even if they were set up for other purposes and tend to lack long-term sustainability, and help them work better. The vast majority of these organizations exist only on paper. They may have lists of members, they may have attendance records, they may have showcase meetings, but they do not function as a group with goals and accountability. Integrated pest management FFSs catalyse these groups to focus on concrete goals: the rice crop and more sustainable production. The IPM training process shows how better to build stronger groups from the individual insights and accomplishments of members. More scientific, informed, profitable and sustainable rice production also helps keep these groups strong.

Science is social, because the cooperative nature of science leads to greater objectivity. Farmers in an FFS together create and test scientific explanations made out of their discoveries. Their day-to-day experiences in the FFS are compared and contrasted in open dialogue, with their foundations of 5, 10, 20 or 40 years of earlier experience growing rice. The actions coming from these scientific explanations are carried out more confidently because of this social reinforcement.

Farmers who live in a village share both its ecological location and a social and political community. Farmers have experience, and must operate in the location and the community as experts even though they do not control either: more powerful groups exert final control. Integrated pest management FFSs are public meeting-places for expressing ideas and opinions. They are forums where farmers and trainers defend observations and debate interpretations. They apply their previous experiences and present new information from outside the community to bolster their arguments. The results of the public FFS meetings are management decisions on what action to take.

The convergence of technical localization with community-based empowerment makes IPM a good starting-point for people sharing thoughts on an equal footing, with the consciously agreed goal to create new solutions. The FFS forum evolved in Indonesia, the Philippines and Vietnam to further promote the communicative actions generated in IPM training. This forum is also protected administratively as the fundamental unit that national IPM programmes use to replicate IPM in other Asian villages.

Integrated pest management field training in village communities has been demonstrated across Asia to be among the best activities for those communities to build up their scientific expertise through theory and experiment; to increase access to that expertise among the members of the community; to sustain encouragement of better decision-making and actions in IPM and other knowledge-intensive technologies; and to increase the effective demand by communities for good-quality technical information to turn into knowledge.

Stronger community organizations and public bureaucracies

If a new variety is recommended by a plant breeding agency or programme, farmers may grow it in a small area. Instead of only measuring the yield, farmers may also measure densities of herbivorous insects and densities of predacious insects. While herbivorous insect density may go down, if predator densities also go down in relation to currently used varieties, the farmer will properly interpret this as a potential risk. It is a risk because those predators can respond to invasions of insects that are not controlled by the inbred resistance of a new variety. A new variety that unnecessarily suppresses insect populations that would feed predators endangers the entire agroecosystem.

Stronger community organizations and private bureaucracies

Even more dramatic can be the impact of new insecticides. Insecticides are routinely so broad in their spectrum of impact on rice-plant arthropod communities that farmers from IPM field schools check the impact of applying an insecticide on predators as much as on potential pests. Farmers have the confidence to do this because they know that the rice plants can tolerate previously frightening amounts of damage without showing a yield reduction. Farmers will therefore not apply an insecticide out of fear or panic but will apply it when it becomes clear that the local agroecosystem has already been so disturbed that the densities of populations of predators are not sufficient to control populations of potential pests. New technology will be assessed in terms of the competence of the agroecosystem.

Trainers and IPM

Trainers of these farmers' groups, who may be employed by government agencies or NGOs, receive intensive (from 200 to over 400 h) field training themselves in residential training of trainers' courses, including preparation both in the technical content of IPM and in the group dynamics of training processes.

Trainers are usually embedded in extension systems that emphasize the mechanical delivery of fixed messages and physical and financial input from a central source. They must learn to become skilled facilitators of groups. That means that the training of trainers, which is much more intensive for IPM than normal agricultural extension, includes a good deal of work in group dynamics and group strengthening. Activities that strengthen groups can include exercises that illustrate principles of mutual reliance and mutual responsibility.

IPM transforms extension systems

As farmers grow stronger in their groups and benefit from the core of IPM knowledge that the groups construct together, they become skilled in demanding more effectively other benefits both from government agencies and from NGOs. Experience over the last decade has shown that, if IPM is reduced to an impact point in a large-scale top-down extension discrimination campaign, then IPM is mortally weakened. The quality of the IPM training experience cannot be maintained unless investment is made in trainers and in farmers' groups, not just to carry a technical message but to demonstrate the value of stronger groups and farmers' initiatives.

Extension workers who get inspired by stronger and more positive interactions with farmers benefit extension systems as a whole. Instead of chasing farmers with packages of inputs and chasing farmers to collect loans given out to enable farmers to use those packages, extension trainers become partners in a joint discovery process. This process carries mutual respect for the persons and experiences of farmers and extension staff. Trainers obtain all of the personal fulfilment that good partnership conveys. This is carried back by rejuvenated extension workers into their extension agencies.

Integrated pest management has triggered a transformation of extension agencies, starting from the local level up to the subnational level, as extension workers learn and confirm that this approach and the style with farmers that grows from it are a more satisfying way to work. They use the same group dynamics methods with their supervisors as they use with farmers. They are able to take initiatives inside the agency in planning, monitoring and evaluating, and ultimately in owning the process of extension. The result on an organization-wide level is similar to results in large-scale industrial organizations in Europe, North America and East Asia (Useem *et al.*, 1992). The more that ownership is transferred to lower-ranking but front-line staff, the more effectively a large organization can respond to a wide variety of location-specific demands.

NGOs are natural partners in IPM

The strong foundation that IPM contributes to self-reliance and environmental conservation makes it acceptable to NGOs. These NGOs usually express strong reservations about agricultural intensification because it has meant increasing dependence on external inputs and reduction in the substainability of the future production base. Integrated pest management allows NGOs to find common ground with agricultural agencies of governments. Non-governmental organizations are stronger in village-level organizations and are able to reach out to farmers' groups that are otherwise unreached, so their interest, support and active participation help get IPM technical content on to the agenda of more and more village organizations. Integrated pest management's primary scientific concern with local ecological variation makes it a powerful technical channel for NGOs.

Non-government organizations in the Philippines, Cambodia, Indonesia, Malaysia, Thailand, Sri Lanka, Vietnam and especially Bangladesh are sponsoring and adapting IPM training with their farmers' groups. This usually means increased cooperation between NGOs and agricultural departments of governments. The stronger expertise and wider experience of NGOs in encouraging farmers' participation have significantly strengthened IPM practice. Asia-wide NGOs with support from donors outside Asia, including Cooperative American Relief to Everywhere (CARE), Save the Children, Plan International, World Neighbours and Cooperation Internationale de la Developmente Perlia Solidaine (CIDSE), have adopted IPM policies within the framework of sustainable agriculture.

How FFSs are followed up: IPM clubs and consolidation

After the first crop season focusing on IPM and rice, FFS groups (who come to resemble farmers' 'scientific societies') have constructed, through their own research, IPM for rotation crops of rice such as vegetables and legumes. They have gone on to experiment with rice–fish culture and considered the use of aquatic resources for the culturing of shrimps, snails, frogs and other aquatic protein. Future work should include experiments and observations to promote soil and nutrient management – based on the same philosophy of conservation that they earlier applied to predators of insect pests.

A next step goes beyond the cropping system of rice. Farmers often have separate cash crops, often on higher lands, and seek to apply the principles of IPM to these crops. In India and Vietnam, IPM has started in cotton systems. Farmers in IPM clubs in Vietnam, facilitated by an NGO, CIDSE, have begun applying IPM to tea. In the Philippines, the participatory approach of FFSs has been extended from rice to coconut in Davao. Some IPM clubs either link with or form the nuclei for credit circles or small credit unions, as farmers' shared technical interests build up opportunities for other mutual enterprises.

A very popular activity for IPM clubs is exchanges with other FFS graduates as they form clubs. These exchanges, organized and paid for by the members, allow farmers to compare their experiences in IPM field practice, in extending principles to other systems and in group building in the larger village community. These exchanges, and the longer-term, large-scale linkages they foster for the future, contribute to better pest management options for communities and governments.

Integrated pest management clubs often seek directly to spread IPM to other groups of farmers in the same or nearby villages. In community IPM planning exercises, farmers use locally adapted managerial planning techniques like Strengths, Weaknesses, Opportunities and Threats (SWOT), Logical Framework (LOGFRAME) and participatory rural appraisal (PRA) to broaden people's participation in decisions about future activities, and draw upon human and material resources within communities to create more expert IPM groups (Indonesian National IPM Programme, 1994).

Research partnerships

This higher level of demand for higher-quality technology makes the farmers better clients for researchers. It also makes them more articulate and noisier clients for researchers. Researchers should welcome this because it means that famers will be able to articulate local needs and be better equipped to participate in directing research. It will, however, require a transformation of traditional agricultural research systems.

Karawang field laboratory, west Java, Indonesia

Dr Hermanu Triwidodo set up a field laboratory for $2\frac{1}{2}$ years. This grew to include not only the group of more than 20 trained farmers in the community, but also 15 students and young researchers. They analysed in detail the spatial distribution, population dynamics, diapause physiology, yield impact and biological control of the white stem-borer of rice. The results were used by farmers to make better decisions. The results were also used by national policy-makers in deciding not to resort to large-scale insecticide applications when the populations of white stem-borer spread alarmingly. The original group of trained farmers grew into a well-linked network of more than 3000 people across two subdistricts able to understand the population dynamic basis for stem-borer and to respond to local immigration of pests by large-scale hand removal of egg masses.

Danang/Hue blast study areas, central Vietnam

Here institution-based researchers have made formal partnerships with two IPM clubs in different communities to explore new options for rice production improvement in a rice blast-prone area. A group of plant pathologists and rice breeders from IRRI joined with the Plant Protection Research Institute and the Agricultural Genetics Institute of Vietnam to approach the Plant Protection Department at national, regional, provincial and district level to identify IPM clubs interested in shared research. Two clubs committed themselves to working for at least 2 years (four seasons) with the researchers, maintaining experimental fields, taking their own measurements, selecting good varieties and planting them in mixtures and field patches. This allows the clubs to assess, and then compare their assessments with the researchers, the performance of alternative gene deployment strategies against rice blast. The first season's results were better than previous years, partly because one variety had come to dominate these districts, and blast resistance in that variety had eroded significantly. The farmers selected about 20 lines from the more than 50 tested, and are replanting them in deployment patterns during the heavier blast seasons. Through the IPM trainers' network, they also gave samples of these 20 lines to two additional IPM clubs and thus expanded the range of environments included. The farmers carried out a number of simulation exercises of uniform versus diverse planting patterns in order to explore the possible consequence of new deployment strategies.

Genetic conservation with NGO researchers, Mindanao, Philippines

A well-established NGO working on *in situ* genetic conservation on three continents, South-East Asian Research Institute for Community Education

(SEARICE), has collaborated for several years with a strong farmers' cooperative in North Cotabato Province. They have identified more than 100 previously uncharacterized local varieties, and are planting them over seasons and across fields to explore their performance. Sharing an organic farming orientation that stresses low or no external inputs, the NGO has concentrated on building strong community organizations with the capacity to conserve local germplasm under the umbrella of the cooperative, which has built its own warehouse to store grain (and take advantage of favourable prices), owns six large trucks to haul grain and manages hundreds of thousands of dollars of members' money and produce every year. The NGO approached the Philippine national IPM programme Kasagan a han ng Sakahan at Kalikasan (KASAKALIKASAN), to arrange for season-long FFS training for several of its constituent organizations. The technical strength of the IPM training made it possible for ecological factors, such as the size and composition of natural enemy communities, to be considered when assessing varieties. In this way the research partnership of the NGO and the farmers' organizations was strengthened by the IPM training.

National IPM programmes

Asian governments use IPM to improve environmental quality within successful development strategies. Economic development is supported through higher food production with lower levels of non-renewable, expensive and polluting inputs. In Asian national programmes IPM strives to meet these goals by empowering farmers to understand and manage their crops as ecosystems. Integrated pest management assists farmers to experiment to improve their own management.

Integrated pest management in rice stresses the capability of farmers for diagnosing pest problems and participating in research to develop solutions. Because IPM recognizes farmers' expertise, national IPM programmes require traditional agricultural extension and crop-protection systems to make a paradigm shift as well. They must transfer responsibility for important decision-making away from government staff to farmers. They must convince themselves that farmers are competent to improve modern technology for local optimization. This paradigm shift starts at the top of Ministries of Agriculture, usually with policy-makers visiting a pilot project site where farmers are carrying out scientific studies, basing their decisions on agroecosystem analysis. The new attitude towards farmers and their capabilities then gradually moves from top to bottom of the agricultural hierarchy, with community IPM convincing frontline field workers and an atmosphere of approval assuring middle managers that the new way of approaching crop protection and production is safe.

National IPM programmes replace investment in chemicals and their associated pest-surveillance systems by investment in people. Indonesia originally obtained most of the funds for IPM training from a United States Agency for International Development (USAID) grant, and in 1993 negotiated a World Bank investment project loan representing about $US8 million per year for 4 years to continue for a second phase a national IPM training programme. The Philippines national IPM programme, KASAKALIKASAN, is a $US1.8 million per year IPM farmer training programme. This programme is funded by the national Department of Agriculture through a combination of monetized agricultural commodity grants from Japan and regular revenue funds from the government. India now supports its national IPM training programme with $US5 million per year of regular government revenue funds. Vietnam's IPM programme has grown from $US50,000 to $US1.2 million per year, largely funded by Australian and Dutch aid, but with over 1100 full-time field staff committed by the Plant Protection Department.

Decentralization and devolution

When local government (e.g. village, town, commune, subdistrict, district or province) is convinced that IPM is a politically viable programme appreciated by local constituencies, IPM programmes are institutionally sustainable. Resources at the disposal of local governments can be allocated to community IPM follow-up or new FFSs. The Philippines national IPM programme, KASAKALIKASAN, has emphasized the proactive involvement of local government units in IPM planning, implementation and funding (Castillo, 1995). Farmers' field schools channel community interest in improved farming, and local politicians respond by greater administrative and then financial support. Local NGOs are also important in Philippine IPM; they carry out FFSs as well as participating in local policy formation. A number of provincial governors and municipal mayors have proclaimed IPM policies for their jurisdictions. These ensure sustained follow-up and rapid expansion of access to IPM.

How Might Biotechnology Help IPM in Practice?

Ground rules for assessing how biotechnology (or any technology) might help

Even if a new genetic, cultural or chemical component of technology could be shown in research trials to reduce pest losses in an appropriate ecosystem, it may not satisfy the standards of good IPM if it does not increase farmers' scientific understanding of the agroecosystems they work in.

To satisfy the standards of good IPM, technology should be understandable and testable in terms of concepts used by IPM farmers, such as population regulation by natural enemies, interplant competition, host-plant resistance, the formation, limitation and reduction of yield by crop physiological mechanisms, pools of crop nutrients, energy flow among trophic levels and flows of crop nutrients from pools and sources to and among sinks. These concepts are demonstrated during FFSs and then used to make and test predictions in the shared experimental fields.

New technology can help farmers extend the range of situations in which they apply their concepts. An example is the analysis of rodent infestations as population changes similar to insect population dynamics.

Good IPM does not take decisions out of the hands of farmers, but instead better informs those decisions. Farmers are rational, sceptical and interested in testing performance rather than potential. In short, farmers are scientific, and in their scientific practice they draw upon the same cognitive structures as institution-based scientists (Amanor, 1994, 1995; Richards, 1995). A good IPM process makes technology more accountable to farmers, and makes agroecosystems more understandable. Bodies of expertise from population biology, genetics and plant physiology that 20 years ago were considered too esoteric to be understood and used by farmers are now applied in daily practice by more than 20,000 village communities in Asia.

This case-study will try to apply standards for good IPM to indicate which biotechnologies may better satisfy them. It will not consider the comparative commercial potential of these biotechnologies, which often seems to be the determining factor in their development. It is possible that educational sectors as well as agricultural sectors may have an interest in supporting their development. The chemical mechanisms of these biotechnologies are also not considered, and fairly traditional, even premolecular, biological tools may provide as good or better service for good IPM practice.

IPM clubs and biotechnology

The first group of biotechnologies that, once tested under realistic field conditions and proved to be accurate, simple to use and affordable, could help IPM practice in rice-growing communities are diagnostic tools. Tools to help identify fungal, viral or bacterial pathogen infections would assist IPM groups to make better observations and decisions. An IPM club could use the diagnostic tools to compare infection levels among varieties, across crop stages, among field conditions (e.g. water status, nutrient status, yield history) and at different times or seasons of the year. Farmers in the IPM clubs could make more accurate, quantitative and dynamic observations with these tools. They could draw conclusions, discuss options and carry out field tests of varietal deployment, time of planting or soil and water management to reduce risks from disease. If the diagnostics help farmers differentiate between pathogen symptoms and nutrient disorders, they will better be able to set up fertilizer trials that reflect field-to-field variation.

Integrated pest management clubs could use these diagnostic tools to assess levels of inoculum in the surrounding habitat, and in following weeks compare apparent risk with field infection. From regular assessments of field infection at important crop stages and comparing these assessments from field to field, IPM clubs could better decide whether an infection actually affected yield.

Diagnostics are often used in response to observed symptoms and may not be available at the time an analysis for action needs to be made. Indicator plants, engineered and bred not only to produce yield but to produce information, could amplify low levels of infection or deficiency into easily observed symptoms. Planted together with the crop, these indicator plants would react to external stimuli by changing colour, showing spots or displaying other exaggerated symptoms. They would give IPM clubs early information on infection and spread of diseases. This would allow farmers to record and understand disease progress curves in their fields. The useful resemblances to insect population dynamics, familiar from FFSs, could then be exploited by IPM clubs in debating alternative management strategies for disease. Indicator plants could allow farmers to assess disease quantitatively as well as qualitatively, compare varieties and nutrient management under local conditions and better select from a wider range of options. For rice-growing communities, particularly in subtemperate climates, an important focus would be rice blast disease, in particular neck blast, which is felt to reduce yield more than the commoner leaf blast.

Similarly, indicator plants could help IPM clubs monitor nutrient status across and within fields. Even better would be plants that showed how nutrients enter plants from the soil environment, assisting farmers to understand how local soil conditions affect crop growth and yield formation. The scale of significant soil variation may be quite small – a few dozen metres – and monitoring that variation could help IPM clubs make better-informed decisions.

Diagnostics could help IPM clubs track the evolution of pest populations. If tools were sensitive enough, then early warning of changes in the genetic composition of local populations could alert IPM clubs to monitor more closely disease progress in the current or succeeding season. If IPM clubs monitor varietal performance and the infections of alternate host plants, they could construct maps of the evolution of local virulence. By understanding local virulence patterns, farmers could better choose varieties for planting and begin to reduce unnecessary selection pressure on pathogen populations.

IPM clubs' linkages and biotechnology

Integrated pest management clubs provide good technical information and analytical capability for the larger community. They are linked through their facilitators, through members and through local officials to other IPM clubs and other villages. As they create more accurate maps of pest populations, these data can be shared with more communities, which can fill in maps from their own experiences and current observations.

IPM clubs, policy-makers and biotechnology

These maps will be more detailed and accurate than any maps now available from routine pest-surveillance systems. These systems typically assign one specialist of variable technical preparation to at least 5000 ha of intensified crop area. In extensive rice systems (e.g. rain-fed India) the area covered by one surveillance officer on one round of 2 weeks exceeds 2000 km^2, not all cropped. It is not possible for such specialists to monitor these large areas. In practice, much information comes from concerned farmers' groups already. If these groups receive IPM training, and then could be provided with more accurate diagnostics, the resulting network of information flowing would be of much higher quality than under present circumstances.

The activities of the field schools, farmers' organizations and local governments in the area of IPM become very useful for national crop-protection services, which all have some version of pest or risk surveillance at the 'above the village' level among local government but within subnational administrative and agroecological zones; the more skilled farmers become in diagnosing and acting on field conditions, the more demanding they become of local agricultural, especially crop-protection, field staff, and the greater and higher-quality becomes the flow of information from the field upwards. In some cases, the exchange of information laterally among localities also increases following this ground-swell upwards.

Higher-quality information generated by this field network of IPM clubs would not only help local communities. Agricultural policy-makers from districts up through national legislatures would benefit. Bureaucracies tend to reward precision and caution, but pest-management policy decisions in real time require accuracy and timeliness. It is more valuable to know what happened in the field yesterday than to be given a prediction of next month derived from last month's or last year's statistics, regardless of the model's precision.

Under current conditions, production intensification programmes tend towards varieties, bred without reference to local pest variation, that can be planted nationally. These programmes often bundle pesticides in the same packages against the same pests when varieties are known to carry host-plant resistance. These contradictory policies not only confuse extension workers and farmers, but they also accelerate the evolution of local insect populations by eliminating biological population regulation by natural enemies. This shortens the usable field life of the varieties concerned, as happened with a number of rice varieties carrying BPH resistance genes (Bph1, Bph2 and probably others).

IPM clubs, researchers and biotechnology

Institution-based researchers would benefit from longer-term relationships with these groups, filling in more and more detailed pictures of changing pest populations. Researchers would be able to test large-scale hypotheses and to provide better, more timely, locally better-adapted technologies to IPM clubs and their larger communities.

More accurate, location-specific information on pest evolution would allow researchers to breed, possibly engineer, select, test and make available more narrowly focused host-plant resistance for village communities. It can be hoped that longer-term relationships with more knowledgeable farmers' groups would help researchers respond to their needs directly. In the short run, this can only be imagined for pilot-scale initiatives. Commercial pressures, common to both seed and agrochemical industries, make it difficult for truly local technologies to survive economic selection processes. At the same time, the pressures of national public-sector seed programmes make local varieties the exception rather than the rule. A strong pilot demonstration of the mutual benefits of longer-term exchange relationships, although requiring at least 5 and probably 10 years of institutional commitment, would help make the case.

A Brief Prospectus for the Future

The Asian rice model of participatory IPM is already being used in crops grown in rotation with rice in Indonesia (soybeans, mungbeans, maize, chilli, garlic, shallots, onion and leafy vegetables); in highland cabbage and coconuts in the Philippines; in cotton in India and China; in brinjal (eggplant) in Bangladesh; and in tea in Vietnam. Non-government organizations have been particularly active in vegetable IPM. In many of these cases, farmers who had finished an IPM rice FFS began to apply concepts in other crops, and then demanded technical backup from extension and crop-protection staff. In nearly all crops, the entomological experience initially drove farmers' interest, and the impact of predators and other natural enemies was the entry point to stronger public experimentation. Perhaps because most of these cases are in the tropics, natural population regulation and the disruptive effects of insecticides are easy to demonstrate.

The first season-long FFSs in Africa began in Dawhenya, Ghana, for irrigated rice in mid-1995 (M'boob and Katelaar, 1995). A national IPM steering committee included extension, crop protection, irrigation and crops directorates under the chair of the Vice Minister. Senior research staff from the national agricultural university are also full members of the steering committee. The attractive aspects of the FFS for African farmers are its trust in farmers' abilities, its concrete, local, specific and timely contents, and the shared field experience with extension workers and researchers alongside farmers.

In order further to expand participatory IPM, in 1994 a task force that included staff of the Food and Agriculture Organization (FAO), UNDP, UNEP, the World Bank, and CAB INTERNATIONAL recommended that an IPM facility be set up. This facility would allow multilateral institutions, governments and other interested partners better to identify opportunities for participatory IPM implementation. The facility would mobilize technical and financial support for pilot field projects by governments, NGOs and farmers' organizations in order that the case for larger and longer support could be made to local and national policy-makers and donors. In mid-1995 FAO and the World Bank initiated the funding and activities of the IPM facility.

Soybean in Brazil 5

Flavio Moscardi and D.R. Sosa-Gómez

Introduction

Soybean (*Glycine max*) is the most important export crop in Brazil (11 million ha producing 21 million t). The annual export value is more than $US3.2 billion. With the rapid increase in cultivated area in the south in the 1970s and subsequent expansion of the crop to new areas in central Brazil (Bonato and Bonato, 1987), farmers had to cope with different species of pests, especially insects.

Several species of insects are associated with soybean in Brazil (Panizzi *et al.*, 1977a; Gazzoni *et al.*, 1981); however, a few of them may be considered as key pests or potentially damaging to soybean. The defoliator velvetbean caterpillar, *Anticarsia gemmatalis*, and the pod- and seed-sucking stink bug complex, mainly represented by *Nezara viridula*, *Piezodorus guildinii* and *Euschistus heros*, are the key soybean pests in Brazil. They are abundant in practically all soybean-growing regions and are responsible for over 90% of the insecticide applications on soybean. Other species are either infrequent at damaging levels or occur at economic levels only in limited areas.

A soybean integrated pest management (IPM) programme, strongly orientated to the insect component, was developed and implemented in Brazil in the mid-1970s, and is considered one of the most significant success stories of IPM implementation for a major annual crop over a wide area (Gazzoni and Oliveira, 1984; Kogan and Turnipseed, 1987). An overview of this IPM programme is presented, with emphasis on insect pests, as well as a discussion of the prospects for its improvement through the use of biotechnological techniques.

Historical Background

Until the mid-1970s insect control in soybean was mostly based on criteria developed by farmers, resulting in excessive numbers of chemical applications at unnecessarily high rates. The most popular chemicals were highly toxic and broad-spectrum, including mixtures such as dichlorodiphenyltrichloroethane (DDT) + parathion and DDT + toxaphene, among others. In 1975, an IPM scheme was tested on farms of Parana and Rio Grande do Sul, the major soybean-producing states in Brazil (Kogan *et al.*, 1977), consisting of: (i) weekly scouting of insect populations and assessment of levels of defoliation and plant growth; (ii) use of action thresholds for key pests; and (iii) application of minimum effective rates of selected insecticides when thresholds were reached. Paired fields of 10–30 ha were compared in each of nine selected farms; one followed the IPM scheme and the other was managed according to farmers' established methods.

The number of insecticide applications was reduced by 78% in IPM areas in relation to farmer-treated fields, with no differential effect on yields found (Kogan *et al.*, 1977). In 1976 results confirmed those of 1975, and the programme was officially adopted for the state of Parana. In 1977 it was adopted by the National Soybean Research Center (CNPSo) of the federally supported Brazilian Organization for Agricultural Research (EMBRAPA), which further coordinated and developed soybean IPM in the country. Close links were established with official extension services and farmer cooperatives to facilitate training of field officers in scouting and decision-making procedures, resulting in rapid expansion of the programme to other soybean producing states (Gazzoni and Oliveira, 1984; Kogan and Turnipseed, 1987).

Specific tactics of technology transfer to extension services and to farmers were devised, as discussed by Gazzoni and Oliveira (1984). These included: IPM demonstration sites for extension workers, who were responsible for training farmers; several types of publications directed to extensionists and farmers; intensive use of mass media, with soybean growers being the main target; implementation of a radio and TV 'alert system'; provision of supporting materials (sampling cloths, recording sheets, slide sets, etc.); and short lectures by researchers to extensionists and farmers. The strong engagement by official extension services and farmer cooperatives was crucial to the success of the programme.

As a result of these activities, a dramatic decrease in insecticide usage in soybean was recorded in Parana. Among assisted farmers, the average number of insecticide applications fell from 5.8 in 1976–77 to 1.8, 1.7 and 0.7/ha in the three subsequent seasons. In the same period, the average number of applications fell to about 2.0 overall among soybean growers in the state, showing that the strategies adopted for IPM implementation had a marked influence in promoting the idea even among farmers not directly assisted (Finardi and Souza, 1980). Similar technology transfer procedures were employed in the state of Rio Grande do Sul and subsequently in other regions, resulting in an estimated 40% of soybean farmers adopting the IPM techniques and over 60% reduction in insecticide usage on the crop.

Research carried out by CNPSo–EMBRAPA and other institutions provided information and advances regarding different soybean IPM components, leading to necessary adaptations, improvement and consolidation of the programme country-wide. An important contribution was a comprehensive survey conducted in different regions of the country to assess the dynamics of major pests and their natural enemies (Correa *et al.*, 1977). Technical bulletins (Panizzi *et al.*, 1977a; Gazzoni *et al.*, 1981; Villas Boas *et al.*, 1985; Oliveira *et al.*, 1988) provided important sources of updated information on the IPM programme, especially to extensionists. Considerable progress has been achieved in developing practical procedures leading to further reduction in the use of chemical insecticides on the crop, which will be discussed later.

Status and Advances in Soybean IPM Components

Economic threshold levels

The availability of practical sampling methods and reliable economic threshold levels (ETL) for key pests was important for IPM implementation and reduction of insecticide usage on the crop. During initial phases of the programme, ETL adopted for lepidopterous defoliators and stink bugs were those developed for the same insects in southern USA (Kogan *et al.*, 1977). These were either confirmed or refined through research under Brazilian conditions (Gazzoni *et al.*, 1981; Villas Boas *et al.*, 1990). Economic threshold levels were also determined for other species, such as *Epinotia aporema* (Gazzoni and Oliveira, 1979) and *Sternechus subsignatus* (Hoffmann-Campo *et al.*, 1990), or are being established for more recent insect pests like the crysomelids *Myochrous armatus* and *Megascelis calcifera*, which have become serious problems in central areas of Brazil.

Biological control

The natural occurrence of many species of parasitoids, predators and pathogens is recognized as an important component of the soybean IPM programme (Correa-Ferreira, 1980; Moscardi, 1984; Moscardi and Sosa-Gómez, 1992). Frequently, these organisms are responsible for maintaining pest populations below ETL, and the adoption of IPM tactics non-disruptive to these agents is considered vital for effective management of soybean insects. One important example is the fungus *Nomuraea rileyi*, which under high humidity in the field practically eliminates the need for insecticide applications against the most important defoliating caterpillars (*A. gemmatalis* and Plusiinae). Brazilian farmers are so aware of its importance that they usually refrain from spraying insecticides on soybean when they notice its occurrence on caterpillars in wet seasons.

Considerable progress has also been made at CNPSo–EMBRAPA towards use of biological agents by farmers in lieu of chemical insecticides, resulting in effective implementation of a programme using a nuclear polyhedrosis virus (NPV) of *A. gemmatalis* (AgNPV) in different regions and, more recently, of the stink bug egg parasitoid *Trissolcus basalis*.

The AgNPV is endemic to Brazil and highly virulent and specific to *A. gemmatalis*. Its effective employment by farmers started in 1980–81 on a pilot basis, being subsequently adopted in the different soybean-producing regions of the country. Research developments towards its large-scale production and use were reviewed by Moscardi and Correa-Ferreira (1985), Moscardi (1989, 1990) and Moscardi and Sosa-Gómez (1992). The AgNPV is produced either in the laboratory on insects reared on an artificial diet or under field conditions, the latter method allowing high production level at low cost. Since 1986, the AgNPV has been processed into a standardized kaolin-wettable powder formulation, before release to farmers through extension services. Currently, the Organization for Farmers Cooperatives of the State of Parana (OCEPAR) and four private companies (GERATEC, NITRAL, NOVA ERA and TECNIVITA) are commercializing the AgNPV under a cooperative agreement with CNPSo–EMBRAPA. In some regions, farmers are instructed by extension officers to collect AgNPV-killed larvae in treated fields, so as to apply the pathogen in larger areas or to store them frozen for use in subsequent seasons as crude preparations.

The simple strategies adopted allowed rapid increase in virus use in Brazil, starting at about 2000 ha in the 1982–83 season, and reaching 1 million ha yearly beginning in 1989–90. Since the beginning of the programme, over 7 million ha have been treated with this microbial insecticide, with savings estimated at about $US70 million, and, more importantly, over 9 million l of chemical insecticides were not sprayed into the soybean agroecosystem. Since the AgNPV is specific to *A. gemmatalis*, natural enemies of different insects are spared, resulting in a gradual increase in the natural mortality of pests in AgNPV-treated areas (Moscardi, 1990). The success of AgNPV use has brought about interest in the development of other entomopathogens as microbial insecticides. Considerable progress has been achieved with the NPVs of *Chrysodeixis includens* and *Rachiplusia nu* and the granulosis virus of *E. aporema*. Fungi, especially *Beauveria bassiana* and *Metarhizium anisopliae*, are currently being investigated at CNPSo–EMBRAPA as potential agents to be used for stink bug control (D.R. Sosa-Gómez and F. Moscardi, unpublished data).

The egg parasitoid *T. basalis* is an important component of the natural mortality of stink bug species (Correa-Ferreira, 1986). Despite its high natural incidence, peak populations of this parasitoid usually occur when stink bugs have already reached damaging levels. Therefore, procedures for production and release of *T. basalis* were developed at CNPSo–EMBRAPA, so as to avoid economic damage to soybean and thus eliminate the need for chemical insecticide applications to control the stink bug complex (B.S. Correa-Ferreira, unpublished data).

The parasitoid is produced in the laboratory on *N. viridula* eggs, obtained though large-scale rearing of the insect, according to the procedure developed by Correa-Ferreira (1985). Parasitized eggs (5000/ha) are glued on cardboard and sent to farmers to be hung on soybean plants in border rows of their soybean fields when stink bug populations are still low (colonization phase).

This programme was moved to the farm level in 1990–91, after it proved successful in a 4-year pilot programme conducted in farmers' fields in Parana (Correa-Ferreira, 1991; B.S. Correa-Ferreira, unpublished data), through cooperative efforts with extension services and farmer cooperatives. The current area under the programme is about 2000 ha. Due to the difficulty in producing *T. basalis* for release in larger soybean areas, regional producing units are being established in several institutions to increase availability of the parasitoid.

Cultural control

Changes in cropping practices may affect the dynamics of insects associated with soybean (Kogan and Turnipseed, 1987). Some of them can be manipulated to reduce pest populations or to avoid the coincidence of peak populations occurring at a time when the soybean is most vulnerable to pest attack.

Growing early-maturing varieties is widespread among farmers in the south, since they normally escape economic damage by stink bugs (Panizzi, 1985). As short-cycle varieties are harvested, stink bugs migrate to late-maturing varieties, reaching high population levels and demanding frequent insecticide applications. The planting of a border strip of an early-maturing variety, in about 5% of the area, has been tested as a trap crop in Brazil. Properly timed insecticide applications have resulted in substantial reduction of stink bugs on the main crop (late-maturing variety) (Panizzi, 1985). Release of egg parasitoids in the trap crop has also been shown to be an effective tactic for management of these insects (Correa-Ferreira, 1991). Although use of trap crops is a potentially useful soybean IPM tactic, it has not yet been adopted by farmers. Proper planning of planting on an area-wide basis is necessary to achieve a synchronized effect of plant phenology in attracting the early stink bug colonizers.

Manipulation of planting dates may also be useful to avoid economic damage by some insects. *Anticarsia gemmatalis* larvae are usually less abundant in late plantings, which can be attributed to a higher larval mortality by the fungus *N. rileyi*. Late planting has also been pointed out as an effective way to avoid high incidence of thrips and consequently of the soybean bud blight virus disease, which is transmitted by these insects (A.M.R. Almeida and I.C. Corso, EMBRAPA–CNPSo, unpublished data).

In the last decade, many southern farmers have shifted from conventional to no-till cultivation, which helps reduce soil erosion. However, no-till soybeans have favoured the build-up of some insects, such as the girdler and stem-borer, *S. subsignatus*, and soil insects, mainly scarabaeid larvae, which have become serious pests in some regions. Crop rotation with maize has been proposed as a suitable tactic to reduce *S. subsignatus* in highly infested areas, as well as periodic shifts from no-till to conventional tillage, since hibernating larvae of *S. subsignatus* and other soil insects would be drastically reduced by ploughing and by adverse microclimatic soil conditions (Hoffmann-Campo, 1989).

Host-plant resistance

Research on development of soybean varieties resistant to insects has been carried out throughout the world, with sources of resistance being found against at least 17 species (Kogan, 1989). In Brazil, efforts have been mainly directed towards developing resistant varieties to stink bugs (Miranda *et al.*, 1979; Rosseto *et al.*, 1981; Rosseto, 1989). Variety IAC 100, resistant to stink bugs but also reported as resistant to crysomelid and lepidopterous defoliators, was released to growers in 1988 (Rosseto, 1989). At CNPSo–EMBRAPA, the breeding programme is incorporating genes for resistance to defoliators and stink bugs from identified resistant genotypes into advanced agronomic germplasm. Some of these have shown high levels of resistance and adequate agronomic characteristics, under heavy stink bug attack, in different soybean-growing regions (C.B. Hoffmann-Campo and D.L. Gazzoni, CNPSo–EMBRAPA, unpublished data). Release of these varieties, associated with other tactics such as biological control, is expected to result in an enormous improvement in the IPM programme, since stink bugs are still controlled almost exclusively by chemical insecticides.

Chemical control

Most chemical insecticides, although considered effective tools to suppress insect populations reaching damaging levels, are toxic to non-target organisms, including humans. Furthermore, their use on soybean often leads to pest resurgence as a result of their adverse impact on naturally occurring biological control agents (Panizzi *et al.*, 1977b; Oliveira *et al.*, 1988; Silva *et al.*, 1988). With the implementation of the soybean IPM programme, research on insecticides was mainly directed towards use of more selective products at minimal effective rates, leading to a gradual improvement in pest control recommendations to the extension services.

Considerable progress has been achieved in the recommendation of insecticides based on their selectivity to predators. Beginning in 1988, this was one of the major criteria adopted by regional entomological committees responsible for annual recommendations of insecticides to be used in soybean IPM. As a result, from 20 active ingredients indicated for *A. gemmatalis* control up to 1987, only eight are presently recommended, including two biological insecticides (AgNPV and *Bacillus thuringiensis*). Recently, it has been found that mixtures of insecticides, at half of the recommended dose, with 0.5% NaC1 in the spray tank, result in adequate control of stink bugs (Corso, 1990). This method is being successfully employed by farmers in over 250,000 ha in Brazil.

Perspectives for Improving the Soybean IPM Programme Using Biotechnology

Characterization of biological agents

Even before the development of engineered bioinsecticides, molecular biology offered new perspectives on the characterization of naturally occurring microorganisms (Kirschbaum, 1985), which is the most important initial step of any microbial control programme. The characterization can be accomplished by morphological, physiological and biochemical methods, as well as by deoxyribonucleic acid (DNA) marker polymorphism analysis, the latter presenting the advantage of representing only genetic variations (Fairbanks *et al.*, 1993). Some entomopathogenic microorganisms occurring in soybean systems, such as the AgNPV and the fungi *B. bassiana* and *M. anisopliae*, have been partially characterized (Pinheiro *et al.*, 1990; Sosa-Gómez, 1990; Zanotto, 1990; Maruniak, 1992; M.S. Tigano, EMBRAPA–CENARGEN, unpublished data).

Genetic engineering techniques provide the means to isolate, purify and amplify portions of DNA, making it possible to establish their size and physical features (relative positioning and genetic distance separating genes). Physical genomic maps can be constructed by using tools such as polymerase chain reaction (PCR), restriction fragment length polymorphism (RFLP) and random amplified polymorphic DNA (RAPD) (Fairbanks *et al.*, 1993), which are recommended to consolidate more reliable classification systems (Billimoria, 1986; Humber, 1990). Also, these techniques make it easier to alter DNA segments *in vitro* to study structure and function relationships and to increase the possibilities of success towards obtaining hybrids with stable and advantageous characters.

Considerable progress has been made towards knowledge of the baculovirus genome, the best example being the NPV of *Autographa californica* (AcNPV) (Blissard and Rohrmann, 1990; Wood and Granados, 1991; Maruniak, 1992). This has been, to a lesser extent, accomplished with the AgNPV (Pinheiro *et al.*, 1990; Zanotto, 1990; Maruniak, 1992), allowing ongoing studies on the similarities between AgNPV geographical isolates as well as on the genetic stability of this virus after successive large-scale use, an important aspect for sustainability of its use in Brazil. In-depth characterization of the most important entomopathogens associated with soybean pests, such as AgNPV, *N. rileyi*, *B. bassiana* and *M. anisopliae*, is needed. This will allow monitoring of these agents through time and space, after their application, and of their performance under different ecological conditions, so that more adapted pathotypes can be selected. Use of biotechnological methods would also be important to refine quality control of AgNPV formulations in use, as well as to determine genomic alterations through time and how these are related to possible changes in virulence.

Some other aspects that need more research or appropriate methods are: (i) determination of which genes are involved in the mechanisms of infection; (ii) alteration of specificity and virulence after serial passages through other host systems (Pavan and Ribeiro, 1989; Morales *et al.*, 1993); (iii) synergistic action between baculoviruses; and (iv) the mechanisms and genes related to possible development of resistance of *A. gemmatalis* to its NPV, since this species has shown a high potential to develop resistance to the AgNPV under laboratory selection-pressure experiments (Abot, 1993; A.R. Abot, F. Moscardi, J.R. Fuxa, D.R. Sosa-Gómez and A.R. Richter, unpublished data).

Use of viruses to control soybean insect pests

Commercial interest in developing entomopathogenic viruses as insecticides is reported by Federici (1990) as lagging behind that of *B. thuringiensis* because 'these viruses kill much more slowly, typically have a narrow host spectrum, and are too costly to produce *in vitro*'. All these 'unfavourable' characteristics are also true for the AgNPV and other baculoviruses associated with soybean insects. However, in spite of these limiting factors, the AgNPV programme has been quite successful, as it has resulted in impressive adoption by farmers in Brazil (10% of the soybean area or about 1 million ha yearly). This success is mainly due to the possibilities of producing the pathogen under field conditions at low cost as well as to the favourable host-plant and pest incidence characteristics for virus application. In most regions, *A. gemmatalis* is the key defoliator on the crop up to the flowering stage, and its major occurrence is temporally distinct from the other key pests, the stink bugs, which are economically important to soybean after the flowering stage. Furthermore, soybean tolerates high levels of defoliation, with economic thresholds being 30–40% defoliation for vegetative and flowering stages (Gazzoni *et al.*, 1981). Therefore, proper timing of AgNPV applications, taking into account the density and age composition of *A. gemmatalis* larvae, allows effective control of this insect while preserving soybean yield potential (Moscardi, 1989; Moscardi and Sosa-Gómez, 1992).

Our knowledge of the molecular biology and genetics of baculoviruses has increased greatly in the last 10 years, allowing the use of biotechnological approaches to increase the speed of kill or host range of these agents (D.W. Miller, 1988; Maeda, 1989; Wood and Granados, 1991; Maruniak, 1992). Several types of foreign pesticidal genes have been proposed for insertion into viral genomes, including insect-specific toxins, hormones, hormone receptors, metabolic enzymes and growth regulators, as reviewed by Wood and Granados (1991). Among the known attempts to increase speed of kill of baculoviruses by these means, most were unsuccessful or did not produce clear results. The first positive report of a foreign gene that enhanced the insecticidal properties of a baculovirus was that of Maeda (1989), who replaced the *Bombyx mori* NPV polyhedrin gene with the diuretic hormone gene from the tobacco hornworm, *Manduca sexta*. The modified virus gave 100% larval mortality at 4 days postinfection, while most larvae infected with the wild-type isolate died by day 5. It has been found that a viral protein, contained in the granulosis virus of *Trichoplusia ni*, enhances infectivity of baculoviruses by rapid disruption of the peritrophic membrane of larvae (Derksen and Granados, 1988). Addition of the virus-enhancing factor gene resulted in more than a tenfold reduction of the median lethal dose (LD_{50}) and a lower LT_{50} in infected host larvae (Wood and Granados, 1991). Field trials with genetically altered forms of viruses have been conducted by the Institute of Virology, Oxford (UK) (Bishop, 1986; Bishop *et al.*, 1988) and by the Boyce Thompson Institute at Cornell University, New York, USA (Wood *et al.*, 1990).

With regard to soybean insects and associated viruses, some attempts have been made, through conventional selection procedures, to obtain viral isolates with an expanded host range. An AgNPV variant has been selected on the sugarcane borer, *Diatraea saccharalis*, which increased 1500-fold in virulence over a series of 20 passages through this species (Pavan and Ribeiro, 1989). Pavan and Ribeiro (1989) also claimed that this variant retained its virulence to the original host (*A. gemmatalis*). This AgNPV was routinely produced on laboratory-reared larvae of *D. saccharalis*, and a product (MULTIGEN) was registered and commercialized in 1989 in Brazil by AGROGGEN, a Brazilian private company, for the control of both *A. gemmatalis* and *D. saccharalis*. However, this formulation showed very poor efficiency against *A. gemmatalis* under field conditions, and its production was discontinued in 1990. Similar attempts were made with the

AcNPV, aiming at developing a single viral insecticide with activity against *A. gemmatalis* and Plusiinae (*C. includens* and *R. nu*) (Morales, 1991; Morales *et al.*, 1993). After five serial passages through these species, highly virulent variants of the AcNPV were obtained for each of the hosts. However, a 'polyvalent' virus isolate could not be obtained, since each of the selected variants was highly active only on the host through which it was serially passed, showing very low activity on the other two species.

Although conventional techniques to select improved viral isolates should not be discouraged, biotechnological methods seem to be an invaluable tool for enhancing the insecticidal properties of the AgNPV and other viruses so as to expand the possibilities of their use in soybean IPM. Significant progress has been attained recently towards a better knowledge of AgNPV molecular biology and its contribution to the improvement of this virus as a microbial insecticide (Maruniak, 1992; M.L. de Souza, EMBRAPA–CENARGEN, unpublished data). Physical maps for AgNPV plaque isolates are being constructed in detail, enabling researchers to locate the virus polyhedrin in the gene (Maruniak, 1992). A fine-structure map of the fragment containing the polyhedrin gene has also been constructed (Zanotto, 1990), which will permit genetic alteration of this biological pesticide, leading to important characteristics such as higher virulence, faster killing and expanded host range.

Current use of insect viruses worldwide is low at present, despite the potential for their large-scale use in many cropping systems. Among the factors limiting the expansion of their use are the technical and economic difficulties with their mass production. This has resulted in low interest by private companies in producing and commercializing viral insecticides (Young, 1989; Federici, 1990). At present, these agents can only be produced *in vivo* for commercial purposes, which is often difficult and costly for many virus–host systems. It is believed that greater interest in viral insecticides would emerge if methods could be developed to produce them cost-effectively *in vitro*. Serum-free media have been developed recently, but the upper limit on fermentation batch size for insect cells, which remains at about 20 l, continues to impede more widespread interest (Federici, 1990). The AgNPV can be produced at low cost under field conditions, for further processing into a standardized formulation by EMBRAPA–CNPSo and five private companies, which has allowed rapid expansion of AgNPV use in soybean in Brazil (Moscardi and Sosa-Gómez, 1992). *In vivo* production in the laboratory is also carried out by two private companies, on larvae reared on an artificial diet, but this method needs further improvement to bring down current production costs. Use of this virus in soybean has levelled off at about 1 million ha annually for the last four seasons due to production limitations, even though the demand for the AgNPV has increased during the same period. Therefore, the development of practical and economical methods for *in vitro* mass production of AgNPV would certainly open new perspectives for expanding the availability of this virus and consequently the area treated with this microbial insecticide in Brazil, as well as in other countries where *A. gemmatalis* is a major soybean pest.

Development of fungi as microbial pesticides

Among the fungi associated with soybean pests, the most outstanding, considering the importance of their hosts, are: *N. rileyi*, species of Entomophthorales (e.g. *Entomophthora gammae*, *Zoophthora radicans*, and *Erynia crustosa*) and *Paecilomyces tenuipes* on defoliating Lepidoptera; *B. bassiana*, *M. anisopliae* and *Paecilomyces* sp. on coleopterous species and stink bugs (Moscardi, 1984; Moraes *et al.*, 1991; D.R. Sosa-Gómez, unpublished data). The entomophthoralean fungus *Neozygites* sp. can also be found at high levels on the mite *Tetranychus urticae*, and may be responsible for reducing populations of thrips on the crop (D.R. Sosa-Gómez, unpublished data). Species of fungi have also been found on nematodes, some of them with potential to be used as microbial nematocides, such as *Monacrosporium ellipsosporum* for strain 3 of *Meloidogyne incognita* (Santos *et al.*, 1992) and *Verticillium lecanii* for the soybean cyst nematode *Heterodera glycines* recently introduced into Brazil.

Considerable research has been done at EMBRAPA–CNPSo on the use of fungi against soybean insects, especially stink bugs (D.R. Sosa-Gómez and F. Moscardi, unpublished data). Some limiting aspects related to the use of these agents as microbial insecticides are: (i) the taxonomic status is not well known for some species, e.g. *Neozygites*; (ii) the high dependency on environmental conditions, especially humidity, for conidiogenesis and conidial germination; (iii) the difficulties of large-scale production for some species and in keeping stability of produced materials.

Due to their complex genome, insect fungi are probably secondary targets for improvement by genetic engineering. However, the following aspects could serve as examples of how biotechnological methods would help improve fungi as pest-control agents.

Reduction of humidity requirements to initiate germination and the infective process on host species

D.R. Sosa-Gómez and S.B. Alves (unpublished data) have observed that some strains of *B. bassiana* are less demanding on humidity for conidiogenesis than others. The relative humidity is important for effective stink bug control in soybean during the first week after field application of fungi (D.R. Sosa-Gómez and F. Moscardi, unpublished data). If humidity is higher than 75%, high levels of infection usually result on *Piezodorus guildinii*, a susceptible species.

Increased persistence of mycoinsecticides under normal storage conditions and in the field

These aspects have been improved for *B. thuringiensis* by means of an encapsulation system in *Pseudomonas fluorescens* (Gelernter, 1990). In the case of fungi, higher persistence of infective units is associated with black-pigmented conidia (Ignoffo, 1992), a character that can possibly be manipulated through biotechnological methods to develop isolates of *Beauveria, Metarhizium, Nomuraea* and *Paecilomyces* with extended storage and field persistence.

Improved methods for production of infective units or mycelium

Some fungi associated with soybean pests, such as *Nomuraea, Entomophthora* and *Neozygites*, are difficult to produce in either the conidial or the mycelial form, which limits their use as microbial pesticides. The finding of more productive mutants, as is the case in *Beauveria* and *Metarhizium*, has not substantially reduced costs. The technology of recombinant DNA will probably increase the chances of solving these problems.

Development of virulent fungal isolates with wider host range

Although they are less specific than viruses, isolates of fungi vary substantially in their virulence to different host species, as observed for *N. rileyi* on species of soybean caterpillars (Moscardi *et al.*, 1992). Therefore, the development of fungal isolates with high virulence to a specific pest complex, such as defoliating caterpillars and stink bugs, is needed if these agents are to be used as microbial pesticides.

Increased conidial adhesion to host

This aspect was shown to be important in the infective process of fungi on soybean stink bugs. Conidial adhesion was higher on *N. viridula* and *P. guildinii* than on *E. heros*, when these species were subjected to the same inoculation method and dosage (D.R. Sosa-Gómez, unpublished data).

Use of **Trissolcus basalis** *against stink bugs*

The most serious problem limiting the expansion of this programme in Brazil is the difficulty in producing enough egg parasitoids to meet increasing demands of farmers. Current *in vivo* production, based on laboratory or glasshouse rearing methods for *N. viridula*, is laborious and time-consuming, allowing production of only a limited amount of the parasitoid for field release. Current strategies, such as the establishment of *T. basalis* regional producing units, will certainly help to increase availability of the parasitoid in other soybean-producing regions, since its use is still restricted to southern states. However, these actions will not be enough to meet the demand for the parasitoid, considering that stink bugs are key pests throughout most of the soybean-cultivated area in the country.

Consequently, development of artificial eggs for large-scale multiplication of *T. basalis* would be an important means of expanding its use in soybean IPM. *In vitro* production methods have been developed for other egg parasitoids, with considerable progress being made with *Trichogramma* spp. (Dai *et al.*, 1988; Li *et al.*, 1988) and, more recently, with *Ooencyrtus nezarae*, a parasitoid of stink bugs (Takasu and Yagi, 1992).

Another important area to pursue through conventional or biotechnological methods is the development of *T. basalis* strains with resistance to chemical insecticides, especially those used mostly in soybean, since these products will be needed to control stink bugs and other pests reaching economic thresholds. The viability of developing strains of parasitoids resistant to insecticides has been demonstrated in California for *Trioxys pallidus*, a parasitoid of the walnut aphid (Hoy and Cave, 1989).

Host-plant resistance to soybean pests

The progress in recent years in introduction of *B. thuringiensis* delta endotoxin into crop plants, as a means of obtaining transgenic plants resistant to insects (Perlak and Fischhoff, 1990), indicates that this technology may represent an important tool to incorporate insect resistance into high-yielding varieties. However, this has raised concern that intensive cultivation of transgenic plants may favour rapid development of resistance by insect populations to *B. thuringiensis* toxins or to other insect-specific toxins incorporated into plants (Castro, 1992; Davidson, 1992). Resistance to *B. thuringiensis* commercial formulations has already been detected in some insect populations subjected to frequent applications of this microbial insecticide (McGaughey, 1985). However, insect-resistant transgenic plants may be useful tools in IPM programmes, if possibilities of resistance by insects are carefully taken into consideration and biotechnological as well as cultural methods are devised to reduce or avoid the appearance of insect resistance (see Gould, Chapter 16, and Roush, Chapter 15, this volume).

Regarding soybean, this technique would be a useful tool as an IPM component, especially for Lepidoptera and stink bug management. However, there are some difficulties related to the manipulation of leguminous crops, including soybean, at the level of cell culture, transformation and plant regeneration (Castro, 1992).

The development of transgenic plants resistant to herbicides is also progressing rapidly, and may be useful for weed management in specific situations. However, from our point of view, this may be a way of sustaining and even fostering the use of chemical herbicides in agriculture, to the detriment of other more environmentally sound weed-management practices.

Conclusion

Soybean IPM in Brazil has progressed substantially since its implementation in the mid-1970s, with economic benefits estimated at $US1.5–2.0 billion in a 15-year period. This represents over 25 times the CNPSo–EMBRAPA budget in the same period for research in all fields related to soybean. Although difficult to quantify, environmental and social benefits are probably even higher, considering that in this period over 90 million l of chemical insecticides were not sprayed into the soybean agroecosystem due to the IPM programme.

In a short time frame, most significant improvements of this IPM programme will probably not result from new biotechnological techniques, but rather from conventional methods and technologies such as: (i) release of new soybean cultivars with resistance to stink bugs and to recently introduced diseases, such as stem canker, caused by *Diaporthe phaseolorum*, and the cyst nematode *Heterodera glycines*, currently considered the most important pest problems of soybean in the country; (ii) improved cultural methods of control for the girdler and stem borer *S. subsignatus* and soil-inhabiting insects; (iii) improved mass production, quality control and formulation procedures for AgNPV and the egg parasitoid *T. basalis*; (iv) selection of more virulent strains of AgNPV and fungi among naturally occurring isolates of these agents; (v) better understanding of factors limiting the epizootiology of fungi and their usefulness as microbial insecticides; (vi) devising proper IPM tactics for crysomelid defoliators in central areas of Brazil; and (vii) development of proper technology-transfer strategies for some regions where insecticide applications are still high due to low adoption of IPM practices. A better knowledge of the impact of interactive pest classes, rather than of single pest classes as currently done, is needed to improve the soybean IPM programme, as proposed by Kogan and Turnipseed (1987).

It is expected that the rapid progress being achieved in biotechnology will result in new breakthroughs applicable to soybean IPM in the near future. The most relevant contributions of biotechnology to soybean IPM will probably first occur in the areas of characterization, quality control and environmental monitoring of microbial insecticides. Other important needs which can be met by biotechnological means include: (i) development of *in vitro* mass-production methods for the AgNPV and the egg parasitoid *T. basalis*; (ii) production of AgNPV isolates with shorter killing time and expanded host range; (iii) development of entomopathogenic fungal isolates that are less demanding on humidity for conidiogenesis and germination; and (iv) transgenic plants resistant to key insects and diseases through incorporation of disease- and insect-specific toxins.

The perspectives and practical value of biotechnological methods to the soybean IPM programme are difficult to assess at the present, since developments in this field are quite recent. As they are made available, these technologies should be carefully evaluated for their usefulness and possible environmental impact, before they are firmly incorporated into routine soybean IPM procedures in Brazil.

India: An Overview 6

Nandini V. Katre

Introduction

India is basically an agricultural society. Most of the agricultural production is by small-scale farmers, many of whom are resource-poor. Less than 20% of food-grain production is through modern industrialized agriculture (Dinham, 1993). Therefore, any integrated pest management (IPM) scheme must include small-scale farmers to generate a healthy agricultural economy. The green revolution, while increasing certain aspects of food production, concentrated on agriculture in the most fertile regions, which constitute less than 10% of all farmland. The use of high-yielding varieties of crops that are fertilizer-responsive and chemical pesticide-dependent has not improved the livelihood of most small-scale farmers. Moreover, these farmers find it difficult to cope with chemical-intensive farming systems as they have no control over resources and lack knowledge on pesticides (Gandhimathi, 1992). The change from traditional systems to modern commercial farming has led to a severe exploitation of scarce natural resources and a breakdown of the balance between human consumption and renewable resources. Consequently, chemical pesticides have adversely affected both human health and the environment.

The green revolution also turned India into a major pesticide producer and consumer, and the agrochemical industry, which began blossoming in the 1970s, continues to flourish in the 1990s. Much of the recent and projected growth is attributed to the changes in government policy, encouraging multinational companies to operate and allowing for future expansion (Verma, 1990). Insecticides account for about 75% of pesticides sold in

India. In agriculture, 45% of pesticides are used on cotton, with rice, plantation crops (tea, coffee, rubber, cardamom) and tropical fruits (mango, banana, etc.) using a considerable amount as well (AGROW, 1992). Benzene hexachloride (BHC) accounts for 50% of the insecticides used, in spite of its environmental persistence and health hazards. Dichlorodiphenyltrichloroethane (DDT) has been recently banned for agricultural use (AGROW, 1989). Nevertheless, other pesticides are sold in about 86,000 retail outlets throughout India. Retailers generally lack training, and this results in a considerable variation in advice to customers regarding pesticide use. Thus, inappropriate choices of pesticides have been linked with incorrect timing and inaccurate application, and have in turn led to an inadequate system of pest control and the excessive use of pesticides. This has led to severe environmental pollution and health problems, affected wildlife and resulted in widespread resistance in pests to the chemicals and in outbursts of secondary pests. For example, the diamondback moth (DBM) (*Plutella xylostella*) and the American bollworm (*Heliothis armigera*) are resistant to all pesticides currently registered in India (AGROW, 1992).

India has millions of small-scale farmers who own less than 4 ha of land. Due to pressure from the government and industry, these farmers use hybrid seeds and chemical fertilizers and pesticides, often without adequate knowledge and training. A disproportionate rise in the prices of these chemical inputs is making farming uneconomical. Most of these farmers manage to survive because of government subsidies and easy credit for chemical fertilizers and pesticides. Price controls on fertilizers have been lifted, and it is anticipated that controls on pesticides will soon be lifted. Many farmers will then go out of business, unless alternatives are sought and implemented.

Integrated Pest Management

The health and environmental effects of pesticides have been largely ignored by the Indian government. However, due to crop losses from pest resistance and to sustained criticism from non-governmental organizations (NGOs), health organizations, consumer groups and some farmer groups, the government has been forced to take an interest in IPM. An ambitious government programme, providing training for farmers and extension workers, is being planned to increase the present 26 IPM centres to 200 throughout India (Dinham, 1993). The definitions of IPM range widely, from a combination of hazardous and safer pesticides, limited use of pesticides on a need basis, to a complete ban on chemical pesticide usage. The Indian government appears to define it narrowly, which has encouraged the agrochemical industry to applaud IPM and increase the manufacture of pesticides, since selective pesticide use in IPM can build long-term stable markets for pesticides. For example, one such industrial pesticide group, the Pyrethroid Efficacy Group, created to counter the pest-resistance problems in cotton, brought out its own extension and communication programmes on IPM for cotton. However, this group's focus is on controlled use of pesticides on cotton, and it recommends use of several highly hazardous pesticides in conjunction with pyrethroids (Dinham, 1993). Thus the government's narrow definition of IPM has generated mistrust among many farmers regarding IPM. To these farmers, IPM appears to be an attempt by the agrochemical companies and the government to salvage something from their investment in the chemical pesticide industry, and only where convenient and economic to substitute other approaches, such as biological control (Pereira, Bombay, 1993, personal communication). The status of IPM in India has been reviewed with special reference to IPM on rice, cotton and potato (Suri *et al.*, 1992). An illustration of the use of pesticides in conjunction with cultural controls is given below.

For agriculture to be truly sustainable in the long term, all chemical pesticides must eventually be eliminated. In many instances, this shift to non-chemical pest control may be gradual, but the appropriate IPM methodologies must incorporate this long-term goal. These non-chemical controls can be achieved by a combination of biological controls, cultural practices and agricultural techniques, such as use of local insect-resistant varieties of crops, late planting, pest monitoring, crop rotation, clean cultivation and multicropping. The main advantage of IPM is that it has made the practice of pest control more precise. However, most programmes are limited by the designs and management strategies of existing agricultural systems (Hill, 1990). Therefore a different approach is necessary to ensure the sustainability of India's agricultural system. If pests are viewed as indicators of badly designed and malfunctioning systems, it is imperative to understand the causes of pest outbreaks and to modify the design and management of systems to prevent them. This chapter focuses on IPM studies using biological controls and other non-chemical methods for controlling agricultural pests.

Case-studies

IPM studies

The case-studies described illustrate the definition of IPM, which combines chemical pesticides at a low level with other methods of pest control, thereby sustaining the market for pesticides. Both studies are field trials demonstrated by government scientists, whose thinking on IPM is reflected in the use of chemical pesticides in conjunction with cultural practices. However, these studies later demonstrated that the best results in controlling pests were obtained when no chemical pesticides were used.

Cabbage

In the first study, near Bangalore, field trials by the Horticultural Research Institute showed that Indian mustard serves as a trap crop for the two major cabbage pests, DBM (*P. xylostella*) and the leafwebber (Srinivas and Krishna Moorthy, 1991). In this IPM study, cabbage was intercropped with Indian mustard, which successfully combated the pests during the rainy season without the need for pesticide applications. Variations on intercropping were tried to obtain the pest-control results. During the winter and summer, the intercropped cabbage required two pesticide applications of 0.05% cartap hydrochloride to control the pests. Spot applications of 0.07% endosulfan and 0.1% phosphamidon were also used to control localized infestation. The results obtained were as follows: the intercropped cabbage recorded 93% marketable heads, whereas those sprayed with weekly pesticide applications (grower's practice) recorded only 20% marketable heads. Moreover, during these weekly pesticide applications, growers tended to mix pesticides from different groups, and these resulted in a rise and fall in pest larval populations and therefore inconsistent produce. The control plot, where there was no intercropping or pesticide applications, yielded no marketable heads.

A methodology free of chemical pesticides was demonstrated by the same researchers in a subsequent IPM study on cabbage. They found that, when a 5% neem (*Melia azadirachta*) seed-kernel extract (water extract) was sprayed on the intercropped cabbage in the field, there was effective control of all pests. When neem sprays were used, there was a significant rise in populations of the resident natural enemy of the pests, *Cotesia plutellae* Kurdjumov. Therefore, neem spray was recommended instead of pesticides on intercropped cabbage.

The low-input requirement of this IPM scheme and the fact that it is impossible to control DBM on cabbages with pesticides around Bangalore have encouraged growers to adopt the IPM technology, and extension services have begun disseminating the technology. However, the extent of adoption of the IPM procedure by the farmers is not known, partly because of some problems of transfer of the methodology. For example, some growers reported difficulty in sowing mustard prior to cabbage as recommended by the IPM technique, since it involves early land preparation and irrigating the mustard rows. To avoid this difficulty, these growers had sown mustard simultaneously with the planting of cabbage, and had also resorted to additional spraying to control early pest incidence. Moreover, the growers often forgot to raise a second mustard row, which was a necessary part of the

IPM technique, resulting in further spraying to control the increased pest incidence. The most readily available pesticides were used for spraying these additional rounds. Thus, these observations indicate that the techniques have to be farmer-friendly and easy to incorporate for IPM to be successfully adopted. In view of these difficulties, the researchers began searching for a long-duration mustard variety, which could allow the mustard to be planted concurrently with the cabbage, and thus make the technology easier to adopt. Cultural acceptance was yet another barrier to adoption of this IPM technology. Mustard was not accepted for intercropping by these cabbage growers since it is of no use to them. The agricultural scientists have developed an IPM method of intercropping cabbage with marigolds. Marigolds have commercial value and presumably will be accepted by the growers. Demonstrations are in progress.

Opportunity for biocontrol

Another strategy for IPM on cabbage is to use biocontrol agents with intercropped cabbage. A local laboratory has identified two parasitoids (*Apanteles plutellae* and *Tetrastichus sokolowskii*) that give effective control of DBM and other pests on cabbage. The release and establishment of these parasitoids on the intercropped cabbage may be the best method for the control of pests. Once these parasitoids become established, there may not be a need for intercropping with the cabbage. However, the growers' indiscriminate use of chemical pesticides would make it difficult for the parasitoids to become established in sufficient density to control the pests.

Lucerne

In this study, field trials were conducted to demonstrate the integration of cultural, chemical and biological control measures for the management of pests on lucerne (Ram and Gupta, 1990). The three major pests are leafhoppers, lucerne weevil and aphids. The cultural practices involved the use of the least susceptible or most resistant variety of lucerne and the selection of the optimum timing for planting. To control the pests, a chemical pesticide (endosulfan, 0.08%) or a biological agent (*Bacillus thuringiensis* toxin, subsp. *berliner*, at 0.84 kg^{-1} ha) was applied 15 days after the first cut. The crop was harvested three times. Pest populations were recorded three times before the first cut and twice before both the second and the third cuts. The combination of the cultural practices with either the chemical pesticide or the microbial agent was effective in controlling all pests. However, the cheapest method, with the best cost/benefit ratio, was the one using the cultural practices alone, without any chemical inputs, thereby illustrating that IPM without chemicals can be effective and more economic.

Natural farming, a non-chemical approach to IPM

A 'redesign' approach to IPM is the truly sustainable approach for the future. This approach begins with cropping systems that are less attractive to pests and therefore minimize pest problems from the start. Such systems seek to solve problems internally, by accommodating and supporting the system's natural homoeostatic processes, rather than relying on the repeated application of increasingly ineffective cures to inappropriately designed, malfunctioning systems. The roots of this strategy are found in cultural methods of farming and pest control.

One success story is from the Friends Rural Centre in Rasulia, central India, which is comprised of 15 small farms. These farmers converted their farms from high-input, chemical-intensive farming to natural farming (Aggarwal, 1989, 1991). The first step was to stop using chemical fertilizers and pesticides. Mexican hybrid wheat and other high-yielding hybrid wheats could not survive without chemicals, because they were not suited to the local habitat. Hardy local varieties of wheat were already extinct; nevertheless, suitable wheat seeds were obtained from a neighbouring state. Rice was more adaptable, with hybrids quickly adapting to the natal system. Many suitable rice varieties were found, because rice is indigenous to central India. Rice and wheat were grown with vegetables and legumes. A combination of controls, such as neem formulations and concoctions using chilli peppers, were used against pests, in addition to trap crops such as sunflower, mustard and cowpea. The natural habitat was soon restored. Subsequently, beneficial organisms and natural enemies of the pests established themselves in the newly created pesticide-free habitat. Crop yields were very good for all crops except wheat. Total production under natural farming was higher than that in previous years under the chemical-input system. As a consequence, net profits increased six- to eightfold, with vast improvements in the health of the soil and of the farmers as well. These results encouraged other farmers in the region to switch to natural farming. Those individual farmers who switched did suffer lower crop yields in the first year. However, from the second year onwards, these farmers were able to feed their families and make a profit on their produce. In earlier years, they had been losing money. Some of the farmers who shifted to natural farming could not sustain their production without some chemical inputs and had to change their farming patterns.

Another example of successful natural farming comes from Bangalore, where a small-scale farmer shifted to natural farming, resulting in higher profits for a family of five and five farm workers (Daitota, 1989). For the past few years, the farmer relied entirely on chemicals on his 3-ha farm. In spite of high production levels, his net returns were negligible due to the high costs of tractor ploughing and chemical inputs. The incentive of lowering costs made the farmer shift to natural farming. It took nearly 12 years to shift completely to chemical-free farming. Initially, the yields were less, as in the case above, but so were the costs. After 4 years, the yields had increased to those obtained with chemical farming, and the output increased every year thereafter, with decreasing input costs and substantial net profits. The pest-control strategies included the use of local resistant crop varieties, diversified cropping, crop rotation and biosprays made from neem leaves, mint, wood ash and tobacco leaves. The farmer also hand-picked the insects from the crops, although it was time-consuming. The annual net income of this farm is now much higher than that of the average farm in the area, and farmers around him have begun to realize the significance of this integrated farming approach.

Several such success stories on natural farming have emerged from different parts of India (Dhavle, 1990). In a recent study, actual on-farm experiences of seven medium-sized farmers engaged in natural farming were compared with those engaged in conventional chemical-intensive farming on plots of the same size. All farms are located in South India, in the states of Tamil Nadu and Karnataka (UNDP, 1992). These natural farms produced yields similar to those from the conventional farms, with fewer external inputs. The experiences of the natural farmers were that pests were not a problem on their farms, again illustrating that a properly designed system can prevent or minimize the outbreak of pests from the beginning. These farmers switched to natural farming on their own initiative. Some large farms have successfully shifted to non-chemical farming, (e.g. the 24-ha Gloria Farm near Pondicherry and a 40-ha plantation near Cochin). Again, pests are not a problem on these farms.

In Maharashtra, ten villages (about 100 farmers per village) have shifted to sustainable agricultural practices, through on-farm experimentation and development. These villagers were part of the comprehensive watershed development programme (COWDEP), carried out over 7–8 years for soil and water conservation. These villages grow a variety of fruits and vegetables and have become prosperous since the implementation of the watershed programme. The farmers are eager to adopt innovative technologies and have reduced their chemical inputs.

Despite the experiences of several small-scale farmers who are successfully farming without chemicals, their experience does not seem sufficient to convince most of the other farmers. The borderline survival and poverty of most small-scale farmers make it difficult for them to shift to natural farming, even though it is profitable in the long term. For example, in the case-study on farm systems from southern India mentioned above (UNDP, 1992), it was observed that the farmers who opted for natural farming had sources of off-farm income that allowed them to bear the risks involved in the process of transition and experimentation. Some degree of experimentation is needed to convert farms to natural farming. In the case of COWDEP, the villages that successfully shifted their farming practices did so through:

(i) the motivation and persistence of the village leaders; (ii) the ongoing external monetary inputs; (iii) the commitment of the local leaders; and (iv) the preponderance of a fairly homogeneous caste structure.

Biocontrol agents in IPM

Among developing countries, India is better placed than most to apply biological pest control and biotechnology to sustainable agriculture. It has a large number of scientists, a well-developed infrastructure and a large agricultural extension system. The Department of Biotechnology (DBT) regards biotechnology as essentially developing products for poor farmers, unlike the green revolution, which relied on costly inputs. Jayaraj (1989) reviewed the use of biocontrol agents on a variety of crops in India.

The implementation of the use of most microbial biocontrol agents, exotic predators and parasites required specialized knowledge as well as some development in technology. The Indian Council of Agricultural Research (ICAR) and the DBT are conducting research on biocontrol agents, and are developing technologies that can be implemented in the field in IPM systems. The main centres for these projects are the Biological Control Centre (BCC), headquartered in Bangalore, and its subcentres in different parts of the country.

The BCCs have screened and released several natural enemies of pests (predators, parasites and pathogens), and about 21 have been established in the field. They maintain insect collections of over 3500 identified species, including most of the agricultural pests and their natural enemies (Singh, 1991, 1992). Substantial progress has been made in India in building infrastructural facilities and in training technical personnel in biological control methods (Jayaraj, 1989). Units for mass production of natural enemies of pests are being built in order to distribute these agents to farmers with information on the advantages of biological control.

Most of the field trials using biocontrol agents have been with parasitoids and baculoviruses. Parasitic wasps have been used to control pests of cotton, sunflower, groundnut and fruit crops in India. Indian agricultural scientists have identified a nuclear polyhedrosis virus (NPV) and a parasitic wasp from the Trichogrammidae that feed on the American bollworm. As many as 30 mass-multiplication units in 11 states have been established to ensure adequate and timely supply of these two biocontrol agents (Sharma, 1991). Further, pest-management packages have been prepared to control the bollworm on various crops, especially cotton, which currently requires large amounts of pesticides. For example, the Bio-Control Research Laboratory in Bangalore has developed an IPM strategy for cotton bollworms that includes all of the following: sex pheromone traps for monitoring or mass-trapping the pests, *Trichogramma* parasitoids to destroy the eggs, predacious chrysopids to devour eggs and newborn larvae and NPV to kill caterpillars.

Several cotton farmers are successfully using this IPM package to control the bollworms. If pest infestation is high an insecticide spray is used with the NPV spray.

In Tamil Nadu a combination of *Trichogramma* and the NPV against the early shoot borer was successfully used to control two major sugarcane pests, the internode borer and the early shoot borer, on 5000 ha. The release of 300,000 adult parasites/ha was effective in controlling these pests, at a cost of Rs60/ha ($US2/ha) (Jayaraj, 1989). Tamil Nadu Agricultural University has successfully tested both an NPV and a granulosis virus (GV) in field trials, against the bollworm, the tobacco cutworm and the red hairy caterpillar (Jayaraj, 1989). A combination of NPV and a trap crop (sunflower) was effective in controlling the fruit borer in tomato fields (Kumar, 1992).

Both baculoviruses, NPV and GV, while providing an enormous reservoir of infection potential, kill the pest larvae slowly. Death occurs several days after infection (Fuxa, 1990). Another drawback of the viruses is that they are sensitive to ultraviolet (UV) light, and therefore the virus formulations must be sprayed in the evening. Despite these drawbacks, the baculovirus from the rhinoceros beetle has been successfully used to control the beetle on coconut in Kerala (Mohan *et al.*, 1989). In this instance, 2 years after the adult beetles infected with virus were released, the virus spread to subsequent generations of adults and larvae in the breeding sites.

Natural enemies of pests, like the parasitoids, once released, have the ability to disperse, thereby increasing the parasitism. Thus, when they are established in their habitats, they need not be continually released. However, in most cases, chemical pesticides are not compatible with the use of natural enemies (Greathead, 1990). All the commonly used pesticides against the *Heliothis* caterpillar in India were found to be toxic to the parasitoid *Trichogramma brasiliensis* Ashmead, and this toxicity persisted long after the use of the chemicals (Paul and Agarwal, 1990). Moreover, most of the natural habitats that supported the natural enemies have been negatively altered due to chemical-intensive agricultural practices. These habitats have to be restored before natural enemies of the pests can be successfully established (Cate, 1990; Altieri, Berkeley, 1993, personal communication). Habitats are also important for preventing and minimizing pests. In addition to habitat requirements, another factor that has an impact on the success of natural enemies for controlling pests is the seasonal modality of pests and their enemies. Sometimes the peak of the pest infestation does not coincide with that of the enemies, and seasonal boosts of the enemies are needed. Methods of clean cultivation prevent the creation of habitats for pests. For example, the mango nut weevil, an important pest on mangoes, was effectively controlled by clean cultivation, such as the destruction of vegetation under the trees and the burning of debris (Dey and Pande, 1987).

DBT and ICAR have organized several meetings, workshops and demonstrations to develop farmer-oriented technology packages and to promote wider awareness of biocontrol among farmers (Kumar, 1992). These 'farmers' melas', as they are called, are designed to promote biological control agents such as NPV, GV, *Trichogramma* and *Chrysopa* in IPM programmes, with the help of extension services. Under a DBT-sponsored all-India project, various crops in nearly 11,000 ha in seven states have been utilized to demonstrate the effectiveness of biological control. The cost of controlling pests by these natural enemies is about Rs120/ha ($US4/ha) whereas the cost for pesticides in a plot of the same size is five times as much, mainly due to the need for several pesticide applications for each crop. These differences in costs could be the incentive for farmers to adopt biological control to cope with pests. However, the lack of information on the incompatibility of pesticides and natural enemies, and inadequate knowledge of the habitat requirements for the biocontrol agents, could lead to the farmers abandoning the use of biocontrol agents after the initial demonstration trials and the withdrawal of government supervision. Thus, it would also be important to include in the farmers' training packages the habitat requirements for the biocontrol agents, the use of cultural controls in addition to biocontrols and the selection of local resistant varieties whenever possible.

Commercialization of biocontrol agents

Currently, most of the focus of research and development (R and D) on biocontrol agents in DBT seems to be on products relevant to small-scale farmers. DBT aims to train farmers and entrepreneurs in methods of producing biocontrol agents as small-scale industries in villages (Kumar, 1992). The plan is to select farmers from various regions to attend the agricultural research centres and train them in the methodologies. The government would also provide the seed money for setting up manufacturing units in the villages and assist in getting credit for the businesses. These units would provide biocontrol agents both for the farmers' needs and for sale.

Although the technology for mass production of biocontrol agents is simple enough to set up in small district units, quality control is important and may be difficult to standardize. Taborsky (1992) has described in detail the small-scale bioprocessing of several microbial pesticides. DBT has set up the Biotechnology Consortium India Ltd (BCIL) to commercialize the production of biocontrol agents (Srinivas, 1992).

Private commerical biocontrol laboratories have been built. For example, in Bangalore, the biocontrol research laboratory of Pest Control (India) Ltd mass-cultures native parasites and predators and supplies them to farmers for control of grape mealy bug, coconut leaf caterpillar, sugarcane borers and other pests. Another private laboratory, the main biocontrol laboratory of the Tamil Nadu Cooperative Sugar Federation, manufactures *Trichogramma* for control of sugarcane borers (Jayaraj, 1989; Nagarajan, 1992). Since the profit margins are low and there are problems with marketing, commercialization of biocontrol agents has not flourished. Moreover, the fact that natural enemies of pests, and sometimes the baculoviruses, can spread from the initial release site and establish themselves in a large area, requiring no further releases (inputs), is a disincentive to large companies that seek continued profits from their products. Nevertheless, when pesticide subsidies are withdrawn there may be more interest in the commercialization of biocontrol agents (Nagarajan, 1992).

Most of the commercialization of biocontrol agents has been undertaken as public-sector projects, since these products are seen as benefiting the small-scale, poorer farmers. Therefore, the large Indian companies are reluctant to invest in these developments. Furthermore, these companies, which have the financial resources, are generally unwilling to trust indigenous biotechnology. This cautious approach is mainly due to the notion that Indian-developed products of biotechnology are inferior to imported ones. Instead, there is great interest in collaborating with foreign firms (Walgate, 1990) to develop products geared towards cash crops and large-scale farmers. The collaboration for the microbial biopesticide *Bacillus thuringiensis* described below illustrates this point. The reluctance of the Indian private sector to become involved may result in much more expensive biocontrol products, made locally by foreign-based firms.

Bacillus thuringiensis *and commercial developments*

Some field trials have been successfully conducted in India with *B. thuringiensis* formulations against a number of pests, including DBM and the American bollworm, and on a variety of crops, such as sugarcane, rice and vegetables (Saxena, 1991). In field trials conducted on cauliflower, *B. thuringiensis* controlled the DBM better than any of the several chemical pesticides tested (Justin *et al.*, 1990). So far, *B. thuringiensis*, mainly subsp. *berliner*, has been used minimally, and only in field trials. Its adoption as a biopesticide by farmers has not been reported.

The reasons for the lack of *B. thuringiensis* use in India until now may be due to: (i) its limited availability, since until recently there were no facilities for production; (ii) the extensive training of farmers that is required for its proper and successful use; and (iii) its high cost, making it available mainly to rich farmers and for cash crops. Moreover, some farmers who rear silkworms have shown a reluctance to use *B. thuringiensis* because of its effect on the silkworm (Padidam, 1991). Unfortunately, not much research has been conducted in isolating, identifying and developing local strains of *B. thuringiensis* and, moreover, the Indian biochemical engineering base has not been fully developed, making the conversion of laboratory processes into commercial products difficult.

With the liberalization of Indian government policies, foreign investment and import of certain technology products will no longer be restricted (Srinivas, 1992). This is expected to increase the availability of *B. thuringiensis*-based products. Ecogen Inc. (a US company) and Gujarat State Fertilizers Company have agreed to develop *B. thuringiensis* products jointly (Ecogen Inc., 1993, personal communication). Sandoz India Ltd has received permission to market its *B. thuringiensis* product in India. Field trials with the Sandoz product have been conducted on cabbage, cauliflower, rice, pulses and tomato (*Chemical Weekly*, 1992b). Sandoz plans to import the finished formulation, but to package and market it locally. The company anticipates that the market in India will be too small for setting up a manufacturing facility. The Sandoz *B. thuringiensis* product has been priced at Rs740/ha/treatment ($US25/ha/treatment), whereas conventional chemicals cost around Rs500/ha ($US16/ha).

Anna University in Madras has created a biotechnology centre that is the first of its kind in India, in that it promotes better university–industry relationships. One of its main purposes is to foster industry-oriented research. In collaboration with Southern Petrochemical Industries Corporation (SPIC), the centre has set up a bioprocessing plant that translates laboratory-scale processes into pilot-scale production. One of its products, Biocide-S (a strain of *Bacillus sphaericus*), has been successfully field-tested. These trials showed that, when formulations of Biocide-S were spread on water surfaces, 100% of mosquito-larval deaths occurred in 24 h and the effect lasted for 30 days. A *B. thuringiensis*-based biopesticide for agricultural use has been developed by Tuticorin Alkali Chemicals and Fertilizers Ltd (TAC), a subsidiary of the SPIC group. This will be the first company to commercialize an Indian-developed *B. thuringiensis* product for agriculture (*Chemical Weekly*, 1992c). Field trials have started with this *B. thuringiensis* formulation on rice and on vegetables, such as cauliflower and okra. Plans to field-test it on cotton, tomato, tobacco and chilli are in progress.

For most farmers, the cost of *B. thuringiensis* may be too high at present to allow significant shifts away from chemicals to *B. thuringiensis* formulations. However, if the subsidies on chemical pesticides are removed, as planned, the price of *B. thuringiensis* will be competitive. Furthermore, as more *B. thuringiensis* products are marketed, there should be a further decrease in price. There is also concern among some that the potential heavy use of *B. thuringiensis* would eventually result in the same treadmill effect as with chemical pesticides (P. Tereira, Bombay, 1993, personal communication). Therefore, it is imperative that only a judicious use of *B. thuringiensis* be promoted, and preferably in an IPM scheme. The lessons learnt from the resistance of DBM *B. thuringiensis* in Hawaii and parts of South-East Asia should serve as a guide to its proper use (Katre, 1990). The DBM has already developed resistance to all chemical pesticides used in India, and extensive use of *B. thuringiensis* may lead to the same result.

The development of pest resistance seems to be directly linked to the persistence of *B. thuringiensis* in the field. If transient foliar applications of *B. thuringiensis* can cause resistance in pests, then persistent production of *B. thuringiensis* in genetically engineered plants may select intensively for *B. thuringiensis* resistance. India has been conducting some research on introducing *B. thuringiensis* into plants. Monsanto has developed a genetically engineered cotton that includes the *B. thuringiensis* gene, and is preparing a draft agreement that would transfer this technology to India (*Chemical Weekly*, 1992a). Negotiations with the Indian government are in progress. The technical programmes under this agreement are anticipated to include the development of methods to transfer the *B. thuringiensis* gene into several varieties of Indian cotton. DBT also plans to introduce the *B. thuringiensis* gene into other crops, such as rice, chickpea, pigeonpea and groundnut. Extensive use of *B. thuringiensis*-containing plants could increase the selection pressure on the DBM to develop resistance, and could drastically reduce the potential that conventional *B. thuringiensis* formulations have in IPM programmes in India, before they have had a chance to succeed.

Opportunity for using B. thuringiensis*-containing cotton*

One approach to the utilization of cotton containing the *B. thuringiensis* gene is to introduce it in areas where pest infestations are heavy, in order to reduce the pest population density. The bollworm is the major pest on cotton. It has now developed resistance to chemical pesticides and can therefore be controlled by *B. thuringiensis* expressed in the engineered cotton. The *B. thuringiensis* cotton variety can be multicropped with non-engineered varieties and resistant varieties whenever possible. Indian varieties of cotton resistant to whitefly and other minor pests have been identified (Suri *et al.*, 1992). In the multicropped system, the emergence of pests can be monitored, and their natural enemies, e.g. the parasitoid and the NPV mentioned above, can be introduced in order to maintain long-term control over pests. The establishment of a sustainable system for cotton production would entail a limited use of the *B. thuringiensis*-containing variety, thereby minimizing the possible evolution of pest resistance. However, it would be cheaper to develop the technology of transferring the *B. thuringiensis* gene into Indian varieties of cotton, in India. Research on trangenic plants is already being conducted in some Indian laboratories. Depending on the level of pest infestation, appropriate IPM systems can be designed using different varieties of cotton intercropped with the *B. thuringiensis*-containing cotton.

New Seed Policy

The issue of foreign-developed genetically engineered plants brings into focus the intentions and goals of DBT. The biotechnology programme in India claims to aim at reducing input costs of small-scale farmers, but imported seeds and technology will be expensive. The Indian government has liberalized its seed import policy and lifted the restrictions on the import of seeds. Now farmers and seed merchants are free to import seeds for planting, and multinational seed companies are free to set up businesses and to import genetically engineered seeds. It is feared that Indian agriculture will become dependent on supplies of seeds from abroad, and the farmers will buy them at higher prices, even though they may be ill-suited to their needs and not be any better than indigenous-bred varieties – this in a country whose plant breeders have developed the highest-yielding sugarcane, potato and grapes, and the world's first hybrid cotton (Walgate, 1990). Furthermore, India is home to a great diversity of crop varieties, and many of the local varieties are hardy, resistant to pests and more suitable for local use. The new seed policy will further widen the disparity between the rich and poor farmers. Nagarajan (1992) expressed concern that the new seed policy and the strategy of selling whole agroinputs to the farmer may create major socioeconomic problems, which may be too difficult to solve. Farmers are also concerned with the policy changes and have demonstrated their objections to these changes. Over 50,000 farmers protested against the changes in government agricultural policies on 3 March 1993 in New Delhi. Cargill Seeds near Bangalore has been the main target of several protests in Karnataka (PIRG, 1993). The farmers were protesting against the Dunkel Draft on Agriculture, which supports the removal of import barriers, the removal of subsidies on inputs and the patenting of plant varieties.

Barriers to Adopting IPM

There is a great deal of research being conducted in India on IPM, as demonstrated in field trials. The scientific literature abounds with papers on this subject, giving the impression that a lot is going on. However, the farmers have chosen to adopt IPM only to a limited extent. In those instances where farmers have successfully adopted pest-management schemes, this has been mostly through their own initiatives and often with the assistance of NGOs.

Despite the government's ambitious plans to institute IPM widely, their policies do not go far enough, and mostly conflict with other government plans that propose increases in the manufacture and consumption of chemical pesticides. Therefore, several health and environmental NGOs and small farming groups have criticized the government for not targeting IPM and sustainable agriculture policies and programmes where they are needed, that is, at the millions of small-scale farmers throughout the country. Furthermore, there is insufficient documentation on natural farming and IPM, especially since most farmers have been exclusively relying on pesticides. The conversion to sustainable agriculture requires substantial knowledge and commitment, and converting to non-chemical agriculture is not easy.

The experiences of farmers who have shifted to non-chemical farming are sometimes only partly relevant to resource-poor farmers, who lack the necessary buffer or off-farm incomes to counter the possible initial lower crop yields during the transition period. For example, in a village in Tamil Nadu, composed of several small farms currently using chemical pesticides, older women in the village had farmed without pesticides. Prior to the use of pesticides, they harvested 20 bags of rice per average plot. Now, they average the same output, but with increased input costs and severe health problems due to pesticide use. These farmers would like to return to pesticide-free farming, but they need to regain their knowledge of alternative pest-control methods, and are unwilling to take the risk unless they have guaranteed income against possible lower yields during the transition period (Dinham, 1993).

Another barrier that impedes the adoption of IPM is that, most often, there is a lack of appropriate research and training, which is reflected in inappropriate services and crop varieties. The relations between farmers and the official research and extension agencies are difficult and inadequate. There are differences in attitude, objectives and communication, and these are difficult to bridge. Also, the farmers' needs and input are usually not considered. There is a tendency to impose agricultural technology packages on the farmers, as was done during the green revolution, without their input. Moreover, women are usually excluded from these technology transfers, in spite of the fact that they constitute a very large part of the agricultural labour force. The success of many projects has depended on the involvement of women (Agarwal and Kumar, 1992).

One major barrier to biological pest control and IPM is the existing, well-established pesticide retail outlets. India has an overwhelming number of pesticide retailers, with a total of 86,538 reported in 1989 (Dinham, 1993), most of which are under the control of the private sector. In order to promote biological control and IPM, a massive retailer education programme is necessary. Most retailers are at present poorly informed even on the proper use of the chemicals they sell. To add to this problem, many retailers are affluent landlords belonging to a higher class within the Indian system, and, since the farm workers are primarily from the underprivileged communities, there is no concern regarding the use of hazardous pesticides in the most improper manner. Furthermore, there is no incentive to switch to IPM. In addition to the lack of complete knowledge regarding the use of pesticides, the fact remains that these pesticide retailers have situated themselves all over the country. Therefore, the sale of biocontrol agents will be in direct competition with that of the pesticides, and will only be successful if the biocontrol agents are cheaper and there is significant involvement of the private sector, which has the know-how to market products competitively. The lack of a high level of input from the private sector prevents the full implementation of India's biotechnology programme, since the complementary inputs from both industry and the public sector are vital to its success.

The success of biological control and IPM depends on the calculated and appropriate use of the available pest-control technology. This very systematic approach may be the biggest barrier to their adoption by farmers. After years of indiscriminate and improper use of chemicals, which required less monitoring and planning, the adoption of a system the requires monitoring pests and using biocontrol agents at specified times may be a serious barrier. The agriculture extension agents, who themselves have to be properly trained in IPM, most often do not interact and advise farmers on a regular basis. Therefore, once the IPM packages are distributed to farmers, they may be left to their own resources. Traditional knowledge of pests may be limited or inadequate when dealing with some new problems, and thus, the entomological limitations of peasant farmers may prevent the appropriate resolution of the problems. Therefore, this scenario could lead to the farmers reverting to the use of pesticides, having found no one to advise them on how to handle the problems. Many biocontrol agents cannot be used with pesticides. Unless farmers clearly understand the relationships of pesticides to biocontrol agents and the differences between the chemicals and the biologicals, improper usage will result in failed IPM systems.

Incentives

The best incentive is the proved improvement in the benefit/cost ratio upon the adoption of IPM. The economic advantages of using IPM have to be well documented in order to persuade the farmers to adopt these methodologies. For example, a government-sponsored IPM programme on cotton in South India demonstrated that the net income could be increased by $US123.08/ha upon adoption of IPM. In this programme, the cotton yield also increased substantially to 2050 kg/ha in the IPM area, compared with 1720 kg/ha in the non-IPM plots (Suri *et al.*, 1992). Furthermore, the government, through its policies and programmes, needs to be fully committed to biological pest control and IPM, without any conflict of interest. The removal of subsidies on pesticides must coincide with the establishment of subsidies on biological controls. Credits for chemical-intensive agriculture must be removed, and instead credits should be given for non-chemical farming and IPM. In addition to 'farmers' melas', sustained interaction and advice from agricultural scientists and extension agents are necessary. There is a need to create models that fit the conditions resource-poor farmers have to cope with, and provisions have to be made to facilitate the shift to IPM. Proponents of natural farming have called for a ban on all hazardous pesticides and the introduction of a system making the other pesticides available on prescription only.

Farmer-to-farmer exchanges with success stories using biological controls and IPM will help in informing other farmers. In addition, selected farmers could be paid to educate and train other farmers in their region on IPM technology. Creation of a farmer network would boost the morale of those practising natural farming and encourage other farmers to do the same. The national network could be hooked up to an international one, which would give additional support. One initiative to create a network was taken in Tamil Nadu in 1990, when a group of small-scale, marginal farmers and NGOs started a low-external-input, sustainable-agriculture network (Gandhimathi, 1992). It is hoped that farmers throughout the country will undertake similar initiatives.

The availability of genetically engineered plants with pesticidal properties, used in well-thought-out IPM schemes, would allow a transition to IPM without losses in crop yields, and would therefore be a great incentive for farmers to switch to pesticide-free agriculture. Incentives need to be created by the government to involve the private sector in order to bring the products of biotechnology into the market.

Cash crops could also provide incentives for shifting to IPM, since many small-scale farmers grow some cash crops. Cash crops such as tea, coffee, cardamon, mango and banana use a significant amount of pesticides at present (AGROW, 1992). Value can be added to these cash crops by switching from the use of chemical pesticides to biological pest-control methods. However, markets have to be developed for these pesticide-free crops. International markets already exist, and may be particularly suitable for plantation crops. Local markets for pesticide-free produce have recently been created in large urban areas, such as Bombay, and these need to be extended to other regions. In many cases, the middleman trader sells the produce for the farmers in the urban areas, and also retains most of the income. Direct access to these markets would generate more income for the farmers and encourage them to take the necessary risks to shift to IPM.

Opportunities for Research and Biotechnology

Basic research is needed to develop cropping systems that are less attractive to pests and therefore minimize pest problems from the start. Some of the key factors that contribute to designing such systems are selecting the best crop varieties and multicropping with trap crops and crops that repel pests, often referred to as cultural or traditional methods. A great deal of this traditional knowledge exists among farmers and needs to be researched and documented.

The cultural methods will not be sufficient in themselves, and they need to be supplemented with innovative technologies. Some of the pests that are problems at present have emerged as a result of modern farming systems, which have used excessive amounts of chemical inputs. The designing of appropriate sustainable farming systems therefore requires more than a revival of traditional practices. Ulluwishewa (1992) studied the indigenous knowledge of pest-control methods practised by paddy farmers in Sri Lanka. This knowledge was remembered only by the elderly farmers. He discovered that these traditional pest-control methods were quite sophisticated, and that they used a wide variety of biological, botanical and cultural controls. Therefore, traditional knowledge of pest-control techniques needs to be studied and documented before the knowledge is completely lost. Selected traditional techniques can then be combined with new technology in the best possible manner, in order to control pests while enhancing agricultural productivity. The new emerging biotechnologies can make a significant difference in both the research and the production of effective biocontrol agents, and will therefore assist in the implementation of integrated biological pest-management systems.

Basic research

Some research is critical to prevent further losses of species and knowledge:

- Collection, screening identification and conservation of indigenous pest-resistant and disease-resistant crop varieties.
- Study of cultural methods of agriculture suitable for use in IPM, such as trap crops.
- Research and documentation of indigenous, traditional knowledge and innovations on pest-control methodology.
- Identification of the useful components in traditional systems for controlling pests.
- Modification of systems with appropriate biotechnology.

Biotechnology

A biochemical engineering and bioprocessing base needs to be developed to manufacture biocontrol agents with the necessary quality control.

Natural enemies

Research is needed in the following areas:

- Improvement in technology for mass production of natural enemies, including parasitoids.
- Enhancement of habitats for natural enemies. For example, the addition of plants with the right kind of flowers can increase the number of parasitic wasps that feed on nectar. Secondly, plants can substitute as secondary hosts for wasps when the pest is not available. Thirdly, biochemicals such as kairomones and synonomes can be used to increase the rate of parasitism and to attract natural predators (Kainoh, 1990).
- Identification of novel natural enemies, e.g. a wasp-like insect, *Trichopria handalus*, which parasitizes the cocoon of the uzi fly, has recently been identified (Kumar, 1992). The uzi fly is responsible for a loss of 30% of India's silk production.

Bacillus thuringiensis *and baculoviruses*

Research opportunities include the following:

- Isolation and identification of local strains of *B. thuringiensis* and screening them for appropriate insecticidal properties. The development of indigenous *B. thuringiensis* strains may result in better pest control than imported strains. For example, in the Philippines an indigenous strain was significantly more toxic to DBM larvae than the imported product (Tryon and Litsinger, 1988). The genetic diversity of *B. thuringiensis* insecticidal genes and their activities has yet to be fully exploited, and therefore provides some unique opportunities for biotechnology. The availability of a large number of diverse *B. thuringiensis* toxins will enable better management of pest resistance, broaden the host range and also allow the design of chimeric toxins.
- Baculoviruses have a narrow host range, and they can be screened for their virulent capacities. These insect viruses can be genetically engineered to make them more efficacious and stable.
- Fermentation and production technology for baculoviruses (NPV and GV), fungal toxins and *B. thuringiensis*-based biopesticides.
- Research and development on formulations of *B. thuringiensis*, NPV and GV that are inexpensive, easy to use and stable under various field conditions (e.g. enhancing the field stability of *B. thuringiensis* such that it is not degraded rapidly, with, for example, controlled-release formulations; and engineering NPV such that they are not UV-sensitive). Polymers have been used to develop sustained-release formulations of proteins and other biologicals (Katre, 1993). Biodegradable polymers can be used to design biocontrol products with the desired release properties. These

controlled-release formulations would last long enough in the environment to kill pests effectively, but not long enough to cause problems of pest resistance. Oil-based emulsions also generate stable and long-lasting formulations. Locally available material, such as coconut husks, can also be used to formulate *B. thuringiensis*, and NPV has been formulated in oil emulsions, in detergents and with jaggery or charcoal.

Small-scale bioprocessing technology

An appropriate technology needs to be developed so that some of the biocontrol agents can be manufactured and formulated by small-scale industries in villages. Farmer cooperatives in Gujarat are operating manufacturing units for biocontrol agents. Taborsky (1992) described the equipment, materials and methods for manufacturing microbial pesticides on a small-scale.

Pheromones and botanicals

Work needs to be done in the following areas:

- Research on improving pheromone traps and identifying novel pheromones; for example, Pawar *et al.* (1988) designed pheromone traps that are effective for attracting pests. Novel pheromones that can trap both sexes of pest species and those that disrupt the mating patterns will be particularly useful for mass trapping.
- Research and development on botanicals with insecticidal properties. Information exists on more than 500 Indian plant species traditionally used as pesticides (Pereira, Bombay, 1993, personal communication). These species can be screened for optimal insecticidal activity and minimal toxicity to non-target species.

Screening wild and other local varieties of crops for resistance to pests

Areas of interest include screening germplasm for resistance to pests, as done, for example, in citrus (Bhumannavar *et al.*, 1988) and in cowpea (Jayappa and Lingappa, 1988), characterization of pest-resistant genes and subsequent insertion of these genes into crops. Cowpea has been used in traditional farming as a trap crop for pests. An insect-resistance gene, which is a trypsin inhibitor, has been isolated from cowpea (Hilder *et al.*, 1987). Introduction of this gene into other plants renders them resistant to pests. India has a variety of genetically diverse crops, which create a large genetic pool for the screening of pest-resistant genes.

Targeted delivery of pesticidal properties

Research possibilities include the following:

- Targeting and controlling the delivery of pesticidal properties – for example, controlled-release formulations and engineering pesticidal factors into local environments and into other microorganisms.
- Genetically engineering crop plants to express specific pest toxins, and to extend the engineering of the toxins to target sites on the plants. Novel *B. thuringiensis* genes cloned from local strains of *B. thuringiensis* or novel pest-resistance genes isolated from local resistant crop varieties can be transferred to the desired plants.

Recommendations

In order for biotechnology to benefit Indian agriculture, it has to meet the needs of all farmers and strengthen the existing capabilities of natural farming. This would provide for long-term sustainability and also produce adequate food and fibre to meet the needs of the expanding population. The failure of the green revolution to achieve equal agricultural development across the country and to provide access to food for the poor must not be repeated by the 'gene revolution'. Despite the initial increases in food production, the green revolution caused a reliance on chemical inputs that resulted in a decline in the real income of farmers. A similar strategy with genetically engineered crops or products like *B. thuringiensis* formulations, which require repeated applications, could make farmers dependent on external inputs, and thus lead them on to a 'biotechnological treadmill'. If so, the unequal agricultural development created by the green revolution will be further exacerbated by the gene revolution, leading to increased rural poverty and social unrest.

Financial institutions, such as the World Bank, can assist in the development of IPM that would be appropriate for the needs of Indian agriculture. Some suggestions follow.

1. Support for developing a biochemical engineering base. The equipment and facilities for fermentation and bioprocessing have to be acquired.
2. Assistance to farmers to provide the necessary support to shift to IPM schemes. Schemes like COWDEP transform entire villages so that they can shift to sustainable agricultural practices and are excellent choices for funding. Such transformed villages will be most open to the introduction of innovative biotechnologies designed to enhance productivity.
3. Training in IPM technologies and their proper utilization. The training should include women.

4. Research on appropriate IPM methodologies and natural farming, incorporating traditional knowledge and indigenous plant varieties whenever possible.
5. Dissemination of information regarding successful adoption of IPM, and support of farmer-to-farmer exchanges.
6. Design of appropriate IPM packages for use with the different types of farming needs. These programmes should incorporate input from farmers.
7. Support of small-scale private industry, to develop and market biocontrol agents. The farmers themselves could be part of these enterprises, and the funds a part of rural development projects. For example, in Tamil Nadu, some farmers and agricultural students have set up production units (K.R. Srinivas, 1993, personal communication). However, procuring finances for such units is difficult, and therefore assistance from donor agencies and financial institutions could help their progress tremendously.
8. Facilitating the creation of markets for pesticide-free produce. Equipment, such as vehicles for transport, and access to markets in urban areas, will encourage farmers to shift to IPM.
9. Identify technological developments relevant to local problems, and facilitate the transfer and adoption of this technology.
10. Ensure that adequate safeguards are in place to protect the environment and its inhabitants from any unwise releases of engineered organisms.

Conclusion

Although several examples of biocontrol in IPM have been presented in this chapter, the implementation of these biocontrols and sustainable farming practices in Indian agriculture has been slow. Nevertheless, there have been steady improvements. In some states, the changes are occurring faster than in others. For example, Punjab, Haryana and Tamil Nadu have good agricultural extension systems, and therefore new technologies are adopted more quickly by farmers. Gujarat has several farmer cooperatives, and these facilitate commercialization and the adoption of the technologies. Maharashtra has had successes with farmers, transforming the agricultural practices of entire villages.

There is an abundance of research on biocontrol agents and biotechnology for crop improvement. However, the transfer of technologies from the laboratory to the land is limited. This is partly due to the lack of manufacturing and bioprocessing facilities, but mostly to economic and political reasons. Chemical-intensive farming has a government-supported infrastructure, while the infrastructure of traditional non-chemical farming has disappeared due to lack of support and recognition. Therefore, the shift to biocontrols and IPM is difficult, with high experimental costs and with most farmers being unwilling to make this shift. Nevertheless, the fact remains that those farmers whose agricultural practices have long-term sustainability, who have managed resources such as soil and water well and who use no or minimal chemical inputs have been the most successful in the long run. Therefore, as more farmers adopt sustainable agricultural practices, including integrated pest-control technologies, other farmers will be encouraged to do the same.

Cassava and Cowpea in Africa 7

Hans R. Herren

Introduction

The concept of sustainability has taken a central place in agricultural research and development (R & D) in view of the rapid degradation of the natural resource base, accentuated by environmentally and economically inappropriate crop production and protection practices, based on non-farm inputs. Sustainable agriculture goes beyond the narrow and short-term goal of increasing crop yields through high-yielding crop varieties and short-term soil fertility increases and pest-control practices. Our new focus on health reflects the approach of considering the crop plant as one of the components of a complex system governed by ecological interactions, the agroecosystem. The peculiarity of these interactions with the other elements of the system, i.e. climate, soil and, particularly, biotic factors, is determining the ability of the plant to grow and produce a yield. If the biotic factors exert a detrimental effect on plant health, then we speak of 'pests', meaning here arthropods, nematodes, pathogens, weeds and parasitic weeds.

Managing plant health implies two distinct processes. First, the noxiousness of a particular organism needs to be assessed by a diagnostic process. Before one makes any attempt to manage a system, one has to have a profound knowledge of its features and functioning. All too often, pest-control projects fail because of the use of a simplistic approach that neglects the nature of the problem. In fact, the crucial step is to separate the causes of a given problem from their expression, i.e. the symptoms. Depending on the complexity of the system under study, this differentiation is often possible only through in-depth ecosystems analysis, investigating the interactions between (e.g. plant–pest–antagonist) and across (e.g. crop–alternative host-plants) trophic levels.

Having characterized the system, we can start the process of defining the most appropriate strategy to avoid or control a particular pest problem. Therefore, the results of the diagnostic assessment are used to tailor prevention and/or control tactics in a multidisciplinary effort. In this context, it is worthwhile to emphasize the prevention aspect, where, for example, the use of clean planting material, sound agronomic practices and resistant/tolerant varieties, together with the strict implementation of quarantine rules, are of primary importance in avoiding pest outbreaks. In contrast to crop protection, where the word 'protection' already implies an intervention-orientated control strategy, often narrowed down to the application of pesticides, plant-health management advocates a holistic approach to solving pest problems occurring in the agroecosystem.

This perspective is particularly suited to agroecosystems in Africa, where small-scale farmers produce a diversity of crops with few resources other than their own labour in a still mostly diversity-rich environment.

Plant-health management aims to maintain good crop productivity within the dynamic balance of forces in the agroecosystem, using a combination of plant breeding and systems management strategies. This ecological approach to crop protection seeks to avoid the need to use environmentally hazardous pesticides, which are a purchased input that constantly has to be renewed. Here, plant-health management differs from integrated pest management (IPM), which integrates genetic resistance, biological and cultural control practices with the judicious use of pesticides. Plant health is the logical and necessary evolution of IPM towards a more sustainable and ecologically and economically sound approach by seeking solutions to the problems and avoiding symptom treatment.

Moreover, the ecological approach works to conserve the efficacy of the pests' natural enemies, by obviating the use of pesticides, which usually eliminate the enemies together with (temporarily) the pests. Without the toxic residue that normally accompanies pesticide use, the integrity of the food chain and of natural resources is preserved in the targeted ecosystem and beyond.

Two case-studies presented in this chapter will help to explain how plant-health management works, from concept to implementation. The first one, the success story of the biological control of the cassava mealybug (CM), illustrates how this holistic approach was first developed and successfully implemented at the International Institute of Tropical Agriculture (IITA), and how this created the basis for an ecologically sustainable cassava plant-protection project (ESCaPP). The same agroecosystem approach was subsequently applied to face the challenges of the cowpea pest problem, a project that is still ongoing and where biotechnology applied to host-plant resistance might be part of the solution.

Cassava: a Case-study

Background

Protection for cassava (*Manihot esculenta*) in Africa, where it was introduced from its native South America, illustrates how ecologically sound plant-health management strategies can be based on knowledge of pest/plant/farmer interrelationships. The need for such management strategies grew during the 1980s as an exotic pest, the cassava mealybug (CM) *Phenacoccus manihoti* Matile-Ferrero (Homoptera, Pseudococcidae), spread in devastating waves over the countries of tropical Africa's 'cassava belt'. CM comes from the same environments in South America as cassava. Its populations do not grow to epidemic proportions there because natural enemies hold them in check.

The first outbreaks of CM were observed in Congo (Sylvestre, 1973; Matile-Ferrero, 1978) and Zaïre (Hahn and Williams, 1973) in 1973, following the illegal introduction of contaminated planting material, violating quarantine regulations, into either Zaïre or Congo. From there, the pest was then accidentally introduced into West Africa (Senegal and Gambia in 1976, Nigeria and Benin in 1979 and Sierra Leone in 1985), where it spread rapidly, at the rate of 300 km/year (Herren *et al.*, 1987b). In East Africa, the CM spread was slowed down by physical barriers, such as the Rift Valley, and the infestation of the highlands was rather patchy in the first few years.

Research approach

Following the rules of classical biological control, the first step was to identify the mealybug in its area of origin, in order to find efficient natural enemies. The first mealybugs believed to be *P. manihoti* that were encountered by the Commonwealth (now International) Institute of Biological Control (CIBC/IIBC) in Guyana, Surinam and northern Brazil yielded parasitoids that failed to reproduce on CM in Congo and Zaïre (Bennett and Yaseen, 1980). Later on, this mealybug was identified as *Phenacoccus herreni* Cox and Williams (Cox and Williams, 1981), which explains why the introduction of its parasitoids against CM was unsuccessful. In 1979, IITA started work on biological control and host-plant resistance against CM. This led to the creation of the IITA Biological Control Programme (BCP), which started an extensive exploration in the neotropics. This included close relatives of cassava in the family Euphorbiaceae in areas of rich species diversity of the genus *Manihot* (Renvoize, 1973), namely in central/western Mexico, Yucatan, north-eastern Brazil, Peru and Paraguay, and covered *Phenacoccus* spp. and other mealybugs (Williams, 1986). In view of the large area to be covered and the time pressure to find a solution to the CM problem, collaboration for the exploration had been arranged with both CIBC/IIBC and the Centro Internacional de Agricultura Tropical (CIAT).

Finally, *P. manihoti* was discovered in 1981 in the Paraguay River basin, with generally very low and erratically occurring populations, together with its most efficient natural enemy, the parasitoid *Epidinocarsis lopezi* (De Santis) (Hymenoptera, Encyrtidae) (Yaseen, 1986). Up to 1987, 18 species of natural enemies were found in the area of origin of CM. Eight species were sent for quarantine, while another two species with interesting traits, the encyrtid *Parapyrus manihoti* Noyes and the syrphids *Ocyptamus* spp., could not be reared successfully.

From South America, these beneficials went to the IIBC quarantine laboratory in the UK. Processing met the requirements established by the Inter-African Phytosanitary Council of the Organization of African Unity (OAU/ IAPSC), which regulates the introduction of plants and animals into Africa and which accredited the IIBC laboratory. The exotic natural enemies were then sent to BCP, all introductions being covered by import permits issued by the national quarantine authorities of Nigeria and, later, Benin. Under the same OAU umbrella, beneficials reared at BCP were and still are being shipped to various African countries upon their request and under cover of their own quarantine permits.

At the same time, studies were undertaken to analyse the African cassava agroecosystem, starting with the growth and yield formation processes of the cassava plant (Schulthess, 1987), in order to evaluate, in subsequent phases, multitrophic interactions and eventually assess the project's impact. The allocation of photosynthates and biomass and the effects of nitrogen under different water regimes were investigated and incorporated in a cassava growth simulation model, which runs on the basis of real weather data (Gutierrez *et al.*, 1988a). The model reflects the wide range of responses by cassava to different environmental factors in the various ecological zones, and it is being extended to include different varieties and multi-cropping. On the same lines, the biology of CM was studied in great detail, both in the field (e.g. Fabres, 1981, 1982; Fabres and Boussienguet, 1981; Fabres and Le Rü, 1986; Schulthess, 1987) and in the laboratory (e.g. Nsiama She, 1985; Le Rü and Papierok, 1987). The quantitative data collected by these studies were subsequently used to develop a CM simulation model, which was superimposed on the cassava model in order to study pest–plant interactions (Gutierrez *et al.*, 1988b).

On the third trophic level, emphasis was given to the inventory of the local predators and parasitoids that switched over to CM when it invaded Africa (Fabres and Matile-Ferrero, 1980; Boussienguet, 1986). Following the arrival of *E. lopezi*, the CM food web was investigated again over the whole continent, and it now comprises about 130 species (Neuenschwander *et al.*, 1987; Biassangama *et al.*, 1989). Only about 20 species are common and seem to have some impact. Because of the reduction in CM populations due to *E. lopezi*, many species became less abundant after the establishment of the exotic parasitoid (Neuenschwander and Hammond, 1988). Due to its establishment, rapid spread and impact on CM (Neuenschwander *et al.*, 1989, 1990), *E. lopezi* became the single most important biotic mortality factor for CM, warranting in-depth studies of its biology and behaviour. The results of these investigations can be summarized as follows: *E. lopezi* has a short generation time and consequently a power of increase about equal to that of its host; it has the capacity for host-feeding, mutilation and – where the host is not killed directly – inhibiting the successful reproduction of CM; it has high specificity and searching capacity for its host, allowing it to survive on very low host populations; and it has a density-dependent aggregation and reproduction on most host population densities encountered in the field. Again, this information was analysed and used to build a new layer on the simulation model, which could now be used to evaluate tritrophic interactions. This was particularly useful in showing the overriding influence of *E. lopezi* in the cassava agroecosystems (Gutierrez *et al.*, 1988b): *E. lopezi* populations in this model are shown to react in a density-dependent manner to their host populations. The simulations predict a strong influence of rains and a relatively low importance of local coccinellids.

An important part of applied research was devoted to the development and implementation of mass rearing and releases, as well as to the follow-up studies needed to measure both ecological and economic impact. These aspects, which had been mostly ignored in earlier biological control projects, needed substantial input to cope with what would become one of the world's largest biological control projects.

The rearing took place in automatic rearing chambers or in units called 'cassava trees', made to a large extent with local material. A comparison of the production costs for *E. lopezi* obtained with different rearing techniques showed that this wasp is produced more economically in 'cassava trees' than in cages or chambers (Neuenschwander and Haug, 1990).

The high quality of *E. lopezi*, as witnessed by its successful establishment all over Africa, is attributed to: (i) the high quality of the insects from South America; (ii) rearing it on the original host and original host plant; (iii) careful technical manipulations and constant supervision in the insectarium; (iv) the good size and the high number of rearing units; and (v) infestation schedules that consider all known facets of the biology of the plant, CM and *E. lopezi* (Neuenschwander and Haug, 1990).

Because of the size of the area to be covered with natural enemies, and frequently difficulties of access, transport was often done by aircraft. Releases were mostly made from the ground; however, for speedy delivery to remote areas an aerial release technique was developed that allowed the ejection of viable natural enemies while flying over cassava-growing areas (Bird, 1987; Herren *et al.*, 1987a).

The clearest demonstration of a parasitoid's impact on its host is usually obtained from data on population dynamics. Seven years of continuous monitoring in numerous fields in two areas of south-western Nigeria revealed that mean CM population peaks never reached the height (means of up to 90 CM/tip) and the duration (7 months with over 10 CM/tip) observed during the first season of release (Hammond *et al.*, 1987). Although occasional sharp peaks of up to 30 CM/tip were registered, it is concluded that *E. lopezi* maintains a high level of biological control of CM in this area. The same data also demonstrate the presence of a significant positive density-dependent reaction by the parasitoid in the field, thus offering one mechanism that explains *E. lopezi*'s efficiency, and giving strong credit to the model's predictions at the same time.

Yield loss experiments indicated up to 84% loss due to CM, with early infestations causing higher losses than later ones (Nwanze, 1982). Such high losses occur if harvest is at the beginning of the rainy season, when the plant mobilizes reserves from the roots for regrowth. Later, losses are partially compensated for (Schulthess, 1987). In a large-scale survey across different ecological zones in Ghana and Ivory Coast, it was shown by multiple regression analysis that the loss due to CM was reduced significantly by an average of 2.5 t/ha in the savannah region in areas where *E. lopezi* had been present for most of the planting season, compared with areas where *E. lopezi* was not yet established (Neuenschwander *et al.*, 1989). This figure gives an idea of the high economic yield of this biological control project. It can be used to recalculate a previous estimate of the cost:benefit ratio of 1:149 (Nørgaard, 1988). This estimate is based on subjective assessments and the assumption of diminishing returns. Since farmers cannot easily switch to another staple and in view of the life strategy of *E. lopezi*, we are of the opinion that returns should be calculated as being constant over time.

From biological control to plant-health management

The sustainable solution to the CM problem for Africa has been biological control, which made use of cautiously selected natural enemies to keep an introduced pest below the damage level.

The success of IITA's biological control campaign against CM has led to development of a large-scale project (ESCaPP), which integrates the advances in host-plant resistance against diseases, biological control against CM and the cassava green mite (CGM) and sound agronomic practices enhancing plant vigour and reducing initial infestation through clean planting material. This project is now the first of three, which will be models for developing and implementing with farmers and extension services environmentally sound and economically feasible plant protection for important food crops in the African farming systems.

Biological control of CM not only involved the classical approach, from foreign exploration and biological studies to mass rearing, release, monitoring and impact studies, but also emphasized the quantification of observations on all trophic levels in a holistic manner. These studies were carried out in the area both of introduction and of origin of CM, but in the latter case with great difficulties in view of the very low CM densities and their heterogeneity in space and time. In this context we must mention the necessity to conserve the biodiversity that will ensure that in the future we shall find the natural enemies needed for the many other exotic and endemic pests. Nature has an abundant inventory of resources for us to discover and make good use of. The constraint in exploiting these resources is the lack of interest to inventory them with adequate annotations and to identify and/or describe the bounty at hand. Also, we believe that we must preserve the genetic integrity of the crops that have a long evolution and adaptation path behind them. The tinkering with cassava to develop varieties lacking the ability to produce cyanogenic glucosides, the precursors of hydrocyanic acid (HCN), are very unwise, since it most probably represents a defence mechanism. In Africa, where the plant is newly introduced, only a few endemic insects are able to feed on it. On the other hand, in the Americas, where insects have coevolved with cassava, they have partially overcome this defence mechanism, and so the relatively large complex of pests, although few, have a major impact.

Although the focused research effort was instrumental in the CM success, it would not have worked out without the simultaneous massive training effort and the development of national capacities in both research and implementation.

Cowpea: A Case-study

Background

No other crop on the African continent suffers such high yield losses due to a plethora of insect pests (Jackai and Daoust, 1986; Singh *et al.*, 1990) as cowpea (*Vigna unguiculata*) does. The key pests for most of the regions where cowpea is an important crop are the bean-flower thrips *Megalurothrips sjostedti* Trybom, the pod-borer *Maruca testulalis* (Geyer) and the pod-sucking bug (PSB) *Clavigralla tomentosicollis* Stål. IITA's long-term research approach to solving pest problems was host-plant resistance. In cowpea this approach was successful against the cowpea aphid, an important pest in dry areas; however, only moderate levels of resistance have been found against other insect pests of this crop.

The Habitat Management Programme of IITA's Plant Health Management Division is now giving emphasis to identifying the reasons for ecological imbalances leading to pest problems in cropping systems in Africa in order to find environmentally and economically sound means to keep pest populations from reaching damaging levels – the so-called ecosystem analysis approach. As opposed to the previous case-study, where both the crop and the pest were exotic, with few interactions with the indigenous fauna and flora, here we are facing a highly complex agroecosystem. Cowpea has its primary centre of domestication in West Africa (Ng and Marechal, 1985), i.e. it is a native crop. The key pests have always been tacitly categorized as indigenous, although no serious attempt had been made to investigate their true origin. Nevertheless, what is probably the major difference between the cassava and the cowpea system is the interactions with the surrounding habitat, and particularly the role of wild and cultivated alternative host plants during the dry season, when no cowpea is present. Given the complexity of the cowpea agroecosystem, together with the uncertain origin of the pests, a control strategy relying on a single intervention tactic would not make a tangible impact in the field. These premises seemed, indeed, to be ideal for testing the systems-analysis approach developed during the CM biological control project.

This is now illustrated for the first of the above mentioned key pests, the bean flower thrips, *Megalurothrips sjostedti*, which is considered to be the first major pest attacking the reproductive structures during plant development (Okwakpam, 1967; Taylor, 1969; Nyiira, 1971, 1973). Early feeding damage on developing flower buds can cause their shedding, leading to complete crop failure in cases of high thrips population levels (Singh and Taylor, 1978; Wien and Rösingh, 1980).

Observed for the first time in 1905 on the African continent (Trybom, 1908), *M. sjostedti* was first recorded as a pest in East Africa (Faure, 1960). Although Taylor (1965) suggested that *M. sjostedti* was the possible cause of 'distortion, malformation, and discoloration of floral parts' of cowpea, the records of its damage in West Africa, the most important cowpea-growing region worldwide, has been a subject of controversy for almost a decade. Okwakpam (1967) indicated that heavy flower shedding could be caused by thrips, but the findings of Booker (1965) and Van Halteren (1971) did not support this hypothesis. More recent studies (Agyen-Sampong, 1978; Singh and Taylor, 1978; Ezueh, 1981), however, emphasized the real importance of *M. sjostedti* as a key pest.

Research approach

To have a clearer understanding of the interactions between *M. sjostedti*, the cowpea plant and the surrounding environment, two parallel evolving research activities were started. The first one was the assessment of factors likely to have a strong influence on the pest status of *M. sjostedti* (Tamò *et al.*, 1993b). The second one dealt with the *M. sjostedti*–cowpea interactions subjected to demographic analysis using computer simulation models (Tamò and Baumgärtner, 1993; Tamò *et al.*, 1993a).

Three factors assumed to be responsible for the pest status of *M. sjostedti* were considered important enough to warrant detailed diagnostic studies. First, the colonization rates of the cowpea fields depend on the occurrence of alternative host-plants that sustain the population of *M. sjostedti* during the off-season. Second, the densities of *M. sjostedti* interacting with fruiting structures indicate that its populations are not effectively controlled by biological factors whose identities are not yet known. Third, the amount of damage is determined by the infestation patterns of the crop at the time of flower-bud development.

During the dry season in southern Nigeria (November to March), Taylor (1974) observed both immature stages and adults on the Fabaceae *Cajanus cajan* and *Centrosema pubescens*, showing that *M. sjostedti* could survive the dry season in the absence of cowpeas, reproducing on these host-plants. Unfortunately, no information was available either for other ecological zones (e.g. the dry savannah zone, where the dry season is more important) or for other possible host-plants.

Although many predators and parasitoid species are known to attack thrips (Ananthakrishnan, 1973, Lewis, 1973), the only antagonists of *M. sjostedti* reported in the literature were the two anthocorid bugs, *Orius amnesius* Ghauri, and *Orius albidipennis* (Reuter) (Ghauri, 1980), and an unidentified entomopathogenous fungus, probably belonging to the genus *Entomophthora* (Salifu, 1986).

Because of the insufficiency of the existing data, it was decided to start a more detailed analysis of the above-mentioned factors. The results, summarized in Tamò *et al.* (1993b), were quite astonishing for what was believed to be an indigenous pest.

Extensive surveys indicated clearly that *M. sjostedti* survives the dry season on a wide range of alternative hosts, all belonging to the Leguminosae, where it was found feeding and reproducing. Among the most important host plants in the south, we encountered mostly annual or biannual plants such as *Pueraria phaseoloides*, *Cajanus cajan*, *Centrosema pubescens* and the *Tephrosia* complex. In the north, *M. sjostedti* was found almost exclusively on savannah trees growing in humid areas, such as *Pterocarpus erinacaeus* and *Millettia thonningii*, and further north mostly on trees growing along water flows, such as the omnipresent *Pterocarpus santalinoides*, where very high breeding populations of *M. sjostedti* were observed causing heavy damage to its flowers.

Different antagonists were observed attacking eggs and larvae of *M. sjostedti*. Their impact, however, was estimated to be too low to prevent pest outbreaks. Two undescribed *Megaphragma* spp. and one *Oligosita* sp., all trichogrammatid egg parasitoids, were recorded for the first time on *M. sjostedti*. The anthocorid *Orius* sp. was the most important larval predator.

No hymenopterous parasitoids could be reared from larvae collected on cowpea and three major alternative hosts, whereas a low percentage of the larvae collected from the flowers of *Tephrosia candida*, an exotic shrub native to India, were parasitized by the eulophid *Ceranisus menes* (Walker), also recorded for the first time in Africa. On the one hand, it is known from the literature that larvae of closely related species, such as *Megalurothrips usitatus* (Bagnall), occurring in South-East Asia, are parasitized by different species of *Ceranisus* such as *C. femoratus* (Gahan), *C. vinctus* (Gahan) (Gahan, 1932; Fullaway and Dobroscky, 1934) and *C. menes* (Walker) (Chang, 1990). This could be a plausible reason why these thrips are not considered important pests there (Kalshoven and Van Der Vecht, 1950; Litsinger *et al.*, 1987; Singh *et al.*, 1990). On the other hand, it is noteworthy that in West Africa the newly recorded *C. menes* attacks *M. sjostedti* on the seldom encountered exotic *T. candida*, but it has never been recorded on cowpea, native to this area. Moreover, there are indications that the parasitism rates are low, and nearly half of the parasitoids cannot develop successfully inside the larvae of *M. sjostedti*. All this suggests that the parasitoid is more likely to be associated with other thrips rather than with *M. sjostedti*, and that it is probably more attracted by *T. candida* than by native plants.

The feeding activity of six larvae of *M. sjostedti* during 5 days induced the shedding of all flower buds of a cowpea inflorescence.

The results of this diagnostic study shed new light on the *M. sjostedti* pest problem and the ways to solve it. The lack of efficient antagonists, particularly larval parasitoids known from closely related South-East Asian *Megalurothrips* spp., and the high damage threshold indicate that *M. sjostedti* is a potential target for biological control. However, further studies are needed to investigate the migration of *M. sjostedti* adults to and from alternative host-plants, in order to reinforce the action of biological control with cultural practices.

In our first case-study we showed how simulation models can be used as research tools in plant health (Gutierrez *et al.* 1988a, b). Similarly, the analysis of the cowpea agroecosystem considers the interactions between the crop plant and associated organisms, such as insect pests, which influence yield formation. Baumgärtner and Gutierrez (1989) emphasized the use of the demographic approach to develop simulation models able to capture the features of these complex interactions. Accordingly, a demographic simulation model for cowpea was developed and validated by Tamò and Baumgärtner (1993). Basically, the plant subunits (i.e. leaves, shoots, roots, peduncles and fruiting structures) were competing for carbohydrates, which the plant acquires according to its demand, modified by water and phosphate availability and the carbohydrate supply under the given growing conditions. In the subsequent work, the populations of *M. sjostedti* were integrated into the cowpea model (Tamò *et al.*, 1993a). This was achieved by considering another type of population interaction, i.e. the destruction of reproductive organs by the thrips and the response of the plant, which affects the pest population dynamics. The model showed that the plant was able to compensate thrips damage up to high pest population densities. Below a critical infestation index, yield formation was mainly controlled by the availability of phosphate and water. The yield responded in a more gradual way to the supply of phosphate than to rainfall. Above the critical index, the impact of the two factors remained small and yield formation was mainly controlled by the thrips population.

Another pest of high impact on cowpea yield, when not already destroyed by the thrips, is the legume pod-borer *Maruca testulalis*. This pest probably has a history similar to thrips, i.e. it is not indigenous to Africa. Our most recent observations (M. Tamò, personal communication) indicate that *M. testulalis* could also be a candidate for classic biological control. Research on the ecology of the legume pod-borer is not yet as advanced as for thrips, but it is certain that traditional breeding strategies will not lead to complete success because of the lack of resistance genes in the cultivated cowpea germplasm. There is, however, scope for good resistance sources in wild relatives of cowpea. To tap into these sources and introduce them into cultivated cowpeas, however, we may need some genetic engineering. Such techniques may also be of use for the other major cowpea insect pests, such as the PSB complex and thrips. As already recommended by Murdoch (1992) and Thottappilly *et al.* (1992), every effort should be made to utilize wild relatives of cowpea with resistance genes to the major insect pests. There is scope to utilize *Bacillus thuringiensis* protoxin, which has shown activity against the pod-borer (Jackai and Rawlston, 1988), but we have our reservations about this method, versus simply applying *B. thuringiensis*. Although cowpea has been found to harbour a trypsin inhibitor gene, now used in tobacco and other crops, its activity is mainly directed toward leaf-feeding insects, as shown by the insignificant damage recorded. It is not the same on the reproductive organs of the cowpea plant, which are attacked heavily by a number of pests. This indicates the need to ensure that any resistance is expressed at different plant part levels – flower, pod and seeds in particular. The same is true for protease, α-amylase inhibitors and lectins, found to have negative effects on insect growth and development. When considering genetic engineering, we must not forget the often severe yield depression that results from such manipulations, thus extending the time for producing varieties with all the necessary agronomic attributes. According to Fox (1992), productivity and fitness – health – are inversely correlated, and he does not believe that a highly controlled technosphere will be more productive in the long term than a less intensively controlled and more natural biosphere.

But we believe that, in view of the high damage level and the complex pest/plant/environment interactions, a solution integrating host-plant resistance with habitat management and biological control will be the most effective and long-lasting. The use of the model for the evaluation of pest-plant interactions under different levels of resistance and of possible biological control candidates and habitat management practices will be an important step in the design and development of an ecologically sustainable plant-health strategy for the cowpea agroecosystem. Studies for both *Maruca* and the PSB along the lines described above for thrips are now under way. The results should be helpful in developing an environmentally sustainable cowpea plant-protection strategy, for which a project has been developed and submitted to donors. First elements of such a strategy have already found their way into farmers' fields. In the Kano area of Nigeria, farmers are adopting local varieties improved in the area of insect and striga resistance through traditional breeding methods. The integration of biological control, improved genetic resistance and management practices will render the pesticide applications unnecessary.

Conclusions

Research for future plant-health management practices needs now to focus more on the entire ecosystem in which the agroecosystem is fully embedded, or the 'holistic approach'. The interactions between the two systems are a fact. If in the past the tendency was to isolate the two systems, it is imperative to study and, as far as possible, re-create the links to stabilize the agroecosystem. The buffer function of the global system should not be neglected; it needs to be strongly promoted. Given the average losses, plant health can raise food production worldwide by 40–50%, with few or no external inputs. The needed research considerably overlaps the research to be undertaken to save our planet and provide for sustainable development.

Coming back to the concept of prevention, georeferenced ecological data encompassing weather, soil, flora and fauna, together with socio-economic data, would provide a decision base (comparable to an expert system) for choosing the most appropriate crop to plant at a particular location. This could, under optimal circumstances, reduce the need for costly corrective interventions, such as the application of pesticides, fertilizers and genetic plant manipulations.

Another aspect that needs to be taken into consideration is the production of quality over quantity. Although in the short term the economic arguments may prevail on plant-health issues, in the long run the farmers might be confronted with better informed, and thereby more concerned, consumers. It is important to stress that the production of better-quality food requires less cropping area for the same nutritive value, thus preserving the natural resources and cutting down on the need for additional inputs.

El Titi (1987) reviews rather classical examples of habitat management strategies that are particularly designed for pest prevention and control, but which also, as the name says, include a more global view of the conservation of natural resources. Among the manipulations of the vegetation, we can mention the use of intercropping, relay cropping and green manure and the management of field margins and fallows. An important practical aspect of these techniques is, on the one hand, the improved conservation of fragile tropical soils and, on the other hand, the generated increase in biodiversity is hoped to have a positive influence on the activity of natural enemies.

The integration of sound agronomic, natural-resource management with plant production and plant-health practices will lead to the much sought-after sustainable, cost-effective and productive agriculture, respectful of the environment and human needs of the generations to come. Rather than investing heavily in 'quick fixes' (with the all-too-well-known backlashes), it is here suggested that we should step back, look at the causes of problems, usually man-made, and work with nature, which holds in trust our needed solutions. Rather than continuing to insist on absolutely 'clean' products, the appearance should take second rank behind quality. Also, it is wrong to think in terms of zero pests, in the way that even biotechnology advocates are selling their products (ideas). It has been shown in numerous instances – cotton, canning peas, rice, etc. – that a few insects actually increase the yield and the quality, through natural thinning of the fruiting organs or tillers.

It is most important to let all concerned know that there are paths other than the chemical or biotechnology ones. Only when all the facts about the different interventions are openly exposed can people make an intelligent choice. Looking back, it is extremely difficult to trust the same people and companies who have brought us more problems than solutions, i.e. the agrochemical industry in particular. Commercial ends of biotechnology are the most worrisome part. Governments will have to spend huge amounts of taxpayers' money to introduce new regulations, watchdogs, etc. According to Gips (1990), biotechnology has already drained billions of dollars worldwide, and brought about a shift away from basic biological research, which is essential to sustainable agriculture already inhibited by insufficient funds. Research in the area of biodiversity and linked disciplines of taxonomy and ecosystem analysis is essential to biologically, economically and socially sustainable agriculture, land use and development in general. The present trend reminds me too much of the years before the 'silent spring'. As Fox (1992) said in the preface to 'Superpigs and Wondercorn', 'The genetic age is now upon us. For better or worse, it will affect our lives and those of generations to come.' I hope, however, that the biotechnocrats will only develop what agricultural scientists, together with biologists, ecologists and sociologists, have found to be needed and that the long-term benefits, not in financial but in ecological terms, and other consequences will be thoroughly assessed.

To rely too heavily on biotechnological solutions for Third World agricultural pest or disease vectors is to continue the South's dependence on the North and to promote capital-intensive, non-sustainable farming and development.

Biological Products for Integrated Pest Management 8

Pamela Marrone

Introduction

A report from the National Academy of Sciences, recently made public, brought up concerns about pesticide residues in children's food. As a result, the US government agencies with oversight for food and pesticides will provide incentives to increase the use of integrated pest management (IPM) and alternatives to chemical pesticides. The California Environmental Protection Agency (CAL-EPA) has appointed a Pest Management Advisory Committee (PMAC) to advise the agency on ways to increase the use of IPM in California agriculture.

Echoing the concern about pesticide residues are food processors. 'All this is driven by food safety. That's why we're moving toward IPM, says Del Monte's manager of pest management programs' (Wilson, 1991). The Processed Tomato Foundation has stated that the tomato industry maintains flexibility in crop-protection alternatives, including the promotion of IPM. As a result of these concerns, microbial pesticides have become crucial components of IPM systems (Wilson, 1991).

Defining IPM

In the early 1970s, pest-resistant insects were becoming widespread, and the appearance of secondary pests and problems with pesticide residues renewed an interest in alternative methods of pest control. Integrated pest management represented a complete change in the philosophy of pest control, away from pest eradication towards pest management – the management of entire populations, not just localized ones (Dent, 1991).

The National Coalition on Integrated Pest Management defines IPM as a pest-population management system that uses all available tactics, such as natural enemies of pests, pest-resistant plants, cultural management and pesticide, in a total crop-production system to anticipate and prevent pests from reaching damaging levels (Leslie and Cuperus, 1993). Integrated pest management is an ecologically based pest-control strategy that is part of an overall crop-production system (Zalom *et al.*, 1992). All appropriate methods from many scientific disciplines are combined into a systematic approach to optimizing pest control. Tactics employed in IPM must be compatible with each other and with social, environmental and economic factors. Pesticides are applied only after all other relevant tactics have been developed or when their need is justified by knowledge of pest biology, established decision guidelines and the results of field monitoring. The concept of treatment threshold (economic threshold) is a key element of IPM systems. It is defined as the population density at which control measures must be applied to prevent an increasing pest population from reaching the point where crop loss exceeds cost of control (Zalom *et al.*, 1992).

This chapter focuses on examples of successful use of *Bacillus thuringiensis* and baculoviruses in IPM programmes and the constraints on broader use of these microbials, and includes some speculation on the potential for engineered microbials and plants in these programmes.

Use of Microorganisms in IPM

Bacillus thuringiensis

Never before has there been such interest in *B. thuringiensis* by industry, as evidenced by the proliferation of small companies (Ecogen, Mycogen, etc.) and large agrochemical companies (Ciba, American Cyanamid) developing *B. thuringiensis*-based products. Because of its environmental safety, its activity on insects resistant to chemical insecticides and its selectivity, usage of *B. thuringiensis* is increasing, particularly for use in IPM programmes.

B. thuringiensis-based microbial pesticides are used extensively in vegetables, particularly in California by growers concerned about residue on harvested products, or to decrease the selection pressure from use of conventional pesticides for control of diamondback moth (DBM) and cabbage looper on cole crops, lettuce and tomato (Zalom and Fry, 1992). 'The quality of *B. thuringiensis* products has improved dramatically, and today's products provide a predictable level of control and are priced competitively with many conventional pesticides' (Zalom and Fry, 1992). Monitoring of pest populations becomes especially important.

Considerable work on the utilization of *B. thuringiensis* in IPM vegetable programmes has been done at the University of California, Riverside (Trumble, 1990). He demonstrated the benefit of the use of *B. thuringiensis kurstaki* in a pesticide rotation to reduce the development of resistance by the celery leafminer *Liriomyza trifolii* (Burgess) (Trumble, 1985). *Bacillus thuringiensis* application suppressed army worms without destroying leafminer parasites.

Moar and Trumble (1987) stressed the importance of understanding some basic parameters of *B. thuringiensis* strains, such as the relative time to mortality and joint action of *B. thuringiensis* plus various other insecticides. This information should lead to more efficient use and successful integration into pest-management programmes.

Some recent work by Trumble (1989) focused on determining the economic benefits of IPM relative to traditional spray programmes. In 1988–89, the control of celery pests was evaluated in standard pesticide treatment plots and in 'low-input' IPM plots, utilizing abamectin, *B. thuringiensis* and methamidophos. The number of pesticide applications was reduced from ten (including two insecticides per application) in the standard plots to four (one insecticide per application) in the IPM plots. The growers netted $US581–654 more in the IPM plots than in the standard plots. This did not take into account the increased costs and potential market value loss of an estimated 3 weeks of additional growing time, which was needed to achieve the same fruit size at harvest in the standard programme as in the IPM programme. Therefore, the actual benefit to the grower was much more. Trumble (1989) concludes that *B. thuringiensis* 'effective against *Spodoptera exigua* (Hübner) can have a strategic role in IPM systems designed for celery'.

In studies on tomato in California, Trumble (1991) compared the economic return to the grower in IPM plots (abamectin and *B. thuringiensis* on an as-needed basis) with the standard insecticide spray schedule (eight applications of methomyl plus permethrin). Although the tomatoes in the IPM approach suffered greater insect damage (7% versus 3%), the net gains were equal to or better than the standard spray schedule, due to the reduction in pesticide costs.

In Mexico, tomato is a $US1 billion crop (J.T. Trumble, personal communication). Pesticide resistance has been a serious problem. Thirty-five applications of mixtures of two to four chemicals are applied each season. Tomato pinworm (*Keiferia lycopersicella* (Walsingham)) is the major pest, followed by *Liriomyza, Heliothis zea* (Boddie) and *Spodoptera exigua*. To demonstrate that it is economic to use IPM for fresh market tomato, Trumble compared an IPM programme (*B. thuringiensis kurstaki*, abamectin, *Trichogramma*, pheromones, endosulfan) with a standard chemical regime and a reduced pesticide regime on processing tomato (Trumble and Alvarado-Rodriguez, 1993). The cost of application plus treatment was

$US511, 1200 and 106/ha, respectively. The amount of marketable fruit production was similar for all treatments in the autumn plantings, but significantly higher in the IPM programme in winter and spring plantings. The net profit for the autumn IPM programme was $US304–579/ha. In the winter and spring, only the IPM approach was profitable.

Hunt-Wesson and Campbell's have introduced IPM programmes such as these in the large tomato-growing area of Sinaloa, Mexico. Yield per hectare has increased, cost per ton has decreased and quality has improved (Moore, 1991). Campbell's has reduced the amount of pesticides used by incorporating *B. thuringiensis* and other non-chemical approaches into most of their vegetable growing (W. Reinert, Davis, 1993, personal communication). They have produced a video detailing their IPM programmes.

B. thuringiensis has been successfully applied to cruciferous crops for many years. An example of its use in an IPM programme for imported cabbage worm on cabbage was reported by Sears *et al.* (1983). Prior to this work, *B. thuringiensis* or chemicals were applied on a 5-day or 14-day schedule. The researchers demonstrated that *B. thuringiensis* applications based on thresholds of eggs or small larvae provided comparable control to chemical treatments. Growers in Massachusetts can grow cole crops without any chemical pesticides by using *B. thuringiensis* for controlling the caterpillar complex (Ferro, 1993b).

The use of *B. thuringiensis* for the Egyptian cotton leafworm *S. littoralis* Boisduval as a potential component of an IPM system was demonstrated in tests by Broza *et al.* (1984). They concluded that the additive effect of natural enemies that were undisturbed by the *B. thuringiensis* treatments resulted in positive results and reduction of treatments (2 versus 4.3) in the *B. thuringiensis* plots versus a chemical spray programme. Timing of the application, after precise surveys, as required by all IPM programmes, was also critical to the successful use of *B. thuringiensis* here.

In Australian cotton, an IPM programme has been in place since 1984, with the primary purpose of preserving the pyrethroids for control of *Heliothis* (Daly and McKenzie, 1986). The cropping cycle is divided into three stages of pesticide application. Pyrethroids (no more than three sprays) are limited to only one stage or 'window' of application during the cotton-growing season. *Bacillus thuringiensis* is used in combination with endosulfan early in the season. In addition, it is recommended that the crop be grown early, unfavourable rotations (chickpea, maize) avoided and the soil cultivated to kill overwintering pupae.

Use of *B. thuringiensis* is increasing in US cotton IPM programmes, as in Australia to manage pyrethroid resistance. The best fit for *B. thuringiensis* is in controlling first-generation *Heliothis*. Significantly higher predator populations were observed when using early-season *B. thuringiensis* followed by pyrethroids in the second generation, versus a standard season-long pyrethroid programme (Halford, 1987).

The diamondback moth, *Plutella xylostella*, which has developed resistance to all chemical classes and also to *B. thuringiensis*, can be managed with IPM. In Taiwan, two larval parasitoids, pheromone traps and *B. thuringiensis* reduced pest population densities on cauliflower and broccoli to levels lower than in neighbouring conventionally sprayed plots (Anon., 1991).

A survey of tree-fruit researchers throughout the USA indicated that the integrated use of pheromone mating disruptants, low doses (one-tenth the recommended label rate) of pyrethroids and the full rate (0.45 kg) of *B. thuringiensis* is a viable IPM programme for leafroller management and is 1–2 years away from commercial implementation (Tette and Jacobsen, 1992).

The peach twig borer (*Anarsia lineatella* Zeller) is a major pest of almonds in California that has been controlled through an IPM programme (Dr Frank Zalom, University of California, Davis, 1992, personal communication). This pest emerges from overwintering hibernacula (frass tubes) in late January or early February. The larvae look for newly emerged leaves to feed on before they bore into twigs, where they are difficult to control. For the past 20 years, dormant oil sprays plus an organophosphate (OP) such as parathion have been used to control the peach twig borer. As the OPs are being phased out because of their adverse effects on hawks in the almond orchards, *B. thuringiensis* has the potential to be used as a replacement. *Bacillus thuringiensis* is used against the borer during the 'popcorn' stage (when 60% of the overwintering larvae have emerged) to petal fall (when 80% of the larvae have emerged), while the larvae are feeding and before they bore into the twigs.

In outlining IPM practices in maize and soybeans, Edwards and Ford (1992) cite the use of *B. thuringiensis* as a crop protectant in maize for European corn borer, and in soybeans for green cloverworm and soybean looper. Naturally occurring predators of mites, aphids and caterpillars, such as *Amblyseius*, *Aphidoletes*, *Chrysopa* and *Hippodamia*, can be preserved through *B. thuringiensis* use. Incorporation of *B. thuringiensis* insect-control protein into a maize endophyte, *Clavibacter xyli*, is expected to be commercialized in the next 5 years, also providing selective control of corn borer and other Lepidoptera pests (Dimock *et al.*, 1988).

Bacillus thuringiensis tenebrionis, active on Coleoptera pests, has been shown to control Colorado potato beetle under field conditions (Ferro and Gelernter, 1989; Jaques and Laing, 1989) and is now being used commercially. In researching information necessary for use of *B. thuringiensis tenebrionis* in IPM systems in Massachusetts, Ferro and Lyon (1991) studied a number of parameters that affected the use of a particular *B. thuringiensis tenebrionis* product. These included larval feeding behaviour, larval size, exposure time and temperature.

Roush and Tingey (1992) reported the development of an IPM system for Colorado potato beetle in New York. The programme includes scouting, border sprays or barriers to trap migrating overwintering adults and sprays of *B. thuringiensis tenebrionis* on small larvae, cryolite on large larvae and endosulfan on adults. The system enhances predators and manages resistance through the use of 'soft' chemicals and *B. thuringiensis tenebrionis*, as well as rotation of pesticides in different chemical classes.

Bacillus thuringiensis has been used operationally for controlling forest caterpillar pests for many years (Cunningham, 1988). The use of *B. thuringiensis* in IPM for arboriculture has shown potential (Bowen, 1991). A major goal of IPM is to provide enhanced tree care while reducing the use of pesticides. Current information indicates that a small fraction of the emitted pesticide active ingredient finds its way to the target pest. Therefore, improved application techniques are essential to the success of an IPM programme for tree care. An example where this has been put into practice is in reducing damage from coneworms and seedbugs in southern seed orchards. Guthion™ and Pydrin™ had previously been used at 1.3 and 0.34 kg active ingredient in 90–110 l of water/ha. Now, 2.2 l of undiluted *B. thuringiensis* (Foray™ 48B) with 185 g of Asana™ XL are used, which is a significant reduction in chemical active ingredient and water volume and a 50% reduction in pesticide cost (Bowen, 1991). *B. thuringiensis tenebrionis* has recently been shown to be a valuable component of an IPM system for the Tasmanian eucalyptus leaf beetle, *Chrysophtharta bimaculata* (Olivier) (Elliott *et al.*, 1993).

Future **B. thuringiensis** *products*

The discovery of *B. thuringiensis tenebrionis* fuelled the search for strains with novel activity. Companies in the *B. thuringiensis* arena (Novo Nordisk, Abbott, Sandoz, Mycogen and Ecogen) boast large collections of several thousand *B. thuringiensis* isolates. Many new crystal types have been discovered with activity on mites, corn rootworm (*Diabrotica* spp.), nematodes, adult flies and ants. Because many of these strains are still being characterized, the potential for use of these new strains in IPM is unknown.

Ecogen, Ciba and Sandoz-Repligen scientists have used various molecular techniques (electroporation, transconjugation, etc.) to develop product combining genes from *B. thuringiensis aizawai* and *kurstaki* strains to increase the activity on key lepidopteran pests such as army worm (*Spodoptera* spp.) and cotton bollworm (*Helicoverpa* spp.). Novo Nordisk used classic mutation to improve *B. thuringiensis tenebrionis*. This strain produces a bigger crystal, which is directly correlated with enhanced field activity. In addition, fusions of genes from baculovirus with *B. thuringiensis* genes are being used to expand the host range of *B. thuringiensis*. These products are in various stages of development and commercialization.

It is well known that *B. thuringiensis* lasts only several hours on plant foliage under typical field conditions due to ultraviolet (UV) degradation, rainfall, etc. In the early 1980s, Monsanto developed a recombinant, plant-colonizing *Pseudomonas* for delivery of *B. thuringiensis* genes, with the objective of improving residual activity and efficacy of *B. thuringiensis* proteins. This concept was further developed by Mycogen into the products MVP™ and M-Trak™. The *B. thuringiensis* bearing *Pseudomonas* is killed (to avoid regulatory hurdles for registering recombinant microorganisms) and sprayed on the crop like other *B. thuringiensis* products. The pseudomonad cell is reported to protect the *B. thuringiensis* protein from environmental degradation, thus providing longer residual activity. Under development by the US Department of Agriculture (USDA) (Peoria) is a *B. thuringiensis* and virus starch encapsulation procedure, which is designed to improve the field life and efficacy of these microbial products.

New genetically engineered and improved *B. thuringiensis* products may provide more opportunities and choices for growers using IPM programmes. The most successful *B. thuringiensis* products will be ones that provide efficacy and consistency competitive with traditional chemical pesticides. Significantly improved army worm (*Spodoptera*) and bollworm (*Helicoverpa*) products, competitive with chemical insecticides in efficacy, will be the most important development in the use of *B. thuringiensis* microbials in agriculture. If this is achieved, *B. thuringiensis* products will capture a larger market share and replace some established chemical products. If not, *B. thuringiensis* strains will continue to be used as they are at present, as supplements and in conjunction with chemical pesticides.

Although genetically engineered microbes designed for broadened host range, increased residual activity and improved efficacy are desired by growers and will lead to increased usage in agriculture, they may not necessarily be optimal for IPM systems. The most desirable products for IPM are selective and have a low risk for resistance development. Engineered *B. thuringiensis* products containing a single *B. thuringiensis* gene may have some of the same disadvantages as engineered crops with one gene. The primary disadvantage is, of course, the potential for resistance development due to increased selection pressure. This may also be true for products engineered for longer field stability and residual activity. However, if the products are used after proper monitoring and integrated with other tactics, resistance should be less of a concern.

Baculoviruses

Although baculoviruses have not had the commercial success of *B. thuringiensis*, they have significant potential for use in IPM programmes. There are a number of advantages associated with the use of baculoviruses, which are ideal for IPM because they do not affect predators and parasites. They are safe for non-target insects, humans and the environment. The host specificity is, therefore, a useful attribute from an environmental standpoint.

Baculoviruses may, in some cases, be the only effective biocontrol agent available for controlling an insect species (Cunningham, 1988) and provide an avenue available to overcome a specific problem, such as resistance.

In pest management, it is important to have a selection of control agents when designing strategies. Viruses are not likely to evoke cross-resistance to chemicals or to each other, so more attention must be given to viral pesticides (Cunningham, 1988). 'B. t. and baculoviruses are complementary ... each can be effective at killing a somewhat different spectrum of insect,' stated Dan Hess of Sandoz Crop Protection, one of the companies pursuing the possibilities of baculoviruses (Anon., 1990). The use of two biologicals could lower the possibility for resistance development. Since widespread practical application of viruses in IPM programmes is at an early stage in US agriculture, the true potential of this approach has not been demonstrated.

Strategies for use

There are four strategies for using viral insecticides (Cunningham, 1988). First, the virus spreads from limited applications and permanently regulates the insect population through classic biological control. Second, an epizootic is established through vertical and horizontal transmission, but reapplication may be necessary since the epizootic is not permanent. Third, a viral inoculum in the environment is conserved and reactivated through environmental manipulation. Fourth, repeated applications are used to control an insect population because there is no horizontal transmission of the virus. The fourth strategy is the most widely used for control of agricultural pests. Since viruses are applied as sprays in the same manner as chemical insecticides and *B. thuringiensis*, this strategy will work well for IPM systems.

Helicoverpa zea Boddie (cotton bollworm) nuclear polyhedrosis virus (NPV) was the first baculovirus to be marketed in the USA. It was developed by International Minerals and Chemical Corp. (IMC), but marketed by Sandoz under the trade name Elcar™ in 1976, after purchase of the IMC's biological products division. However, interest in Elcar declined with the introduction of pyrethroids, which are effective, inexpensive, broad-range insecticides. The properties of the virus for fit in IPM systems have been well studied (Ignoffo and Garcia, 1992). Elcar has been tested recently by the USDA Agricultural Reasearch Service (USDA-ARS) for control of *H. zea* in field borders, ditch banks and road edges. The objective is to control *Heliothis* larvae while feeding on wild geranium and other weeds prior to larval movement into neighbouring cotton fields. The virus caused a 64% reduction in adult emergence in 1989 trials (Ignoffo and Garcia, 1992).

Currently, the largest use of baculoviruses is in Brazil, where *Anticarsia gemmatalis* Hübner (velvetbean caterpillar) NPV protects 5.9 million ha of soybeans. In Europe, a number of companies, including Kemira Oy (Finland), Oxford Virology and Imperial Chemical Industries (ICI) (UK), Solvay (Belgium), Calliope (France) and Hoechst (Germany), either have viral products or are developing products for the insecticide market. Viral products include *Cydia pomonella* L. (codling moth) granulosis virus (GV), *Neodiprion sertifer* (Geoffrey) (European pine sawfly) NPV, *Spodoptera exigua* (Hübner) (beet army worm) NPV and *Autographa californica* (Speyer) (alfalfa looper) NPV. United States companies actively involved in baculovirus research are Sandoz, American Cyanamid, Ciba, Crop Genetics International and DuPont.

In North America, the effort in baculoviruses has been led mainly by governmental agencies (Cunningham, 1988; Otvos *et al.*, 1989; Podgwaite *et al.*, 1991). The US Forest Service has registered NPVs to fight *Lymantria dispar* (L.) (gypsy moth), *Neodiprion sertifer* (European pine sawfly) and *Orgyia pseudotsugata* (McDunnough) (Douglas fir tussock moth) in forestry. The Canadian Forest Service holds registrations for *O. pseudotsugata* NPV and *Neodiprion lecontei* (Fitch) (redheaded pine sawfly) NPV. Louis Falcon (University of California, Berkeley, personal communication) has demonstrated the use of codling moth GV in pear, apple and walnut IPM in California and Washington. In a pilot project with several organic growers, virus is one component of an orchard management system. This includes orchard sanitation, codling moth monitoring using pheromone traps, spray timing based on accumulated heat units and proper virus application and coverage. Organic growers pay $US74/ha-treatment of virus and apply five to ten times per season, in contrast to monthly sprays (3 times per season) of Guthion™ at $US18.50–24.70/ha-treatment. According to Falcon, total chemical input (insecticides, fungicides, acaricides, bactericides) cost approximately $US890/season/ha. Total cost of all pesticides (fungicides, miticides, insecticides) when using the virus for codling moth is $US780/season/ha. In the virus-treated orchards, natural enemies can survive to control mite pests, thus eliminating the need for miticides, which are required in Guthion-treated orchards.

Although Falcon's programme has been successful for organic growers, it has not been utilized by mainstream fruit producers because of the effectiveness of Guthion. With the recent development of Guthion-resistant codling moth populations after 20 years of use, an incentive for the adoption of Falcon's programme may develop.

Application of baculovirus for control of beet armyworm has been well studied in greenhouse systems (Smits *et al.*, 1987b). Parameters such as dose–response, larval feeding behaviour, application techniques (Smits *et al.*, 1988), timing and strain (Smits *et al.*, 1987a) have been integrated into recommendations for operational use of virus with other control methods.

On head lettuce in California, a beet armyworm NPV was field-tested for 3 years and compared with chemical insecticides (Gelernter *et al.*, 1986).

Results indicated comparable control to methomyl and permethrin. Research in California is now focusing on data necessary to bring NPVs into operational use with vegetables such as lettuce, tomato and celery.

Future baculovirus products

At the present stage of baculovirus products development, however, baculoviruses have several limitations associated with their use as insecticides in IPM programmes. A major limitation is the slow rate of kill, resulting in feeding damage. Reduction of kill time will rely on improvements in formulation and application in the immediate future. However, this limitation can be managed in the short term by employing baculoviruses in combination with other insecticides through IPM.

Although viruses are less expensive to produce *in vivo* than *in vitro*, the cost still exceeds that for *B. thuringiensis*. Viruses are formulated to be applied in the same fashion as *B. thuringiensis* strains. However, for extensive use in IPM, improvements in formulation and application technology are needed. In formulation, knowledge of stability and shelf-life is required in order to optimize storage and distribution. In application, droplet size, density and dosage, as well as the tank mix (e.g. stickers, UV protectants), need to be optimized.

Another limitation is that baculoviruses can be highly host-specific, which may reduce their commercial potential. However, the host specificity is viewed positively from the environmental and IPM standpoint. Two viruses with relatively broad host ranges are *Autographa californica* (alfalfa looper) NPV and *Syngrapha falcifera* (Kirby) (celery looper) NPV, each of which kill over 30 insect species. The celery looper virus is reported to have commercial potential in cotton IPM systems (Wood, 1992).

In the long term, the development of recombinant baculoviruses that can kill a range of insects in 48–72 h will be the focus. To increase the ability of baculoviruses to kill early, research to insert specific genes into the baculoviral genome is under way. These proteins can be toxins or disrupters of larval development. The proteins being tested for exploitation include *B. thuringiensis*, juvenile hormone esterase, prothoracic trophic hormone (PTTH), melittin, trehalase, scorpion toxin and mite toxin. The knowledge of the molecular biology of viruses has also promoted interest in modifying and improving baculoviruses with regard to host range and virulence.

Development of engineered viruses shifts the registration process to a more difficult venue. The US Environmental Protection Agency (EPA) has not yet formulated guidelines for viruses engineered with new insect-control proteins. The products most likely to be commercialized first will be ones with insect hormones or other well-characterized proteins. More likely to encounter regulatory and public-perception roadblocks are engineered products containing scorpion venom proteins and itch-mite venom protein.

While these proteins may be selective to insects and thus very desirable for IPM programmes, their use in IPM systems will depend on demonstrated safety to mammals and other non-target organisms.

The growth and success of baculoviruses as commercial insecticides will depend on reducing production costs, developing practical, effective formulations, optimizing field performance, overcoming regulatory obstacles and educating users and the public. There are a number of opportunities in the baculovirus area, but an increase in research and development (R & D), focusing on improving production, formulation and application technologies in conjunction with genetic engineering for faster kill and broader host range, will be necessary to enable the development of more economic and efficacious products.

Plant Resistance to Insects

There are numerous review articles on plant resistance to insects (Harris, 1980; Kogan, 1982; Hedin, 1983). Resistance as a pest-management factor has achieved some outstanding results (for example, grape phylloxera, wooly aphid, Hessian fly, wheat stem sawfly) (Kogan, 1982). Desirable features of pest-resistant plants are specificity, cumulative effectiveness (effect on the pest is compounded in successive generations), persistence and harmony with the environment, ease of adoption and compatibility with other IPM tactics (Kogan, 1982). Disadvantages of pest-resistant plants for pest-management systems include the long time for development (3–15 years), genetic limitations (the lack of available resistance genes), evolution of resistant pest biotypes and conflicting resistance traits.

Transgenic plants, whether engineered to contain an insecticidal protein, such as an endotoxin protein from *B. thuringiensis*, or a chitinase gene to control root-rot pathogens, appear to have the same advantages and disadvantages as traditional resistant crop varieties. For example, cotton engineered with a *B. thuringiensis Cry IA(c)* or *(b)* protein is indeed selective for Lepidoptera, increases the persistence of *B. thuringiensis* for season-long control, is compatible with the environment in much the same way as microbial *B. thuringiensis* can be easily adopted by farmers, and is compatible with other tactics, such as the use of natural enemies, chemical pesticides for other pests, etc.

Likewise, the disadvantages are the same. The time period needed to develop transgenic plants, including the search for the appropriate gene(s), tissue culture selection, backcrossing, etc., is still as long as traditional plant-breeding methods. Although much progress has been made in the discovery of new genes for introduction into plants, the ability to introduce the genetic material has surpassed the discovery of new genes to engineer. For insect control, only *B. thuringiensis* genes and certain protease inhibitor genes show potential. These genes are effective against certain Lepidoptera and Coleoptera, with no effectiveness on sucking insects. In cotton, for example, transgenic plants with *B. thuringiensis* genes will provide growers with another alternative for bollworm, but the need for products that control whiteflies, mites and lygus bugs remains.

The third limitation, occurrence of resistance biotypes, is a key issue for integration of transgenic cotton containing *B. thuringiensis* genes and probably other transgenics with useful pest-resistance traits. The issue of resistant pest biotypes and which strategies to use to delay resistance development will not be dealt with here, but is reviewed in Chapter 17 by Whalon. Gould (1988a) is concerned about the incompatibility of transgenic plants with IPM systems. He states that 'the successful engineering of highly resistant crops could lead to the elimination of IPM techniques that aim at using intense, pest suppressive measures only when pests are likely to cause economically important damage'. Obviously, strategies (as described in Chapter 17 by Whalon) designed to limit the development of resistant biotypes based on pest population dynamics are critical and necessary for optimal use of transgenic plants in IPM systems.

The current transgenic plants, containing single *B. thuringiensis* genes are 'first-generation' plants and will be followed by more sophisticated second- and third-generation plants with greater flexibility for use in IPM systems. These include plants with inducible and tissue-specific expression systems as well as multiple genes. Future IPM programmes will have a combination of genetically engineered and modified *B. thuringiensis* microbial products, several types of engineered plants and traditional *B. thuringiensis*.

Constraints on Using Engineered Biopesticides

The constraints on the use of engineered pest-control agents in IPM, such as recombinant microbes and plants, are the same as for the adoption of IPM strategies in general. The general opinion in the academic community is that, as long as effective and inexpensive synthetic chemicals are available, growers do not have an incentive to adopt IPM (George Kennedy, North Carolina State University, 1991, personal communication). It is easier to use a continuous chemical strategy. Integrated pest management programmes require extensive grower education, and dealers stock only certain chemicals, which may not always be the best choice for the IPM programme. An example Kennedy cites is Colorado potato beetle control in North Carolina. Continuous sprays of pyrethroids have proved to be profitable; therefore growers have no compelling reason to switch to an IPM programme that includes *B. thuringiensis*. Resistance development is not a concern because it is presumed by growers that the next new chemical is just around the corner.

Constraints on the use of biologically intensive IPM are summarized by Glass (1992). These are categorized as technical, implementation, training, regulatory, policy and economic. Some key technical constraints include lack of understanding of crop and pest biology interactions, lack of field-orientated and IPM research, inconsistent efficacy of biological agents and lack of funding. A main barrier to implementation is the perception that IPM is riskier and more expensive than conventional controls and the preference for immediate solutions. Some key regulatory constraints that have an impact on the use of genetically engineered products are diversity and uncertainty of interstate and international regulations, lack of clear EPA guidelines and unreasonable crop-destruction requirements. Lack of a consistent national IPM policy is seen as the fundamental reason why IPM efforts have remained stymied until lack of available pesticide products forced its use.

Frank Zalom (University of California, Davis, 1992, personal communication) has outlined the elements of successful IPM programmes. There must be: (i) demonstrated need (regulatory or economic); (ii) research: a body of knowledge on applied ecology of the pest(s), crop, beneficials and control alternatives; (iii) commercial development: testing via extension and compatibility with other practices and materials; and (iv) implementation: willing growers, pest-control advisers, extension, marketing and sales representatives. Most importantly, a demonstrated success will pave the way for adoption of microbials and other new technologies such as transgenic plants in IPM.

A report prepared for the CAL-EPA Department of Pesticide Regulation (Benbrook and Marquart, 1993) provides a detailed analysis of the constraints on IPM in California agriculture. The authors make several suggestions as to how IPM can be increased in California. One suggestion is to integrate IPM practices into pesticide product labels and permit requirements. Regulation should be used to encourage innovation, and pest-control advisers should have a special licence category for IPM-trained specialists. Over the next year, the CAL-EPA PMAC will also develop solutions to the constraints on IPM.

Conclusion

In the short term (1–5 years), microbial use in IPM systems will continue to increase, and will consist largely of naturally occurring, transconjugant, genetically altered (via classical mutation), and dead, engineered products. The IPM systems will incorporate living recombinant *B. thuringiensis* strains and transgenic plants in the next 5–10 years. Engineered viruses are likely to be integrated after the year 2000. The advantages that engineered microbials and plants will bring are improved efficacy and speed of kill, broader host range and increased residual activity.

Biotechnology provides new tools that can be applied to systems where answers to problems are needed, just as traditional plant breeding or organism selection is a tool to improve crops and microbes. The products of biotechnology are unlikely to completely replace other pest-control means such as chemicals and classic biocontrol agents. Therefore, the products of biotechnology should be seen as options that can be integrated into pest-management systems with other existing alternatives. Sound strategies based on our knowledge of pest population dynamics are critical for the maximum use of the new microbial pesticides and pest-resistant plants. In development of these strategies, a key point is that a pest-control tactic developed via biotechnology may not be any different from one developed by traditional means.

Acknowledgements

The author thanks G. Kirfman and D. Edwards for reviewing the manuscript.

Novel Arthropod Biological Control Agents

9

MARJORIE A. HOY

Introduction

There are many definitions of biotechnology, but most include the concept of technologies that employ biological systems. The term as often used implies that recombinant deoxyribonucleic acid (DNA) techniques are employed. Until recently, arthropod natural enemies have not been subjects of biotechnology, but this is beginning to change. This chapter reviews the potential for biotechnology to improve the efficacy of arthropod natural enemies (parasitoids and predators) in integrated pest management (IPM) programmes. I shall focus this review on the use of molecular techniques to genetically improve arthropod natural enemies. However, biotechnology, in the sense of using molecular techniques, could also provide solutions to a number of basic and applied problems, as outlined in Table 9.1.

For example, maintaining quality in laboratory-reared arthropods is considered to be difficult due to possible genetic changes caused by inadvertent selection, inbreeding, genetic drift and founder effects (Stouthamer *et al.*, 1992; Hopper *et al.*, 1993). Until recently, monitoring genetic variation in arthropod natural enemies has been limited to the use of enzyme electrophoresis, but many natural enemies such as parasitoids (Hymenoptera) lack substantial variation in these proteins (Packer and Owen, 1992). New DNA-based methods for monitoring genetic variation are now available, including mitochondrial DNA analysis and DNA sequencing.

Portions of this article have been adapted and reprinted by permission of the publisher, from Hoy, M.A., *Insect Molecular Genetics*, chapter 15. Copyright 1996 Academic Press, Inc.

The use of the polymerase chain reaction (PCR) to identify DNA markers, particularly markers identified by a random sample of the genome (such as random amplified polymorphic DNA (RAPD)-PCR, offers a highly efficient method for detecting genetic changes in arthropod populations (Arnheim *et al.*, 1990; Williams *et al.*, 1990; Hadrys *et al.*, 1992). The RAPD-PCR method is also of potential value for identifying and monitoring establishment and dispersal of specific biotypes of arthropod natural enemies (Chapco *et al.*, 1992; Edwards and Hoy, 1993).

Another potential application of biotechnology is the development of cryobiological methods for preserving embryos of arthropod biological control agents. Currently, arthropod biological control agents can be maintained only by continuous rearing or by holding specimens in diapause (for those species that can diapause). This is expensive and can lead to loss of colonies, as well as to genetic drift or contamination of the colonies. Mazur *et al.* (1992) recently demonstrated that embryos of the fruit fly (*Drosophila melanogaster*) could be preserved in liquid nitrogen and then be thawed to develop into viable and fertile adults. If cryopreservation can be adapted to other arthropods, a significant saving in rearing costs can be achieved. More importantly, valuable collections of arthropod natural enemies could be maintained indefinitely.

Biotechnology could provide a solution to one of the most critical issues that limits the use of arthropod natural enemies: the difficulty and high cost of mass-rearing them for either classical or augmentative releases. If high-quality and inexpensive artificial diets for predators and parasitoids were available, biological control programmes would no longer be restricted by inefficient mass production methods. This area of research has received relatively little attention (Thompson, 1990).

Until recently, genetic improvement of arthropod natural enemies was achieved by traditional genetic methods: artificial selection, or hybridization of different strains to achieve heterosis (Hoy, 1990a). Beckendorf and Hoy (1985) suggested that recombinant DNA techniques could make genetic improvement of arthropod natural enemies more efficient and less expensive, because once a gene has been cloned it can be inserted into a number of beneficial species. If the manipulated populations need not be reared for long periods of time, there is less likelihood that laboratory selection and inbreeding will occur. One of the significant benefits of recombinant DNA techniques may be that it will be easier to maintain 'quality' in transgenic arthropods.

The ability to manipulate and insert genetic material into the genome of *Drosophila* has been used to develop a fundamental understanding of genetics, biochemistry, development and behaviour (Lawrence, 1992). Genetic engineering of arthropods other than *Drosophila* has been attempted, with limited successful results (Walker, 1989; Handler and O'Brochta, 1991;

Hoy, 1994). However, new techniques to achieve stable transformation of insects are being developed.

Table 9.1. Some potential applications of biotechnology to arthropod natural enemies.

Objective(s)	Potential method(s)	Reference(s)
Alter genome of natural enemies	Various	Ballinger and Benzer, 1989 Handler and O'Brochta, 1991
	Maternal microinjection	Presnail and Hoy, 1992
Alter sex ratio of parasitoids	Unknown now	Slee and Bownes, 1990 Bownes, 1992
Cryopreservation of colonies	Modify method for *Drosophila*	Mazur *et al.*, 1992
Develop genetic maps	PCR	Ashburner, 1992 Sobral and Honeycutt, 1993 Tingey and del Tufo, 1993
Develop information on genome organization	RAPD-PCR	Hunt and Page, 1992 Tingey and del Tufo, 1993
Identify biotypes of arthropod biological control agents	PCR, several methods RAPD-PCR	McPherson *et al.*, 1992 Ballinger-Crabtree *et al.*, 1992 Black *et al.*, 1992 Chapco *et al.*, 1992
Improve artificial diets	New approach needed?	Thompson, 1990
Monitor establishment and dispersal of arthropod biological control agents	RAPD-PCR Anonymous single-copy nuclear DNA PCR	Edwards and Hoy, 1993 Karl and Avise, 1993
Monitor genetic changes in arthropod colonies	RAPD-PCR	No examples yet
Parentage analysis	RAPD-PCR	Scott *et al.*, 1992 Scott and Williams, 1993

PCR, polymerase chain reaction; RAPD, random amplified polymorphic deoxyribonucleic acid; DNA, deoxyribonucleic acid.

Traditional genetic manipulation projects typically have three phases: conceiving and identifying the problem, developing the genetically manipulated strain and evaluating and implementing the new biotype (Hoy, 1990a). The use of recombinant DNA techniques may alter this sequence because risks associated with releases into the environment need to be resolved (Hoy, 1995).

What Genetic Manipulations of Beneficial Arthropods Might be Useful?

Selections for resistance to pesticides, lack of diapause or enhanced temperature tolerance have been successful, although most projects involved selection for resistance to pesticides (Hoy, 1990b). Pesticide-resistant predators and parasitoids have been evaluated in the field and are being implemented in several IPM programmes (Hoy, 1990b). Genetic improvement has proved to be practical and cost-effective when the trait(s) limiting efficacy can be identified, the improved strain retains its fitness and methods for implementation have been developed (Headley and Hoy, 1987).

Traits primarily determined by single major genes, such as pesticide resistance, are most appropriate for manipulation at this time because methods for manipulating and stabilizing traits that are determined by complex genetic mechanisms are not yet available. Genetic improvement can be useful: (i) when the natural enemy is known to be a potentially effective biological control agent except for one limiting factor; (ii) the limiting trait is primarily influenced by a single major gene; (iii) the gene can be obtained by selection, mutagenesis or cloning; (iv) the manipulated strain is fit and effective, and (v) the released strain can be maintained by some form of reproductive isolation. Typically, applications of pesticides are employed to reduce populations of susceptible natural enemies, allowing the resistant population to establish and persist. Alternatively, genetically manipulated strains can be released into greenhouses, or in new geographic regions, so that the population is maintained in isolation with the desired trait intact. It is not clear whether genetically manipulated natural enemy strains can interbreed with native populations and if field selection will result in a new 'hybrid' population that carries the introduced attribute (Hoy, 1990a).

Farmers in many developing countries cannot afford pesticides for managing their crops, but pesticide-resistant natural enemies could still be useful to them. For example, mass-rearing of foreign predatory mites (Acari: Phytoseiidae) for releases to control the cassava green mite, *Mononychellus tanajoa* (Acari: Tetranychidae) in Africa is being conducted in Benin by the International Institute of Tropical Agriculture (Herren and Neuenschwander, 1991). Cassava is an exotic staple crop in Africa, and serves as a valuable food during the 'hungry period' for millions of people. The release and establishment of several exotic predatory mite species over a continent-wide area is hampered by the difficulty in rearing huge numbers of phytoseiids economically. Current rearing methods rely on mass production in greenhouses/rearing rooms. However, if pure cultures of phytoseiids could be maintained outdoors on cassava plants infested with the cassava green mite, then rearing would be more economical. If the exotic phytoseiids were resistant to pesticides, they could survive applications of pesticides to the cassava to eliminate contaminant species such as generalist insect predators. Approximately 62 million individuals of *Metaseiulus occidentalis* were reared on 0.2 ha of soybean (Hoy *et al.*, 1982). Because the population of *M. occidentalis* was resistant to carbaryl, extraneous phytoseiid species or insect predators were eliminated from the plot by applications of carbaryl.

Steps in Genetic Manipulation by Recombinant DNA Methods

Genetic improvement by recombinant DNA techniques involves several steps (Fig. 9.1). A successful outcome generally requires that we have a thorough knowledge of the biology, ecology and behaviour of the target species. Identifying one or more specific traits which, if altered, would potentially achieve the goals of the project is a critically important first step. Next, suitable genes must be identified and cloned and appropriate regulatory sequences must be identified so that the inserted gene will be expressed at appropriate levels in the correct tissues and at a relevant time.

It is likely that the construct containing the gene of interest and the regulatory sequences will be evaluated by means of transient transformation. If the cloned gene and its regulatory sequences function well, they can be inserted into the germline. This may require a modification of the current methods for transformation, taking into account the specific aspects of the target species' morphology, life cycle or behaviour. Once the construct has been incorporated into the genome and demonstrated to be stable, expressed appropriately and stably transmitted to progeny, the relative fitness of the transgenic strain should be evaluated. These steps can take place in the laboratory.

If the laboratory tests indicate that the transgenic strain is relatively fit and the trait is stable and appropriately expressed, the transgenic strain(s) should be evaluated in small field plots to confirm their efficacy and fitness. Before any field tests can occur, however, regulatory issues relating to the safety of releasing transgenic beneficial arthropods must be resolved (Hoy, 1990a, 1992a; Table 9.2).

Research is essential in all aspects in this sequence (Fig. 9.1). Inserting cloned DNA into pests or beneficial arthropods could be accomplished by several different techniques (Table 9.3). The effects of the inserted DNA could be transient and short-term or stable and long-term. If the inserted DNA is incorporated into the chromosomes in the cells that give rise to the germ line, the foreign genetic material should be transmitted faithfully and indefinitely to successive generations. Cloned DNA can be isolated from the same or other species, and it is technically feasible to insert genes from

Table 9.2. Risk issues associated with releasing transgenic arthropods.

1. *Each transgenic species should be evaluated on a case-by-case basis*
Is the transgenic species a pest or a beneficial arthropod? It seems likely that permission to release transgenic arthropod natural enemies that are host-specific will be more readily received than permission to release a transgenic mosquito or tick that could potentially vector a human or animal disease or a transgenic Mediterranean fruit fly that is a pest of agricultural crops (US Department of Agriculture, 1991).

2. *Has host range/preference of the manipulated arthropod been altered?*
Host/prey specificity is usually important to ensure that arthropod biological control agents control the target pest. Likewise, changes in host specificity of a vector insect or agricultural pest should be carefully evaluated. Host range or preference should be documented through laboratory or greenhouse no-choice tests.

3. *Has the geographical range of the manipulated arthropod been altered?*
Temperature/relative humidity tolerances and diapause attributes often restrict geographical distribution of arthropods. Changes in responses to these abiotic factors should be demonstrable with growth chamber and laboratory tests if the transgenic strain's responses are compared with the responses of unmanipulated strains.

4. *Is the genetic alteration stable?*
Laboratory tests should determine whether the trait is transmitted faithfully to progeny. Likewise, genetic evaluations should demonstrate that the inserted gene(s) are maintained in their original insertion site. For that reason, it may be a good idea to avoid the use of transformation methods, such as transposable-element vectors, that could result in movement of the inserted DNA. Even a transposable-element vector lacking the transposase gene could possibly move if a transposase were supplied by a 'helper' transposable element that was 'native' to the engineered species. Likewise, it may be important to demonstrate that the inserted gene cannot be transmitted to pest species, particularly if it is a gene such as a pesticide resistance gene (Tiedje *et al.*, 1989; Hoy, 1992a).

DNA, deoxyribonucleic acid.

microorganisms into arthropods and have the DNA transcribed and translated. However, DNA coding sequences isolated from microorganisms must have promoters (controlling elements) and other regulatory DNA sequences derived from a higher organism attached so that the gene will be expressed in arthropods. These regulatory sequences determine when a gene will be

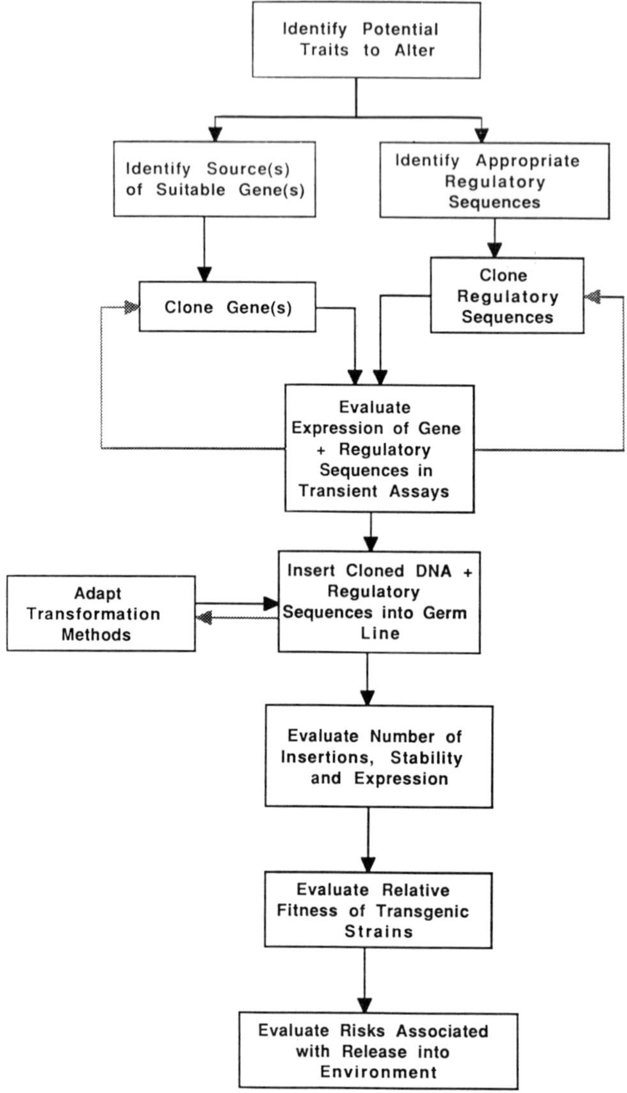

Fig. 9.1. An outline of some of the steps involved in genetic engineering of beneficial arthropods up to the point when they are ready to be released into the environment. DNA, deoxyribonucleic acid.

Table 9.3. Some potential methods for stably transforming arthropods other than *Drosophila*.

Technique	Example(s) available	Reference(s)
Artificial chromosomes	None in arthropods; feasible with yeast and mice	Schedl *et al.*, 1992
Baculovirus vectors *Autographa californica* *Bombyx mori*	Primarily for protein expression in larvae or cell cultures; lethal to infected host unless additional genetic modifications are conducted	Iatrou and Meidinger, 1990 Miller, L.K., 1988
DNA delivered by microprojectiles	Transient expression only in *Drosophila* embryos	Balderelli and Lengyel, 1990
Electroporation	Transient transformation of *Drosophila* only	Kamdar *et al.*, 1992
Maternal microinjection	*Metaseiulus occidentalis*	Presnail and Hoy, 1992
Microinjection of eggs	Three mosquito species; P element apparently not functional	Miller *et al.*, 1987 McGrane *et al.*, 1988 Morris *et al.*, 1989
P-element vectors	*Drosophila* species only	Handler and O'Brochta, 1991
Soaking dechorionated eggs in DNA solution	*Drosophila*	Walker, 1989
Sperm as vectors of DNA	*Lucilia cuprina* *Apis mellifera* (DNA bound externally only?)	Atkinson *et al.*, 1991 Milne *et al.*, 1988
Transfection of cultured cells	*Aedes albopictus*	Fallon, 1991

Table 9.3. *continued*

Technique	Example(s) available	Reference(s)
Transformation of insect symbionts	Bacterial symbiont of *Rhodnius prolixus* engineered; symbionts inserted into symbiont-free insects were transmitted to successive generations and *Rhodnius* survived antibiotic treatment	Beard *et al.*, 1992
Transplant nuclei and cells	*Drosophila*	Zalokar, 1981
Transposable-element (TE) vectors from target species	None at this time, but TEs are known from several, including *Anopheles gambiae* and *Bombyx mori*	Michaille *et al.*, 1990 Besansky, 1990 Robertson, 1993
Yeast recombinase (FLP)-mediated recombination on specific target DNA sequences (FRT)	*Aedes aegypti*	Morris *et al.*, 1991

DNA, deoxyribonucleic acid.

transcribed, at what level, in what tissues and how long the messenger ribonucleic acid (RNA) can be used for translation. The status of these components of genetic manipulation of both pest and beneficial arthropods is reviewed.

Potential Germ-line Transformation Methods

Most research on stable transformation methods has been accomplished with *Drosophila melanogaster* (Table 9.3). Initial efforts to genetically engineer *D. melanogaster* were rarely successful until the P transposable element was genetically manipulated to serve as a vector to carry exogenous genes into the chromosomes of germ-line cells (Rubin and Spradling, 1982; Spradling and Rubin, 1982). This pioneering work has elicited immense amounts of research on fundamental analyses of gene structure, function and regulation in *Drosophila* and has given us a broad understanding of how the flies develop (Lawrence, 1992). Many genes have been identified, isolated and cloned from *Drosophila*. At present, only a few of these genes appear to be potentially useful in genetic manipulations of either pest or beneficial arthropods.

Since the pioneering research of Rubin and Spradling (1982), P-element vectors have been investigated as possible vectors for other arthropods, and have been used effectively with *D. similans* and *D. hawaiiensis*. Other insects, including three mosquitoes and the Mediterranean fruit fly, *Ceratitis capitata*, have received microinjected DNA cloned into P-element vectors (Miller *et al.*, 1987; McGrane *et al.*, 1988; Morris *et al.*, 1989), but the rate of transformation was low (less than 0.1% of the microinjected embryos), and there is no evidence that the process of transformation was P-element-mediated. P-element-mediated transposition may be limited to *Drosophila* species (Handler and O'Brochta, 1991), because there is no firm evidence that integration of any exogenous DNA in an insect outside the genus *Drosophila* has been P-element-mediated. As a result, a variety of other methods for achieving transformation have been considered and evaluated (Table 9.3).

Transposable elements are commonly found in all organisms whenever they have been sought, but they have been less well studied in other arthropods (Berg and Howe, 1989: Perkins and Howells, 1992). It is possible that species-specific transposable elements could be isolated and genetically modified for use as vectors in specific insects. However, this is neither a rapid nor an inexpensive process and this approach may be limited to those arthropod species that are of major economic importance. Furthermore, because transgenic arthropods being released into the environment should be stably transformed, such transposable-element vectors should ideally be incapable of additional movements subsequently. Thus, issues of risk assessment should be considered in designing a genetic manipulation project involving 'native' transposable-element vectors (Hoy, 1992a, 1995).

Microinjecting DNA carried in P-element vectors into *Drosophila* eggs is a well-developed technique (Santamaria, 1986). These microinjection methods had to be modified for mosquito eggs, and slightly different injection methods were required for different genera (Miller *et al.*, 1987; McGrane *et al.*, 1988; Morris *et al.*, 1989) and for honey-bees (Milne *et al.*, 1988). Presnail and Hoy (1992) found that eggs of the phytoseiid predator *Metaseiulus occidentalis* were extremely difficult to dechorionate and dehydrate and that the needle tip had to be modified. It appears that *Drosophila* microinjection methodology will have to be adapted empirically to each insect species and will not be feasible with all. Variables to consider include whether to dechorionate or not, whether to dehydrate and for how long, at what age/stage to inject, what holding conditions to implement after injection and what size and shape of needle to use. It may be possible to microinject exogenous DNA into insect embryos without using any transposable-element vector with stable transformation occurring at a low rate (Walker, 1989).

Early preblastoderm eggs present within adult females of the predatory mite *Metaseiulus occidentalis* were microinjected by inserting a needle through the cuticle of gravid females. This technique, called 'maternal microinjection', resulted in relatively high levels of survival and stable transformation without the aid of a transposase-producing helper plasmid (Presnail and Hoy, 1992). The *lacZ* reporter gene regulated by the *Drosophila hsp70* promoter was expressed in larvae developing from the injected eggs and in subsequent generations. Stable transformation was confirmed in the sixth generation by heminested (PCR) amplification of a region spanning the *Drosophila*/*E. coli* DNA sequences inserted into the mite.

Maternal microinjection of *M. occidentalis* is less laborious than microinjection of eggs. Survival rates of injected females were comparable with rates of microinjected *Drosophila* eggs. The transformation rate was approximately one-tenth the efficiency of P-element-mediated transformation of *Drosophila*, but comparable to techniques employed for species in which transformation is achieved without a P-element vector.

Maternal microinjection may provide a 'universal' DNA delivery system for arthropod species. It should be possible to adapt needle diameter and tip structure so that it can be inserted into the region of the ovary/ovaries of many arthropods. Injection may be facilitated by inserting the needle into membranous regions between sclerotized segments, and preliminary dissections to determine the precise locations of the ovaries should allow more precise insertion of the DNA (Presnail and Hoy, 1992).

Before the development of P-element vectors, exogenous DNA was introduced into *Drosophila* embryos by soaking them in DNA solutions after dechorionation (Walker, 1989). However, the method was not much used because of low uptake (< 2%), variable phenotypes and the difficulty in establishing stably transformed lines of flies. Most experiments used total genomic DNA, and Walker (1989) speculated that soaking embryos in specific cloned gene sequences could produce higher rates of stable transformation. Whether this method can be used for other arthropod species remains to be determined.

The 'gene gun' has been used successfully to transform major crop plants, yeast and cultured cells. Its use with arthropod eggs is limited, but Balderelli and Lengyel (1990) obtained transient expression of DNA in *Drosophila* embryos. The authors suggest this method may, with some modification, be suitable for stable germ-line transformation. Whether other arthropod eggs can be dechorionated and transformed has not been determined. This technique may be particularly useful for species that deposit large numbers of eggs. It would not be advantageous for species such as parasitic Hymenoptera, which deposit their eggs into the body of their insect host, because obtaining large numbers of eggs by dissection would be extremely tedious. However, if an artificial ovipositional medium were available for the parasitic wasps, it might be possible to obtain reasonable numbers of eggs at the appropriate developmental stage.

Other possibilities for transformation include soaking embryos in a DNA solution; *Drosophila* can take up and transiently express this DNA (Kamdar *et al.*, 1992). In yeast, artificial chromosomes have been constructed which behave much like the natural ones do, and it may be possible to develop artificial chromosomes for arthropods. Recently, transgenic mice were obtained by injecting a yeast artificial chromosome (YAC) into fertilized mouse oocytes (Schedl *et al.*, 1992). These mice carried the YAC DNA and expressed the YAC-encoded tyrosinase gene, so that the albino mice were pigmented. The YAC integrated into the mouse genome and the presence of yeast telomeric sequences apparently did not reduce the efficiency of integration. Artificial chromosomes may be particularly useful for situations where it is desirable to insert a number of genes that are linked.

Zalokar (1981) reported methods for injecting and transplanting nuclei and pole cells into eggs of *Drosophila*. Thus, it might be possible to genetically transform insect cells in cell culture, isolate the nuclei and transplant them into the region where the germ-line cells (pole cells) will develop in embryos.

It is relatively easy to genetically engineer microorganisms. Because many arthropods, particularly blood-feeding species, contain symbiotic bacteria that supply nutrients that are essential for their hosts, it is possible to use bacterial symbionts as vehicles for expressing foreign genes in arthropods. Ideally, the exogenous genes in the symbionts would enhance the fitness of their beneficial arthropod hosts. Beard *et al.* (1992) demonstrated that genetic engineering of insect symbionts is feasible by transforming a bacterial symbiont, *Rhodococcus rhodniia*, of the Chagas' disease vector *Rhodnius prolixus*. The symbiont was genetically engineered to be resistant to an antibiotic and the resistant symbionts were transmitted to insects lacking any symbionts. Beard *et al.* (1992) speculate that the symbiont could be used to express a gene product within the insect gut that would negatively influence attachment or development of the pathogen.

Cultured insect cells can be induced to take up exogenous DNA by electroporation, liposomes, laser micropuncture and several types of microinjection, and the transformed cells can be used to evaluate the expression of genes and promoters (Walker, 1989; Fallon, 1991). However, at present, adapting these methods appears to be an empirical process for each species.

Baculoviruses have double-stranded, circular DNA genomes contained within a rod-shaped protein coat. Baculoviruses infect insects, and several have been used as biological pesticides. *Autographa californica*, *Bombyx mori* and *Lymantria dispar* nuclear polyhedrosis viruses have been exploited as vectors to carry exogenous DNA into insect cells (L.K. Miller, 1988; Iatrou and Meidinger, 1990; Yu *et al.*, 1992). Because insect cells or larvae die from their infection, baculoviruses are not suitable for producing stably transformed insects, but, if non-lethal baculoviruses are developed, they could be used as vectors for stable transformation.

A gene coding for a yeast recombinase, FLP, has been found on a plasmid isolated from the yeast *Saccharomyces cerevisiae*. This plasmid also carries two inverted recombination target sites (FRT) that are specifically recognized by the FLP recombinase. FLP recombinase will catalyse recombination of the DNA between the FRT sites in the plasmid, inverting the sequences between them. FLP will catalyse both intramolecular and intermolecular recombination, and the possibility thus exists that FLP-mediated recombination could be used to insert foreign DNA into a specific site in an arthropod chromosome after it had been engineered to have the FRT sites. This would allow insertion of foreign DNA into the same site in a strain each time and could eliminate some of the position-effect influences on gene expression associated with random insertion of exogenous DNA. Morris *et al.* (1991) showed that FLP-mediated, site-specific intermolecular recombination occurred in microinjected embryos of the mosquito *Aedes aegypti*, although the results did not allow them to determine whether the mosquitoes were stably transformed. Establishing whether this technique can be used for developing stable transgenic arthropod strains requires additional work. This system could provide a rapid method of inserting different DNA sequences into a specific chromosomal site (where the FRT site is). However, because a stable FRT site in the genome is necessary and is integrated into the arthropod genome through non-specific recombination, different lines will have to be evaluated to determine which is the best for allowing expression of the foreign genes. It may be that the FLP system would be best suited for those species undergoing intensive and long-term genetic analysis and manipulation.

One of the drawbacks of the P element is that insertion of genes is uncontrolled and it is difficult to modify existing genes in specific ways. Recent efforts to insert genes in specific target sites using the P element have been successful with the X-linked *white* locus in *Drosophila* (Sentry and Kaiser, 1992). Other transposable elements may be used as potential tools for targeted gene insertion in beneficial insects in the future.

What Genes are Available?

Theoretically genes can be isolated from either closely or distantly related organisms for insertion into other arthropod species. It may also be possible to isolate a gene from the species being manipulated, alter it and reinsert it into the germ line. Assuming that a transformation method is available so that either transient or stable transformation can be achieved, the major issue then becomes whether the exogenous gene is expressed appropriately and effectively. Expression requires an appropriate promoter and other regulatory elements.

Many genes have been cloned and inserted into *Drosophila* by P-element-mediated transformation, but few of the cloned genes are of value for genetic manipulation of beneficial arthropods, because most projects were directed at understanding gene regulation or developmental processes. Some genes are useful for identifying transformants, including microbial genes, such as neomycin or G418 resistance, chloramphenicol acetyltransferase (CAT) and β-galactosidase. Relatively few genes cloned from *Drosophila* could be used directly for transforming beneficial arthropods, but they could serve as probes for homologous sequences in other arthropod species. Cloned genes could also be modified by *in vitro* mutation to achieve a desired phenotype.

For the near future, resistance genes will probably be the most available and useful for transforming arthropods (Table 9.4). Potentially useful resistance genes that have been cloned include: (i) a parathion hydrolase gene (*opd*) from *Pseudomonas diminuta* and *Flavobacterium*; (ii) a cyclodiene resistance gene (γ-aminobutyric acid A (GABA$_A$)) from *Drosophila*; (iii) β-tubulin genes isolated from *Neurospora crassa* and *Septoria nodorum* that confer resistance to benomyl; (iv) an acetylcholinesterase gene (*Ace*) from *D. melanogaster* and the mosquito *Anopheles stephensi*; (v) a glutathione-S-transferase gene (GST1) from *Musca domestica*; (vi) a cytochrome P450-B1 gene (*CYP6A2*) associated with dichlorodiphenyltrichloroethane (DDT) resistance in *Drosophila*; and (vii) the amplification core and esterase B1 gene isolated from *Culex* mosquitoes that are responsible for organophosphorus insecticide resistance (Table 9.4). Metallothionein genes have been cloned from *Drosophila* and other organisms that appear to function in homoeostasis of copper and cadmium and in their detoxification (Theodore *et al.*, 1991). Perhaps these genes could provide resistance to fungicides containing copper in arthropod natural enemies. In many crops, including Florida citrus, fungicides may have serious negative impacts on beneficial arthropods such as phytoseiid predators.

Multidrug resistance genes, *mdr* or *pgp*, in mammals become amplified and overexpressed in multidrug-resistant cell lines, resulting in cross-resistances to a broad spectrum of compounds, including those used in cancer chemotherapy. The multidrug resistance genes code for a family of membrane glycoproteins that appear to function as an energy-dependent transport pump. Two members of this multigene family were isolated from *D. melanogaster* and these genes (*Mdr49* and *Mdr65*) could provide resistances to a number of exogenous chemicals (Wu *et al.*, 1991). For example, *D. melanogaster* strains that were made deficient for *Mdr49* were viable and fertile, but had an increased sensitivity to colchicine during development. Whether the insertion of multidrug resistance genes would provide a useful increase in tolerance to chemicals that arthropods might encounter in the environment remains to be determined.

Preliminary results suggest that microbial genes conferring resistance to pesticides can function in arthropods. The *opd* gene, isolated from *Pseudomonas* and conferring resistance to organophosphorus pesticides, has been inserted, using a baculovirus expression vector, into cultured fall armyworm (*Spodoptera frugiperda*) cells and larvae (Dumas *et al.*, 1990). Phillips *et al.* (1990) also transferred the *opd* gene into *D. melanogaster*. The *opd* gene was put under the control of the *Drosophila* heat-shock promoter, *hsp70*, and stable active enzyme was produced and accumulated with repeated induction. It is likely that this gene could be used to confer resistance to organophosphorus pesticides in beneficial arthropod species, as well as serving as a selectable marker for detecting transformation of pest species. Increased freeze resistance in frost-susceptible hosts may be made possible by gene transfer. Antifreeze protein genes cloned from the wolf-fish (*Anarhichas lupus*) have been expressed in transgenic *Drosophila* (Rancourt *et al.*, 1990, 1992), using the *hsp70* promoter and yolk polypeptide promoters of *Drosophila*. Although additional work is required, the results suggest that subtropical or tropical species of arthropod natural enemies could become useful in or adapted to a much broader range of climates.

Altering longevity of certain arthropods might be beneficial, and research on mechanisms of ageing may provide useful genes in the future. A cloned catalase gene inserted into *D. melanogaster* by P-element-mediated transformation provided resistance to hydrogen peroxide, although it did not prolong the lifespan of flies (Orr and Sohal, 1992).

As basic research progresses, other traits that might be important or useful to introduce into beneficial insects will become obvious. Shortening developmental rate, enhancing progeny production, altering sex ratio, extending temperature and relative humidity tolerances and altering host or habitat preferences could enhance biological control (Hoy, 1976). However, it is not simple to document that changes in one or more of these attributes would actually improve the performance of a biological control agent.

Importance of Appropriate Regulatory Signals

Whether a coding region is transcribed and translated in a specific tissue is determined by a number of regulatory sequences in the DNA, including promoters and enhancers. The stability of messenger RNA is influenced by polyadenylation (poly(A)) signals at the 3' end of the RNA, which can influence the amount of protein produced. It is crucial to obtain expression of the inserted gene at appropriate times, levels and tissues. Another factor that may be important in maintaining the inserted DNA in the transgenic

Table 9.4. Some cloned resistance genes possibly useful for genetic manipulation of beneficial arthropods.

Gene (abbreviation) (resistance)	Source(s)	Reference(s)
Acetylcholinesterase (*Ace*) (pesticide resistances)	*D. melanogaster* *Anopheles stephensi*	Hall and Spierer, 1986 Fournier *et al.*, 1989 Hall and Malcolm, 1991 Hoffmann *et al.*, 1992
ß-Tubulin (benomyl resistance)	*Neurospora crassa* *Septoria nodorum*	Orbach *et al.*, 1986 Cooley *et al.*, 1991
Catalase (H_2O_2 resistance)	*D. melanogaster*	Orr and Sohal, 1992
γ-Aminobutyric acid A ($GABA_A$) (dieldrin resistance)	*D. melanogaster*	Ffrench-Constant *et al.*, 1991, 1993
Cytochrome P450-B1 (DDT resistance)	*D. melanogaster*	Waters *et al.*, 1992
Esterase ß1 amplification core (organophosphate resistance)	*Culex* species	Mouches *et al.*, 1986, 1990
Glutathione-S-transferase (DmGST 1-1) (DDT resistance)	*D. melanogaster*	Toung *et al.*, 1990
Glutathione-S-transferase (MdGST1) (organophosphate resistance)	*Musca domestica*	Wang *et al.*, 1991 Fournier *et al.*, 1992
Metallothionein genes (*Mtn*) (copper resistance)	*D. melanogaster*	Theodore *et al.*, 1991
Multidrug resistance (*Mdr49* and *Mdr65*) (colchicine resistance)	*D. melanogaster*	Wu *et al.*, 1991
Mercury resistance	*Streptomyces lividans*	Sedlmeier and Altenbuchner, 1992

Table 9.4. *continued*

Gene (abbreviation) (resistance)	Source(s)	Reference(s)
Neomycin phosphotransferase (*neo*) (resistance to kanamycin, neomycin, G418)	Transposon Tn5	Beck *et al.*, 1982
Parathion hydrolase (*opd*) (parathion, paraoxon resistance)	*Pseudomonas diminuta* *Flavobacterium* sp.	Serdar *et al.*, 1989 Dumas *et al.*, 1990 Phillips *et al.*, 1990 Mulbry and Karns, 1989

H_2O_2, hydrogen peroxide; DDT, dichlorodiphenyltrichloroethane.

line over time is the presence of origins of replication (DePamphilis, 1993). If exogenous DNA is inserted into a region of the chromosome far from a site where an origin of replication occurs naturally, the exogenous DNA could be lost over time because it is not replicated (Benbow *et al.*, 1992).

Because regulatory sequences may vary from species to species, the source of regulatory sequences chosen for cloning may be as important as or even more important than the source of the protein-coding sequences (Fig. 9.1). Furthermore, some regulatory sequences allow genes to be expressed only in particular tissues or in response to particular stimuli (such as heat shock), while other genes are expressed in most tissues most of the time. If it is important that the inserted gene function in a tissue- or stimulus-specific manner, it is essential to identify tissue- or stimulus-specific promoters. Currently, the number of suitable regulatory sequences available for genetic manipulation of arthropods is limited. The heat-shock (*hsp70*) promoter from *Drosophila* is commonly used as an inducible promoter. It is the strongest promoter known in *Drosophila* and appears to function in all cells. Heat-shock proteins are present in all organisms subjected to high temperatures and, while the number of these proteins varies from organism to organism, all produce a 70 kDa protein. It is likely that the *Drosophila hsp70* promoter can be used whenever an inducible promoter is required that will function in all cells. However, induction of the *hsp70* promoter may be different in different species. For example, the mosquito *Anopheles gambiae* was transformed with a plasmid containing the *hsp70* promoter of *Drosophila* attached to a microbial neomycin resistance gene, which also confers resistance to the antibiotic G418 (Miller *et al.*, 1987). Transgenic mosquitoes expressed the *neo* gene at a low level in adults at 26°C and a heat shock for 15 min at 37°C enhanced the level of expression. Recently, Sakai and Miller (1992) found that survival of transgenic larvae exposed to G418 was increased after heat shock at 41°C, which is higher than the temperature (37°C) typically used to induce genes in *Drosophila*. McInnis *et al.* (1990) found that three heat shocks produced higher survival rates in Mediterranean fruit flies (*Ceratitis capitata*) transiently transformed with *neo* and treated with geneticin.

Other commonly used regulatory sequences from *Drosophila* are the actin 5C promoter, the α1-tubulin promoter and the metallothionein (*Mtn*) promoter. Angelichio *et al.* (1991) compared the ability of four promoters in cultured *D. melanogaster* cells and found that the actin 5C and the metallothionein promoters generated comparable levels of RNA and protein. The α1-tubulin promoter generated about fourfold lower levels and the fibroin promoter had no detectable activity in these cells. The effects of three poly(A) signals were also evaluated to determine their impact on stability of the transcribed messenger RNA. Angelichio *et al.* (1991) compared the poly(A) signals of the SV40 early region, the SV40 late region and the *Drosophila* metallothionein gene. The SV40 late poly(A) constructs yielded protein levels that were three- to fivefold higher than the SV40 early construct. The metallothionein poly(A) and SV40 early constructs produced nearly equivalent levels.

Chromosome replication requires that origins of replication be located at intervals along each chromosome. Until recently, origins of replication on arthropod chromosomes were not available, but those involved in amplification of chorion genes (ACE3) in *D. melanogaster* have been identified and cloned (Carminati *et al.*, 1992). During genetic manipulation of pest or beneficial species, it may be useful to insert ACE3 or similar elements along with the exogenous genes to ensure that replication of this region of the chromosome occurs, in order to increase the stability of the introduced DNA in the transgenic strain.

Identification, cloning or genetic modification of promoters and other regulatory sequences may increase the precision with which desired proteins are transcribed and expressed in transgenic arthropods. Research to understand the structure and function of regulatory sequences for use in transgenic arthropods should have high priority. Project goals will dictate what type of regulatory sequences are most useful. In some cases, a low-level constitutive production of transgenic proteins will be useful, whereas in other cases high levels of protein production will be required after inducement by a specific cue. Researchers will have to evaluate the trade-offs between high levels of protein production and the subsequent impact on relative fitness of the transgenic arthropod strain, based on the specific goals of each programme.

Identifying Stably Transformed Arthropods

After inserting the desired genes, the next issue is how to detect whether the exogenous gene has in fact been incorporated into the germ line. Because transformation methods are, at least so far, relatively inefficient, a screening method is needed to identify transformed individuals. This process is relatively simple in *Drosophila*, where there is a wealth of genetic information, including visible markers that can identify transgenic individuals. Most pest or beneficial arthropods lack such extensive genetic information.

Identifying transformed individuals could be achieved by using a pesticide resistance gene, such as the *opd* gene as the selectable marker. Another option is to use the neomycin (*neo*) antibiotic resistance gene, which functions in both *Drosophila* and mosquitoes and is less likely to provoke concern about risks of releasing transgenic arthropods into the environment. Another marker is the β-galactosidase gene (*lacZ*) isolated from *E. coli* and regulated by the *Drosophila hsp70* promoter, which has been expressed in both *Drosophila* and the phytoseiid predator *Metaseiulus occidentalis* (Presnail and Hoy, 1992). If an appropriate selectable marker is not available, identifying transformed lines can be accomplished with PCR and subsequent analysis by Southern blot hybridization or an immunological procedure.

Risks Associated with Releases of Transgenic Arthropods

Risk assessments will be somewhat different for pest and beneficial arthropods. Until recently, most practitioners of biological control asserted that biological control of arthropod pests or weeds by arthropod natural enemies was environmentally safe and risk-free if carried out by trained scientists. However, questions about the safety of classic biological control have been raised, particularly where environmentalists are concerned about the preservation of native flora and fauna (Howarth, 1991), and the era of accepting classic biological control as environmentally risk-free appears to have passed (Harris, 1985; Ehler, 1990; Hoy, 1992a). Protocols for evaluating the risks associated with releasing parasitoids and predators that have been manipulated with rDNA techniques do not currently exist but will probably include, as a minimum, the questions or principles outlined in Table 9.2 (Tiedje *et al.*, 1989; Hoy, 1990a, 1992a, 1995).

The suggestion by Raymond *et al.* (1991) that there has been a worldwide migration of *Culex pipiens* mosquitoes carrying naturally amplified organophosphorus resistance genes suggests that dispersal of arthropods can be rapid and extensive. Another risk issue involves the possibility that horizontal transfer of genes may occur between one arthropod species and another (Houck *et al.*, 1991). The P element appears to have invaded *D.*

melanogaster populations within the last 50 years, perhaps from a species in the *D. willistoni* group. Houck *et al.* (1991) showed that P elements may have been transferred between *Drosophila* species by the semiparasitic mite *Proctolaelaps regalis*. Horizontal transfer of P elements from *D. willistoni* to *D. melanogaster* must be a very rare event, requiring that two *Drosophila* females of different species lay their eggs in proximity so that a mite can feed sequentially on one and then on the other (in the correct order). The mite must carry the P element to the recipient egg, which must be in a very early stage of embryonic development, the recipient embryo must incorporate a complete copy of the P element into a chromosome before it is degraded by enzymes in the cytoplasm, the recipient embryo must survive the feeding by the mite and the adult that develops from the embryo must transmit the P element to its progeny. If each event is rare and the combined probability is multiplicative, then the probability that horizontal gene transfer between different arthropod species will occur must be vanishingly small.

Interspecific transfer of another transposable element (*mariner*) has been suggested as an explanation of their presence in the drosophilid genera *Drosophila* and *Zaprionus* (Maruyama and Hartl, 1991) and in the genome of the lepidopteran *Hyalophora cecropia* (Lidholm *et al.*, 1992). While the interspecific transfer of *mariner* is suspected only on the basis of DNA sequence similarities and no specific vector has been identified, the data are consistent with the hypothesis that transposable elements can move between different species and orders of arthropods (Kidwell, 1992; Robertson, 1993).

Evidence from sequencing DNA isolated from bacterial endosymbionts of mosquitoes, Coleoptera and *Drosophila* suggests that the symbionts may have been horizontally transferred between these species (O'Neill *et al.*, 1992). Bacterial endosymbionts of insects are involved in many examples of cytoplasmic incompatibility, in which certain crosses between symbiont-infected individuals lead to the death of embryos or distortion of the progeny sex ratio. An analysis of the 16S ribosomal RNA (rRNA) genes specific to prokaryotes from *Culex pipiens*, *Tribolium confusum*, *Hypera postica*, *Aedes albopictus*, two populations of *Drosophila simulans* and *Ephestia cautella* indicated that their symbionts are all closely related to each other (O'Neill *et al.*, 1992). O'Neill *et al.* (1992) speculated that cytoplasmic incompatibility is due to infection with a specialized bacterium that infects a wide range of different arthropod hosts and that the symbiont has been acquired more than once by different insects. Preliminary surveys with a DNA probe indicated that additional insects, including *Corcyra cephalonica*, *Sitotroga cerealella*, *Diabrotica virgifera*, *Attagenus unicolor*, *Rhagoletis pomonella*, *Rhagoletis mendax* and *Anastrepha suspensa*, carry the symbiont, although cytoplasmic incompatibility has not been demonstrated in these species.

If horizontal transmission of DNA (or microorganisms) between arthropods occurs, even if exceedingly rarely, there is no guarantee that genes inserted into any species are completely stable. Naturally occurring horizontal transmission of DNA between species may have provided some of the variability upon which evolution has acted, but the extent and nature of this naturally occurring gene transfer are just being determined. Thus, releases of transgenic arthropods will have to be evaluated on the basis of their probable benefits and potential risks.

Experience indicates that the probability that a new organism will become established is small (Williamson, 1992). Discussions of risk include questions about survival, reproduction and dispersal of transgenic species and their effects on other species. Questions are also asked about the inserted DNA, its stability and its possible effect on other species should the genetic material move (US Department of Agriculture, 1991; Table 9.2). Historical examples of biological invasions or classical biological control demonstrate the lack of predictability, the low level of successful establishment and the importance of scale, specificity and the speed of evolution (Ehler, 1990). Transgenic organisms could pose risks because they will be released in large numbers. Williamson (1992) speculated that the greater the genetic novelty, the greater the possibility of surprising results, and recommended using molecular markers to begin to understand dispersal and the interactions between species.

Conclusions

One factor hindering progress in the genetic manipulation of beneficial arthropods is the lack of a 'universal' transformation system. The availability of a transformation method that would provide a rapid and general system for introducing exogenous DNA into species for which little genetic information is available would revolutionize the genetic engineering of arthropods (Presnail and Hoy, 1992a). If it is necessary to identify and engineer a specific vector for each target species, then these techniques may be limited to a very few species of great economic importance because of the high cost and lengthy time needed to develop the vector system.

Currently, we lack an example that demonstrates that transgenic arthropod biological control agents can be effective in a pest-management programme or that they can control a pest population. For many years, genetic manipulation of arthropod natural enemies was considered to be impractical for pest-management programmes (Hoy, 1976), and this limited the resources devoted to this tactic. The demonstration that a laboratory-selected strain of predatory mite could provide cost-effective control of spider mites in an agricultural crop (Headley and Hoy, 1987) provided an impetus to this research tactic in biological control. It is important to demonstrate that a transgenic beneficial arthropod can be effective in regulating pest populations in the field and have no negative impacts on the environment. Until this has been achieved, adequate resources and funding will be difficult to obtain because it is considered to be high-risk research.

Because the potential risks of releasing transgenic arthropods into the environment have not been resolved, it may be proper to first consider releasing a relatively risk-free example. This might involve the release of a transgenic beneficial arthropod that is carrying either a non-coding segment of exogenous DNA or a gene such as β-galactosidase (Hoy, 1992a,b). A transgenic strain of the phytoseiid predator *M. occidentalis* carrying the *lacZ* construct would be relatively simple to evaluate. It is an obligatory predator, has a low dispersal rate and is unlikely to become a pest (Hoy, 1992a,b). Ideally, the transgenic *M. occidentalis* could be released into a site in Florida, where it is unlikely to become permanently established. Risk assessment of transgenic arthropods, as with transgenic crops and microorganisms, adds a significant cost in both time and resources to the project. It has taken years for companies to get to the point where transgenic crops could become commercially available, and it remains unclear how successful they will be in the market-place. Thus, it will be important to conduct benefit–cost analyses when transgenic arthropods are used in pest-management programmes.

Significant, exciting and unpredictable advances are being achieved in molecular biology and genetics. As a result of rapid advances in molecular genetic techniques and knowledge of basic developmental mechanisms, it is difficult to anticipate the opportunities for genetically manipulating beneficial arthropods over even the next few years. Despite these anticipated advances, additional research is required if we are to gain an understanding of the attributes other than resistance to pesticides that we might manipulate. Furthermore, getting a transgenic arthropod into the field will be an awesome challenge, requiring risk assessments, detailed knowledge of the population genetics of the target species (Caprio *et al.*, 1991) and coordinated efforts between molecular and population geneticists, ecologists, regulatory agencies and pest-management specialists.

Marker-assisted Plant Breeding

10

REBECCA J. NELSON

Introduction

The use of host-plant resistance (HPR) is an inexpensive and environmentally sound approach for the control of many crop pests. It has been used to effectively control many insect pests and diseases (hereafter collectively termed 'pests'), reducing farmers' reliance on chemical inputs for crop protection. In this chapter, the role of HPR in integrated pest management (IPM) is first considered, and the ways in which molecular marker technology (MMT) can contribute to the more effective use of resistance is discussed.

How does HPR fit into the IPM concept? The use of HPR is a powerful and attractive tactic that can hardly be overlooked in a crop-protection programme (Maxwell, 1991). There may, however, be some practical and philosophical issues to consider in the use of HPR. For pests for which there are adequate sources of available HPR, the primary practical problem encountered in exploiting resistance is that it may be unstable. That is, resistance often becomes ineffective within a few years of its deployment, presumably due to shifts in the structure of pest/pathogen populations. This may be due to the presence of compatible pest subpopulations at the time of varietal deployment or to the evolution of novel pest genotypes in response to the selection pressure exerted by the host genotype.

Although it is unlikely to pose a health hazard, the indiscriminate use of HPR may be likened in some ways to the abuse of pesticides. One of the fundamental principles of IPM is that each intervention should be employed as little as possible, with the aim of encouraging the development of a stable system. The advocates of IPM have stressed the importance of encouraging farmers to assess pest populations and to utilize control measures only when warranted by high levels of pest pressure. It has been argued that the use of HPR is essentially indiscriminate, since the farmer must choose the variety planted before the pest situation can be evaluated, and thenceforth selection pressure is exerted on the pest population whether or not there would have been economic losses (Gould, 1988a). For insect pests, a strong form of resistance may reduce the levels of the pest to such an extent that the pest's natural enemies decline to dangerously low levels.

While plant genotypes cannot be adjusted over the course of the cropping season, resistant genotypes do not have to be selected indiscriminately. There are a large number of different resistance genes available for many host–pest systems, and these genes vary in important ways. Defence mechanisms may be classified as conditioning avoidance, tolerance or resistance (Parlevleit, 1983). In this chapter, genes involved in plant defence will be referred to as 'resistance genes'. Resistance genes may differ in their specificity or in the pest/pathogen subpopulation(s) that they affect. They may also differ in the magnitude of their effects, as measured by the extent of disease/damage to the plant or as measured by an aspect of pathogen/pest biology. For instance, genes for resistance to insect pests may be described in terms of the effect that they have on insect survival, reproduction and behaviour. Host genes may also affect pest populations indirectly, e.g. through effects on parasitoids and predators. Thus, host genotypes that encourage natural enemies should be selected and resistance mechanisms that are toxic to them avoided (Maxwell, 1991).

The choice of resistance genes is not an 'either–or' matter: different genes can be combined in a single plant genotype, and multiple plant genotypes can be deployed in a single cropping system. These different 'gene deployment' strategies can produce crop genotypes with different properties in relation to the degree of resistance expressed and the effect of the resistance on the pest population. These properties in turn affect the stability of resistance and the impact of resistance on other components of the ecosystem.

For resistance to be utilized in the most 'enlightened' way possible (that is, in a manner that is compatible with the ideal approach to integrated crop management), it is necessary to have a sound knowledge of the properties of individual resistance genes and of the properties of different gene combinations, both within individual genotypes and in different mixtures of genotypes. In addition, it is necessary to have tools that make it possible to efficiently determine the resistance genotypes of crop varieties and of individual plants in a breeding programme.

Molecular marker technology is a set of tools that can greatly enhance the efficiency and effectiveness of the genetic analysis of resistance and the selection of combinations of resistance genes in a crop-improvement programme. This same technology can also make it possible to analyse the structure of pest populations, and thus to better understand the effects of resistance genes and genotypes on pest subpopulations. Both (complementary) types of application are reviewed in this chapter.

What is MMT?

Molecular marker technology (MMT) is a set of tools afforded by the techniques of molecular biology which allow differences in the genetic material of different organisms to be examined. Molecular markers (also known as deoxyribonucleic acid (DNA) markers) can be used in the same ways that other genetic markers, such as morphological and biochemical markers, are used. Because molecular markers detect differences in DNA sequence, however, they are abundant and applicable across virtually all organisms. Because much of an organism's DNA does not encode proteins, the vast majority of DNA polymorphisms are selectively neutral.

There are two principal classes of molecular markers currently in use for gene mapping and tagging, namely restriction fragment length polymorphisms (RFLPs) and random amplified polymorphic DNA markers (RAPDs). These methods differ in the way in which DNA sequence polymorphisms are detected. In RFLP analysis, DNA samples are digested with restriction enzymes. Fragments of different sizes are separated by gel electrophoresis and transferred to membranes. Specific DNA fragments are detected by hybridization with labelled probes. The probes are usually cloned DNA fragments that have been linked to radioisotopes or to other molecules that can be detected. Depending on the application, RFLP analysis can make use of single-copy probes (DNA sequences present in one location in the genome) or repetitive probes (DNA sequences present in multiple copies in the genome). In analysing the clonality and population structure of an asexually reproducing pathogen, repetitive DNA probes are convenient markers, because they allow multiple loci to be evaluated simultaneously (Hamer *et al.*, 1989; McDonald and Martinez, 1990; Kistler *et al.*, 1991; Goodwin *et al.*, 1992; Leach *et al.*, 1992).

Another class of DNA markers is generated through the use of the polymerase chain reaction (PCR) (Saiki *et al.*, 1988). Using PCR, particular fragments of DNA are amplified to large quantities, considerably facilitating detection. The amplified portions of DNA are defined by short, synthetic DNA molecules (primers) that match the ends of the DNA to be amplified. Specific primers (normally 16–20 bases long) based on the DNA sequence data for the target region can be used to amplify specific, defined DNA fragments. The PCR-based markers can be utilized in much the same way that RFLPs are used (e.g. Karl and Avise, 1993). In some cases RFLP markers are actually converted to PCR-based markers. This has been done for both single-copy RFLPs (Williams *et al.*, 1991) and repetitive elements used for DNA 'fingerprinting' (Palittapongarnpim *et al.*, 1993).

It was Williams *et al.* (1990) who developed the RAPD procedure, in which 10-base primers of arbitrary sequence are used to amplify discrete DNA fragments using genomic DNA as template. The basic principle is that, by virtue of the short primer sequence, there is a high probability that two priming sites will occur in the genome in inverted orientation and in close proximity. The intervening region is amplified, resulting in different sizes of DNA fragments, which can be resolved by gel electrophoresis. The amplified DNA fragments behave as simple Mendelian markers. The main advantage of RAPD analysis is that no prior knowledge of DNA sequences is necessary to design the primers. Hence, abundant genetic markers can be generated with almost any organisms of interest. A variation of the random primer–PCR technique called DNA amplified fingerprint (DAF), which apparently yields more scorable loci with higher resolution, was developed by Caetano-Anolles *et al.*, 1991.

The use of RAPD for genetic analysis has recently been reviewed by Tingey and Del Tufo (1993). It has been used extensively for tagging resistance genes in crop plants (e.g. Martin *et al.*, 1991; Michelmore *et al.*, 1991; Paran *et al.*, 1991), and also for DNA fingerprinting of microbes (e.g. Crowhurst *et al.*, 1991; Goodwin and Annis, 1991; Hadrys *et al.*, 1992; Mazurier *et al.*, 1992; Bernardo *et al.*, 1993; Tibayrenc *et al.*, 1993) and crops (Halward *et al.*, 1991). There is evidence for parity between RAPD and RFLP-based typing (Bernardo *et al.*, 1993; Tibayrenc *et al.*, 1993).

Molecular marker technology can be used for genetic linkage analysis and for analysis of genetic similarity among individuals. The process of locating genes of interest on a genetic map via linkage to molecular markers is referred to as gene mapping. Classical 'phenotypes' and molecular marker 'genotypes' are evaluated for the same individuals (or related progenies) in a segregating population and the data are analysed to determine if any of the markers cosegregate with the target phenotype. If a marker is sufficiently close to a gene of interest, the gene can be selected indirectly through the use of the marker. Marker data can also be used for numerical taxonomy, to allow the genetic relationships among individuals to be inferred.

Although the types of markers that are currently in routine use in research laboratories are leading to useful insights into crop and pest genetics, there is a need for improvement in the methodologies before these can become useful for analysing the vast numbers of samples needed for some types of application. For instance, for MMT to be useful in a crop-breeding programme, it may be necessary to analyse large numbers of plants with large numbers of markers. For some types of studies on pest populations, again, very large sample sizes may be required. Although current methodologies may be unsatisfactory in some regards, molecular marker techniques are evolving extremely rapidly, and procedures have been getting more efficient all the time. We can expect that the methodologies will increasingly 'meet the user', becoming less expensive and easier to use (Rafalski and Tingey, 1993). Some recent developments include the wide adoption of the RAPD technique and other PCR-based markers, the development of a procedure to detect polymorphisms between plants based on sampling of half-seeds without DNA extraction (Chunwongse *et al.*, 1993), non-electrophoretic methods for detecting polymorphisms, greater amounts of accessible polymorphism through the use of microsatellite-based probes (Love *et al.*, 1990; Serikawa *et al.*, 1992), availability of lyophilized enzymes and thermocyclers that can accommodate large numbers of samples.

Strategies for the Deployment of HPR, and the Role of DNA Markers

Molecular marker technology should serve, rather than drive, an effort for resistance breeding and deployment. That is, one should determine the desired type of resistance and deployment strategies that are most likely to provide adequate, stable resistance, and then utilize the appropriate tools to achieve these objectives, rather than utilizing resistance genes on the basis of convenience for a gene-tagging programme. Molecular marker technology can help in the utilization of HPR in two ways: first, in the research phase, in analysing pest population biology and the genetics of resistance; and, second, in the practical phase, in which the desired genotypes are being constructed and utilized.

In general, stability is one of the desirable features of HPR. What is durable resistance, and what factors affect the stability resistance? 'Durable resistance' has been defined as resistance that remains effective while a cultivar possessing it is widely cultivated for a relatively long period of time (Johnson, 1981). This definition does not state or imply anything about the degree of expression, the genetic control of resistance, its mechanism or race specificity; it is a property of resistance that can, by definition, be recognized only retrospectively (Johnson, 1981). The useful lifetime of resistance is influenced by characteristics of the host genotype, as well as environmental factors that affect the host–pest interaction, including characteristics of the pest population and crop management practices (Bonman and Mackill, 1988). To design and select crop genotypes that are likely to have durable resistance it is clearly important to understand how these factors affect durability.

The level of resistance expressed is among the host characteristics that affect the durability of resistance. Genes that condition complete resistance may be termed 'major genes', and genes that lead to a reduction in disease in the context of a compatible interaction may be termed 'minor genes'. It is often pointed out that strong resistance is associated with a strong selective advantage for pest types that can overcome the resistance. Selection pressure cannot, however, be gauged exclusively from the level of resistance. The number of genes conditioning the resistance, their quantitative effects and their specificities with respect to pest subpopulations are critically important. For instance, a cultivar expressing complete resistance to a pest is much more likely to 'break down' (become susceptible in the field, presumably due to a shift in the composition of the pest population from non-virulent to virulent) rapidly if resistance is conditioned by a single gene than if it is conditioned by multiple genes, with redundant effects. In the latter case, a pest strain with the ability to overcome a single gene will give that strain little or no selective advantage. For the rice blast system, as with others, resistance associated with single major genes has been generally found to be non-durable (Bonman and Mackill, 1988). Cultivars with durable resistance have been identified, however, and durability has been associated with minor genes.

How the genetic control of resistance contributes to durability has been the subject of some discussion in the literature. Van der Plank (1968) suggested that polygenic resistance should be more stable than monogenic resistance. Green and Campbell (1979) observed that wheat cultivars showing durable resistance to stem rust in Canada 'have more genes for resistance and different gene combinations' compared with the cultivars with short-lived resistance. Quantitative, polygenically inherited resistance ('partial resistance'; Parlevleit, 1988) is generally considered the most likely to be stable (Leonard and Mundt, 1984). Modelling studies have shown that a host genotype with multiple genes with minor effects would behave as largely race-non-specific and therefore be stable, even if the minor genes were race-specific (Parlevleit and Zadoks, 1977; Jenns and Leonard, 1985). These and similar observations have led to a strategy for management of resistance by combining ('pyramiding') multiple resistance genes into a single plant genotype (Nelson, 1978). One hypothesis holds that cultivars carrying multiple resistance genes may tend to show greater durability of resistance because mutation to multiple virulence in a pest population is unlikely ('the probabilities hypothesis'; Wheeler and Diachun, 1983; Schafer and Roelfs, 1985). Mundt (1990, 1991) reviewed the available data on this issue and concluded that, while there is inadequate information to allow a final judgement of the probabilities hypothesis, existing evidence suggested that the utility of gene pyramids may be due to the effectiveness of specific gene combinations rather than simply the number of genes. Certain combinations of resistance genes may be particularly difficult for pathogen populations to overcome because the corresponding combinations of virulences may reduce pathogen fitness (Parlevleit, 1981). These gene combinations would be particularly desirable to identify and exploit in a breeding programme.

What can be done with molecular markers? There are two general ways in which MMT is being employed for the enhancement of resistance-gene deployment for stable crop protection. The first is for phylogenetic analysis and population genetics of pest organisms. The second is for genetic analysis, gene mapping and gene tagging in the host plant. Molecular markers offer a range of genetic markers that can be used to detect differences among pathogen/pest variants, to measure diversity of populations, to determine the phylogenetic relationships among variants and to analyse the differences among populations in different regions or on different hosts.

For the genetic analysis of the host, the primary objective is to determine the number and nature of the genes involved in conditioning resistance. In many cases, cultivars with desirable forms of resistance have multiple resistance genes. Particularly when individual resistance genes have only small effects on disease, it is extremely difficult or impossible to identify the genes using the techniques of conventional genetics. Molecular marker technology makes it possible to individually identify all genes that show a significant effect on disease, and to determine their chromosomal location. In addition to allowing efficient gene mapping, a molecular marker located very close to a gene can act as a 'tag', which can be used for indirect selection of the gene in a breeding programme.

In addition to the identification of particular loci in a selection programme, resistance breeding may involve other steps for which MMT can be useful. Molecular markers are useful for characterizing diversity within and between accessions in germplasm collections (Andersen and Fairbanks, 1990). This may aid in the efficient selection of germplasm for screening. 'Linkage drag' can be minimized by the use of RFLPs in selecting against extraneous genetic material from resistance donors such as wild relatives of crop species (Young and Tanksley, 1989; Paterson *et al.*, 1991b). If bulk breeding methods are employed to accumulate resistance genes, MMT may be useful in monitoring changes in allele frequency during population improvement.

Wolfe (1983) distinguished between the inherent durability of a cultivar genotype (due to race non-specific resistance, gene combinations or other forms of durable resistance) and the systems durability of an overall deployment scheme for resistance. Various strategies for incorporating host diversity into the crop could yield a more durably resistant system. These include the deployment of individual cultivars to give heterogeneity in time and/or space, and the use of heterogeneous cultivars (varietal mixtures or multilines). Molecular marker technology may be useful for monitoring changes in pest gene/genotype frequencies with experimental or agricultural interventions such as different deployment strategies.

Analysis and manipulation of crop pests

An understanding of pest population structure is important for each of the several steps involved in the identification and utilization of HPR (Gould, 1983). Identification of sources of resistance is the first step in resistance breeding. If selected pest strains are used for screening potential resistance donors, the choice of strains used during the screening will determine which resistance genes will be identified. The use of diverse strains will improve the likelihood that useful sources of resistance are identified. If screening is conducted in the field, the structure of the pest population at the field site(s) will affect the selection process. Entries that are resistant at a particular field site may be susceptible to pest subpopulations not represented in that population. Conversely, entries may be susceptible to certain pest subpopulations and thus not be selected, but may actually carry valuable resistance that is effective against other pest subpopulations.

Once potential sources of resistance are identifed, they are then analysed and particular sources are selected for use in the breeding programme. During the evaluation phase, the resistance spectra of genotypes and genes relative to pathogen subpopulations should be determined, and this requires that the structure of the population be understood. When the breeding programme has been undertaken, the performance of breeding lines is evaluated in the field and promising lines are selected. As during the screening phase, it is important that the field resistance of the breeding materials are evaluated at sites where the pest populations best represent those found in farmers' fields. The finished varieties are eventually released to farmers and cultivated according to one of various possible deployment strategies. Again, the optimal deployment strategy utilized will depend on the population biology of the pest.

'Pest population structure' refers to the amount of variation in the population, the ways in which the variation is partitioned in time and space and the phylogenetic relationships among variants (Leung *et al.*, 1993a). Variation can be produced through the processes of mutation, recombination and migration. Existing variation is acted upon by selection and drift (McDonald *et al.*, 1989). Among the forces that shape pest population structure, selection is the one that can most obviously be manipulated through breeding, deployment and management of the crop. It is also important to understand the roles of the other forces, however. For instance, the role of recombination in generating diversity varies among pests, and this has profound implications for analysis and management of pest populations (e.g. Burdon and Roelfs, 1985). The extent of pest migration will affect the rate at which virulent forms of the pest will spread.

For practical purposes, the most important pest characteristic to assess is the ability of pest types to attack and damage particular crop genotypes.

Accordingly, the conventional way of characterizing pest population structure has involved the identification of pathogen 'races' and insect 'biotypes'. While virulence and fitness information is critical and cannot be replaced by other types of markers, genes affecting these traits may be subject to strong selection, and data on these genes are expensive to obtain on a per-locus basis. Used in combination with virulence data, neutral genetic markers such as those afforded by MMT are extremely useful tools for developing an understanding of pest population structure and dynamics. Concepts, methodologies and progress in analysing the population structures of phytopathogenic bacteria and fungi were recently reviewed by Leung *et al.* (1993a). Molecular marker technology has also been used to study the population structures of some insect pests (e.g. Ehtesham *et al.*, 1993; G. Roderick, University of Hawaii, personal communication; Huynh, 1993). Case-studies on two rice diseases are presented to illustrate the utility of marker information on pest populations.

A case-study: bacterial blight of rice

For the bacterial blight (BB) pathogen, *Xanthomonas oryzae* pv. *oryzae*, a series of repetitive DNA elements have been isolated from the genome of the pathogen and used for DNA fingerprinting of various collections of pathogen strains. First, a subset of 97 strains from the collection maintained at the International Rice Research Institute (IRRI) was analysed using a repetitive element selected from a library of genomic clones (Leach *et al.*, 1990, 1992). Based on the banding patterns of the different strains, the diversity of the pathogen population was assessed, and a dendrogram depicting the relationships among strains and pathotypes was produced.

The relationship between phylogeny and phenotype was further examined (Nelson *et al.*, 1994). First, the extent to which the inferred relationships among strains/pathotypes were dependent on the probe used was assessed. A set of five probes was used, including four transposable elements (Nelson *et al.*, 1994) and a gene involved in specifying avirulence on rice hosts carrying a particular resistance gene (Hopkins *et al.*, 1992), to analyse 150 strains of the pathogen. The diversities estimated for the different probes were similar across the entire collection analysed, but were dramatically different for some of the individual pathotypes. The phylogenies inferred based on the different probes were generally similar, though not identical. From the consensus phylogeny that emerged, it appeared that the relationship between phylogeny and virulence phenotype was not simple. Some lineages contained strains representing more than one pathotype (presumably due to divergent evolution), and some pathotypes appeared in more than one lineage (presumably due to convergent evolution). The cases of apparent convergent evolution were further examined by testing individuals of the same pathotype on near-isogenic lines (NILs) carrying resistance genes not present in the standard differential set used for pathotyping. This led to the identification of two (possibly three) new races of the pathogen, corresponding to the pathogen lineages.

Analysis of 1208 isolates of the BB pathogen systematically sampled from 11 sites along a 310-km transect in northern and central Luzon in the Philippines revealed a phylogenetic structure similar to that found for the previous analyses of smaller samples (E.Y. Ardales and R.J. Nelson, unpublished data). This suggests that, by combining the available results of DNA fingerprint and pathotypic analyses, we have developed a fairly clear picture of the existing variation of the pathogen in the field. Nine strains representing the known diversity of the pathogen are now in use for screening potential sources of resistance, and these strains have been used to characterize the resistance spectra of all the known resistance genes available in near-isogenic lines (NILs) (R.J. Nelson and T.W. Mew, unpublished data).

The more intensive sampling also provides additional information about the spatial distribution of the pathogen. Hierarchical analysis of genetic variation showed substructuring both within and between sites. For 50 of the 55 site-by-site comparisons, there was highly significant differentiation between sites. For six of the nine sites for which multiple fields were analysed, significant differentiation was observed between fields within a site (Nelson *et al.*, 1993). This differentiation suggests that the population structure of the BB pathogen is affected by a low level of gene flow and/or a high level of host selection. The former would be consistent with the expectation that novel pathogen variants able to attack a resistant variety would spread relatively slowly. Field experiments utilizing molecular markers are now under way at IRRI to analyse gene flow and selection and to compare the impact of various strategies for the deployment of BB resistance on pathogen population structure.

Studies of the population structure of *X. oryzae* pv. *oryzae* in other countries are currently under way through various international collaborations (e.g. Adhikari *et al.*, 1995; Zhang *et al.*, 1995). The results of these studies indicate substantial differences in the pathogen populations present in different regions, both in phylogenetic and pathogenic terms. This observation underscores the general need to analyse pest populations on a local basis for effective local use of HPR for pest management.

A case-study: rice blast

Populations of the blast fungus *Pyricularia grisea* (syn. *P. oryzae*; teleomorph *Magnaporthe grisea*) have also been analysed extensively by DNA fingerprinting. This pathogen is notorious for extreme variability and for rapidly overcoming resistant varieties (Ezuka, 1979; Ou, 1980, 1985; Chin, 1985). Although isozyme markers failed to detect substantial polymorphism among blast-fungus strains (Leung and Williams, 1986), RFLP analysis using repetitive DNA probes has revealed abundant polymorphism among pathogen strains (Hamer *et al.*, 1989; Hamer, 1991). In addition, RFLP maps of the fungus have been constructed using single-copy DNA markers and a repetitive DNA element (Skinner *et al.*, 1990; Valent and Chumley, 1991; Romao and Hamer, 1992).

The dispersed repetitive DNA element MGR586 isolated from the blast fungus (Hamer *et al.*, 1989) has been used in several studies of the pathogen's population structure. One long-standing question in this area has been the role of weed-infecting populations of the fungus on the rice-blast epidemiology (Teng *et al.*, 1991). Using MGR586 for RFLP analysis of an international collection of blast-fungus isolates taken from rice and weed hosts, Hamer *et al.* (1989) found that isolates infecting rice and weeds were clearly differentiated in terms of the number of hybridizing bands. Borromeo *et al.* (1993) analysed a collection of isolates from rice and weed hosts from the Philippines, using MGR586 and several other types of markers, and concluded that, while populations infecting rice and some common rice-field weeds share a common ancestry (and common mitochondrial genotype), there is little ongoing gene flow between the pathogen populations infecting rice and weed hosts. These observations suggest that weed-infecting populations of the fungus do not in general pose a hazard to the rice crop, and that they may in fact be considered for use as biological control agents for weeds.

Studies using DNA fingerprinting of rice-infecting strains of blast fungus have indicated that the pathogen populations consist of clonal lineages. The numbers of lineages and the numbers of pathotypes per lineage appear to vary considerably between geographical regions. When MGR586 was used for DNA fingerprint analysis of 18 isolates of *M. grisea* from the USA, a close association between the hybridization banding patterns and race groupings was observed (Levy *et al.*, 1991). These observations suggest that fungal pathotypes are relatively stable.

Much more complex pictures of blast-fungus population structures have emerged from studies in the Philippines and Colombia, however. Borromeo (1990) isolated the repetitive element PGR613 from the genome of *M. grisea* and used it to type 156 isolates from different locations in the Philippines. Lineages of the pathogen defined using this probe corresponded to those defined by MGR586 (Borromeo, 1990). Although some lineages showed characteristic reactions to some hosts, consistent differences between lineages were not observed on most hosts, and no clear association between lineage and pathotype was evident (Borromeo, 1990; B.A. Estrada, J.M. Bonman and R.S. Zeigler, IRRI, 1992, personal communication). Further studies on the relationship between lineage and pathotype in the Philippines, in which 234 isolates were tested on 21 hosts, have indicated again that lineages appeared to have consistent patterns of interaction with some hosts but not others. That is, some hosts were resistant to all the members of a pathogen lineage tested, while other hosts were resistant to some isolates but susceptible to others.

Analyses of blast-fungus populations in Colombia, conducted in collaboration between researchers at the Centro Internacional de Agricultura Tropical (CIAT) and Purdue University, also revealed multiple pathotypes for many lineages (Levy *et al.*, 1993). Analysis of the pathotypes found within each lineage (Correa and Zeigler, 1991, 1993) revealed that they were closely related, suggesting step-by-step changes in virulence within lineages. Further, over 90% of the pathotypes were found in only one lineage. A close relationship between lineage and host origin was detected (Levy *et al.*, 1993).

Thus, while the relationship between lineage and the pathotypes of individual isolates is quite complex in some locations, it is possible to construct a 'composite' pathotype of each lineage, such that only hosts that are resistant to all tested members of a lineage are considered resistant to that lineage. If appropriate cultivars are selected for assessment of the virulence spectra of pathogen lineages, this assessment simultaneously provides useful information on the resistance spectra of host germplasm. Based on the composite pathotypes, resistance genes conditioning resistance to the full spectrum of pathogen lineages can be selected for a gene 'pyramiding' programme.

For blast, evaluation of resistance is often conducted in field 'hot spots', sites conducive to disease development. Natural pathogen populations are used for screening, with susceptible host cultivars used as 'spreader rows' to enhance the production of inocula. The effectiveness of the screening process depends on the diversity of the pathogen populations in the field. As part of a larger collaborative effort to analyse blast-fungus populations at upland screening sites (Leung *et al.*, 1993b), DNA fingerprint analysis was used to measure the diversity of natural populations of the blast pathogen at two IRRI screening sites. Over 1800 pathogen isolates were collected from a set of diverse rice germplasm and analysed by RFLP, using the probe MGR586 (Chen *et al.*, 1995).

Several interesting points emerged from this study. First, among the 25 lineages detected at the two screening sites, some dominated the collection while other lineages were very rare. Some lineages present may not have been detected at all. Rice varieties and lines tested at the screening sites are probably not effectively exposed to the rare types. Second, although the two screening sites are only about 40 km apart, they share relatively few lineages. This differentiation could be due to differences in host selection operating at the two sites (different types of rice germplasm are screened at the two sites), to environmental differences and/or to the geographical separation.

Analysis of host–lineage associations gives a preliminary view of the virulence spectra of the lineages and of the resistance spectra of the host genotypes. Some hosts were susceptible to several lineages. Certain other cultivars appeared susceptible in the field, but were found to host isolates of only a single lineage. This suggests that these cultivars possess broad-spectrum resistance in spite of their susceptible appearance. After the lineage resistance of these cultivars is confirmed in greenhouse tests, these cultivars could be donors for a gene-pyramiding programme.

Because the composition of the pathogen population at a screening site may be strongly affected by any selection exerted by the cultivars used as spreader rows, it is important that these be carefully selected to generate a diverse inoculum. Marker analysis of host–lineage relationships provides a basis for selecting a combination of cultivars that will amplify a wide range of pathogen subpopulations. Several lineages were found on only a single host, and in some cases these lineages were uniquely capable of attacking that host. Inclusion of these cultivars in the spreader rows may ensure more effective screening efficiency. The initial diversities of the pathogen populations in the screening nurseries were calculated, using Nei's (1987) diversity index with RFLP typing data. Improved spreader-row compositions were selected based on the host–lineage relationships observed. The impact of this change will be evaluated.

Analysis and manipulation of resistance

As discussed above, gene pyramiding is one of the main strategies for improving inherent resistance. Gene combinations may include major and/or minor genes. Although many such gene combinations have been produced by conventional breeding methods, it is often difficult to select for multiple resistance genes because the action of one gene may obscure the action of another. The action of minor genes is difficult to measure at best, particularly in small field plots (Van der Plank, 1968; Parlevleit, 1979), and cannot be detected in the presence of major genes. It is difficult to identify minor genes individually or to estimate the number of minor genes present in a genotype by conventional means. Although the value of such genes in conferring stable resistance has long been recognized, this type of resistance has been underutilized because of inherent technical difficulties associated with minor effects. With the advent of MMT, quantitative traits can be dissected systematically, and individual loci contributing to them can be identified and mapped.

Breeding methods aimed at accumulating minor genes have been devised (e.g. Robinson, 1976), but these methods tend to underutilize major genes, and do not prevent the loss of minor genes during subsequent breeding procedures. Marker-aided selection offers great advantages for gene pyramiding. Other-wise confounding phenotypic interactions do not interfere with marker-based identification of resistance loci. Molecular marker technology offers particularly revolutionary potential for the exploitation of quantitative resistance, or resistance conditioned by multiple minor genes (Paterson *et al.*, 1991b; Bonman *et al.*, 1992).

How should genes be selected for a pyramiding programme? Because of the considerable time and expense involved in incorporating resistance genes, it is important that the most useful gene combinations be utilized. One approach to this is to characterize individual genes and to select combinations that provide the most broad-spectrum and overlapping resistance. A complementary approach is to characterize resistance in cultivars that are considered to have shown durable and non-durable resistance, and to try to infer something about the properties and particular gene combinations that contribute to durability. With the use of MMT, it is possible to analyse the genetics of resistance in greater detail than was previously possible. Linkage maps of DNA markers have been constructed for at least 23 crop plants (Paterson *et al.*, 1991b), including the major cereal crops.

A case-study: resistance to the rice blast fungus

Resistance to the blast fungus *Pyricularia grisea* has long posed a serious challenge in rice improvement. Resistance, particularly when based on single major genes, has generally been overcome in one or a few years in blast-prone environments. The average longevity of resistance in Japan is less than 3 years (Kiyosawa, 1982). Some cultivars, however, have shown more long-lasting resistance (Bonman and Mackill, 1988). In several cases, durable resistance to blast is believed to be associated with complex inheritance (Toriyama, 1975; Notteghem, 1985). Analysing five improved cultivars and four traditional cultivars for resistance, Yu *et al.* (1987) found at least seven genes conditioned resistance against the three isolates tested, with each cultivar having one or two genes effective against each isolate. Many rice improvement programmes aim to incorporate multiple resistance genes, including genes for partial resistance, into rice varieties (Ikehashi and Khush, 1979).

Systematic genetic studies of blast resistance led to the identification of 13 dominant resistance genes in Japan (Kiyosawa, 1981). Subsequent studies have revealed several more resistance genes (e.g. Yu *et al.*, 1987; Mackill and Bonman, 1992; Wang *et al.*, 1994), although the allelic relationships between all identified loci have not always been analysed in detail. At least one gene was identified as recessive (Yu *et al.*, 1987). The availability of an RFLP map of rice (McCouch *et al.*, 1988; Causse *et al.*, 1994; Kurata *et al.*, 1994) has accelerated the mapping of resistance genes. At least seven major genes for blast resistance have been mapped with respect to RFLPs (Yu *et al.*, 1991; L. Zhu *et al.*, Academia Sinica, Beijing, 1991, personal communication; J. Tohme, CIAT, Cali, Colombia, 1993, personal communication; Wang *et al.*, 1994), and a number of minor genes have also been mapped (Wang *et al.*, 1994).

At IRRI we are taking two interrelated approaches to understanding and manipulating blast resistance. First, we are analysing the genetics of resistance in cultivars that have shown durable resistance. In this approach, we make use of fixed recombinant populations and molecular markers in an effort to understand the resistant genotype and to dissect the whole into its constituent parts. The results of this effort should be useful in guiding our breeding strategy, and will also produce a set of specific 'gene tags' that can be utilized in constructing resistant varieties. Second, we are developing a set of NILs in which single blast resistance genes are isolated in a common genetic background. With this approach, we are able to study and compare the characteristics of individual genes. Once individual genes are characterized, gene combinations likely to contribute to durable resistance will be selected for pyramiding in the breeding programme. Further, the NILs provide a set of defined differential hosts for analysing pathogen virulence.

To better understand the genetic basis of durable blast resistance, a set of crosses was undertaken to produce recombinant inbred (RI) populations that could be used for RFLP mapping of major and minor genes in a group of upland rice cultivars considered to have shown long-lasting resistance (J.M. Bonman and D.J. Mackill, IRRI, 1989, personal communication). Analysis of the first of these crosses was recently completed (Wang *et al.*, 1994). A population of F7 RI lines was generated from a cross between the japonica cultivar Moroberekan, a traditional West African upland cultivar considered to have durable resistance (Bonman and Mackill, 1988), and the indica cultivar CO39. Each of the 281 RI lines were probed with 156 RFLP markers. The RI population was evaluated for qualitative resistance, using five pathogen isolates, and for quantitative resistance, using a single isolate in 'polycyclic' tests, in which multiple infection cycles were allowed to proceed. The associations between markers and phenotypic effects were analysed, using linear regression analysis and interval analysis (Paterson *et al.*, 1988; Lander and Botstein, 1989). Two genes conferring complete resistance to isolate PO6–6, designated *Pi-5*(t) and *Pi-7*(t) respectively, were identified and mapped on chromosomes 4 and 11. Preliminary evidence indicates that Moroberekan also carries other major genes effective against different isolates. Nine regions of the genome were identified by both interval analysis and linear regression analysis as having quantitative effects on resistance to isolate PO6–6. These were considered to be minor genes or putative quantitative trait loci (QTLs) for blast resistance. Thus, the resistance genotype Moroberekan is evidently quite complex, consisting of multiple major and minor genes. This conclusion is consistent with the view that durable resistance is associated with genetic complexity.

To be useful in a breeding programme, QTLs must behave in a predictable way and be robust in their effects. Unfortunately, QTLs identified in one genetic background and under one set of environmental conditions are not necessarily effective in other crosses or environments (Paterson *et al.*, 1991a). To investigate the potential utility of the QTLs identified for blast resistance, the RI population derived from Moroberekan/CO39 was tested at a field site in the Philippines for 2 years and at a site in Indonesia for 1 year. The results of the field testing indicated that the QTLs identified in greenhouse tests using a single isolate were generally effective in reducing levels of disease in the field at both locations.

In addition to analysing the genetic constitution of complex resistant genotypes, detailed analyses of individual resistance genes are being conducted using NILs. A set of NILs carrying different blast resistance genes was developed by several generations of backcrossing and selfing, with selection for the resistance phenotype to each of several isolates at each generation (Mackill and Bonman, 1992). Several other NILs have been added to this set by marker-aided selection from the RI population used for mapping resistance genes from Moroberekan. DNA marker data allows the level of isogenicity of the NILs to be evaluated. The lines may be further ‘cleaned up’ by employing marker-aided selection to reduce or eliminate extraneous regions of introgressed donor DNA detected by RFLPs.

An additional pair of NILs was developed for a blast resistance gene transferred from a wild species of rice. Wild relatives of rice are a rich, largely untapped, source of pest resistance genes (Heinrichs *et al.*, 1985; Sitch, 1990). Resistance to grassy stunt virus was transferred from *Oryza nivara* (Khush, 1977), and resistance to two insect pests was incorporated from *O. officinalis* (Jena and Khush, 1990). The grassy species *Oryza minuta* is resistant to a number of pests (Sitch *et al.*, 1989; Amante-Bordeos *et al.*, 1992; Reimers *et al.*, 1993). Resistance to blast and BB was transferred from this species to an elite breeding line of rice (Amante-Bordeos *et al.*, 1992). Blast resistance in the introgression line was shown to be conrolled by a single gene (Amante-Bordeos *et al.*, 1992), which is incompletely dominant (P.J. Reimers and R.J. Nelson, unpublished data). The line was highly resistant to all of the many blast pathogen isolates tested, as well as to field populations of the pathogen in test sites in Bangladesh, Colombia, Indonesia, the Philippines and Thailand (Amante Bordeos *et al.*, 1992; Reimers *et al.*, 1994; D. Chen and R. Nelson, unpublished data).

Subsequent molecular marker analysis of the introgression line and a backcross population derived from it showed that very little genetic material from the wild species had been transferred to the rice line. More than 100 RFLP markers distributed an average of 25 cM apart were tested, but failed to reveal any introgressions. Thus, the introgression line is nearly isogenic to the elite breeding line to which the resistance gene was transferred, and the two can be considered a pair of NILs.

To uncover the region carrying the resistance gene from *O. minuta*, 540 RAPD primers were used to analyse the genome (P.J. Reimers and R.J.

Nelson, unpublished data). On average, each primer amplified 13 bands. If an RAPD band is equivalent to a locus (Williams *et al.*, 1991), approximately 7020 loci were thereby tested. Of these, four loci linked to resistance were identified. These markers will be used to assess the relationship between this gene and other known resistance genes, and could be used to track the resistance gene in a breeding programme.

The availability of NILs carrying major and minor genes offers the opportunity to answer several long-standing questions concerning the behaviour of major and minor genes for disease resistance. For instance, does a major gene condition some resistance against compatible isolates (or does a 'defeated' major gene provide 'residual' resistance against isolates that have overcome the major gene), or are such isolates equally fit on lines with and without the defeated gene (Pedersen and Leath, 1988)? Are individual minor genes for resistance specific or non-specific in their effects? Three of the genomic regions identified as carrying minor genes affecting blast in the analysis of Moroberekan/CO39 corresponded to loci previously shown to be linked to major genes for blast resistance (Yu *et al.*, 1991; L. Zhu, Academia Sinica, Beijing, 1991, personal communication), raising the question of whether these genomic regions carried both major and minor genes, or if a single locus could have both major and minor effects on disease expression.

To answer these questions, quantitative inoculation experiments were conducted using NILs carrying four major genes and one minor gene (D. Chen, R.S. Zeigler and R.J. Nelson, unpublished data). The lines carrying major genes showed inconsistent reactions to the pathogen isolates tested. Although reduced lesion numbers were sometimes seen when the lines carrying the major genes were inoculated with compatible isolates, these lines sometimes showed similar or greater numbers of lesions relative to the recurrent parent. For the lines carrying the minor gene, the numbers of lesions were consistently reduced relative to the recurrent parent. Pending further experiments of this sort, we tentatively conclude that 'defeated' major genes do not show consistent residual resistance, and that a putative QTL conditions relatively 'horizontal' incomplete resistance.

Several genes involved in the non-specific plant defence response have been cloned and characterized (reviewed by Collinge and Slusarenko, 1987; Rigden and Coutts, 1988; Lamb *et al.*, 1989). We are now attempting to determine if any of these known defence genes correspond to the minor genes for resistance located on the molecular genetic map of rice chromosomes. If this effort is successful, it may be feasible to efficiently clone particularly effective alleles of defence-response genes from cultivars that have shown durable resistance, and to use these for genetic transformation of rice lines. This process could be much more rapid than transfer of minor genes by conventional or marker-aided breeding.

A case-study: resistance to the bacterial-blight pathogen of rice

Breeding for resistance to the BB pathogen *Xanthomonas oryzae* pv. *oryzae* has not been as difficult as for the blast pathogen. Although the breakdown of resistance has sometimes been spectacular (Ezuka and Sakaguchi, 1978), several examples of durable resistance are known. The Japanese variety Nongken 58 has shown resistance for more than 24 years when widely planted in China (Lee *et al.*, 1989). Cisadane is considered to have durable BB resistance in Indonesia. Cultivars carrying the *Xa*-4 gene for resistance have been grown for 20 years in the Philippines without the occurrence of widespread epidemics, although compatible races have become prevalent in the country (Mew *et al.*, 1992). The durability of field resistance of these cultivars has been attributed to quantitative resistance present in the cultivars (Lee *et al.*, 1989). Genetic and molecular genetic analyses support the suggestion that Cisadane and *Xa*-4-carrying varieties also carry multiple minor genes for resistance (Koch and Parlevleit, 1991; N. Huang, S.R. McCouch and R.J. Nelson, unpublished data).

Efforts to detect markers closely linked to genes for resistance to the BB pathogen have taken advantage of the availability of ten sets of NILs (Ikeda *et al.*, 1990; Ogawa *et al.*, 1990), each carrying a single gene for resistance. One of the genes, *Xa*-21, was introgressed from a wild relative of rice, *Oryza longistaminata*. Segregating populations have been used to confirm cosegregation between RFLP markers and *Xa*-1, *Xa*-2, *Xa*-3, *Xa*-4, *Xa*-5, *Xa*-10 and *Xa*-21. In the case of *Xa*-1, *Xa*-5 and *Xa*-21, 'gene tags' useful for selection in a breeding programme are now available. In these cases, markers are linked within 1 cM of the target gene and the probability of a crossover event separating the marker from the gene of interest is remote (McCouch *et al.*, 1991; Ronald and Tanksley, 1991; Yoshimura *et al.*, 1995).

Physiological and biochemical studies have been conducted, using some of the BB NILs, to investigate the mechanism of resistance conferred by these genes (Reimers and Leach, 1991; Reimers *et al.*, 1992). Each gene was analysed to determine the spectrum of resistance by inoculation, using a set of pathogen isolates (R.J. Nelson and T.W. Mew, unpublished data) considered to represent the diversity of the bacteria in the Philippines and in Japan. Based on these studies, gene combinations of particular interest for rice improvement have been tentatively identified.

To test the utility of markers as selection tools, and to study the behaviour of gene combinations in a particular genetic background, pairs of genes were combined by marker-assisted selection, and confirmed when possible by phenotypic selection. The markers provide immediate information about which plants carry a target gene and whether it exists in a homozygous or a heterozygous state. In other cases, markers allow accurate identification of genotypes in which two or more genes have been combined in one individual, but where the phenotype of the individual is indistinguishable from those carrying fewer genes. Based on this capability, we are testing the hypothesis that polygenic forms of resistance can be effectively constructed by combining genes that were originally identified as conferring qualitative or single-gene resistance. Unexpected interactions between genes that have been well characterized individually can also be evaluated using this approach. Molecular marker technology provides many new opportunities for manipulating genes in a plant breeding programme once those genes have been identified through the gene-tagging process.

In field experiments NILs carrrying individual resistance genes and pairs of resistance genes are being used to evaluate the effectiveness of various deployment strategies for the control of BB, and to determine the effect of these strategies on the structure of the pathogen population (C.C. Mundt, IRRI, 1993, personal communication).

Conclusion

Host-plant resistance is a valuable tool for the control of pests and pathogens. Because strong forms of resistance can be associated with ‘boom and bust’ pest cycles, HPR has been undervalued by some IPM practitioners. HPR need not, however, be extreme, unstable and at odds with an IPM strategy.

Once the type of desired resistance is determined and the existence of such resistance has been established, MMT offers powerful tools for the genetic analysis and genetic manipulation of resistance. The same methods that are used for genetic analysis and selection of HPR are also applicable for analysis of pest and pathogen populations. These tools have been widely applied for research purposes. The first applications for crop improvement are now under way. As the methodologies are rapidly becoming more efficient and user-friendly, they may soon contribute substantially to the integrated management of crop pests.

Modern Plant Breeding: An Overview 11

IVAN W. BUDDENHAGEN

Introduction

Plant breeding, biotechnology and integrated pest management (IPM) implementation and research require substantial funding. What drives the system is the *perception* of the cost/benefit ratio in relation to doing something. Please note I emphasize perception because I doubt if anyone really knows the reality of the cost/benefit ratio of any particular activity, and bandwagons often distort reality and greatly influence perception. Resources in biotechnology are very high; backing new IPM efforts has considerable momentum; and the old activity of plant breeding is struggling to survive.

Good field research on plant diseases, entomology and even agronomy is less common today than before the interest in IPM and biotechnology became so intense. In this chapter, the interrelationships and the interdependence of biotechnology, IPM and plant breeding are discussed.

Plant Breeding and Breeding for Resistance

What is the present reality? First, most plant life has never been 'bred', except by natural evolutionary forces. Look at our forests, grasslands, weeds, shrubs and all our natural plants and crops. My thesis is that most crops have hardly been 'bred' – and, if they have been bred a little bit, they have hardly been touched on 'resistance'. Think of all the crops where quality has essentially precluded any work on resistance – temperate and tropical fruits and nuts and many vegetables and ornamentals. Think of grasses and pasture plants, such as clovers. A little resistance improvement over natural levels may have accumulated by default, while selecting for yield.

So, in spite of all the publications on crop breeding and genetics, most of the effort and investment on breeding has been on a few field crops, usually those where returns are so low that the simple expedient of spraying pesticides has been considered too costly. So will investing in biotechnology really do much for resistance when existing economics and other factors have failed to provide good breeding programmes that include a good resistance focus for most crop and forest species? The answer is: probably not, but with some specific exceptions.

Natural Breeding Systems and Propagation Methods in Relation to IPM

How do breeding systems and propagation methods affect resistance levels and ease of improving resistance?

Different crops have different breeding systems (Simmonds, 1991) and these greatly affect how easy or how difficult it is to breed them for resistance. There are inbreeders, outbreeders and clonal or partly clonal species. Humans can easily make hybrids economically out of outbreeding crop. For inbreeding species, it is much more difficult, but, if F_1 hybrids can be cloned, their advantages are easily exploited. If the natural seed-cloning present in some grasses (apomixis) could be more broadly extended into major crops it would be a major boon, and a good place for much more biotechnology investment.

The key point is that the level of resistance of a crop variety establishes the degree of importance of its pests and pathogens. Damage levels will fluctuate with weather fluctuations and to some extent with agronomic manipulations, but the genetic level of resistance is fundamental. The less resistance, the more damage; the more damage, the more attempts at control with pesticides; the more pesticides, the more imbalance; the more imbalance, the more pressure to look at alternatives to pesticides to reestablish balance – i.e. IPM approaches.

The IPM approach has devolved into IPM projects, which their proponents advocate as leading to empowerment of farmers to be observant in their fields, to recognize and preserve natural enemies, to allow some damage instead of spraying and thus lessen pesticide use, which will enable more balance. More balance is good and thus IPM may be considered to be good to the extent that it improves balance. At present, IPM is implemented only for a few major crops and for major insect pests. If the existing varietal susceptibility remains constant (i.e. no change in variety and no pressure or funds added to develop a more resistant, better locally adapted variety are made), then this 'farmer empowerment' type of IPM will continue, as will support for such IPM projects. Integrated pest management is needed because of poor plant breeding. The need for advocacy, funds and improved strategies of breeding for improved disease or pest resistance lies outside this concept of IPM.

Thus, there is a fundamental problem with IPM and what it has come to stand for. As the proponents say, it is not a technology, it is a process; and then this process is variously described to fit the case or the audience. When pressed, the proponents of IPM will say it includes varietal resistance and all other factors that can be useful to reduce pests. When pressed, they say yes, it includes diseases and weeds. But IPM as practised really bypasses the questions of whether we can develop better host–plant resistance and what it will take to do it. In each local situation we may also ask: How good and thorough is the research in entomology and plant pathology on the ecology and biology of the system? How thoroughly has anyone done a local agroecosystem and pathosystem analysis? In my experience in many parts of the world, not nearly sufficient local work has been done. Researchers today do little field work. Yet pests, pathogens and weeds, which, by their very existence, generate funds for basic laboratory research, do so by their existence as field and postharvest problems. Yet their ecology is complex and little studied or understood within the local field situation. It should be obvious that neither IPM as currently advocated nor biotechnology will redress this fundamental problem.

Why is There This Fundamental Problem?

There are historical roots to the problem, one very old and one created recently by the green revolution. Long before pesticides were discovered, crop diseases and pests were handled by sensible agronomic/horticultural practices. These practices were developed locally by farmers and horticulturists and differed in different places. As science came to be applied to agriculture in the 1800s, divergence occurred, with the differing trends in education in different regions. In the USA, with the growth of land-grant colleges, there was a practical trend connecting knowledge, agriculture and farmers. Depredations still occurred and, with the expansion of intercontinental commerce and plant introductions, new-encounter and re-encounter epidemics, such as mildew on grape and blight on potatoes, initiated the pesticide age for plant diseases. Good horticultural practices (rotations, sanitation, etc.) were not forgotten for disease control. For at least a century, pathologists did not recommend excessive pesticide use, and plant breeding for resistance in field crops with coevolved pathogens grew in importance. So farmers had IP(athogen)M for a long time. Organic fungicides specific for a single metabolic-inhibiting step (after the Second World

War), along with single vertical gene resistance, revealed the power of pathogen evolution to pathologists, probably earlier than to entomologists, who relied too heavily (after dichlorodiphenyltrichloroethane (DDT)) on the idea of insecticides alone to solve insect problems.

Insect resistance breeding was slower to develop and pesticide misuse increased, so crash IP(est)M had to start. Integrated pest management started in the USA with cotton, which had received inordinate insecticide applications, generating insecticide resistance, loss of control and negative environmental effects. Gradually people in crop agriculture relearned principles of evolution and they were stimulated at first to restudy insect-pest biology.

A different historical trend was developing in the Third World, where the long-known fertilizer/agronomic/crop yield principles were belatedly applied to crop agriculture in the tropics in the 1960s. Unfortunately, they were applied through a crop breeding focus, where great emphasis was put on the production of new varieties of rice and wheat that would cover millions of hectares. International institutes were established, with world mandates, thus lacking focus and with 'broad adaptation' as the guiding light in varietal development. This varietal broad-adaptation concept, applied globally, with centralized control on funding, strategy and research itself, essentially removed from the agenda the focus on studying or understanding the local environment, the local pests, pathogens and weeds and their interaction with old or new crops, varieties or agronomic cropping practices. It also removed the idea of breeding new varieties appropriate to local conditions. Thus, the two historical forces converged in the Third World, generating pesticide misuse, pest resistance to insecticides and inadequately bred varieties. Hence the need for IPM for the Third World, where its main emphasis and funding now is.

IPM in Perspective

A little more discussion is needed to place IPM in perspective in relation to both plant breeding and biotechnology. Investments made to intensify or raise agricultural production will continue. This will increase environmental disturbance, which will in turn increase opportunities for pests, pathogens and weeds. Thus, increasing efforts on production necessitates increased and rational agroecosystem management, of which a major part will be 'pest management'. The idea is sound that, for a balanced, economic and least environmentally negative management of pests, an integration of all possible techniques and approaches is required. It is important, however, that IPM be understood as an integral part of crop and agroecosystem management. There is a need to invest in agroecosystem management – which includes developing locally adapted varieties, durably resistant to local pests and pathogens. The historical root of IPM as an approach to reduce the excessive use of insecticides has restricted the understanding and acceptance of IPM as a holistic and analytical part of agroecosystem management, even by its main proponents, entomologists.

A holistic approach to IPM must be concerned with the three major and different biological inhibitors of crop production: pests, pathogens and weeds.

Box 11.1: Pesticide abuse: insects on bananas in Central America.

As commercial banana production in Central America became more concerned with high-quality fruit and high productivity in the 1960s, complete control of insect pests became a field objective. Fruit blemishes due to insects were to be stopped and any insect leaf-feeding that might reduce yields was to be prevented. Broad-spectrum insecticides were used routinely whenever a field superintendent felt he/she might lose some fruit. At first, insects were controlled, but soon, as the natural parasites and predators of these indigenous, originally non-banana insects were eliminated, it took more sprays to be effective and insecticide use as well as insect damage soared. Many insect and mite species that had been unknown as banana pests began to cause serious widespread damage. Insecticide poisoning occurred in the environment.

In 1974, management in a banana company in Panama finally agreed with recommendations of entomologists, and halted insecticide sprays. Very quickly, insect and mite damage declined as predators and parasites resurged. Insecticide use is now minimal in banana plantations throughout Central America. Insecticides are used mainly as insecticide-impregnated plastic bags to cover the fruit, and spot treatments of larvae using microbial insecticides.

The basic cause of the problem was not a changing of varieties, but a tightening of quality requirements, together with inadequate appreciation of ecological and evolutionary principles. The use of non-selective insecticides in a crop that can tolerate moderate levels of foliar damage without suffering losses in quality or quantity of fruit was a misguided approach from the outset.

The whole ecosystem and human cultural system are part of this relationship, but the focus must be on management, which affects crop relationships with pests, pathogens (disease) and weeds. The holistic approach must be concerned with the basic reason for the importance of the first two: the genetic make-up of the crop variety itself in relation to existing and potential levels of resistance/susceptibility to local pests and pathogens. Thus, any investment to intensify crop production must start with agroecosystem analysis, pathosystem analysis and analysis of crop varietal resistance and the potential for improving varietal resistance.

Although descriptions of IPM often mention varietal resistance (varietal improvement) as part of IPM, it should be recognized that in hardly any IPM programmes is this fundamental component of potential pest and pathogen depredation funded or an integrated part. Existing and proposed IPM programmes emphasize farmer intervention in existing fields and with existing varieties, with little attention paid to the role of new varieties.

Two examples of what I call 'holistic IPM' emphasize that a holistic view reveals two basic and quite different types of potential interventions to mitigate pest/pathogen depredations. These are: (i) pesticide abuse (Box 11.1): application of techniques to reduce already excessive use of pesticides (mainly insecticides); and (ii) anticipatory breeding (Box 11.2): application of strategies and tactics to avoid the need for excessive use of pesticides and still maintain biologically induced losses at a low level.

Box 11.2: Anticipatory breeding: maize streak virus disease in Africa.

Maize streak virus, transmitted by a leafhopper, causes periodic epidemics on maize in tropical Africa. The diverse maize germplasm from tropical America is uniformly susceptible to this indigenous African virus and its vectors. Disease occurrence was so erratic that neither farmer selection nor maize-breeder selection in a routine maize-breeding programme resulted in usable resistance. Entomological studies of vector ecology and simple studies of virus epidemiology enabled techniques to be developed for mass rearing of the vector and uniform challenge of diverse maize germplasm and quantitative assessment of the virus reaction. Resistance was found and selection criteria were used which developed durable resistance, effective across Africa. National breeders and entomologists were trained in the use of the methods so that resistance could be developed in local materials suitable for different African maize environments. A major threat to African maize production was halted, not by pesticides or farmer training but by 'preventive' breeding based on application of pest/pathogen ecological knowledge in a practical and multidisciplinary breeding programme.

Most examples cited are of the first type, and for many people IPM is only considered in this context. It is the simplest, most straightforward and easily explained approach. Less appreciated is that the need for this type of IPM follows after inappropriate plant breeding, coupled with changed agronomic practices, or just inappropriate and indiscriminate application of pesticides in ignorance of ecological and evolutionary principles.

Biotechnology and Plant Breeding

As more organic chemicals were used to control pathogens and pests and as more single resistance genes were used in developing new varieties, plant pathologists, entomologists and breeders had to face up to pest/pathogen evolution against both pesticides and plant resistance genes. They did this in slightly different ways; so we have horizontal/vertical resistance concepts, which have had some influence on plant breeders handling disease resistance; and for insects, with their more complex biology, much recent work on adaptation and on various potential adaptation mechanisms in relation to plant resistance mechanisms, and on population replacement studies.

The two disciplines of plant breeding and entomology are now, fortunately, much closer in thinking and this can now proceed in concert with the advent of biotechnology applied to plant breeding and genetics – the attempt to improve resistance generally and to make it more durable. It seems that, insect ecologists notwithstanding, those pursuing insect resistance through biotechnology still underrate evolutionary potential. Pathologists, with their long history in studying host/parasite biochemistry and genetic interaction and resistance breakdowns, are less sanguine and are still searching for realistic approaches in biotechnology for fungal/nematode/bacterial pathogens. The simpler viral pathogens seem more tractable to biotechnology-based approaches (see Beachy, Chapter 14, this volume).

A key problem is that biotechnologists do not really understand plant breeding and vice versa. There are exceptions, but they are too few. Biotechnologists want genetic precision and knowledge. Breeders want a new and better variety. This difference is fundamental to their thinking and their methodologies. Progress would be much more rapid in getting better varieties for farmers if there were a marriage of these two fields. Plant breeders have been handling resistance for a century without seeing the genes and usually with very little genetics. The genetics and the physiology of the interaction were all retrospective. Biotechnology now offers the promise of more precise targetting of resistance genes for incorporation in breeding programmes.

Most plant pathologists do not understand breeding and the needs of breeders either, and that does not help the positive interaction required for getting new and better varieties that are durably resistant. Hence, there has been much altercation over horizontal/vertical resistance and little progress on how to actually go about obtaining durable resistance. These problems will not be reduced by biotechnology, but will be increased, since biotechnology-based breeding will work, at first, on major genes. All the problems of enhancing heritability values that breeders now face will continue. Making progress in minimizing environmental variance when making selections among progeny is the key to making better varieties. This needs doing with careful judgement of the right environment and of environmental stresses where selection is practised, and with the right challenge by the appropriate pests and pathogens. All of this is equally important, whether one is dealing with 'biotech inserted genes' or with 'natural crossing inserted genes'. Thus biotech will change plant breeding in the practical world much less than most people think. I believe most plant breeding will be changed very little by biotechnology. Much of plant breeding is an art, and resistance is often just a small part of looking at the beautiful product of the reshuffling of thousands of genes as a new plant is created to be what it is, as it unfolds itself in its particular environmental space of the breeder's plot. How much that environment will be represented in, say, 100,000 or 1 million ha is an important point that cannot be resolved in the laboratory. I have discussed many of these issues in a series of papers (Buddenhagen, 1977, 1983a–d, 1987, 1991, 1992). These issues are also relevant to biotechnology's potential input to plant breeding.

We should remember that most plants are resistant to most pathogens. How can we exploit this amazing phenomenon through biotechnology? To me, that is the key question and should be a major focus of investment. Plant breeders now cannot really exploit this point because they can mix genes only from closely related plants that do not have this generalized non-host resistance.

Plant breeders deal in large numbers, applying a shotgun approach. An F_2 population from a cross may have 1000 plants to interact holistically with their environment to enable selection of the best of a vast array of mixed-up genes. Could we obtain the same through biotechnology if we went after it with a large number of plants with a shotgun mix of deoxyribonucleic acid (DNA) pieces from non-hosts, rather than by the approach of isolating, sequencing and making a single transformation of a single gene and its promoter? What inhibits this approach? Why not try to figure out how to do it? There are, of course, many reasons why it might not be possible. But maybe, by looking at the blocks one by one (transformation difficulties, expression difficulties, regeneration difficulties, etc.), one could reduce them to the point of enabling shotgun breeding with biotechnological methods. Who can predict what might be revealed? Who would finance such efforts? Certainly not the private sector, with its need for short-term commercial acceptance. In the public sector are there those with a longer vision? Unfortunately there are too few. Thus the structural difficulties of our research system probably outweigh the possibilities.

Conclusion

Backing basic research and innovative approaches to the use of biotechnology in developing crop resistance to pathogens will be needed for eventual solutions to currently intractable problems. Much of this should be in the public sector, where immediate dollar pay-off pressures are less severe. Even now, innovative ideas are appearing that look very interesting. For example: (i) the two-component system for non-specific resistance proposed by Cornelissen and Melchers (1993); also see de Wit (1992); (ii) the 'plantibodies' approach advocated by Schots *et al.* (1992); and (iii) somatic hybridization in citrus and in potato (Helgeson, 1992). Other innovations are under development, and with further backing of new ideas much can be expected from biotechnology. Control of the non-viral pathogens especially needs further effort, since little progress has been made to date. For viruses, much has been accomplished already, and pay-off is close enough to generate considerable investment, in both the public and private sectors. As concluded by Wilson (1993), and paraphrased here, successful virus disease is a result of a very subtle blend of virus and host proteins working together very precisely – almost any unregulated superimposition of interfering protein or nucleic acid intervention will disrupt this subtle blend and give resistance (see also Harrison, 1992).

The selection of where to apply a promising new innovation should not be left entirely with the private sector, which requires large financial pay-off. Many speciality crops are scarcely bred or cannot be bred and they are most in need of a biotechnology approach. Banana/plantain is a key example of such a crop. Often they already receive a large quantity of pesticides, and have their appeal to the private sector. Thus a strong case can be made for backing both basic innovative research and its application, in universities and other public-sector organizations. This should lead to a more balanced approach to biotechnology applied to crops.

Weeds are pests largely ignored by IPM proponents, and now by agronomists, since herbicide field people and companies are dominant in weed control. This is a mistake, as weeds are most vulnerable to wise crop management in a cropping systems context. Biocontrol of weeds receives little support and now the private sector is busy inserting herbicide-resistance genes in a few major crops (Te Beest *et al.*, 1992). Although this approach receives considerable opposition from the environmental groups, there may well be merit in this innovation, even for developing countries. There is a possibility that less herbicide would be used if crops had herbicide resistance. Moreover, use of herbicides may be much more benign to the environment in the tropics than alternative cultivation methods that result in much more soil erosion and degradation.

Biotechnology efforts to make endophytes more useful in reducing fungal or nematode pathogen damage also need backing, as does the area of rhizosphere organisms (Andrews, 1992; Sikora, 1992). Finally, in relation to crop production in the long term, broadening nitrogen fixation to new plant families should remain a major target.

Long-term investment and vision are needed.

Insect-resistant Crop Plants 12

David A. Fischhoff

Introduction

The past few years have seen the development in several laboratories of transgenic crop plants engineered for resistance to insect pests. These have included cotton, maize and tomato plants resistant to lepidopteran insects and potato plants resistant to Colorado potato beetles (*Leptinotarsa decimlineata*). These plants have resulted from the convergence of two areas of molecular biology. First is the ability to engineer a variety of plant species to contain and efficiently express heterologous genes. The range of crops that have been shown to be capable of genetic transformation now includes many of the major food and fibre crops, and the history of this technology over the past decade suggests that essentially all major crops should be amenable to transformation in the near future.

The second enabling result is the isolation and characterization of single genes that encode insecticidal proteins. Most known insecticidal proteins, and most of those that have been tested for utility in plants, are derived from *Bacillus thuringiensis*, a ubiquitous Gram-positive organism. *Bacillus thuringiensis* strains produce a crystalline protein inclusion containing an insecticidal protein. In their microbial form, *B. thuringiensis* strains are the active agents in a variety of microbial insecticide products that have been used for over 30 years. Based on this history of use, it is well established that *B. thuringiensis* proteins are highly active insecticides with a novel mode of action that involves receptor-mediated binding to and eventual lysis of the midgut epithelium, causing a relatively rapid cessation of feeding and leading within hours or a few days to insect death. *Bacillus thuringiensis*

 Biotechnology and Integrated Pest Management (ed. G.J. Persley)

proteins are highly specific, acting only on a limited spectrum of insects, typically restricted to a single order or two, and often they are highly selective for insects within that order. At the same time, *B. thuringiensis* proteins have no activity against other organisms, such as non-target insects, including beneficial insects and predacious insects, other invertebrates, such as earthworms or spiders, and higher animals, including fish, birds and mammals.

The genes encoding *B. thuringiensis* proteins have been isolated and characterized from many strains. These *B. thuringiensis* genes, called *cry* genes (for 'crystal'), have been categorized into families, based on sequence homologies and insecticidal spectrum. Recent years have seen a great expansion in the number and types of isolated *cry* genes. The earliest type characterized, and the most common in *B. thuringiensis* strains, are *cryI* genes, active on lepidopteran pests. At this point there are over 12 known types of *cryI* genes, which differ in their potency against a given lepidopteran insect; many *cryI* genes have a spectrum that includes a large variety of agronomic pests. Other lepidopteran active proteins come from *cryII*, some of which also have dipteran activity. *CryIII* contains coleopteran active genes, which have been characterized for activity against Colorado potato beetle in particular. *CryIV* is a diverse gene family active against dipterans, especially mosquitoes. New families of genes include examples with dual lepidopteran and coleopteran activity, those with reported activity against plant-pathogenic nematodes and several novel types reported in patent applications but not yet fully characterized.

For development of insect-resistant transgenic plants, attention has initially focused on lepidopteran resistance, with *cryI* and *cryII* genes, or on coleopteran resistance, with *cryIII* genes. These genes were among the first to be characterized and their activity spectrum includes many of the most important insect pests, so that these plants could provide an economically attractive pest-control agent. For example, the *cryIA*(c) gene is active against essentially all of the major lepidopteran pests of cotton in the USA and many other cotton-producing areas around the world. As such, plants expressing this gene could provide an alternative for most, if not all, of the chemical insecticide applied to this important category of cotton pests. Similarly, the *cryIII* genes in potato can provide effective and selective control of Colorado potato beetle, a primary pest of potato in North America and Eastern Europe.

Transgenic expression of *cry* genes in plants has proved to be a significant technical challenge, but one that now appears to be solved. In their wild-type forms, *cryI* genes isolated from *B. thuringiensis* are poorly expressed in plants or, in some cases, not expressed at all. Plants that do express these wild-type genes are typically resistant to only the most sensitive lepidopterans, such as *Manduca sexta* (tobacco hornworm). We have developed a novel method for increasing the expression of *cry* genes, involving modifications to the coding sequence, which allows for efficient expression in plants. *Bacillus thuringiensis* proteins from modified genes can typically be expressed at levels of about 0.1% to more than 1.0% of plant protein (Perlak *et al.*, 1991). These are levels at which plants can provide essentially complete control of the insects that are sensitive to the given protein. This approach has been elaborated by ourselves and others and shown to work on all classes of *cry* genes tested and in all plants engineered with these modified genes so far. The list of major crops in development utilizing modified *cry* genes includes *cryI* genes in cotton, maize and tomato, *cryII* genes in cotton and maize and *cryIII* genes in potato. Early-stage experiments with other genes and other crops are being done in laboratories worldwide.

Current Status of Insect-resistant Transgenic Plants

Cotton

Cotton plants engineered for resistance to lepidopteran pests are among the most advanced in terms of product development (Perlak *et al.*, 1990). We have tested plants expressing *cryIA* genes in the field for more than 4 years. These tests have been done in the USA in all of the major cotton-production regions. Plants have been analysed for insect control efficacy compared with traditional chemical insecticides, and they have also been analysed for agronomic characteristics, such as growth and yield. The major result of these tests has been to demonstrate that insect-resistant cotton can effectively control the major lepidopteran pests of cotton, including *Heliothis virescens* (tobacco budworm), *Helicoverpa zea* (cotton bollworm), *Pectinophora gossypiella* (pink bollworm) and several others, such as cabbage looper, salt marsh caterpillar, cotton leaf perforator and beet army worm. This control is comparable to that achieved in the field with typical lepidopteran control agents, such as pyrethroid insecticides. Cotton plants with *B. thuringiensis* proteins have insect damage ratings comparable to or less than those achieved even with higher than normal applications of pyrethroids (e.g. weekly applications regardless of infestation level). These plants provide essentially complete control of lepidopteran pests even when control (unsprayed) plots show very high levels of damage (greater than 70% square and boll loss). Importantly, this protection from damage extends to the level of yield. Insect-resistant cotton plants have yields comparable to or higher than the same variety treated with traditional insect-control programmes.

For these cotton plants, product development work is proceeding on three parallel paths that should lead to commercial introduction in the USA in the next few years. First, the gene is being transferred from the initially transformed variety into an advanced cotton germplasm via traditional breeding (backcrossing). This will allow the introduction of insect-resistant plants in cotton varieties that have all of the agronomic and fibre traits necessary for efficient and economic cotton production by the farmer. Second, data to support product approval by government regulatory agencies are being developed and submitted. In the USA, the primary regulatory body is the Environmental Protection Agency (EPA), and we have worked with EPA to define the regulatory requirements for plant pesticides, especially those plants expressing *B. thuringiensis* proteins. The data supporting product approval include a broad spectrum addressing issues of human, animal and environmental safety. *Bacillus thuringiensis* proteins are already approved for use in microbial formulations with an exemption from the requirement of a tolerance, and our expectation is that the *in planta* delivery of these proteins will have the same overall favourable picture. Third, product development involves the evaluation and testing of approaches to best implement *B. thuringiensis* cotton plants into the whole cotton production and pest-management programme. Our aim is to develop and implement approaches that will utilize *B. thuringiensis* cotton as a central piece of a cotton pest-management system. The focus is to develop schemes that maximize the utility and durability of this insect-control product. These aspects are addressed in detail below for insect-resistant cotton as a case-study.

Maize

In addition to cotton, two other insect-resistant crops are in advanced stages of development. Maize plants expressing *cryI* genes have been tested for a few years by Monsanto, Ciba and others. The results, while less complete than for cotton, indicate that these plants provide control of *Ostrinia nubilalis* (European corn borer). This is an important pest of US maize and is very difficult to control with chemical agents because its habit is to spend most of its life inside the maize stalk, where it is relatively inaccessible to chemical sprays. Plant expression of *B. thuringiensis* proteins is a method that can, in this case, overcome the technical barriers to insect control with other agents, and thus address an unmet need in pest management.

Potato

Potato plants expressing *cryIII* genes have been field-tested for more than 3 years. These plants show complete control of Colorado potato beetle in the field in all of the major potato-growing areas of North America (Perlak *et al.*, 1993). Because potato is vegetatively propagated, the varieties used in transformation are the same as those used commercially. We have focused on *B. thuringiensis* genes in Russet Burbank, the predominant North American variety. Under conditions where natural potato-beetle infestation is high enough to cause more than 80% defoliation, leading to reduced tuber yields, the insect-resistant plants show no damage. In all tests, these plants performed better than the current commercial chemical insecticide standards. Recently (in 1995) potato plants expressing the *cryIIIA* gene became the first genetically modified insect resistant crop to gain full commercial approval in the USA.

Results

The laboratory and field results with insect-resistant cotton, maize and potato lead to three general conclusions. First, the plants can be highly effective in controlling the targeted insect pests as well as or better than current practices. They thus provide a substitute for current chemical insecticides that are directed at the same pests. This high efficacy is due to the potency of the *B. thuringiensis* proteins themselves at the achieved levels of plant expression, plus the timing and localization of delivery of these proteins. For all insects, *B. thuringiensis* proteins are most efficacious against neonate larvae. Plant expression delivers the active *B. thuringiensis* protein directly to the neonates immediately after hatching, when they are most susceptible. Second, the *B. thuringiensis*-expressing plants can be more effective than alternative methods of control for certain pests. This is clearly seen for European corn borer, which can be difficult or impossible to control with topically applied pesticides. With plants, the *B. thuringiensis* protein is delivered to the larvae as they feed and move into the plant at their most sensitive stage. Another example is the pink bollworm in cotton. In some field trials of *B. thuringiensis* cotton plants, chemical application was not at all effective in preventing boll damage by pink bollworm because this pest enters the boll (by way of feeding) where it is much less susceptible to spray applications within 24 h or so of hatching. However, *B. thuringiensis* protein is present in the plant tissues that the larvae consume as they enter the boll, and so we have observed a very high level of control compared with chemical insecticide application. Third, the reduction of chemical insecticides used in plots of *B. thuringiensis* plants leads to: (i) higher levels of beneficial and predatory insects; and (ii) lower levels of other pests whose populations tend to increase following some chemical insecticide treatments. Because most field trials so far have been on relatively small plots (often much less than 0.4 ha), these observations of shifting population dynamics are especially encouraging and are expected to be even more significant when *B. thuringiensis* plants are cultivated on larger areas.

Implementation Plans for Insect-resistant Cotton

The ideas and plans proposed here for the implementation of insect-resistant cotton plants into the overall programme for cotton pest management and cotton production are based on introduction of these cotton plants for use in the USA, but many of the same ideas will apply directly to the use of *B. thuringiensis* cotton in other parts of the world. This is because many of the same, or closely related, pests occur on cotton in all areas. Also, in areas where other lepidopteran pests are seen, their natural history is often similar to that of some US pests.

It is important to recognize that the plans for implementation of insect-resistant plants will be crop- and insect-specific, and each example will have to be tailored to the individual cropping system. At the same time, many of the general principles on which these proposals for *B. thuringiensis* cotton are based will apply to other crop/insect combinations because some of the features of the *B. thuringiensis* proteins in plants, such as efficacy and selectivity, are conserved. Importantly, the goals of implementation remain the same in all cases, ensuring the utility and durability of this novel type of pest-control agent.

The proposals for *B. thuringiensis* cotton outlined here are based on examination of pest-control practices and technical literature, coupled with experimental observations on *B. thuringiensis* plants and proteins. In addition to our scientists, we have involved dozens of external advisers and consultants. Finally, we have taken into account current and emerging trends in cotton production and seed production to develop plans that should be acceptable to those involved in cotton production, such as seed companies, growers, crop consultants and extension agents. While these proposals are presented in considerable detail, the specifics are still being examined in field trials, and we shall continue to adjust and modify these plans based on field results both before and after commercialization. As seed supplies expand, additional information from commercial-scale cotton production will provide valuable input from academic experts, extension agents and growers.

Benefits of insect-resistant cotton

The use of insect-resistant cotton will provide important benefits to the cotton grower and to society and the environment. First and foremost, insect-resistant cotton offers an alternative to chemical insecticides for controlling lepidopteran insects. The use of insect-resistant cotton has now been shown to control these pests and the crop damage that they cause with an efficacy equal to or better than that of the current control methods, while significantly reducing the application of chemical insecticides directed at these pests. This substitution will have direct benefits to the grower, such as less time and effort spent on lepidopteran control and reduced farmer exposure to chemical insecticides. It will also have beneficial impacts on the environment due to the reduction of chemical insecticide applications. Insect-resistant cotton is also likely to produce secondary benefits in pest control as an indirect result of the reduction in use of chemical insecticides.

Chemical insecticides such as pyrethroids are relatively non-specific and have the effect of killing beneficial predatory and parasitic insects. Because insect-resistant cotton is not active against these beneficial insects, we expect to see, and have already seen, beneficial populations rise significantly in fields planted with the *B. thuringiensis* cotton. This should provide additional control of the lepidopteran target pests and some control of non-target pests, such as mites, which increase as problems as their natural predators are removed. Insect-resistant cotton is also capable of controlling populations of budworm and bollworm, which are losing their sensitivity to chemical insecticides, thus filling a need that is likely to grow in coming years in all areas of the world.

To achieve all of these benefits, it is important that insect-resistant cotton be implemented and managed properly. In this respect, insect-resistant cotton is not different from other pesticides. There are two aspects of this management. The first is the development of pest-management techniques that allow the farmer to maximize the ability of these plants to control target lepidopterans and non-target pests. In essence this is the development of a total insect-management package that will be centred around a new tool, insect-resistant cotton. The second is the development of appropriate strategies to maximize the product durability and the utility of insect-resistant cotton. Part of this is the development and implementation of strategies that will delay or prevent the potential development of insect resistance to the *B. thuringiensis* protein in cotton. Because both management aspects can affect the way in which insect-resistant cotton is used by the grower, these two types of management for insect-resistant cotton, total insect management and insect resistance management, are necessarily interconnected.

Although we do not see resistance management as an issue peculiar to insect-resistant cotton, it has become a topic of discussion. As a result, in the sections that follow we shall focus attention on resistance management as one component of lepidopteran pest management. Our scientists have worked for several years on laboratory studies of insect resistance, and with outside collaborators we have examined nearly every suggestion that has been made for resistance management in insect-resistant cotton. As the discussion below demonstrates, we have concluded that promising, viable and effective strategies for resistance management in insect-resistant cotton are available and can be implemented. These strategies have been developed in consultation with the advice of experts, taking into account research and an understanding of cotton seed production and agronomic practices. In addressing these options, it is evident that insect-resistant cotton offers some unique options in pest management and resistance management that are not available to traditional pesticides. Access to these new options is an additional benefit of insect-resistant cotton.

Pest and insect resistance management

As part of a package to provide economic control of insect damage in cotton, we envision insect-resistant cotton as providing a central focus around which other insect management practices will be developed (Fischoff, 1992). In many areas lepidopteran pests are the primary damaging insects of cotton, so the use of insect resistant cotton to control these pests will be a major portion of total insect control. In addition, by substituting the modified cotton plants for chemical pesticides directed at lepidopteran pests, insect-resistant cotton will certainly lead to other changes in overall insect management. Many of the details of overall pest management in insect-resistant cotton can only be determined by multiyear, large-scale field tests designed to incorporate insect-resistant cotton into current growing practices. Such field trials are now beginning, and will provide the data needed for a full package of insect control prior to and after commercialization of insect-resistant cotton. These trials will involve collaboration between our company, cotton seed companies and academic and extension entomologists. These ongoing studies examine insect population dynamics in *B. thuringiensis* cotton fields, the effect of insect-resistant cotton on beneficials and on non-lepidopteran pest species and the establishment of economic thresholds for lepidopterans in insect-resistant cotton.

We foresee the use of insect-resistant cotton as a primary tool in a total insect pest-management package as being fully consistent with the goals of integrated pest management (IPM). Insect-resistant cotton is a useful tool in IPM because it:

1. is insect-specific, affecting only a few targeted pest species;
2. is active only against insects feeding on the plant and thus doing damage;
3. will reduce use of chemical insecticides;
4. will be combined with pest-control approaches targeted at non-lepidopterans; and
5. will lead to an increase in beneficial insects.

Because IPM and resistance management are interconnected, it is important to develop both of these approaches for insect-resistant cotton in tandem.

Combination of insect-resistant cotton with chemical insecticides

One aspect of the use of insect-resistant cotton for integrated insect management in cotton is the continued use of chemical insecticides. Some chemicals will continue to be used in cotton for non-lepidopteran pests, and these chemicals need to be chosen to fit an IPM approach for overall cotton insect control. These chemical insecticides will need to be chosen so as not to have a negative impact on predatory insects that play a role in both resistance management and IPM. However, this combination with chemicals, while part of a total package of insect control, is not a complete resistance-management option for insect-resistant cotton. Chemical insecticides can reduce the population size of any insects selected for resistance to *B. thuringiensis* but probably cannot alter the gene frequencies within this population. Alternatively, insect-resistant cotton should have a positive impact on current cotton chemical controls by helping slow pyrethroid resistance development and prolonging the life of these important agricultural chemicals. Chemical insecticides will continue to play a role in an IPM package for total cotton insect control. It is likely that new families of chemical insecticides, and perhaps some new biologicals, for use in cotton will be introduced during the next decade, which will open new possibilities for overall pest management.

Resistance management for insect-resistant cotton

As described above, part of managing the implementation of insect-resistant cotton for lepidopteran control is the design and implementation of appropriate strategies to delay or prevent the potential development of insect resistance to *B. thuringiensis* in cotton plants. Described below are approaches that should effectively manage resistance development in insect-resistant cotton. It is important to note that these strategies are being tested and validated in the field, and they could be modified based on additional data. They are summarized briefly and then expanded in greater detail in the next section.

Recommended resistance-management strategies

Short-term strategies (implemented at commercialization)

- High-dose expression of *B. thuringiensis* in cotton plants to control insects heterozygous for resistance alleles.
- Refugia as hosts for sensitive insects provided through non-insect-resistant cotton or non-cotton hosts.
- Agronomic practices that minimize insect exposure to *B. thuringiensis*.
- Integrated pest management (as described above).
- Monitoring of insect populations for susceptibility to *B. thuringiensis*

Medium-term strategies (implemented 2–5 years after commercialization)

- Continue all short-term strategies.
- Combination of two genes within the same plant, both of which are active on budworm/bollworm/pink bollworm and other targeted lepidopterans

but with different sites/modes of action.

Long-term strategies (more than 5 years after commercialization)

- Continue all short- and medium-term strategies.
- Incorporation of host-plant resistance (HPR) traits into insect-resistant cotton as they are proved effective.
- Incorporation of novel proteins that provide effective control of lepidopteran pests

Details of recommendations

High-dose expression

High-dose expression for resistance management is based on three assumptions:

1. Resistance will most probably be controlled by one major locus with recessive resistance alleles.
2. Insects developing resistance to *B. thuringiensis* will be rare initially and will almost always mate with susceptible insects, giving rise to heterozygous progeny.
3. The vast majority of the heterozygous progeny will be disabled or killed by insect-resistant cotton with the same dose as the homozygous susceptible larvae.

The high-dose expression strategy uses plant expression of *B. thuringiensis* in quantities sufficient to kill those insects heterozygous for resistance to *B. thuringiensis*. This resistance strategy fits well with the fact that high-dose expression is essential for commercial efficacy of insect-resistant cotton because of the range of sensitivity to *B. thuringiensis* in cotton insect targets (e.g. at least a tenfold difference between budworm and bollworm). High-dose expression is also necessary to maintain consistent control across environments and genotypes. We are evaluating and developing the high-dose expression strategy, and testing aspects of its predictions in both laboratory and field experiments.

Refugia for sensitive insects

Refugia provide a refuge for sensitive insects within a population so they will not be exposed to *B. thuringiensis* and not be selected for resistance. As a resistance-management technique, refugia are based on the concept that control failure due to resistance is a population genetics phenomenon. Control failures are observed when the frequency of resistant insects in the population reaches a critical level. Refugia supply susceptible non-selected individuals to the general population. With adequate refugia, the frequency of resistance genes will be very low and spread only slowly through the population. Refugia are an important component of our insect-resistant cotton resistance-management strategies.

Refugia can be provided either within the crop or outside it. The refuge can also be planted specifically as such or exist naturally. In all of these approaches, the effectiveness of the refuge is based on those insects that survive on the refuge crop rather than its total acreage. This is an important point because, if the refuge is chemically treated, the refuge population is reduced and the amount of acreage required is increased. The following examples of refugia can be utilized for insect-resistant cotton.

Refuge outside the crop: non-insect-resistant cotton This type of refuge will exist in all cotton areas not covered by insect-resistant cotton. This area will be substantial in the early years after introduction and could supply a sufficient refuge for several years. As insect-resistant cotton is more widely grown, this refuge will be reduced. Consequently, over time, reliance on non-insect-resistant cotton-fields for refugia may not be adequate.

Refuge outside the crop: non-cotton hosts The cotton bollworm has many non-cotton hosts, including other crops in all locations, which may provide an adequate refuge. Tobacco budworm has fewer alternatives and pink bollworm has none. In some locations cotton may be the only host for at least one insect generation per season. The use of *B. thuringiensis* microbials or transgenic *B. thuringiensis* plants on other crops will also have an impact on their utility as a refuge for insect-resistant cotton. Obviously, this option has to be evaluated carefully based on pests and growing regions.

Refuge within the crop: non-insect resistant cotton In certain cases a possible solution is to provide an 'in-crop' refuge of non-insect-resistant cotton. For this in-crop refuge, the choices are: (i) mixed seed lines; or (ii) non-insect-resistant cotton in the same field. The optimum refuge area required will need to be determined.

Mixed seed lines (*B. thuringiensis* and non-insect-resistant cotton within the same bag) have a certain appeal due to the 'automatic' implementation. A possible problem with mixed seed arises from larvae that survive on a non-insect-resistant cotton plant and migrate to the modified cotton, where they will be less sensitive to *B. thuringiensis* because of size. This could compromise insect control and increase selection pressure for resistance. The likelihood of this occurring needs to be determined experimentally before this strategy can be implemented.

There may also be economic and logistic problems with implementing a mixed seed strategy. However, we are interested in determining the viability of the mixed seed approach. It is clear that field research is required to determine the percentage of non-insect-resistant cotton needed as a refuge, and what the impact of this percentage will be on overall yield, quality and production economics.

Another in-crop refuge could be non-insect-resistant cotton planted specifically by the farmer. Besides providing a refuge, such planting of separate indicator rows of non-insect-resistant cotton could potentially make scouting easier. Field research is needed to determine the optimum type of planting regime.

Agronomic practices

Certain agronomic practices may need to be recommended for insect-resistant cotton. In particular, plough-down dates need to be recommended to eliminate unnecessary insect exposure to *B. thuringiensis* from cotton regrowth. These will need to be determined on a regional basis.

Monitoring insect resistance

Insect resistance monitoring is an important component of an insect resistance-management strategy. A baseline frequency must be developed in order to know when resistant genotypes have increased within the population. Baseline information should be collected on all *B. thuringiensis* products (engineered plants and *B. thuringiensis* microbials). This information must be developed on regional bases over several years so that susceptibility changes in populations can be identified.

Multiple insect control traits

A set of strategies for the medium and long term focuses on combining multiple insecticidal agents. The rationale is essentially the same for all of these: expose the insects to two or more active agents with distinct modes of action at the same time, and the probability of any one insect being selected for resistance to both agents simultaneously is extremely low.

Combination with a second insect resistance gene

A second gene within the same plant possessing a different mode of action will significantly reduce the frequency of resistant individuals. Population models indicate that other alternative uses of a second gene such as seed mixture or using single genes in rotation are not as effective as two genes within the same plant. Assuming initial gene frequencies for *B. thuringiensis* resistance are low, initial introduction of a product with a single *B. thuringiensis* gene should not negatively compromise a second gene because the single gene product will be planted on limited acreage in the first few years. In the medium term, the best choice of second gene is an unrelated *B. thuringiensis* gene. We are currently pursuing this option with alternative *cryI* and *cryII* genes and by examining novel *B. thuringiensis* genes as they become available. In the long term, the use of novel, non-*B. thuringiensis* insecticidal genes holds great promise. This area is under active study and holds great long-term promise; our programme has already yielded some new insecticidal proteins, such as the enzyme cholesterol oxidase.

Combination with host-plant resistance traits
This is a long-term strategy to be implemented by seed companies or public breeders. Host-plant resistance traits used in combination with insect-resistant cotton need to be insecticidally effective and not to have a negative impact or cotton quality. We have funded academic work on HPR to help set direction on HPR traits that alone or in combination are useful in protecting the plant from lepidopteran insects. Cotton seed companies are interested in incorporating these traits if they are effective and have no negative effects on cotton yield or quality.

Summary of recommended plans

Insect-resistant cotton will offer great benefits in overall insect control in cotton. Insect-resistant cotton will be incorporated into an IPM programme for cotton pest control. Work on this has begun and will continue in the next few years prior to commercialization. With proper management and implementation, potential insect resistance to *B. thuringiensis* will not be a technical or commercial problem that will limit the value or efficacy of insect-resistant cotton or its lifespan. We and others in the academic and industrial community have developed a package of strategies that should effectively manage potential insect resistance. The details of this programme and its incorporation into the IPM programme will also be further developed and optimized in the field in coming years.

Many aspects of the use of insect-resistant cotton in IPM and the incorporation of resistance-management strategies are unique to its position as an insect-resistant plant, as compared with traditional chemical or microbial insecticides. For example, the use of refugia and the incorporation of multiple resistance traits through molecular biology or through plant breeding are aspects that are ideally suited to insect-resistant plants. This ability to utilize new methods in IPM and in resistance management is another beneficial feature of insect-resistant cotton.

Future Prospects

The use of *B. thuringiensis* proteins and genes to produce insect-resistant transgenic plants is now well established, and the plants are recognized to provide useful levels of insect control against many important pests. In determining which additional crops might be engineered to express *B. thuringiensis* proteins, consideration of their utility in pest control based on local conditions is important. Equally important will be the development of appropriate strategies to integrate these plants into IPM and integrated resistance management (IRM) programmes, again based on local cropping and production conditions. This will of necessity be a case-by-case analysis. Our expectation is that it should be possible to effectively manage and utilize *B. thuringiensis* expressing plants in many situations.

The pace of *B. thuringiensis* protein and gene discovery and characterization continues to increase. As new genes are discovered and tested, new pesticidal activities will become available from *B. thuringiensis*. Some of these new genes are likely to be useful in generating second- and third-generation products in the current crops, by providing genes useful for enhancing both pest management and resistance management. Other new genes will open up new pest-control opportunities, for insects not currently considered targets of *B. thuringiensis*.

In the future our expectation is that pest-resistant transgenic plants of all types, not just insect-resistant, will be generated using biotechnology. This will depend on the discovery and development of genes with useful pesticidal properties. That this is already becoming a reality is evident from the recent discovery of other potentially useful insecticidal gene products, such as protease inhibitors, lectins and enzymes, and from the development of genes for fungal or bacterial disease control in plants. Based on this we expect transgenic plant pesticides to be a major component of future pest-control systems in agriculture. The types of strategies considered for implementation of *B. thuringiensis* cotton that we have outlined here will help provide a starting-point for the effective implementation of these future plant products.

Cotton in Australia 13

W. James Peacock, D.J. Llewellyn and G.P. Fitt

Cotton (*Gossypium hirsutum* L.) production is now Australia's second largest cropping enterprise, annually grossing more than $A1 billion (thousand million). A major input into this predominantly irrigated farming system is chemical pesticides, with up to $A100 million/year being spent on insect control (Fitt, 1994). The major pests are two lepidopteran species, *Helicoverpa armigera* (cotton bollworm) and *H. punctigera* (native budworm), and most of the control measures are directed at the eggs and larvae of these two species. Although both species are important, differences in aspects of their biology and population dynamics make *H. armigera* the more serious pest (Fitt, 1989). *Helicoverpa armigera* is a polyphagous feeder attacking other crops, such as sorghum (*Sorghum bicolor*) and grain legumes, and its population distribution and density are closely tied to cropping activities in cotton-growing areas. It goes through a number of generations during the growth of the cotton crop and then into a winter diapause at the end of the season. Insects exposed to pesticides therefore remain locally to emerge in large numbers in the next season. They are then resprayed with the same chemicals, resulting in a strong selection pressure for insects resistant to most of the common control chemicals. *Helicoverpa punctigera* is also polyphagous and an important early season pest on many summer crops, including cotton, but is more easily controlled by conventional pesticides. Adult *H. punctigera* are more mobile in relation to the cropping areas than *H. armigera*. Large populations are generated in inland Australia, feeding on native vegetation, and migrate towards the major cropping areas in spring to attack our exotic crop species. Populations in cropping areas are therefore continually replaced by immigrants, ensuring that resistant insects rarely or never arise. This difference in distribution can be accounted for by the differences in dispersal behaviour of the two species following emergence. Both species are capable of long-distance migration on high-altitude winds. However, *H. armigera* tends to remain in areas where host plants are available, moving only short distances (5–10 km). *Helicoverpa punctigera*, on the other hand, appears to be an obligate migrant, moving long distances each generation.

Mites are not normally serious pests in cotton as their populations are maintained at subeconomic levels through natural predation by thrips and other predators. Populations can flare as a result of some external factor, usually a pesticide such as a pyrethroid applied to control *Helicoverpa*, which disrupts this control by affecting the population density of the predators. Early-season control of lepidopteran pests must therefore take into account impacts on other pests, since chemical control of mites is considerably more difficult as there is currently only one key miticide (propargite) used on cotton and selection of resistant mites is a serious possibility. Mirids (e.g. green mirid, *Creontiades* sp.) cause damage to terminals and young flower buds, and are generally considered pests of cotton, but their role in the crop, like that of thrips, is considerably more complex since they are also predators of all stages of mites and the egg stages of *Helicoverpa* species. They enter the cropping system before *Helicoverpa*, but their populations are generally controlled to some extent by the same pesticides as those used for *Helicoverpa* species, and they are currently only sporadic pests in some fields in some seasons. This may change as pesticide usage changes with new technological developments in pest control.

This complex temporal and spatial pattern of different pest species and the interdependence of their population densities as a consequence of the timing and spectrum of pesticide usage is something that has to be reckoned with when introducing any new control measures, such as transgenic insect-tolerant plants or even a new pesticide, into the farming ecosystem. The necessity for a total integrated approach to pest control was highlighted by the large-scale introduction of synthetic pyrethroids to the industry in the early 1970s and the rapid build-up of *H. armigera* that were resistant to all of the synthetic pyrethroids (Gunning *et al.*, 1984). Unfortunately, this resistance build-up was clearly due to overuse of the pesticide in a situation where, with hindsight, it could have been managed to better advantage in the longer term.

The Emerald region in Queensland generated the first significant levels of resistance after only a few years of pyrethroid usage, the reason being that the subtropical situation of the industry in this region allowed year-round cropping and that pyrethroids were used not only for cotton but for other crops, such as sorghum. Selection pressures were therefore continuous and the *H. armigera* populations responded accordingly. Once the resistance problem was recognized, the whole of the cotton-cropping industry, and related industries, voluntarily adopted a recommended management strategy that has probably slowed down the build-up of resistance-gene frequencies within the insect populations. Basically, the strategy required that pyrethroids be restricted for use in a defined window of time that covered just a single generation of the moth in the field. In some situations the use of piperonyl butoxide (PBO) with pyrethroids increased insect control by suppressing at least one of the insect resistance mechanisms to pyrethroid insecticides. This chemical has no toxicity on its own but inhibits a mixed-function oxidase that has evolved in the insects to detoxify pyrethroid pesticides. In 1990–91 PBO was introduced for commercial use, but pyrethroid resistance levels continue to rise in the *H. armigera* population and the resistance situation still remains critical in Australia.

In the periods before and after the pyrethroid window of the summer-crops resistance-management strategy, alternative chemicals are recommended, but again with the inevitable results. We now have, for example, significant levels of resistance to the pesticide endosulfan. Apart from the toxic organophosphates, endosulfan is basically the last line of chemical defence available to the industry until new synthetic pesticides are developed, and this appears unlikely for the next 4–5 years. Farmers, at the instigation of the chemical companies, are increasingly turning their interest to insecticides based on *Bacillus thuringiensis* spores and crystals (Aronson *et al.*, 1986) and on other soft options that avoid broad-spectrum insecticides. In earlier years farmers were mostly disdainful of this biologically soft and specific pesticide, but, with nothing else available and with improved formulations, *B. thuringiensis* is now becoming a serious component in our pest-management system. Its management in a way that minimizes the probability of resistance is critical because it represents one of the few alternatives left. However, it is all the more critical because this same protein insecticide is the basis of some of the genetic engineering strategies that will probably be brought into play over the next few years with the development of transgenic cotton varieties expressing the *B. thuringiensis* protein.

A transgenic addition to the pest-management armoury is clearly important and probably critical to the longevity of the Australian cotton-production industry. One of the principal reasons for hope with a transgenic insecticide is that it will come with a vastly increased knowledge of the mechanism of action of the particular insecticide on the target pest species (Van Rie *et al.*, 1990) and a greater awareness, learned from past failures, of the importance of resistance management (Gould, 1988a, 1991; Roush, 1989). There is the knowledge that resistance can develop to *B. thuringiensis* insecticidal proteins, both in the field and in the laboratory (Stone *et al.*, 1989), and some understanding of the mechanisms of that resistance (Van Rie *et al.*, 1990), but there is also an expectation that, with the appropriate level of biochemical and population-dynamics knowledge, transgenic plants expressing *B. thuringiensis* may be developed and successfully deployed in an equilibrated integrated pest management (IPM) system.

In Australia we have initially used the Monsanto *B. thuringiensis* constructs (Perlak *et al.*, 1990) and have reached the stage of successful field trials with cotton containing a truncated *CryIA(b) B. thuringiensis* gene (see Hofte and Whiteley, 1989, for *B. thuringiensis* gene nomenclature). The protection afforded in the field against lepidopteran species by the *B. thuringiensis* gene in Australian cultivars has been as expected, and high mortality has been observed for both important species of *Helicoverpa*. *Helicoverpa punctigera* is even more susceptible than *H. armigera*, in line with previous observations with *B. thuringiensis* sprays. Our observations suggest that the transgenic plants are also resistant to other, more minor, moth species. However, further trials will assess the impact of the transgenic plants on all the insect components of the cotton-cropping ecosystem, both pest and non-pest species. New cultivars based on the *CryIA(c)* gene developed by Monsanto (Perlak *et al.*, 1991) are also being produced, both by direct transformation of commercial Australian cultivars (Cousins *et al.*, 1991), and by conventional backcrossing programmes, and these new transgenic plants show even greater mortality to *Helicoverpa* species. Field trials on this new material were to begin in 1993, subject to regulatory approval.

In the first Australian field trials, *B. thuringiensis* cotton was surrounded by non-transgenic plants to trap transgenic pollen that might be dispersed from the site by insect vectors. The trap-row plants were unsprayed and subject to attack by all the normal cotton pests, including *Helicoverpa* species. Surprisingly, very little damage was seen on these non-transgenic plants, presumably as a result of the very high populations of beneficial insects that established early in the season. These insects appeared to suppress *Helicoverpa* populations through predation and parasitism. Pest populations did not rise until late in the season, when an accidental pesticide spray drift from a neighbouring commercial field decimated beneficial insect populations. The potentially significant contribution of beneficial insects in the control of pest species highlights the importance of developing the soft option for pest control as part of the overall management strategy with transgenic *B. thuringiensis* cotton, and this will form the basis for further studies with these transgenic cultivars.

In designing strategies for the deployment of transgenes in our cultivars, many factors will have to be taken into account. Other chapters deal with the possibilities of using insect refugia to dilute out resistance genes, tissue-specific expression of the *B. thuringiensis* transgene and other levels of control of expression of such genes. Our own efforts, apart from devising effective field deployment strategies for *B. thuringiensis* cotton, are directed towards providing additional and different transgenes coding for molecules that have quite different pathways of action, that use different receptor molecules and that will hopefully have little chance of being subjected to cross-protection from any resistance built up by selection with the *B.*

thuringiensis transgene and/or (just as importantly) the *B. thuringiensis* sprays. A survey of a large number of proteinase inhibitors has identified one from the giant taro of the Pacific islands as having particular promise against our *Helicoverpa* pests. We have made good progress in isolating this gene and introducing it in a suitable construct into our cotton cultivars. Other enzymes and proteins are also showing some promise as insecticidal genes for transgenic plants and are being developed concurrently with our *B. thuringiensis* programme. We are particularly excited about the potential for the use of a plant-based production of antibodies, so-called plantibodies, directed against specific components of the *Helicoverpa* feeding system. This same approach might be useful, too, for the sucking pests, which are certainly going to need to be addressed in the transgenic mode of control once overall pesticide usage declines in cotton cropping. Other avenues of protection, including the expression of insect viral components and manipulation of secondary metabolites within the plant, are also being considered.

The propensity of *H. armigera* for rapid resistance development to conventional chemical pesticides may possibly be reflected in a similar ability to develop resistance to *B. thuringiensis*-based insecticides or transgenic plants. An extensive baseline survey of existing levels of tolerance in *Helicoverpa* species to *B. thuringiensis* has begun, and monitoring for resistance will continue as *B. thuringiensis* becomes more established as an insecticide in the Australian cotton industry and as we move closer to commercial release of transgenic plants. As with conventional insecticides, it will be important to be able to distinguish the population densities of the two closely related *Helicoverpa* species at the egg and larval stages, to assist with management decisions on control options. CSIRO has recently developed an antibody-based identification kit, based on a simple colour-detection reaction of squashed larvae or eggs, which will allow growers or consultants to distinguish the two species. This is playing an important role in the pyrethroid resistance-management programme and will be just as critical for *B. thuringiensis* cotton used in conjunction with conventional chemical sprays.

One of the critical components of an IPM system in cotton is the contribution made by the plant itself through some innate characteristics (host-plant resistance). One key plant character introduced into Australian cotton by conventional breeding was the okra leaf shape (Thomson, 1987). This different leaf shape has particular advantages for pest tolerance, to both mites and *Helicoverpa* species, as well as providing an open leaf canopy for adaptation to the humid summers that can occur in the production areas in inland northern New South Wales and south-central Queensland. The open canopy is of advantage, too, in the application of chemicals, and the okra-leaf character in itself has been shown to contribute savings of one or two insecticide sprays per season in many localities of our cotton-production area. There are other morphological characters being investigated that show promise for better insect control. The removal of the nectaries (nectariless character), removal of hairs (glabrousness) and finally the frego bract character are all possibilities under investigation (Thomson, 1987). We have cultivars with various combinations of these characters nearing commercial use and have begun breeding programmes to introduce *B. thuringiensis* genes into these cultivars. We are also conducting analyses to determine, particularly in the wild cottons present in Australia, whether there are other chemical characteristics comparable to the well-known contribution of gossypol to insect tolerance, which might be introduced into our commercial cultivars along with the genes to be introduced by genetic engineering.

The Australian cotton industry is one of the most successful agricultural production industries in Australia, with remarkable increases in its total production and improvements in its yield during the last 20 years. In 1973 production was approximately 40 kilotons and in 1992 it was over 1600 kilotons. Yields in the same period went up from around 800 kg of lint/ha to almost 1800 kg/ha. Nevertheless, the industry still faces serious problems because of our failure to put into place an effective IPM system that could be sustained over many seasons. However, as we have indicated above, there are many lines of improvement available with advances in technology. We are confident that, with the combination of a vigorous research system and a receptive industry with a desire to develop the appropriate strategies, there is every hope of developing a well-controlled pest-management system that will be able to be operated in a sustainable way.

Virus-resistant Transgenic Plants

14

Roger N. Beachy

Introduction

Plant virus diseases are responsible for significant crop losses throughout the world. For example, it has been estimated that viruses in potato cause annual losses of between $US60 and 120 million in the USA alone. Annual loses in field crops are generally placed at between 5 and 10% in the USA with significantly higher losses in other parts of the world. Crop-yield losses are especially significant in tropical regions, where insect vectors survive year-round and where alternate hosts provide year-round reservoirs for both virus and insect vector.

In the USA and Europe, research in plant breeding has led to the development of crop varieties that possess high levels of disease resistance against specific viruses. Many resistance genes are not durable, but others are highly durable and have been deployed in agricultural regions for 15–20 years. For example, the *Tm-2* gene for resistance against tomato mosaic virus (ToMV) has been used in both greenhouse and field situations for more than 15 years, following its introduction by seed companies in Holland. *Tm-2* has found its way into tomato varieties around the world and has been a highly durable gene. Recently, however, strains of ToMV that overcome the *Tm-2* gene have been isolated.

There are, on the other hand, numerous examples in which breeding for virus disease resistance has been notably unsuccessful. For example, the disease caused by papaya ringspot virus (PRV) has largely eliminated commercial production of papaya in many parts of the world, due to the lack of genes for resistance. Perhaps more importantly, the geminiviruses have emerged as significant pathogens in many parts of Asia and Latin America and recently in the southern USA. These agents comprise a genome of single-stranded deoxyribonucleic acid (DNA), and they are borne by whiteflies. Whiteflies have been difficult to control by chemical means, and little or no genetic resistance has been found to control the disease. These characteristics have contributed to the increasing importance of geminiviruses in agriculture.

In the mid-1980s, a number of research laboratories began experiments to prevent the replication and disease-causing cycle of plant viruses through the use of transgenic plants. A number of different approaches were taken to interrupt the disease cycle, as shown in Fig. 14.1. This illustrates virus entry into the cell through a wound caused by either mechanical damage or by insects. After the virus is placed within the cell, an as yet uncharacterized reaction causes the release of viral capsid (coat) proteins (CPs) from the viral genome, releasing the genome for translation and replication. During replication, complementary strands of the viral genome are synthesized by a replicase encoded either wholly or in part by the viral genome. As this enzyme replicates the genome, it produces new mRNAs, messenger ribonucleic acids which use the host translation machinery to produce the viral proteins. Some of these proteins may be important for the acquisition and spread of the virus by the insect vector. Other gene products may be proteases, which cleave a virus preprotein to release active viral protein(s). Some proteins are responsible for modifying or creating the channels that enable the virus to move from cell to cell. Virus mRNAs produce additional CP which is responsible for encapsidating the replicated virus and the accumulation of new virus particles.

Since 1986, novel transgenes have caused the interruption of the virus replication and life cycle in a number of ways, as indicated in Fig. 14.1. These will be briefly reviewed here, with emphasis on studies with the virus CP.

Developing Virus Disease Resistance in Transgenic Plants

The first example of interference with virus infection and disease was reported by Powell *et al.* (1986), who described transgenic plants that express the CP of tobacco mosaic virus (TMV). Plants that express the CP gene are resistant to infection by the strain of TMV from which the CP gene was isolated and to closely related strains. This resistance was subsequently described as CP-mediated resistance (CP-MR). The effects of the expression of the gene were to reduce the number of sites where the infection occurred, and to reduce the rate of spread of virus from the inoculated leaf to upper leaves. This resulted in decreased severity of infection in plants that accumulate CP. It was subsequently shown that this approach can be effective for the control of a large number of different classes of viruses. In each of these cases, the CP gene provides resistance against the virus from which the CP sequence was isolated and against closely related strains, but not to different groups of viruses. In some cases, there was significant resistance against distantly related serotypes (reviewed by Fitchen and Beachy, 1993).

To date, CP-MR has been reported to be effective against viruses in at least 11 different classification groups, and in a variety of dicotyledon and monocotyledon crop species, including potato, tomato, papaya, squash, cucumber, melon, alfalfa, sugar beet, rice and maize.

An alternative approach to conferring resistance has been to express portions of the viral genome that encode the viral replicase. It has been shown in several laboratories that resistance against at least several classes of viruses can be engendered by expressing a gene encoding either the full viral replicase or a portion thereof (Carr *et al.*, 1992). Such resistance can be extremely high, i.e. near immunity. The expression of these viral sequences apparently interferes with the replication process. Replicase-mediated resistance, while potentially very strong, has been reported to be very narrow in nature (Mueller *et al.*, 1995). Thus a gene encoding replicase is effective against the virus strain from which it was isolated, but not against closely related strains.

Attempts to interfere with virus replication through the application of antisense RNAs have achieved mixed results. Genes that encode sequences complementary to viral RNA have been shown to be effective in providing resistance to viruses whose genome comprises DNA, such as the geminiviruses (Stanley *et al.*, 1990). However, antisense genes have been less effective against viruses whose genome is single-stranded RNA. In the case of the geminiviruses, whose replication cycle occurs in the nucleus, replication is especially sensitive to inhibition by expression of antisense genes.

In contrast, replication of viruses that replicate in the cytoplasm, such as those whose genomes comprise single stranded (+)-sense RNA molecules (representing more than 95% of the plant viruses known) is less susceptible to inhibition by antisense RNA (e.g. Powell *et al.*, 1989). There are, however, some notable exceptions, e.g. antisense strategies have been somewhat effective against the luteoviruses.

Other strategies are predicted to yield likely candidate genes for interfering with virus diseases, but have yet to be successfully demonstrated. Among these is the possibility of interfering with the proteins that are necessary for acquisition of the virus by the insect vector. The so-called 'helper-component proteins' are essential for virus acquisition. It is likely that several research teams are attempting to interfere with this process by expressing mutant acquisition proteins that block the involvement of the virus helper component in acquisition, or by expressing mutant CPs that

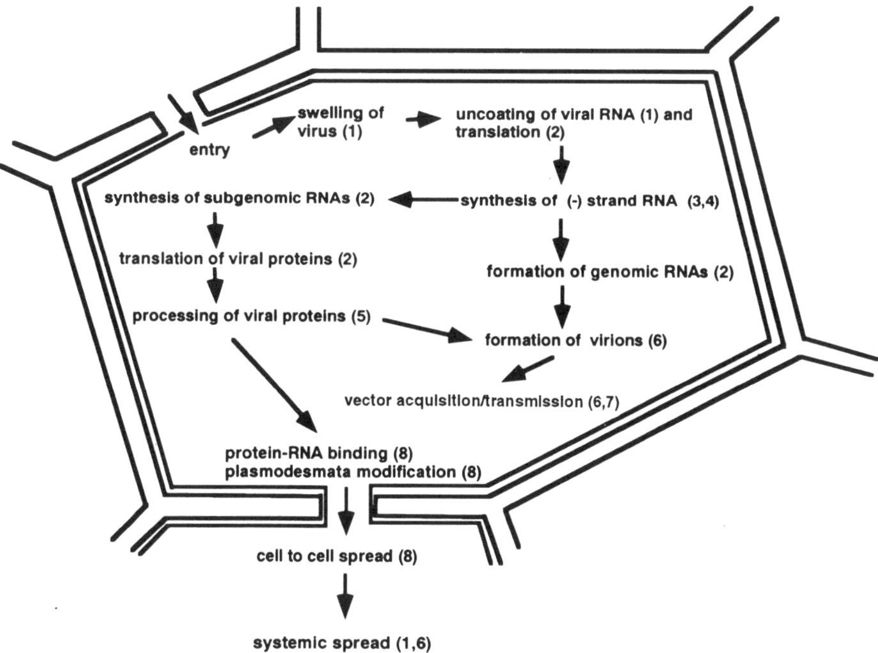

Gene products that may interfere with different steps of virus infection in transgenic plants:

1 coat protein
2 (-) sense RNA (+ribozyme)
3 (+) sense RNA (+ribozyme)
4 defective replicase
5 modified protease
6 defective coat protein
7 defective transmission factors
8 defective movement protein

Fig. 14.1. A generalized virus life cycle and gene products that may interrupt the cycle.

block transmission of the virus.

Other groups are attempting to block acquisition of virus by the expression of proteins such as antibodies or other competitors to block the sites in the insect to which viruses bind.

A promising type of disease resistance has recently been described by several research groups. This strategy involves attempting to block the spread of virus from cell to cell in the host plant, thereby limiting the accumulation of virus throughout the plant. This function is controlled by one or more virus-encoded protein(s) that modify the function and/or structure of the natural cell-to-cell communication channels between plant cells. For example, TMV encodes a 30 kDa movement protein that modifies the function of the plasmodesmata. This modification is apparently essential for the movement of the virus from cell to cell.

Lapidot *et al.* (1993) demonstrated that the expression in transgenic plants of a gene that encodes a dysfunctional movement protein from tobacco mosaic tobamovirus can interfere with the movement protein produced by infecting tobamoviruses, thereby limiting the spread of the virus from cell to cell. While this dysfunctional movement protein does not confer immunity to the transgenic plant, it does significantly interfere with the spread of the virus and the systemic effects of the disease. Furthermore, the modified dysfunctional movement protein used in this study is equally effective against all tobamoviruses thus far tested, and may be expected to interfere with the spread of other viruses that use similar mechanisms for cell-to-cell spread (Cooper *et al.*, 1995). Thus a dysfunctional movement protein is likely to provide the most wide-ranging type of resistance yet described. Similar approaches are most probably being taken by other research groups to block the movement of the geminiviruses and other important plant viruses (Beck *et al.*, 1994).

In summary, a number of experimental approaches have been taken to block the cycle of virus infection and replication in transgenic plants through the use of genes that are derived from the targeted viral pathogen. Approaches that are most likely to be successful will include those that prevent the spread of virus and its accumulation in infected cells or tissues. An effective resistance could be engendered by decreasing the accumulation of virus, thereby limiting the likelihood that the virus could be acquired and transmitted to adjacent plants by insect vectors. This would have the effect of reducing the rate of spread of the virus in the field, thereby reducing losses to the farmer.

Field Trials of Virus-resistant Transgenic Plants

One year after the first report of transgenic virus-resistant plants (Powell *et al.*, 1986), a field study was carried out with transgenic tomato plants that expressed a gene encoding the CP of TMV. In this first field test of virus resistance, the plants were evaluated for their agronomic characteristics and for resistance against inoculation with TMV. The results demonstrated that the CP gene was effective under field conditions, and that it significantly reduced the impact of infection on virus yield. Non-transgenic plants suffered a 30–35% decrease in yield, whereas plants that expressed the TMV CP gene showed little or no significant difference in yield compared with control plants that were not inoculated (Nelson *et al.*, 1988).

Since that time, a number of other experiments have been carried out with these and other transgenic plants, with equally encouraging results (e.g. Gonsalves *et al.*, 1992). In such field studies, the greatest degree of resistance has been provided by CP gene sequences that are closely related to the CP sequence of the virus from which the gene was taken. This implies that it is important to isolate CP genes from strains of virus similar to those that are endemic to the targeted region. Experiments carried out with potato demonstrated that plants that express CP genes for both potato virus X and potato virus Y were resistant to both viruses under field conditions.

Other field studies carried out over the last few years include those with transgenic cucumber plants that are resistant to cucumber mosaic virus and to several polyviruses, potato plants that are resistant to potato leaf-roll luteovirus, and tomato plant varieties that carry resistance against cucumber mosaic virus and potato virus Y. Only a few of the many field tests carried out to date have been reported in the literature.

As indicated above, studies are under way with transgenic plant lines that carry viral sequences other than those that encode CPs; some of these have included initial field trials. For example, it has been shown that there are high levels of disease resistance against tobacco etch virus (TEV), a polyvirus, in transgenic plant lines that express a segment of the TEV CP gene but fail to encode the CP. At the current time, none of the alternate strategies are in commercial development.

The level of resistance that has been engendered by the expression of CP genes can be from high to low. However, the general impression gathered from studies of CP-MR is that transgenic plants have significantly lower levels of virus accumulation as a result of expressing a CP gene than do those plants that do not.

To date, there have been no published reports on the epidemiologic impact of the transgenic virus CP. However, informal discussion with those involved in the field tests indicated that the reduced rate of virus accumulation in transgenic plants reduced the frequency of virus transmission by insect vectors. Thus, even though genetic transformation generally does not produce immune plants, it will have a significant impact on the spread of virus under field situations.

Concerns about the Use of Transgenic Plants in Agriculture

Tepfer (1993) discussed the pros and cons of the use of viral genes in transgenic plants under field situations. In this and other articles, researchers have pointed out that the use of virus-derived sequences to confer resistance raises important scientific questions. First, there is the possibility that CPs can transcapsidate other viruses, and make it possible to transmit virus to plants that are usually not exposed to the transcapsidated virus. This should be expected, based upon the nature of certain virus CPs and the lack of specificity of nucleic acid molecules that can be encapsidated. Furthermore, it has been known for some time that many plant virus diseases result from mixed infections. Under these conditions, mixed encapsidation can and often does occur. It is unlikely that transgenic plants represent a situation that is significantly different from that of mixed infection.

The second concern is that an mRNA expressed as a transgene could recombine with viral genomic RNAs, causing the derivation of novel virus sequences. The likelihood of such intermolecular recombination cannot be entirely ruled out. In fact, it has been demonstrated that in the case of some viruses, e.g. brome mosaic virus, intermolecular recombination between virus strains occurs. Virus strains recombine, however, depending upon the degree of sequence relatedness. Certain other viruses apparently do not recombine. Again, the pertinent question is not whether or not this can be expected to occur, but rather, if it does occur, what are the biological implications?

The third concern voiced is that the presence of a virus resistance gene would cause the pathogen to undergo a change to overcome resistance. This has been a traditional problem with breeding for disease resistance, and may also happen in CP-MR or other types of non-classical disease resistance. For that reason, it is the opinion of many scientists that the best types of resistance may be those that do not lead to immunity, but rather those that provide a relatively broad and yet durable type of resistance. Furthermore, should infection lead to the selection of virus strains that overcome resistance, it can be anticipated that the variant could become the source of a new gene for resistance.

Conclusions

Genes derived from fragments of virus genomes are effective transgenes for the control of plant virus diseases. The use of such genes has been shown to be effective in controlling diseases in a variety of field trials, and it is expected that plant varieties that have incorporated CP-MR will be released as improved cultivars in the near future. The rate at which such plants will be adopted for commercial use will certainly be driven by the success of the resistance and the ability to withstand repeated exposure to disease agents over multiple years and in multiple locations. Nevertheless, it is anticipated that this technology will be efficiently deployed in a variety of crops in a number of locations round the world before the end of the century.

Note in Proof:
In 1995 the Asgrow Seed Company (Kalamazoo, MI, USA) commercialised a variety of yellow squash that is resistant to two different viruses. Resistance was due to the expression of two transgenes that encode two different virus coat proteins.

Can We Slow Adaptation by Pests to Insect Transgenic Crops?

15

Richard T. Roush

Introduction

Strains of *Bacillus thuringiensis*, or more specifically the δ-endotoxins that they produce, have been used as biological insecticides for control of several lepidopteran and dipteran pests since the 1960s. Unfortunately, even the most recently developed *B. thuringiensis* products often have poor to mediocre efficacy against at least some targeted pests under field conditions compared with alternative insecticides, and they are relatively expensive, a particularly important consideration for the use of *B. thuringiensis* in developing countries. At least part of this expense is due to the costs of quality control. The production of consistently effective *B. thuringiensis* products is not an easy task, and quality control of such products has been a problem recently in the USA (B.A. Federici and W.F. Moar, 1993, personal communication) and Israel (M. Wysocki, 1993, personal communication). The *β*-exotoxins produced by some poorly characterized strains can adversely affect predators and parasites (Flexner *et al.*, 1986). Further, *B. thuringiensis* cannot be used effectively against a number of pest species that might otherwise be sensitive, such as stem-borers, because of their feeding habits. Consequently, at least the current formulations of *B. thuringiensis* seem unlikely to reduce the use of conventional insecticides, except in those areas with widespread resistance or if more effective insecticides are banned for environmental reasons.

One tactic demonstrated to improve the effectiveness of *B. thuringiensis* is genetic transformation of plants. The genes that produce the active ingredients of *B. thuringiensis* (δ-endotoxins) have been transferred to and

 Biotechnology and Integrated Pest Management (ed. G.J. Persley)

expressed in tobacco, tomato, cotton (Gasser and Fraley, 1989; Benedict *et al.*, 1993), rice, maize (Warren *et al.*, 1992; Koziel *et al.*, 1993a) and potato (Perlak *et al.*, 1993). Cotton, maize and potato already provide excellent control of targeted pests. This approach revolutionizes the use of *B. thuringiensis* by effectively increasing both its persistence and coverage of the plant, allowing the potential for use of *B. thuringiensis* even against stem-borers. This could, in turn, drastically reduce the use of 'hard' chemical pesticides in crops, as well as reducing the use of fossil fuels and the soil compaction that is incurred with sprays of synthetic or biological insecticides. Thus, transgenic plants may be one of the most important recent developments in crop management. Given the reports that one often hears about the amounts and types of insecticides used on vegetables in developing countries and the limited extent to which vegetables are processed prior to consumption, vegetable crops seem to be an important target for this kind of genetic improvement. Given the increasing ease with which plants are being transformed and the increasing availability of codon-optimized *B. thuringiensis* constructs, this may be a locally accessible alternative to the quality-control problems associated with *B. thuringiensis* fermentation.

Unfortunately, there are also concerns that the benefits of genetically transformed plants will be short-lived (McGaughey and Whalon, 1992). As much as 1640-fold resistance to *B. thuringiensis* has been found in localized populations of the diamondback moth (*Plutella xylostella*) (DBM) from Hawaii, Florida and Asia (Tabashnik *et al.*, 1992a; Tabashnik, 1994). As I shall discuss later in this chapter, at least some of these strains can severely damage *B. thuringiensis* crucifers. Laboratory selection programmes have generated resistance in Indianmeal moths (IMM) (McGaughey, 1985), one cotton bollworm (*Heliothis virescens*) (Stone *et al.*, 1989; Gould *et al.*. 1992) and the Colorado potato beetle (CPB) (Whalon *et al.*, 1993b), although none of these strains has been documented to prosper on *B. thuringiensis* plant genotypes. On the contrary, the *B. thuringiensis*-resistant CPB fails to develop successfully or reproduce, even on genotypes with very low expression of *B. thuringiensis*, and will be discussed later in this chapter.

Although some molecular biologists have assumed that resistance to *B. thuringiensis* can be easily overcome by simply modifying the structure of the *B. thuringiensis* gene, it was similarly once thought that we could stay ahead of pesticide resistance by making new pesticides. Unfortunately, pests have been able to adapt more quickly than chemists (National Research Council, 1986), and the same seems likely to happen for *B. thuringiensis*. In *B. thuringiensis*-resistant IMM, cross-resistance extends broadly to most *B. thuringiensis* isolates tested (McGaughey and Johnson, 1987) and was very stable even in the absence of selection (McGaughey and Beeman, 1988). It has subsequently been shown in both IMM and DBM that resistance extends only to one toxin type, as a result of decreased binding affinity (van

Rie *et al.*, 1990; Ferre *et al.*, 1991), suggesting that it is not strains but toxin types that are important for toxicity.

Further, there seem to be relatively few toxin types that are effective against any given species of pest. In the case of DBM, for example, only the *CryIA* and *CryIC* toxins seem to be very effective (A.M. Shelton and J. Tang, unpublished data). Subsequent selection in IMM showed that resistance could readily be extended to *B. thuringiensis* strains (carrying alternate toxins) to which the original resistant strains were susceptible (McGaughey and Johnson, 1987). Selection on *H. virescens* for resistance to *CryIA(c)* has produced a single strain with broad cross-resistance to a range of other *B. thuringiensis* types, including *CryIB*, *CryIC* and *CryIIA* toxins (Gould *et al.*, 1992). Besides the problem of cross-resistance to multiple toxins, resistance to any single toxin appears to be more problematic than originally thought. In the case of DBM, resistance to *B. thuringiensis* sometimes decreases gradually when selection pressure is removed, but resistance is quickly restored when selection pressure is reintroduced (Tabashnik, 1993; Shelton, unpublished data).

Thus, it may be futile to develop *B. thuringiensis*-transformed crops unless we also develop approaches to extend their usefulness. At the same time it has been argued that *B. thuringiensis*-transformed plants, when planted in large acreage, may cause rapid and widespread resistance so that *B. thuringiensis* foliar sprays will no longer be effective and will be useless in a rotational scheme of insecticide management.

As will be discussed later, the assumption that transgenic plants will cause resistance faster than sprays is not necessarily true, although identifying those cases where plants are better than sprays may be complex. None the less, it is certainly likely that the increased use of *B. thuringiensis* toxins that will result from transgenics will result in the more rapid evolution of resistance. This forces us to consider what we expect of *B. thuringiensis*. If our goal is to prevent resistance to *B. thuringiensis* the best way is to leave it on the shelf. I suggest instead that we should aim to use *B. thuringiensis* in the manner most effective to achieving societal goals of reducing pesticide use and increasing or at least stabilizing crop yields. In this context, the increased efficacy of transgenic plants gives much to recommend them. In any case, the influence of transgenic *B. thuringiensis* plants will have long-term and profound consequences for the continued use of *B. thuringiensis* sprays and the future uses of biorational insecticides.

While current thinking focuses on transgenic *B. thuringiensis* plants, other insect control genes are also being incorporated into plants and some (e.g. proteinase inhibitors) have provided significant levels of insect control (Hilder *et al.*, 1987). Thus, even though I will concentrate on *B. thuringiensis* in this chapter, *B. thuringiensis* should probably be considered only a model for transgenic resistance factors. I emphasize *B. thuringiensis* merely because we have a lot more experimental data for it than for any other transgene.

Hopefully, what we learn about managing resistance to *B. thuringiensis* even where we fail, will help to establish sound principles for managing resistance of other plant-incorporated insecticides.

Principles of Resistance Management for Transgenics

To manage resistance, it is obviously necessary to minimize 'selection pressure', itself a somewhat ambiguous notion. Selection works in a heavily non-egalitarian way, by discrimination between genotypes. For example, doses of toxins that do not make life hard for susceptible individuals, either by killing them or by reducing their reproductive output, do not select for resistance. (Unfortunately, except for plant varieties that are so much more vigorous that they can withstand insect attack, weak expression of genes cannot do much to manage the pests.) On the other hand, doses that are so high that they kill all individuals in the population, including the most resistant genotypes, do not select for resistance either, because no one has been favoured by discrimination. However, it is often difficult to achieve this high level of kill, and especially dangerous to try to do so in most circumstances with classical insecticides. Further, the slow decay of toxin residues means that there will almost certainly be a time period where discrimination works strongly in favour of resistance (Tabashnik and Croft, 1982). If one is going to avoid discrimination, one wants to have tight control over exposure such that the insect is exposed only to non-selecting doses at the high or low end but not in between.

There are at least four possible ways to minimize discrimination in favour of resistance by transgenic plants: (i) modify the expression of the genes in each plant such that they are expressed only when and where needed through tissue-specific, temporal-specific or inducible promoters; (ii) use varieties that express *B. thuringiensis* genes only moderately strongly, so that not all susceptible individuals are killed; (iii) include within the cropping system a percentage of plants that do not produce *B. thuringiensis* (a 'refuge'); and (iv) use plants carrying mixtures of *B. thuringiensis* toxins.

Specific promoters

The first option for managing resistance to genetically transformed plants was advanced by Gould (1988a,b). Most current varieties of *B. thuringiensis* transformed plants express the genes continuously throughout growth of the plant. An alternative approach that would moderate selection is to use specific gene promoters that would express the *B. thuringiensis* genes only: (i) in the most important tissues ('tissue- or structure-specific' expression), such as the bolls in cotton; (ii) in critical growth periods ('temporal-specific' expression) (Gould, 1988a,b); or (iii) when environmentally induced, perhaps by the spraying of an environmentally benign chemical (Williams *et al.*, 1992). These systems are not necessarily exclusive; a temporally specific promoter may be effectively structure-specific if it is turned on only when needed (e.g. late in the season and affecting only the top of a plant.) Unfortunately, the molecular biology of such expression systems seems highly complex. Thus, at the very least, they are likely to be costly in both time and money to develop. Perhaps more importantly, the systems may be difficult to control in a way that manages resistance.

Inducible and temporal-specific promoters

My collaborators and I have started to test whether the residues in inducible delivery systems can be tightly controlled. By manipulating light intensity on different plant parts in a variety of tobacco transformed with a light-inducible promoter, we have found that we can easily prevent the apparent production of a *B. thuringiensis* toxin (by covering growing plant parts), but that, once the leaf has grown in light (and toxin has been produced), covering the plant part does not cause a rapid decay of the residue (T. Metz, R.T. Roush and E. Earle, unpublished data). On the contrary, it appears that, once the toxin has been made, the residues continue to linger at a high level until the cell walls break down with leaf decay. Our observations are consistent with more detailed analyses by Carozzi *et al.* (1992), indicating that older tissues have generally higher levels of toxin than younger tissues.

Although this system we have studied is inducible, it illustrates the problems that might be suffered by temporal expression as well. A number of targeted crops, including potato and cotton, can suffer moderate to severe defoliation at some times of the year without yield loss. Unfortunately, for both cotton and potato and most other crops, it is the end of the season that is least vulnerable (e.g. Ferro *et al.*, 1983); if the residues do not decay much after the early-season promoters shut off, temporal-specific promoters may not help much.

A related problem with environmentally inducible promoters, at least some of those that have been developed so far, is that they are 'leaky'; that is, they give great expression when induced, but some expression, perhaps at selecting levels, even when they are not (John Ryals, July 1993, personal communication). Promoters that must be induced with chemical sprays also have a host of potential sociological problems. First, it seems pretty clear after the assault in the USA on Alar, a material that was neither a pesticide nor a strong carcinogen but was characterized as both, that the public has spoken loudly and clearly; they did not say that they only wanted safe chemicals sprayed into the environment, they stated pretty clearly that they do not want any chemicals sprayed into the environment. It is not obvious why a chemical inducer would escape such wrath, not just in the USA but eventually in many other countries. Second, given that one of our current problems in managing pesticide use and pesticide resistance is that farmers are risk-aversive, what would deter a grower from using the inducer on a weekly schedule, negating its benefits?

Tissue- or structure-specific promoters

Gould and Anderson (1991) and Gould *et al.* (1991b) have demonstrated that one of the potential advantages of a tissue- or structure-specific approach is that insects might be deterred from feeding on the most economically important plant structures without being killed, thus lowering selection pressure. In the field, transgenic cotton did not cause any consistently significant change in the behaviour of *Heliothis virescens* larvae (Benedict *et al.*, 1993). In laboratory and field tests on CPB and *Manduca sexta*, we have not found strong feeding deterrent (choice experiments show no more than 40% of the insects avoid the transgenic leaves) to transgenic potato or tobacco (respectively), which raises questions as to how general the benefits might be (R.T. Roush and N. Carruthers, unpublished data).

Further, to be fully effective, expression in the transgenic structures may have to be fairly high. Otherwise, physiologically resistant larvae may still attack their preferred (and presumably more nutritious) parts of the plant, resulting in greater fitness. In the case of IMM, for example, when fourth-instar larvae are given a choice between untreated diet and diet treated at a concentration that kills susceptible but not resistant larvae (100 mg/kg of diet; McGaughey, 1985), only 22% of the susceptible larvae are found in the treated diet after 2 days. In contrast, 53% of resistant larvae are found in the treated diet, a difference that was statistically significant ($p < 0.05$; R.T. Roush and W.H. McGaughey, unpublished data). When the concentration was increased to 1000 mg/kg, all larvae avoided the treated diet.

Alternatively, all first-instar larvae might be deterred and driven off to susceptible tissues, but become more tolerant as they get older (Zehnder and Gelernter, 1989) and may then return to preferred tissues. Early-instar larvae can recover from short-term feeding on *B. thuringiensis* (Abdul-Sattar and Watson, 1982). In addition to the feeding damage that could occur, resistant larvae might be able to return to preferred tissues sooner, conferring a reproductive advantage.

Another problem is designing promoters that will provide expression high enough to protect all important tissues without affecting the tissues that can be sacrificed. Much is now known about tissue-specific promotion of genes in plants (Edwards and Coruzzi, 1990), but most of the promoters appear to be either too specific or not specific enough for our needs. In cotton, for example, both bolls and terminal meristems must be protected (Hopkins *et al.*, 1982; Wilson and Waite, 1982; Ramalho *et al.*, 1984). There are promoters that affect both meristems and flowering parts in crucifers, but at least some of these promoters also leave product in stems (Medford *et al.*, 1991), which may or may not be relevant to some pests. It is not yet clear whether the levels of expression achieved by such promoters will be adequate in crucifers or other plant species to protect against key pests. In contrast, the PEPC promoters used in maize provided high levels of expression in green tissue and much lower expression in kernels (Koziel *et al.*, 1993a). While this may make maize products more acceptable to the consuming public, the level of expression in the kernels is still four to six times the LC_{50} for European corn borer (*Ostrinia nublialis*) (calculations from data in Koziel *et al.*, 1993a). Damage to maize kernels is tolerable for maize grown for feed, but this high a level of expression suggests that the ear will not prove to be a very hospitable site for susceptible corn borers.

While their use may be complex, it would be premature to dismiss temporal, inducible or structure-specific expression. There is much to be learned about how genes are controlled in plants, and therein lies considerable potential to develop resistance-management strategies ideally suited for specific circumstances. On the other hand, because of the technical difficulties inherent in perfecting the expression needed, it seems unlikely that such plants will be available in the near future or that the approach will be readily generalized across a wide range of pest taxa or commodities.

Moderate expression

The second general approach for reducing the intensity of selection for *B. thuringiensis* resistance would be to use varieties that express *B. thuringiensis* genes only moderately strongly, so that not all susceptible individuals are killed on any given plant. This is functionally equivalent to using lower concentrations of insecticide in each application, the 'low-dose' approach to managing resistance. In general, models suggest that a disadvantage of the low-dose approach is that one must allow a fairly significant proportion of the treated individuals to survive if resistance is to be significantly delayed (Curtis, 1991), which is not always practical. This point can be illustrated with computer simulations, using dose–mortality data from studies of *B. thuringiensis*-resistant DBM. For example, a concentration of *B. thuringiensis* applied to a leaf surface that killed 80% of susceptible larvae killed only 60% of F_1(presumed heterozygous) larvae; the LC_{90} for the susceptible larvae killed only 70% of the F_1 (Tabashnik *et al.*, 1992b). Using these data to set mortality values for the model predicts that resistance should occur in less than about 30 generations. When only 80–99% of susceptible larvae are killed, resistance occurs in less than about 30 generations, and it is hard to imagine that growers would willingly tolerate much poorer control than 80%. Only when all homozygous susceptible insects (SS: 100%) and greater than 95% of the heterozygotes are killed does resistance begin to be significantly delayed (Fig. 15.1). Resistance in the field in Hawaii, from which these DBM were studied, evolved after about 50–100 treated generations (Tabashnik *et al.*, 1992a), a result which can be achieved with this model by lowering the initial gene frequency or (perhaps more reasonably) raising the portion of larvae untreated with *B. thuringiensis* to 30–40%.

In the only cases where the low-dose approach has been successfully applied to resistance management, it is incorporated with the use of other alternative controls, such as predators, which can further reduce pest populations (Roush, 1989). However, as noted by Gould *et al.* (1991a), natural enemies may accelerate the evolution of resistance in a transgenic system if the natural enemies are more effective against susceptible individuals than against resistant ones. We have artificially manipulated larvae of CPB to mimic resistant (fed non-transgenic plants) and susceptible larvae (fed transgenic plants). In every one of ten replicates, the sick susceptible larvae were much more heavily attacked by predators than were the healthy 'resistant' larvae (R.T. Roush, C. Richael and W.M. Tingey, unpublished data). Warren *et al.* (1992) similarly observed that parasitism tended to be higher on transgenic than on non-transgenic tobacco.

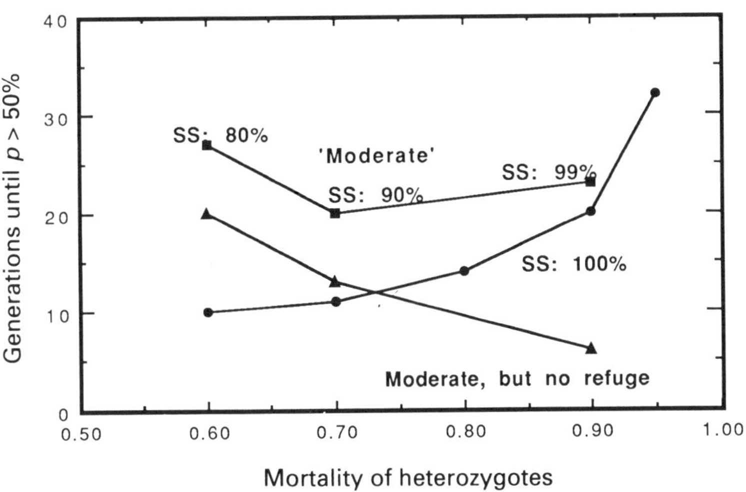

Fig. 15.1. Moderation in use to manage resistance to *B. thuringiensis*. Results are given in terms of the number of generations until the frequency of the resistance gene (*p*), initially at 10^{-6}, exceeds 50%. The numbers near the curve show the approximate percentages of mortality of susceptible (SS) insects from data on resistant and susceptible diamondback moth (e.g. at a concentration that killed 90% of susceptible larvae, only about 70% of F_1larvae were killed; Tabashnik *et al.*, 1992a). In the middle curve, all SS larvae were killed. Except for the curve noted at the lower right, 10% of the population was assumed to escape exposure each generation.

Another suggested advantage of a moderate-dose approach is that it may be sufficient to slow development of the insects and thereby reduce the total number or earliness of generations such that the pest is less likely to be a problem (Ferro, 1993a). We have studied this with a clone of potato that kills only half of the neonate CPB larvae. Development time was slowed only about 10% (R.T. Roush, C. Richael and W.M. Tingey, unpublished data), not enough to be of much benefit.

On the other hand, a serious and perhaps fatal problem for the use of moderate expression is that it appears to be difficult to adjust concentration of *B. thuringiensis* such that only moderate mortality is reliably obtained. The average mortality in our moderately expressing potato plants was 50%, but the range was 20–80%, expression that would probably be too unreliable to satisfy growers. A second problem for a moderate-expression approach is illustrated with *Heliothis* on a crop that faces a complex of several pests. In such cases, a variety that will effectively compete with pesticides will need to be capable of controlling all of the targeted pests in the complex. In other words, the concentration of *B. thuringiensis* must be targeted against the least susceptible pest, implying that some species are going to suffer a much higher dose. Finally, although a moderate dose may be better buffered against selection for a major gene, it may select more effectively for minor genes.

Refuges

Approaches that leave refuges for susceptible insects have been important for the management of resistance to chemical pesticides (National Research Council, 1986; Roush, 1989), and appear to be promising for pest-resistant plant varieties produced by classical breeding (Gould, 1986a,b). In theory, the refuge approach would be especially useful where the expression of the *B. thuringiensis* gene is extremely high, causing greater than 95% mortality of heterozygous insects (Fig. 15.2), which is equivalent to a 'high-dose' or 'high-kill' approach in the management of pesticide resistance. With the high-dose approach, resistance is delayed because many resistant heterozygotes, the most common carriers of resistance (and perhaps even many resistant homozygotes), are killed by the dose of toxin used. Although most effective when the plants show high expression relative to survival of the heterozygotes, a refuge will always be beneficial in diluting resistance (Fig. 15.2).

Although successful in laboratory experiments and in models, high-dose or high-kill tactics do not appear to be generally promising for the management of resistance to synthetic pesticides because of environmental concerns and because of the lethal effects of pesticide residues on the inward migration of susceptible insects (Tabashnik and Croft, 1982; Roush, 1989). However, these problems can be overcome by genetically transformed

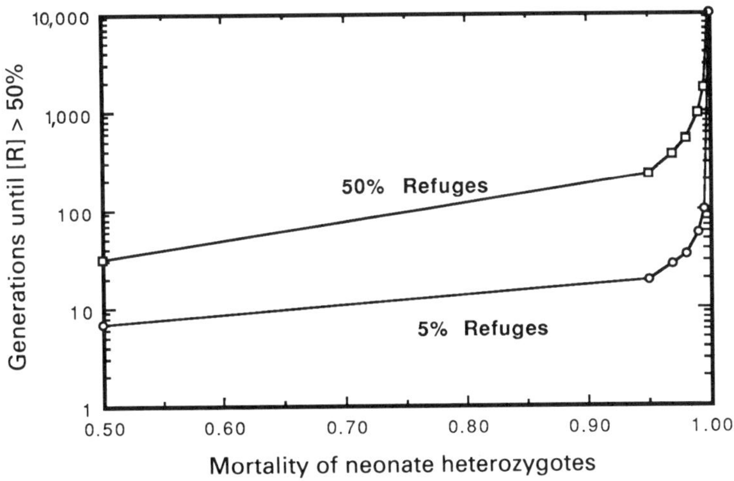

Fig. 15.2. Effect of mortality (from feeding on transgenic plants) of *B. thuringiensis*-resistant heterozygotes on the evolution of resistance ([R]). Results of a simulation model assuming a single locus, random mating, no selective mortality of resistant homozygous larvae and initial resistance allele frequencies of 10^{-6}. When there is 100% mortality of susceptible larvae feeding on the transgenic plants, the upper and lower curves illustrate that the time to resistance evolution is longer as a greater fraction of the population develops on fully susceptible plants (in the refuge).

plants: *B. thuringiensis* residues are considered safe, and a stomach poison would not deter the mating of susceptible and resistant individuals among the pests. The larvae are usually most susceptible when first hatched, making possible a relatively high dose. Coverage is essentially complete. However, in order for the plants to effectively slow resistance, there must be a relatively large number of susceptible individuals within the population to dilute the resistance of any *B. thuringiensis*-exposed survivors, especially resistant homozygotes. This in turn requires that there are non-transformed host plants within or neighbouring the crops, close enough for resistant insects to be likely to mate with susceptible counterparts.

The effectiveness of refuges, moderate doses and sprays has been modelled experimentally in the laboratory (R.T. Roush and W.H. McGaughey, unpublished data), using IMM. In these experiments, selection was conducted on a 'synthetic' strain, one composed mostly of susceptible individuals, with a small fraction of resistance genome (at a 2% frequency) hybridized into the strain through backcrossing to ensure that there was genetic variation for a selection response. In IMM, a concentration of 100 mg *B. thuringiensis*/kg diet killed about 95% of susceptible larvae, 60–80% of heterozygous larvae and less than 10% of resistant larvae. Increasing the concentration just fivefold to 500 mg/kg kills greater than

99% of all susceptible and heterozygous larvae and up to 50% of resistant strains (McGaughey, 1985). When monitored at a single diagnostic concentration of 200 mg/kg, resistance developed within five or six generations to treatments that modelled moderate expression in plants without refuges or spraying, but not to a treatment that modelled a mixture of 90% resistant plants showing high expression and 10% susceptible plants (Fig. 15.3).

Sprays vs. transgenic plants

These experiments suggest, perhaps contrary to intuition, that transgenic plants may actually delay resistance more effectively than sprays. *Bacillus thuringiensis* sprays share a problem with chemical insecticide sprays; the level of control achieved of heterozygotes is relatively poor compared with that of a plant that can provide high expression. In fact, based on the history of the evolution of resistance in the field (Tabashnik *et al.*, 1992a), genotypic fitnesses estimated from data with foliar preparations (Tabashnik *et al.*, 1992a, b) and recent results with transgenic plants, we can infer that plants could outperform sprays in managing resistance in DBM.

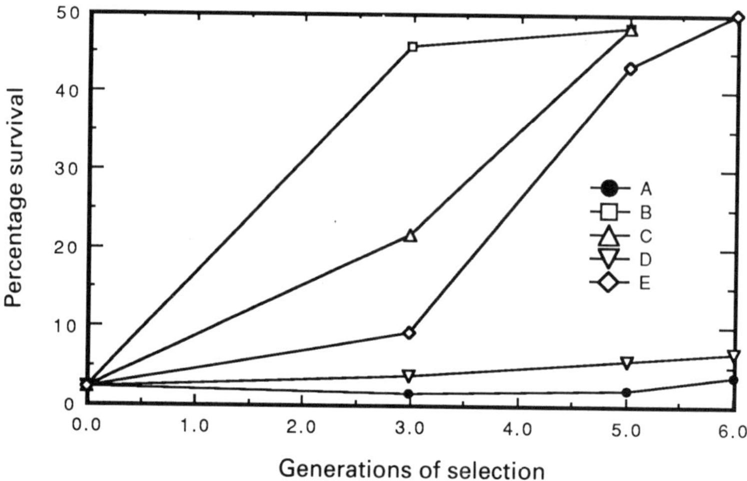

Fig. 15.3. Effect of *B. thuringiensis* concentration on the evolution of resistance in the Indianmeal moth, *Plodia interpunctella*. A, Untreated control; B, 100 mg/kg diet each generation (modelling moderately high continuous expression in plants or routine spraying with no refuges for susceptible insects); C, 500 mg/kg diet every other generation (a control for selection at this high dose); D, 90% of the population treated at 500 mg/kg diet and 10% untreated (to mimic a seed mixture with plants showing high expression); E, 60% of the population treated at 500 mg/kg diet, 30% at 100 mg/kg diet and 10% untreated (to mimic sprays with non-uniform coverage and residue decays such that half of the treated population received dose).

We have produced broccoli (*Brassica oleracea*) transformants expressing a *B. thuringiensis* gene at a sufficient level to kill susceptible DBM larvae, but unable to kill a population that developed resistance to foliar sprays in the field (T. Metz, E. Earle, R.T. Roush and A. Shelton, unpublished data). These results justify concerns about insect resistance to transgenic plants, but the plants were generated to serve as a model crop plant to test theories about the management of resistance. Thus, the genotypes used have not been intentionally modified to maximize expression, which might control even these resistant insects. None the less, the F_1 larvae all died when tested as neonates. Even at only a 5% refuge (half that used for the simulations of sprays) and only 99% mortality of heterozygotes, the transgenic plants would have controlled the insects for 110 treated generations (Fig. 15.2), longer than predicted by the same model for sprays (Fig. 15.2) or observed in the field. The issues inherent in providing the necessary refuge will be discussed below, but at least some strategies can enforce a refuge even where this is difficult for sprays.

Sprays may also suffer another disadvantage compared with transgenic plants: they may be vulnerable to a wider diversity of resistance mechanisms. We have tested a strain of CPB that shows at least 60-fold resistance to *B. thuringiensis tenebrionis* spray preparations (Whalon *et al.*, 1993a) against transgenic potato clones that show low levels of expression, apparently less than five times the concentration needed to kill all susceptible beetles (e.g. clone 13 of Adang *et al.*, 1993). When sprayed at a high concentration, 100% of field-collected larvae die, but more than 70% of the resistant strain survives. On the other hand, adults of the *B. thuringiensis*-spray-resistant strain cannot lay eggs when fed on transgenic plants, nor can larvae develop to pupation (R.T. Roush, C. Richael and M.E. Whalon, unpublished data). Thus, it appears that the mechanism(s) in the spray-resistant strain are ineffective against the plants. This may be because the *B. thuringiensis* toxins in sprays must be processed before they can bind to and disrupt the gut membrane, whereas the plants produce only active toxin (McGaughey and Whalon, 1992).

Seed mixes vs. refuges

Perhaps the only real controversy over the need for a refuge with transgenic crops is whether it should be outside the crop or inside the crop as a seed mix. Where only the larvae feed and do not move from plant to plant, a seed mix is ideal. It ensures that every farmer has a refuge with developmental phenologies and insect-control practices consistent across both host-plant types, and, as long as the fraction of susceptible seed or natural pest infestation is typically low, there would be little or no yield loss without additional management. It also ensures that the susceptible insects develop in close proximity to the resistant insects, which will help to ensure random mating.

On the other hand, if the larvae move around and feed on different plants, one net effect may be to reduce the proportion of the population in a refuge (a result of 'suicidal dispersal'). Second, heterozygous larvae that could have been killed as neonates may now survive if they move from a non-transgenic to a transgenic plant at a later, less susceptible life stage. This would increase the relative fitness advantage of heterozygotes (i.e. more than 5% of the heterozygotes on the transgenic plant could survive). The result of these factors is that resistance can occur more quickly with a seed mix than with a similar proportion of susceptible plants outside the crop. Using a model that assumed that larvae would move between plants twice in their lifetimes, Mallet and Porter (1992) showed that, under certain conditions, using a seed mix when a refuge was also present could cause resistance more quickly than if only the refuge was relied upon. Their paper has been widely interpreted to show that seed mixes could be worse than pure stands of transgenic plants, but the pure stands were always modelled to include a refuge.

As noted by Mallet and Porter (1992) and shown in more detail by Tabashnik (1993), there is a rather limited set of conditions under which seed mix can speed resistance in the presence of a refuge. First, the mortality of heterozygotes must be very high, close to 99%, when they stay on transgenic plants. Second, at least 20% of the larvae must move between plants. Third, when larvae move, unless neonate heterozygotes show better than about 5% survival on the transgenic plants, the older heterozygotes must show a significant advantage over both susceptible homozygotes and neonate heterozygotes. Even under these conditions, the worst case shown by Mallet and Porter (1992) is that, where resistance would evolve to a pure stand with a 10% refuge in 779 generations, adding a seed mix would cause resistance in about half the time, 419 generations. While one would always like to double the time to resistance, 419 generations is at least 50 years for most insect pests. The added benefits pale before the risk of not using a seed mix, where resistance might evolve in just a few generations if the refuge disappeared or failed to produce insects.

The effects of movement and survival of heterozygotes can be illustrated with data from DBM (Table 15.1) and computer simulations of the effects of interplant movement when the larvae move only once in a lifetime. When mortality of the neonate heterozygotes is less than 98%, there is not much difference between the seed mix and the refuge (Fig. 15.4). The combination of a 10% seed mix with a refuge and limited survival of late-instar heterozygous larvae (10%, a little higher than observed for DBM) outperforms a refuge alone, even in the face of 20–50% larval movement. When mortality of neonate larvae is very high (e.g. when only 0.01% of the neonates survive) and the refuge is provided only outside the crop (SM =

Table 15.1. Diamondback moth survival on transgenic broccoli.

Strain (N = number tested)	% Survival
SS neonates (N = 1000)	0
SS late instar (N = 1000)	0
R (field) neonates (N = 325)	90
F_1 neonates (N = 170)	0
F_1 middle instars (N = 150)	2
F_1 late instars (N = 920)	8
F_2 neonates (N = 1000)	21

0), the time until the resistance allele frequency reaches 50% can be quite long (779 generations). In contrast, where there is a seed mix and the survival of late-instar larvae is high (RSM = 0.5) or there is a lot of

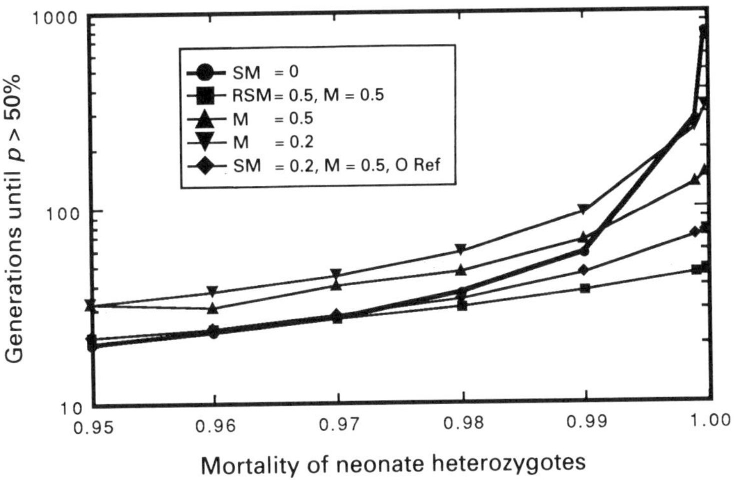

Fig. 15.4. Effects of interplant movement on resistance evolution in Lepidoptera when the level of expression is always sufficient to control susceptible larvae of any instar. Following Mallet and Porter (1992), the initial frequency of the resistance allele, p, is assumed to be 10^{-4}. Except as noted, the survival of late-instar heterozygous larvae that move from a transgenic to a non-transgenic plant (RSM) is assumed to be 10%, using values found for F_1 diamondback moth (DBM) larvae. Further, except as noted, the percentage of the population that is growing outside the transgenic crop (Ref) and the percentage of susceptible plants in the transgenic crop (SM) is 10%.

interplant movement of the larvae (M = 0.5), the time to resistance is much shorter. Thus, the main effect of a seed mix is to negate the benefits of extremely high mortality of neonate heterozygotes.

How commonly such high mortality of heterozygotes may be achieved is not yet known. Although the results with DBM (Table 15.1) show that high mortality can occur, less sensitive insect species may tax the level of expression that is possible in the plants. Certainly, whenever susceptible insects survive exposure, we can expect that resistant homozygotes will do a little better, but we shall not know how well until resistance evolves. Further, since laboratory-selected strains are often unrepresentative of those that occur in the field (Roush and McKenzie, 1987; Roush and Daly, 1990), we may never really know the survival of heterozygotes until resistance evolves in the field. However, on a more optimistic note, all of the simulation results described above suggest that seed mixes will do fine as long as less than 20% of the larvae are moving between plants. Even if tedious to do so, interplant movement can be measured before the evolution of resistance in order to avoid seed mixes where they are likely to have serious problems.

In the case of CPB, similar simulations strongly suggest that, due to the feeding-suppressive and even lethal effects of plants on the adults, seed mixtures will be an essential feature of any resistance-management programme, as will be discussed later.

Feasibility of seed mixes

One concern about the seed-mix approach is the practicality of mixing seed. Potato has been of particular concern. Due to the bulk of potato seed pieces, it has been argued that growers would be unable to mix seed efficiently. However, mixing can occur inadvertently during transfers at several steps during commercial production if the seed potatoes are simply dumped into storage together. We have tested this by persuading a grower to mix two potato varieties with different flower colours at cutting on his farm, which is several steps fewer than the mixing that would occur if the seed producer mixed the varieties in his storage facility. The grower simply dumped one 'box' of the odd variety on top of a load of the standard variety. At flowering, we took 100 3-m samples randomly through the 16 planted rows and calculated a Morisita's index ($I\sigma$) for dispersion (Southwood, 1966). Our value of 1.84 was not random ($P < 1\%$), but the marker plants were not strongly clumped either. We believe that this suggests that mixing potato seed is not a real problem in a typical commercial setting (R.T. Roush and W. Tingey, unpublished data), and similar solutions can probably be found in other systems.

Optimal seed mixtures or refuge sizes

Intuitively and in simulation models, the higher the proportion of susceptible plants, the greater the delay of resistance. On the practical side, however, a high level of susceptible plants may result in considerable yield loss, increased pesticide use or, as discussed below, faster resistance, contrary to our objectives. This would be especially serious if females selectively oviposit on non-transgenic plants. At least in our experiments, both female moths and CPB lay their eggs at random with respect to whether or not plants express *B. thuringiensis* (R.T. Roush, P. Beckley, N. Carruthers and W. Tingey, unpublished data).

For seed mixes at least, one standard for the optimization of the relative densities of susceptible plants might be to use the highest percentage that does not cause a yield loss or allow plants to be so stripped that larvae will be encouraged to disperse. We have tested this in CPB and found that, even under densities at least four- to eightfold higher than would be tolerated by growers, there was no yield loss or significant defoliation, even where 45% of the potato plants were non-transgenic (yields in t/ha pure transgenic plots, 47.8; 30% non-transgenic, 48; 45% non-transgenic, 45.5; 100% non-transgenic, 7.6). However, it was clear that the 45% non-transgenic plots allowed enough survival of beetles for their population densities probably to increase over the years, which suggests that, in the absence of some other control tactic, yields would eventually suffer. Further, as first pointed out by Ferro (1993a) specifically with respect to CPB, densities so high that the susceptible plants are defoliated may actually accelerate resistance by causing starvation and mortality in the susceptible refuge. In contrast, plots with 30% susceptible plants allowed only about as many beetles to develop in the plots as arrived there (about 110), and would thus be expected to hold population densities stable over time.

Thus, for insect pests that develop more or less exclusively on crops targeted for *B. thuringiensis* the ideal seed mixture may be less than that which will allow population growth. For pests like cotton bollworm with a wide range of hosts, this approach would not apply, and the simple criterion of the maximum quantity of susceptible hosts that avoid yield loss would be more appropriate.

Extinction or resistance?

Except for Ferro (1993a), models published to date for the management of transgenic insecticidal crops have not explicitly included pest population growth. When population growth is considered for pests that feed essentially only on transgenic hosts (as is often the case for both CPB and DBM), it is at least theoretically possible that transgenic plants coupled with low relative densities of untransformed hosts will cause local extinction. However, especially if the pest population grows at a rapid rate, the refuge/seed-mix percentage may have to be very small to prevent population growth. If a pest population grows at a rate of 40-fold per generation in the absence of plant protectants, one can only afford about 2% of the population on non-transgenic hosts each generation to prevent population replacement. Such low frequencies of susceptible hosts may cause rapid selection for resistance. Under what conditions will resistance occur before extinction?

A related problem is that all of the models published thus far for transgenic plants take a deterministic approach to resistance, assuming essentially infinite population sizes and complete random mating. However, resistance is most likely to occur where there is a 'clutch' of resistant insects among populations that are strongly subdivided by reason of distance or patchy habitat. Thus, I developed models that could be called 'directed deterministic'; I tried to look at just those patches that would be most likely to produce resistance if a stochastic approach were adopted.

To study both extinction and subdivision, I used two equations from population ecology. The first defines the genetic population size in terms of the area under which random mating can occur. The neighbourhood effective population size, N_e, depends strongly on S, the mean displacement, which is the distance between the birth sites of individuals and their offspring. It also depends on pest density, d, as follows (modified from Futuyma, 1979):

$$N_e = 4\pi S^2 d$$

The model is conservative in that once the number, N, of individuals in an area covered by the local population (or 'deme') is set, it cannot increase except through its own reproduction (i.e. there will be no in-migration of susceptible insects), intensifying the potential and rate of resistance evolution.

If the population size is constricted by mortality on the plants, the frequency, P, of mated females may decline, due to the inability of males to find them at low densities, and can be derived from equations described by Hopper and Roush (1993) to give

$$P = 1 - e^{-mld}$$

where m is the distance at which males always find and mate with females, l is the distance traversed by males during the generation and d is the pest density (half of which are assumed to be males). I also assumed that the level of homozygosity might be increased by inbreeding, but found that this did not have a strong influence on the rate of evolution.

Inspection of these equations might confirm one's intuition that extinction is most likely to occur when the mate-detection distance and distance covered in search are low, as these contribute strongly to the failure of females to mate. Resistance, in contrast, is most likely to occur when mean displacement is low, subdividing the population enormously and reducing N – in effect, reducing the number of susceptible individuals that can dilute resistance.

Specific values for these parameters are generally poorly understood for any pest insect, so I leaned to causing resistance rather than extinction in choice of values for the model. I shall focus first on the results of this model for host-specific Lepidoptera. Although lepidopteran males may start to orientate to females at more than 100 m (Carde, 1986; Perry and Wall, 1986), most researchers on insect pheromones suggest that mate finding probably does not reliably occur until the males get much closer (C. Linn and S. Ramaswamy, 1993, personal communication). Schneider *et al.* (1989) cite 15 m for *Heliothis virescens*. For the simulations shown here, I used 10 m.

Mark–release–recapture data give us some indication of mean displacement and distance covered by males in search, but probably underestimate search (since they do not record deviations from a straight line) and may overestimate mean displacement. Individuals may leave many offspring between release and recapture, and mean displacement as defined here will underestimate the rate of resistance evolution when individuals mate before undergoing longer-range dispersal. Schneider *et al.* (1989) show that dispersal is at least 18 km in the spring for males of *H. virescens*. Showers's (1993) data suggest dispersal of at least 800 m for European corn borer. To encourage resistance rather than extinction in Lepidoptera, I generally used 200 m as the distance covered by males searching for females and 100 m for mean displacement.

To produce a local clutch of resistant individuals, the most likely source of a resistance outbreak, I included a cohort of ten homozygous resistant (RR) and 20 heterozygous (RS) larvae. The initial frequency of the R allele was 10^{-4}. Survival of RR, RS and SS larvae on transgenics was assumed to be 100%, 0.01% and 0, respectively; other assumptions were that there was no interplant movement (i.e. either there were no susceptible hosts within the crop or there was no dispersal of larvae), that the initial larval pest density was 10/m and that the mate-detection distance was 10 m. Except as noted in Fig. 15.5, females produced 40 offspring each (R_0 = 40), and 2.5% of the insects grew on hosts outside the crop (refuge).

When the fraction of eggs laid in the refuge was 10%, the pest population was not held in check (Fig. 15.5, curve marked A). When the refuge was maintained at 2.5%, resistance caused control failures at generation 54 (B). When the initial local population size was increased 100-fold by increasing the mean displacement tenfold to 1000 m (to complement this, male search was also increased tenfold), control failures occurred in 178 generations (C). When the refuge portion of the population was decreased to 2% compared with B, control failures occurred in 29 generations (D). In contrast, the same reduction relative to C caused extinction in 34 generations (E).

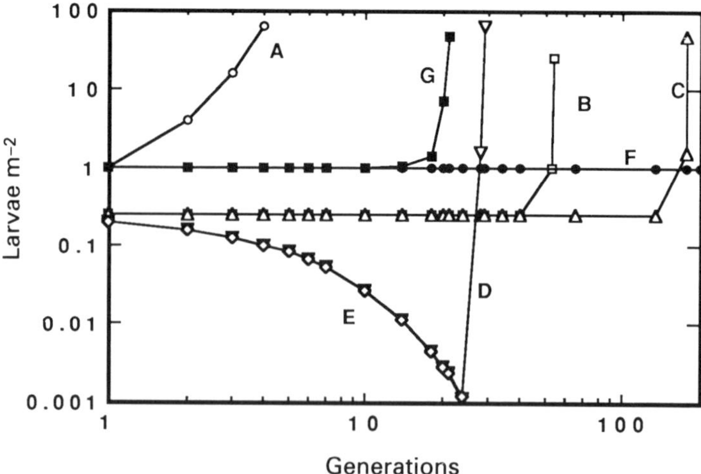

Fig. 15.5. Effects of refuge size and population subdivision on the evolution of resistance and subsequent population growth in Lepidoptera. The details of the simulations are explained in the text.

Resistance may have occurred in D only due to a gross underestimate of the dispersal of any crop-specific moth. However, we should not gamble on our poor understanding of lepidopteran dispersal and mating biology, but should genuinely integrate transgenic plants with other control tactics. When other pest-management tactics, such as classical resistance or cultural controls, are used to reduce R_0 from 40 (curve A) to 10, the population does not grow out of control and resistance takes more than 200 generations (F). Curve G is a sobering reminder that demonstrates why extinction has not often been observed with the use of classical insecticides. Here the survival of RS is only 5%, probably much less than for most classical insecticides. Even with large, expansive populations (mean displacement and male search set at 10,000 and 20,000, respectively) for dilution, resistance occurs quickly, in 21 generations.

In many ways, CPB is among the most difficult for resistance management in transgenic crops. The adults seem to move less than most moths (Roush and Tingey, 1992), and mate detection appears to occur visually and only over a short range (Szentesi, 1985). Not only do transgenic potato plants suppress feeding and oviposition by CPB adults, but adults die much more quickly on the plants than on susceptible controls (Perlak *et al.*, 1993), factors not incorporated by Ferro (1993a). On the other hand, the concentration of *B. thuringiensis* in at least some clones is 50–100 times that needed to kill larvae and prevent oviposition by adults (Perlak *et al.*, 1993; R.T. Roush and W.M. Tingey, unpublished data), so the dose is very high. Because of the rather sedentary nature of CPB, it has been possible to estimate most of the relevant parameters for simulations parallel to those made above for host-specific Lepidoptera. The simulations included a cohort of 25 RR and 50 RS larvae, presumed to be the offspring of a pair of heterozygous parents, as the most likely source of a resistance outbreak. The initial frequency of the R allele was 10^{-5}. In parallel with the strongly recessive inheritance of resistance to *B. thuringiensis* in DBM, it was assumed that survival of RR, RS and SS larvae on transgenics was 100%, 0% and 0%, respectively, and that heterozygous adults had only 1% of the fitness of resistant homozygotes, but that 50% of late-instar heterozygotes survived when they moved from a susceptible to a resistant plant. Consistent with experimental observations, it was assumed that the distance at which males could detect females was 0.1 m, that the distance actually travelled by males searching for females was 200 m, that 10% of the beetles moved between fields and 20% of the larvae between plants, but, to be conservative, that the mean displacement was only 30 m. Suppression of adult feeding in the crop was dependent on the proportion of susceptible plants, and adult males were assumed to be only half as badly afflicted at any given seed mix as the females. Any refuges outside the crop were managed without *B. thuringiensis* strains and on an economic threshold.

In the absence of susceptible plants, control failures due to resistance occurred in seven generations (Fig. 15.6, curve A). When 20% of the plants in the transgenic crop were susceptible, extinction occurred in seven generations (B). When the fraction of susceptible plants in the crop was increased to 30%, resistance occurred in 15 generations (C). If the mean displacement

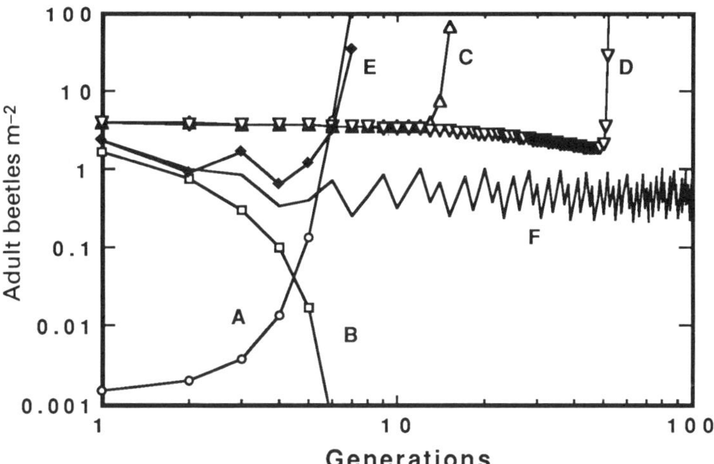

Fig. 15.6. Importance of percentage of susceptible plants in the crop and population subdivision for the evolution of resistance and population growth in the Colorado potato beetle, as explained in the text.

was increased to 1000 m, control failures occurred in 52 generations (D). Without a seed mix, even a 30% refuge allowed resistance in seven generations (E). Thus, because of the lack of long-range mate detection and the need to feed while searching for mates or to lay eggs, a seed mix is necessary to ensure that susceptible adults can mate with resistant individuals surviving in the transgenic crop. However, the most effective strategy appears to include both a seed mix (to provide the opportunity to mate) and a refuge (to provide a lot of susceptible adults), as illustrated with F, where a 20% seed mix and 10% refuge outside the crop delayed resistance to more than 200 generations.

Pyramiding

The fourth potential tactic is mixtures of toxins, which look promising in simulations when there is an untreated refuge and when heterozygotes for both genotypes suffer high mortality when the toxins are used independently (Gould, 1988a,b). Toxin mixtures are more forgiving when mortality of the heterozygotes is not extremely high, but high mortality of susceptible insects that feed on the transgenic plants is still needed to provide maximum benefits (Table 15.2). In the best case, the two toxins would be very different. Where there are genes that confer cross-resistance to several toxin types, as found by Gould *et al.* (1992), resistance can occur more quickly than if the second toxin is held in reserve until the first fails. However, the disadvantage at risk appears to be no more than twofold (i.e. both toxins are

Table 15.2. Generations until resistance increases until control of the population is lost to resistance for one toxin used alone or two toxins pyramided in the same plant, assuming that resistance to each toxin is controlled by a different gene (i.e. there is no cross-resistance). For simplicity, it is assumed that the mortalities of the RS and SS genotypes are the same for each gene and that the initial frequency of resistance is 10^{-6}. It is also assumed that 10% of the eggs from each generation are laid on non-transgenic hosts and that these larvae do not move between transgenic and non-transgenic hosts.

	Survival (%) of		Generations to resistance
	RS	SS	
One gene	30	5	15
	30	0	11
	1	0	112
	0.1	0	778
Two genes	30	5	95
	30	0	< 44,000

lost in the same time that one would have been lost if used alone). In contrast, a properly deployed pyramid can cause an enormous delay of resistance (e.g. Table 15.2).

Conclusions and Needs

Transgenic plants offer a powerful new tool for managing pest populations. As seems to be the case with all powerful new tools, it is up to us to use them wisely. The inclusion of non-transformed host plants close to the crop to ensure the availability of susceptible insects is likely to be an effective way to slow the evolution of resistance to these crops. From a tactical standpoint, this would best be achieved through the use of a refuge, but growers may be reluctant to do this. Unless governments can mandate and enforce such refuges (thereby exercising their obligation to look after the common good), the next best choice is a seed mix, which will do almost as well as a refuge so long as interplant movement of larvae is less than 20%. The refuge will work somewhat to very much better when two or more toxin genes are pyramided in the same plant.

This discussion emphasizes that we need to understand aspects of pest biology that have not concerned us much in the past, such as interplant movement of larvae, as well as others that have always vexed us, such as dispersal and mate-finding in adults. Quantifying adult behaviour will probably not get any easier, but we can plan around it by continuing to emphasize larger rather than smaller refuges and the integration of transgenic plants with other control tactics, including cultural practices, predators and parasites that slow the growth of pest populations. Further, we need to understand a lot more about the molecular biology of plants if we are to attempt strategies more sophisticated than refuges. Ultimately, we shall need large-scale experiments in the field. One can only hope that we take full advantage of the introduction of *B. thuringiensis* transgenics to learn how to better manage other genetically modified plants that will follow.

Acknowledgements

My work has benefited from the contributions of many collaborators, who will be coauthors elsewhere of much of the research mentioned here, including: W. McGaughey, W. Tingey, A. Shelton, E. Earle, T. Metz, J. Tang and M. Whalon, and my assistants P. Beckley, N. Carruthers, C. Richael and M. Burgess. I also thank B. Tabashnik for sharing unpublished manuscripts and K. Hopper and F. Gould for discussions. Calgene, Hybritech and Monsanto have provided clones and constructs. This work was supported in part by the US Department of Agriculture (USDA) grant No. 91–37302–6199.

Deploying Pesticidal Engineered Crops in Developing Countries

16

Fred Gould

Introduction

How do we get the most benefit from genetically engineered pesticidal crops in poor and developing countries? I shall deal with a variety of issues regarding the development and deployment of pesticidal crops that I feel are far from being resolved and could benefit from more discussion. As a tool for framing some of the issues, I shall focus on three major crops of poor and developing countries – cotton, rice and maize – which differ in aspects of their production, pest complexes and socioeconomic characteristics.

There appear to be at least five diverse perspectives on pesticidal engineered plants:

1. Pesticidal plants are a one-shot replacement for other pest-control methods.
2. Insects may adapt to the first pesticidal plants, but by that time other toxins will be engineered into the plants as replacements.
3. The development of pesticidal plants requires a top-down technological approach and is therefore doomed to failure, at least for the small farmer.
4. Insects are capable of adapting to pesticidal plants but resistance-management strategies can be used to significantly slow the rate of adaptation.
5. Insects are likely to adapt to pesticidal plants, but while the plants are still working they offer a means to limit insecticide use and to allow natural biological control agents to re-establish, making it easier for farmers to wean themselves from future reliance on insecticides.

The first viewpoint is naïve, given the history of insect adaptation to insecticides (Georghiou, 1990). Although some insects may not adapt to a pesticidal crop, this is not a reasonable general expectation. I feel that there is some merit to each of viewpoints 2 to 5, but I also feel that it would be arrogant and irresponsible to hold tenaciously to any one of these viewpoints for all major cropping systems in poor and developing countries.

Given the many forces that are pushing toward the use of engineered pesticidal crops, it is important to determine achievable goals that will allow farmers to use these crops for long-term benefits. The best goal for one crop may be to let private companies bring in the technology and use public resources to assist farmers in making the transition to biological and cultural control while the pesticidal plants still offer a temporary shield from the most disruptive pests. In another cropping system, there may be potential to use resistance-management strategies in concert with a fall-back position in case resistance does develop. The worst thing we could do is give a farmer false confidence in these plants that lasts for 5–10 years and then falls apart – without an immediate replacement. That approach could be worse than leaving the farmer on the pesticide treadmill.

In most areas, maize, cotton and rice have a number of important pests. It will often be impossible to optimize the development and deployment of pesticidal plants to offer stable protection from all of these species. The hardest decisions in allocating resources may be in determining which pest(s) should be targeted for sustained or temporary control with pesticidal crops. These decisions will hinge on ecological and social factors, including whether the goal of a programme is to increase or to stabilize the farmer's yield.

In this chapter I assume that our major goal is to develop and deploy engineered crops in ways that lead to reasonable short-term crop protection without compromising long-term stable crop production. My focus is, therefore, on both ecological and evolutionary aspects of crop protection. Although I am not an expert in economics or sociology, I shall deal with these subjects in a general way because they have such a significant impact on ecological and evolutionary aspects of pesticidal plants.

The first section of this chapter deals with the general issues and the concepts involved in assessing the problems and potentials of alternative strategies for designing and deploying pesticidal crops. The second section deals with some operational (i.e. practical) factors that determine whether a certain general approach to development and deployment of pesticidal maize, cotton or rice in a specific location is reasonable. These practical concerns may be humbling and challenging for the enthusiastic, good-willed, molecular biologist. A research programme funded by the International Rice Research Institute (IRRI) and the Rockefeller Foundation is briefly described at the end of the chapter as one approach for entomologists and molecular biologists to work together in addressing some of these concerns.

General Issues and Concepts

Genes for insect pest protection

Why are so many public and private research groups working with Bacillus thuringiensis *genes?*

If we look at the overall efforts to use biotechnology in the management of insect pests of crops, most of the work is focused on toxins coded for by genes of the bacterium *Bacillus thuringiensis*. Even before the biotechnology revolution, *B. thuringiensis* was the most successful microbial insecticide. From a commercial perspective it was basically the only successful microbial insecticide.

Perhaps the most important characteristic of *B. thuringiensis* for plant molecular biologists is that toxins produced by *B. thuringiensis* are proteins. Proteins are the direct products of transcription and translation of deoxyribonucleic acid (DNA). To get a plant to make a novel protein generally requires only the insertion of a single gene with the DNA code for the protein and a promoter sequence of DNA to turn on the gene.

To get a plant to make a novel non-protein toxin is much more complex. It requires that genes be inserted into the plant that code for protein-based enzymes. These enzymes, in turn, run a biosynthetic pathway that leads to production of the toxin. In rare cases only a single enzyme may be required to produce a non-protein toxin from a compound already present in the target plant, but in most cases many enzymes are required, and they must interact properly with each other to produce high quantities of toxin.

Most plant-produced toxins are not proteins, so plants have not been a general source of transferable toxin genes. One exception is proteinase inhibitors, which are plant proteins that inhibit the digestion of other proteins. Unfortunately, most of these inhibitors must be produced at very high levels to be effective (see Peacock *et al.*, Chapter 13, this volume).

Many vertebrates and invertebrates produce toxic proteins. Under natural circumstances these proteins are injected into the victim or are produced by a parasite while it is inside the host. Unfortunately, these proteins are generally not designed to withstand degradation in the digestive conditions of the insect midgut. Experience with scorpion-venom proteins indicates that ingestion by insects is far less effective than injection (Lois Miller, personal communication).

There is an active search for non-*B. thuringiensis* proteins that are orally active, but whether this will lead to many new products in the future is uncertain. Monsanto has recently applied for patents on a boll-weevil-active protein that is unrelated to *B. thuringiensis* toxins, so there is at least some hope (J.P. Purcell, poster at the Second International Conference on Insect Molecular Biology).

Apart from being the only widely effective proteins currently available, *B. thuringiensis* proteins have a history of safe use as formulations of the sporulating bacterium itself. This gives companies and regulatory officials an easier time in assessing health and environmental impacts.

Although it would be foolish to think that *B. thuringiensis* toxins will continue to be the sole focus of plant engineering for insect management, they are, at least currently, the only show in town. I shall therefore use *B. thuringiensis* toxins as a model for addressing important questions regarding: (i) approaches for interfacing toxin-expressing plants with integrated pest management (IPM); (ii) pest adaptation to the toxins in different cropping systems; and (iii) broader impacts of these engineered crops on agriculture in poor and developing countries. When other new toxins are expressed in plants, the same basic questions will need to be asked again, and they will be easier to answer if we already have experience from studies of *B. thuringiensis* toxins.

Taxonomy of some B. thuringiensis *toxins*

The species *B. thuringiensis* produces many toxins. A brief overview of the way these *B. thuringiensis* endotoxins are categorized is given in Fig. 16.1. Amino acid sequence similarity ranges from a high point of at least 80% for all members of the *CryIA* group to a low of about 25% between the *CryII* and the *CryI* group. As indicated in Fig. 16.1, each group has activity for a set of species within an insect order. However, there are exceptions, such as a *CryIII* toxin that affects some Coleoptera (beetles) and some Lepidoptera (caterpillars), and a *CryII* toxin that affects some Lepidoptera as well as some Diptera (flies). More exceptions are likely to surface in the future.

Within a group of toxins there is far from perfect overlap in biological effects. For example, *CryIA(a)* is hardly toxic at all to *Heliothis virescens* but is highly toxic to *Bombyx mori* (two lepidopteran species). *CryIA(c)* is highly toxic to *H. virescens* but is hardly toxic at all to *Bombyx mori*. The specificity of *B. thuringiensis* toxins is both a blessing and a curse, as will be seen later.

While the current toxins available are effective on many major insect pests, some groups of pests, such as aphids, grasshoppers and some beetles, are unaffected by these toxins. There is an ongoing search for novel *B. thuringiensis* toxins that kill such insects.

Expression of B. thuringiensis *genes in crops*

Initial attempts to engineer crops to express *B. thuringiensis* toxins were limited: (i) to crops that could be transformed with *Agrobacterium*; and (ii) to insertion of natural, *B. thuringiensis*-toxin DNA sequences obtained from *B. thuringiensis* strains. Although some of these attempts produced highly toxic plants, most did not. The development of particle bombardment techniques increased the range of crop species that could be transformed. Furthermore, the development of synthetic DNA sequences that can more efficiently code for *B. thuringiensis*-protein production in plants has led to crops with extremely high *B. thuringiensis* toxin titres.

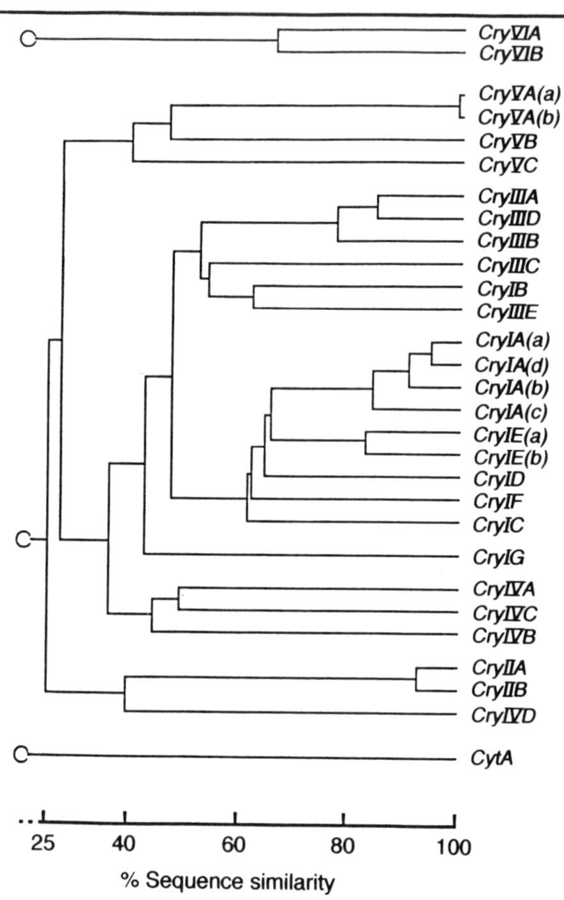

Fig. 16.1. Possible evolutionary relationships between *B. thuringiensis* toxins. The Fitch–Magoliash algorithm was used to compare amino acid sequences of 29 domains (aa 1 to approximately aa 640). From Feitelson *et al.* (1992).

In most cases, expression of *B. thuringiensis* genes has been driven by constitutive promoters, such as the 35S promoter, which result in *B. thuringiensis* toxin production in almost all plant parts. Recent work at Ciba-Geigy has produced maize plants in which *B. thuringiensis* expression is driven by tissue-specific promoters (pollen-specific and green-leaf tissue-specific).

Two technical stumbling-blocks that still need to be addressed in some crops are: (i) the difficulty and time required to regenerate fertile plants from transformed tissues; and (ii) the difficulty in working directly with élite lines of some crops.

Potential for Insect Resistance to *B. thuringiensis* Toxins

Just because we can find a toxin that is a protein and can put it in an appropriate plant does not mean that we have the problem of insect-pest protection solved. Insects have proven their capacity to overcome, by evolution, the effects of many toxins (Georghiou, 1990; Gould, 1991). Diamondback moth (DBM) (*Plutella xylostella*) populations in many areas of the world have already developed resistance to *B. thuringiensis* spray formulations (Tabashnik, 1993). The task of putting a *B. thuringiensis* gene in a plant may turn out to be less formidable than determining how to use this new type of plant in a way that leads to sustained crop protection from insects.

Slowing down B. thuringiensis *resistance in crop pests*

Environmentalists as well as agricultural scientists have voiced considerable concern over the possibility that improper design and deployment of engineered crops will lead to only temporary protection from insect pests and will put agriculture right back on the pesticide treadmill.

One approach for avoiding rapid pest adaptation to *B. thuringiensis* and other insect toxins employs the tools of population genetics and population ecology. I shall offer an overview of this approach because a general understanding of the concepts involved can help us see the strengths and weaknesses inherent in applying these tools to solving pest problems in developing countries.

For many decades, scientists have been using Fisher's (1930) fundamental theorem of natural selection to help them increase the rate at which they are able to genetically alter crops and animals. Fisher's important statement was that the rate of evolutionary change is proportional to the additive genetic variance for fitness in the population. Without getting too far into theory and precision, this theorem can be approximately restated as follows: within a population of individuals, the more variation there is for a trait that can be passed on from one parent to its offspring, and the more closely the trait is associated with the number of offspring produced, the faster the trait will evolve in the population. Whether the trait is litres of milk per day or grain yield per tiller, the success of a breeder depends on how much useful variation exists in the population of organisms that he or she is working with, and how good he or she is in artificially manipulating the matings and choices of offspring to increase the important components of variation discussed by Fisher (and others).

Thousands of articles and books have reported the results of theoretical and empirical studies devoted to maximizing inherited variation in economically important traits, and to minimizing non-inheritable variation in those traits so that a better correlation could be produced between the inherited component of variation for an economic trait and the number of successful offspring a domesticated organism is allowed to produce. All this has been for the sake of speeding up evolution in the barnyard or experimental field plot.

We want to do just the opposite. We want to decrease the rate of evolution in wild pest populations so they do not overcome the effects of toxins in our crops. To accomplish this, we need to decrease the correlation between the inherited component of variation in the pest's toxin-resistance traits and the number of successful offspring produced. This is not easy because we generally expect to find a strong positive relationship between how adapted an individual is to the toxin and the number of offspring produced.

We have two options: (i) decrease the difference in the number of offspring produced by toxin-adapted and unadapted individuals within the population; and (ii) decrease the amount of 'adaptedness' that one parent can pass on to its offspring. (Emphasis is on one parent with resistance genes while the other parent is susceptible, because, when a toxin is first introduced, the frequency of individuals with resistance genes is so low that it is unlikely that two individuals with resistance genes will find and mate with each other.) All genetic models and approaches for slowing pest adaptation entail the use of one of these options or a combination of both options.

In order to develop appropriate population genetic models that predict the most feasible option(s), there is a need for at least qualitative knowledge about the inheritance and frequency of *B. thuringiensis* resistance genes. I shall therefore briefly review what is known about *B. thuringiensis* resistance genes before going on to discuss models and strategies.

Resistance to **B. thuringiensis** *toxins in crop pests*

The genetics of resistance to *B. thuringiensis* has been examined in a number of species, and Tabashnik (1993) and McGaughey and Whalon (1992) have recently reviewed much of the literature. The most thorough genetic studies have been conducted with the Indianmeal moth (IMM) (*Plodia interpunctella*), DBM and tobacco budworm (*Heliothis virescens*).

In IMM and DBM, where resistance is related to a change in protein-based receptor molecules of midgut epithelial cells, resistance appears to be a mostly recessive trait (McGaughey and Beeman, 1988; Van Rie *et al.*, 1990; Ferre *et al.*, 1991; Tabashnik *et al.*, 1992a, b), i.e. offspring most closely resemble the susceptible parent. This makes sense because the heterozygotes are likely to continue producing about one-half normal receptors, which would theoretically translate to about twofold resistance. In practice the level of resistance in the F_1 hybrids is generally somewhat higher than twofold. R. Roush (personal communication) has indicated that higher than expected resistance in heterozygotes may be related to pharmacokinetic properties of the system.)

In at least one case, resistance to *B. thuringiensis* toxins by *H. virescens* appears to be unrelated to changes in midgut receptors (Gould *et al.*, 1992). Resistance in this species has been shown to be either additive or partially recessive. Recent studies of one strain with 25–50-fold resistance to *CryIA(c)* indicate that more than one locus is involved in this resistance (F. Gould and A. Anderson, unpublished data). In another strain of *H. virescens* with over 3000-fold resistance, a single locus appears to be responsible for most of the resistance (Gould *et al.*, 1995; Heckel *et al.*, in prep.).

In IMM and DBM, resistance appears to be specific to a few related toxins (Van Rie *et al.*, 1990; Ferre *et al.*, 1991). However, one case of *H. virescens* resistance that evolved in response to a single toxin has led to adaptation to other distantly related *B. thuringiensis* toxins (Gould *et al.*, 1992).

Work on DBM indicates that resistant individuals are less fit than normal individuals and that resistant strains will revert to susceptibility (Groeters *et al.*, 1993). Although a direct fitness cost has not been associated with the 3000-fold resistance to *B. thuringiensis* in *H. virescens*, reversion has been documented in this strain (F. Gould, unpublished data).

Although there is ongoing work in our laboratory to directly estimate the initial frequency of *B. thuringiensis*-resistance genes in the field, current estimates are indirect, being based on the number of individuals that have been needed to select resistant strains in the laboratory. Based on the work of Gould *et al.* (1992) and McGaughey and Beeman (1988), Tabashnik (1993) concluded that initial gene frequencies were higher than 5 in 1000. This is much higher than the initial frequencies assumed in most insecticide resistance models. If these estimates are correct, the issue of resistance to *B. thuringiensis* toxins is indeed a crucial one. Recent work in our laboratory (Gould *et al.*, 1995) supports this estimate.

Models to predict rate of pest adaptation

One- and two-locus population genetic models as well as quantitative genetic models have been used for over 15 years developing general strategies to slow the rate of evolution of insecticide resistance (Roush and Tabashnik, 1990) and pest adaptation to crops with natural defensive chemistry (Gould, 1986a). Although quantitative genetic models that assume polygenic inheritance are highly appropriate for assessing adaptation to some insecticide spray regimes, they are less useful when the toxin is produced at very highly efficacious levels by the crop plant. Under these circumstances the most troublesome resistance traits in the pest are likely to be those controlled by one or two genes with major effects. Thus single- and two-locus population genetic models are appropriate for many questions posed about crops that express *B. thuringiensis* toxins.

Most of the models developed for predicting pest adaptation to toxins are deterministic models that assume diploid, sexually reproducing, randomly mating organisms. These assumptions are generally reasonable but there are exceptions. For example, Caprio and Tabashnik (1992) examined insecticide evolution in small subdivided populations with gene flow. Results of their stochastic model differed from results of deterministic models in predicting non-linear effects of increased gene flow. Gould (1986b) examined a special case of Hessian fly adaptation to resistant wheat in which the insect had paternal gene loss and subdivided populations. Gould *et al.* (1991c) also briefly explored the differences between strategies that would be best for facultatively parthenogenic insects and obligately sexual insects. Follett *et al.* (1993), who recently developed a specific model for Colorado potato beetle (CPB) (*Leptinotarsa decemlineata*) adaptation to pyrethroids, pointed out the need for more specific models.

In the case of *B. thuringiensis*-producing plants, there is definitely a need for specific models or at least for models that have subroutines to handle specific ecological attributes of a pest. Because of the intense selection pressure exerted by some resistant plants, deviations from the assumptions of general models can be caused by small changes in movement, mating, cropping patterns, etc.

In the next sections I shall first describe some of the resistance-management strategies that, based on general models, show promise for at least some insects. I shall then examine insect pests in tropical maize, rice and cotton in more detail.

General approaches for the use of **B. thuringiensis** *genes*

Most of the strategies for engineering and deploying crops with *B. thuringiensis* genes in ways that will slow down the evolution of resistance can be placed in five general categories:

1. High levels of constitutive expression of single toxins in all plants.
2. High levels of constitutive expression of two or more toxins in all plants.
3. Spatial or temporal mixtures of plants with high levels of constitutive expression of one or more toxins and other plants with no toxin expression.
4. Low levels of expression of single toxins interacting with the pests' natural enemies.
5. Targeted gene expression.

We can categorize the five strategies as follows: for the most part, strategies 1 and 2 aim at decreasing the 'amount' of adaptedness that a single resistant or partially resistant parent can pass on to its offspring. Strategy 4 aims to decrease the difference in fitness between resistant and susceptible individuals. Strategies 3 and 5 attempt to do both. The utility of each of these strategies can be measured based on how much longer it is expected to last compared with a homogeneous planting of a variety expressing a moderate dose of toxin that kills between 95 and 99% of susceptible individuals (Fig. 16.2a).

Strategy 1

The single-toxin, high-dose approach derives its power from the assumption that high expression of a single toxin is so effective that only an incredibly small fraction of susceptible or partially resistant individuals will survive. 'High dose' has been defined as 25 times the amount of toxin needed to kill 99% of susceptible insects (Gould, 1994). Even if a single resistance allele in homozygous condition (RR) can confer immunity to the toxin, strategy 1 can be effective as long as both the homozygous susceptible (SS) and heterozygous (RS) individuals have very low and almost identical fitness. If these conditions hold, genetic models indicate that, at gene frequencies below 1 in 1000, the few RR individuals will be likely to mate with susceptible individuals and produce offspring that are heterozygous. These heterozygous offspring will have fitness about as low as the totally susceptible individuals. In terms of Fisher's fundamental theorem, the single parent (RR) is not able to pass its resistance on to its offspring.

Strategy 2

The use of two or more genes at high expression levels works much like strategy 1, with a little more complexity added. Assuming that two toxin genes act independently and that there is no cross-adaptation, an individual that is immune to one toxin (R1R1) will still have low survival (about 0.1%), as long as it does not have alleles for adapting to the second toxin (S2S2).

Only individuals homozygous for resistance to both toxins (R1R1, R2R2) will have high survival. When the allelic frequency of each of the two resistance alleles is 10^{-3}, the frequency of these doubly resistant individuals is 10^{-12} in a randomly mating population. The chance that two such individuals will exist is low, and the chance that they will mate with each other is even lower. Thus, it is unlikely that such a resistant individual will ever pass on a large component of its adaptedness to its offspring.

Strategies 1 and 2 look great at a superficial level, but they break down even if there is what might seem to be a tiny difference between the fitness of totally susceptible individuals and those that are heterozygous for one or both alleles. When we look at figures showing that the totally resistant individuals have a fitness of 1.0 and the highest fitness of any of the other genotypes is 0.0002, we may be impressed. However, we must realize that if the heterozygote's fitness is 0.0002 and the totally susceptible individual's fitness is 0.0001, the heterozygote is twice as fit and it will pass on this fitness to one-half of its offspring. As seen in Figs 16.2 and 16.3, this causes strategies 1 and 2 to begin collapsing.

Strategy 3

Strategy 3 does a marvellous job (theoretically) of making up for the Achilles' heel of strategies 1 and 2. Assume that we mix 10% totally susceptible plants with the plants from strategy 1. Assuming no cost to resistance, the RR individuals have a fitness of 1.0 on all plant types. The SS individuals have a fitness of 0.0001 on *B. thuringiensis* plants and a fitness of 1.0 on the susceptible plants. Because the frequency of *B. thuringiensis* plants is 0.90 and the frequency of non-*B. thuringiensis* plants is 0.10, the overall fitness of the SS individuals is $(0.0001 \times 0.90) + (1.0 \times 0.10)$ or 0.10009. The fitness of the heterozygote is $(0.0002 \times 0.90) + (1.0 \times 0.10)$ or 0.10018. Dividing 0.10018 by 0.10009 gives 1.000899. Therefore, in this mixed planting, the heterozygote is 1.000899 times more fit than the susceptible individual. In strategy 1, the heterozygote was 2.0 times more fit. The 10% mixture almost totally negates the fitness differences between these two important genotypes. A 10% mixture has an even stronger effect in the multiple-gene case and can therefore, at least theoretically, slow evolution to a crawl.

Strategy 3 also decreases the difference between the fitness of totally adapted individuals and that of partially adapted individuals. In strategy 1 the difference in fitness between RR and RS individuals was 1.0 divided by 0.0002, or 5000-fold. With a 10% mixture, this becomes 1.0 divided by 0.10018 or 9.98-fold.

Mixtures have great advantages as long as two assumptions about the pests are met: first, there must be approximate random mating; second, the difference in fitness of pest genotypes in the mixture must be directly related to the ratio of the two types of plants. That is to say, the simple equations given above hold true. The concern of many entomologists is that for some species it will be hard to devise a spatial mixture in which both of these assumptions hold true (Roush, Chapter 15 this volume; Mallett and Porter, 1992; Gould, 1994).

We must ask whether, in a mixture, all of the heterozygotes that hatch on a *B. thuringiensis* plant have a fitness of about 0.0002 and all of the heterozygotes that hatch on a non-*B. thuringiensis* plant have a fitness of about 1.0. As long as the larvae are restricted to the plant upon which they hatched, this makes sense. However, if larvae are able to move from one plant to another and the adjacent plant is of a different type, the above

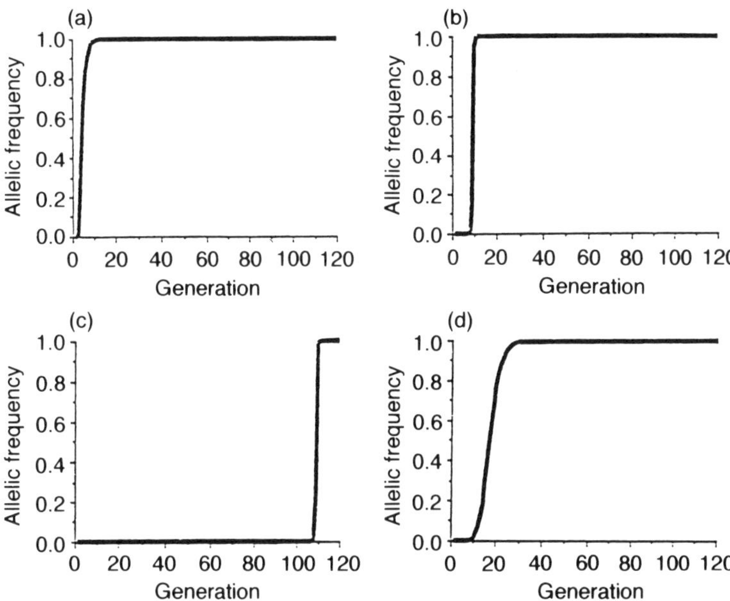

Fig. 16.2. Sustainability of moderate, high and low expression of single *B. thuringiensis* genes when 100% of the plants express the toxin. (a) Moderate expression where the fitnesses of genotypes are as follows: SS = 0.01, RS = 0.505, RR = 1.0 (this is the probable result of having no strategy for sustainability). (b) High expression when there is complete recessiveness and fitnesses are: SS = 0.0001, RS = 0.0001, RR = 1.0 (strategy 1 at its best). (c) High expression when fitnesses are SS = 0.0001, RS = 0.0002, RR = 1.0 (strategy 1 when the assumption of complete recessiveness is relaxed). (d) Low expression when fitnesses are: SS = 0.40, RS = 0.70, RR = 1.0 (strategy 4 when only direct effects of plants on survival are considered). In all cases, initial frequency of R is equal to 1 in a million. SS, homozygous susceptible; RS, heterozygous; RR, homozygous resistant. From Gould, 1994.

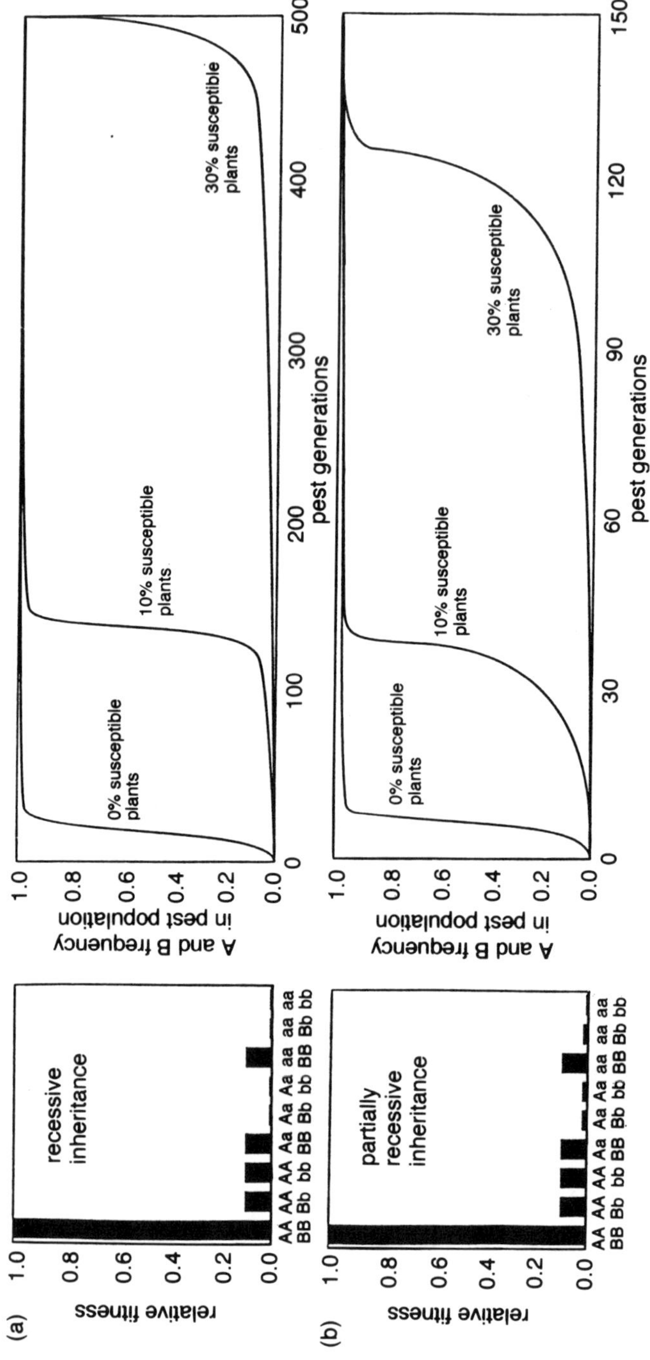
(a)
relative fitness
1.0
0.8
0.6
0.4
0.2
0.0
recessive
inheritance
AA BB
AA Bb
AA bb
Aa BB
Aa Bb
Aa bb
aa BB
aa Bb
aa bb
A and B frequency
in pest population
0% susceptible
plants
10% susceptible
plants
30% susceptible
plants
0
100
200
300
400
500
pest generations
(b)
relative fitness
partially
recessive
inheritance
AA BB
AA Bb
AA bb
Aa BB
Aa Bb
Aa bb
aa BB
aa Bb
aa bb
A and B frequency
in pest population
0% susceptible
plants
10% susceptible
plants
30% susceptible
plants
0
30
60
90
120
150
pest generations

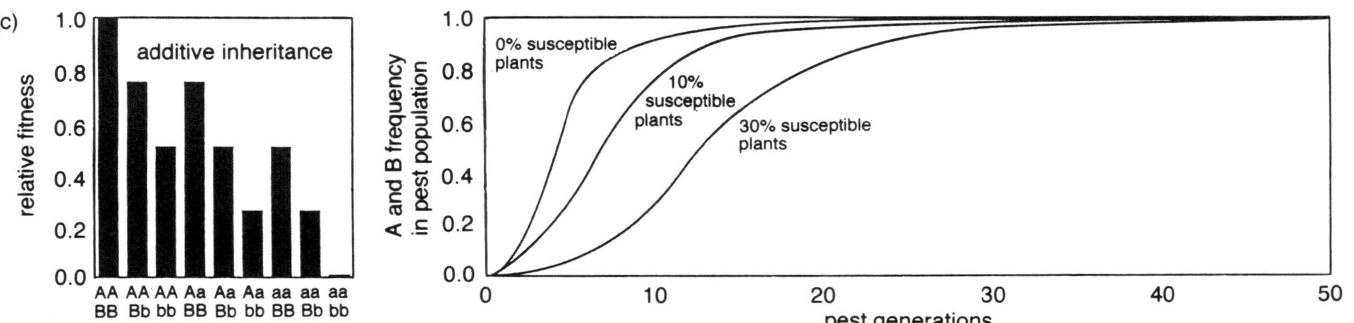

Fig. 16.3. Strategies 2 and 3. Results of simulations from Gould (1991), in which there are two toxins and two independently assorting genetic loci carrying resistance alleles. The A and B alleles confer resistance, whereas the a and b alleles do not. The effect of using a mixture of some plants with two toxins and others with no toxins is most dramatic when resistance is inherited as a completely recessive trait (a). In this case, only the AABB genotype has high fitness when the pest must live on a host plant that produces both toxins (all genotypes are assumed equally fit on plants with no toxins and larvae do not move between plants). When 100% of the plants produce the toxins, resistance evolves in about 30 pest generations; with 10% susceptible plants it takes almost 150 generations, and with 30% it takes over 500 generations. The benefits of mixtures are less extreme when individuals with a single resistance allele are slightly more fit than totally susceptible individuals (0.02 versus 0.01): see (b). If resistance is additive the benefit is even less (c) but is still significant. Note that each of the figures has a different time-scale. These simulations used an initial frequency of resistance alleles of 0.01 and did not assume a very high level of mortality in susceptibles due to single toxins. The positive impact of mixtures is more pronounced when there are lower initial frequencies and higher mortalities of susceptible individuals.

relationship may break down. For example, if an RS larva grows to third instar on a non-*B. thuringiensis* plant, then moves to a *B. thuringiensis* plant for its fourth instar and then moves back to the original non-*B. thuringiensis* plant for its final instar, what will its fitness be compared with an SS larva that had the same movement history? If the average fitness of the RS individual is 0.30 and that of the SS individual 0.10, and if this type of movement history is common, the effectiveness of the mixture strategy is decreased. We obviously need to gather more information on the relative fitness of such genotypes in plots planted to seed mixtures before endorsing a seed-mixture strategy.

Recent work conducted in my laboratory and in the field, using *B. thuringiensis*-expressing maize plants from Ciba-Geigy, indicated that about 90% of newly hatched European corn-borer larvae left *B. thuringiensis*-expressing maize within 36 h, while only 50–70% of such larvae left normal maize plants. This amount of movement could decrease the utility of seed mixtures, but we need to know more about the comparative behaviour and survival of SS and RS larvae before we can come to any conclusions. The issue of movement and larval fitness will be discussed later.

One way to avoid the problem of larval movement between *B. thuringiensis* and non-*B. thuringiensis* plants is to create a mixture at the field-to-field level instead of the plant-to-plant level. However, at this spatial scale, the model assumption about random mating becomes suspect. If the spatial scale of the mixture is at 5 ha, one 5-ha field out of every ten such fields is planted to non-*B. thuringiensis* plants. The question is what proportion of RR adults that emerge in a given field will mate with SS individuals coming from the non-*B. thuringiensis* field, and what proportion of RR adults will mate with other RR adults from the same *B. thuringiensis* field. Answers to this question will have to be addressed on a case-by-case basis. This may mean that such estimates will have to be made in more than one location for a single pest species. (Of course, the utility of the field-to-field mixture is negated if insecticides or other control measures are used to decimate the insects in the 'refuge' field.)

In some cases, general knowledge of a pest insect species may give us simple answers to questions about both larval and adult movement. For example, we can assume that once a Hessian fly larva starts feeding on a plant it cannot easily move to another plant based on its biology and morphology. This is probably also true for rice gall midges. We know that adults of some *Helicoverpa* and *Spodoptera* species move long distances and are good local fliers as well, so they are likely to cross field boundaries in search of mates. Unfortunately, as discussed below, we have fewer answers for many important pests.

Strategy 4

The use of low levels of *B. thuringiensis* toxin interacting with natural enemies was once thought to be an ideal solution to the resistance problem. It was assumed that low levels of toxin in the plant only slowed the growth of the immature stages and caused little mortality (i.e. 10–20%). This strategy was therefore expected to result in only minor differences between the fitness of RR and SS individuals. Such an outcome is possible, but in many cases the low-dose strategy could lead to large fitness differences between RR, RS and SS individuals. Potential causes for differences are as follows: (i) the slower-developing larvae do not reach a diapause stage before the crop matures or the weather becomes unsuitable; (ii) the resistant individuals are able to complete more generations per season than the more susceptible individuals and therefore leave many more offspring; and (iii) predators and parasites cause a higher percentage mortality of susceptible individuals than they do for more resistant individuals (see Gould *et al.*, 1991b; Johnson and Gould, 1992). Furthermore, asynchronous development of resistant and susceptible individuals could lead to non-random mating. Strategy 4 could work very well in some cases, but details of the fitness of adapted and unadapted individuals must be understood before such a strategy is used.

Strategy 5

Targeted gene expression has some things in common with strategy 3, which is the use of mixtures. While strategy 3 can give pest individuals a choice of neighbouring plants that differ in toxicity, one form of targeted gene expression, tissue-specific expression, can give pests a choice between toxic and non-toxic plant parts. There are many promoter sequences available that turn on structural genes only at certain times and places. Some tissue-specific promoters turn genes on only if they are in pollen cells, others only in root tip cells. If a pest feeds on a number of plant parts but only causes economic injury based on damage to specific tissues, it might be possible to express the toxin only in the tissues that need protection. In many cases these are fruiting structures or growing tips of the plant (meristematic tissues). If high doses of multiple toxins were only expressed in some tissues and if no toxin or extremely low levels of toxin were present in other tissues, such a strategy could have promise.

As with the seed-mixture strategy, however, random movement of larvae from one structure to another (accompanied by feeding) could cause problems with this strategy. If instead of random movement there is directed movement, with larvae avoiding toxic plant parts, the disadvantage could be negated. This type of directed movement has been seen in tobacco budworm on tobacco plants with and without *B. thuringiensis* production (F. Gould *et al.* unpublished data). Unlike strategy 3, the tissue-specific expression strategy could lead to a reduced number of pest generations being exposed to toxin. This would be likely if only fruiting structures produced the toxin and these fruiting structures were present in only one of the generations when a pest was found on the crop.

Another approach for decreasing the number of generations when a pest was exposed involves time-specific or phenologically specific expression. It may be possible to find promoters that cause expression of their structural genes only when a plant changed from its vegetative to its reproductive phase. This might be more likely in plants with determinant growth and those with clear photoperiod responses. This approach will be discussed below in some detail.

A final targeted approach for expression involves use of inducible promoters. There are many promoters that start transcription only when plants are exposed to environmental cues, such as heat shock, water stress or insect or pathogen attack. If a gene was turned on every time an insect took a bite, the result would be similar to constitutive expression, but, if the gene turned on in response to a threshold of damage, this approach would have merit, especially for a pest that had only periodic outbreaks. Ciba-Geigy now has a *B. thuringiensis* gene linked to a promoter that turns on expression only when an exogenous compound is sprayed on the plants. This approach could have merit under specific economic and political systems.

Some of these strategies (e.g. strategy 1) are more risky than others, but, no matter which of these strategies is being considered, we need to understand how the strategy is likely to interact with the ecological, behavioural and genetic attributes of the target pests.

The following sections of this chapter will give a general assessment of some of the attributes of specific pests, cropping systems and social systems that could have an impact on which of these strategies are likely to be most useful and realistic. Given the complexity of choosing between resistance-management strategies, one may be tempted to give up on using any strategy and instead use the first *B. thuringiensis* plants that are available. Therefore, it is important to remember that, even when strategies 3–5 do not work very well, they are almost always expected to lead to significantly more sustained efficacy of the resistant plants than the use of pure varieties that express enough toxin to kill between 90 and 99% of susceptible insects.

Cropping Systems and Pests in Developing Countries

Contrasts between systems

This section offers some general observations on maize (*Zea mays*), cotton (*Gossypium hirsutum*) and rice (*Oryza sativa*) systems, which could serve to initiate more detailed assessments.

Contrasts between ecology of pest complexes

Rice

Most of the major insect pests of rice that are currently targeted for control by *B. thuringiensis* toxin-expressing plants use rice as their only or major host (e.g. striped stem-borer, yellow stem-borer, leaf-folders). The general ecology and behaviour of some of these pest species are at least partially understood, but mating and movement patterns have not been a focus of study. Efforts to control leaf-folders and stem-borers with broad-spectrum insecticides have often caused outbreaks of brown planthoppers, because these insecticides kill natural enemies (Kenmore, Chapter 4, this volume). Therefore, a more specific control tactic that does not disrupt pest enemies could have positive, system-wide ramifications.

The leaf-folders cause clear visual damage to the rice plant but a number of studies indicate that they rarely cause any yield loss (K.L. Heong, IRRI, personal communication). In many cases, farmers may spray these insects because they fear rare outbreaks.

Striped and yellow stem-borers attack rice during two stages of plant growth: the tillering stage and the reproductive stage. Studies have shown that rice plantings can completely negate the effects of stem-borer damage to the tillering stage by compensatory growth (Rubia, 1990). It is not clear that irrigated rice can compensate for a significant amount of stem-borer damage during the reproductive stage (but deep-water rice may be able to compensate).

A number of entomologists feel that almost all rice pest outbreaks are pesticide-induced. The opinion has been expressed that the most important rice pests to control with *B. thuringiensis*-producing plants are leaf-folders, for, although they generally cause no significant yield loss, they trigger the use of pesticides more than do stem-borers. This is because the leaf-blade feeding damage caused by leaf-folders is so much more apparent to farmers than the internal damage caused by stem-borers. Although IRRI has not opted for this approach, it is worthy of discussion.

Peter Kenmore (Bellagio, 1993, personal communication) recounted a pest-control problem of a similar nature to that of the leaf-folder. He indicated that, in Bangladesh, the government uses an 'air force' of spray planes to control hispa beetles, which look foreboding on the rice plants but generally do not decrease yields. Farmers are worried about potential damage, so they are generally in favour of the spraying. Kenmore suggested that, if a rice variety could be developed that killed the hispa beetles, there would be no rationale for widespread spraying. This would decrease disruption of natural enemies and would stop the recurrent fish kills caused by the pesticides.

It is clear that a low density of insect pests is not usually a problem in a rice field. From an IPM perspective it might be possible to keep all of the lepidopteran pests of rice at reasonably low levels if even one-half of their rice host plants were toxic. With one-half of their hosts causing nearly 100% mortality, the rate of pest population increase would be cut in half (assuming no strong density-dependent or inversely density-dependent population regulation). Only large-scale field experiments could determine what percentage of the plants would need to be toxic in order to reduce the pest status of these lepidopteran pests over a long time period. Because stem-borers lay eggs in clusters and some neonates move to adjacent plants, it will be important to assess the impact such movement has on the long-term utility of a seed-mixture strategy.

Although the major lepidopteran stem-borers are closely related, work at IRRI (described below) indicates that *B. thuringiensis* toxins that are highly effective on one stem-borer species are not always effective on a second stem-borer species. It will be important to choose carefully among *B. thuringiensis* toxins if there is a goal of creating rice varieties that are toxic to the stem-borer complex and/or stem-borers and leaf-folders. There has been no detailed screening to find *B. thuringiensis* toxins that are effective on dipteran, coleopteran or homopteran pests of rice.

Maize

Maize, or corn, is grown throughout most of the world, and its pest complex shifts with location. Major in-field pests include Lepidoptera, Coleoptera, Hemiptera, Homoptera and Diptera. Focusing on the Lepidoptera, the European corn borer is a sporadic but important pest in temperate areas. Armyworm and other species of stem-borers are more important in the tropics and sub-tropics.

While some maize pest problems are pesticide-induced, others are more probably the result of natural cycles of the pest, climate and/or area-specific cultural practices. It is difficult to get good data on pesticide-induced outbreaks, but we at least know that insect-induced yield losses were sometimes severe in the prepesticide era and that, in areas of Africa where insecticides are not used, insects still cause significant crop losses. (This is also true in some areas of the USA where foliar insecticides are not used.) The view that most maize pest losses are not simply a result of insecticide use stands in contrast to the prevailing view regarding tropical rice pests.

Maize plants can compensate for some foliage loss but they do not tiller the way rice does. Maize plantings are therefore highly sensitive to damage that disrupts the growth of stems.

Considerable research has been done on temperate maize pests. In terms of mating and larval movement, there is some good information on European corn borer, rootworm and armyworm species. Not surprisingly, there is less research on tropical pests of maize. For many tropical maize pests, we even lack basic information on host range.

The 1992 Annual Report of the International Institute of Tropical Agriculture (IITA) (Anon., 1993b) gives us a good sense of the species of pests that are considered threats to African maize production, but also gives a strong sense of how little we know about these pests. Just in terms of stem- and cob-boring Lepidoptera, the IITA report lists seven species. These species are biologically distinct, coming from three families of Lepidoptera: Noctuidae, Pyralidae and Tortricidae. Most of these species are pests only of maize in Africa. The impact of each of these species on maize yield is only now being surveyed in a systematic manner. As indicated earlier, *B. thuringiensis* toxins that are effective on one species of caterpillar may have no effect on another species of caterpillar, so it is not clear that a *B. thuringiensis* toxin that is optimal for the European corn-borer (Pyralidae) would be effective on these African insects. We already know that *CryIA* toxins are active on European corn borers but have poor action against armyworm species (Noctuidae) that are major pests in Central America and in other tropical areas. Before we know which species of Lepidoptera is/are most important in a given region of Africa, we are certainly in a weak position for choosing which *B. thuringiensis* toxin to use.

Cotton

As with maize, many cotton pests vary with region. The boll weevil, which is restricted to certain parts of the western hemisphere, is a good example of a region-specific pest and is also one of the few major pest insects that are specialists on cotton.

Most cotton production systems are severely affected by the Heliothine family of moths, but the western and eastern hemispheres have different species of this bollworm complex. None of these species are cotton specialists, but in areas of intense cotton production a very high percentage of a regional bollworm population may feed exclusively on cotton at certain times of the growing season. Only some problems with bollworms can be traced directly to the use of broad-spectrum insecticides (Bradley, 1993).

Armyworms are a major problem in tropical cotton. In Egypt the most important armyworm is the Egyptian cotton leafworm, *Spodoptera litoralis*, which is resistant to many insecticides and could be a clear target for *B. thuringiensis*-producing cotton. Although the toxin in current commercial development for cotton, *CryIA(c)*, is not very effective on armyworms, it may be used for convenience in Egyptian cotton and could lead to rapid resistance development if the high dose of *CryIA(c)* for Heliothine moths is a moderate dose for this *Spodoptera* species. Other *Spodoptera* species are important pests of cotton in Central and South America.

It is clear that for all three crops the geographical diversity of the pest complexes suggests that a single *B. thuringiensis* toxin will not be optimal in all locations.

Crop breeding and seed production in these crops

At present rice and cotton seed are generally produced as inbred lines, whereas maize hybrids are used in most growing regions (but there are exceptions as in some areas of Africa). In the future this situation may change. There are current programmes to produce and use high-yielding hybrid rice (e.g. in China), and as genetic engineering enters the mainstream it is likely that some companies may try to protect their germplasm by developing engineered hybrids of many crops that do not produce high-quality fertile seed in farmers' fields.

Currently, the difference between hybrid and inbred crops is quite important. Producers of hybrid seed can get premium prices for high-yielding seed because farmers are obliged to always buy seed from the seed producers. With inbred crops, if the price is too high the farmer has the option of saving seed from her/his own fields. This has had a major impact on commercial development of engineered crops.

Maize

Maize is a major target for biotechnology, primarily because it is used as a hybrid and because it is a major crop in developed countries, where farmers are already paying premium prices for seed.

A number of commercial seed companies with subsidiaries in developing countries have temperate-zone maize lines that express *B. thuringiensis* toxins that are active against European corn borer. It is easy to envision these companies moving *B. thuringiensis* genes into tropical maize varieties without optimizing them for local maize pests. If such varieties were marketed broadly, resistance problems could emerge quickly. Because multinational corporations have expressed a concern over the problem of *B. thuringiensis* resistance, it is possible that they will take a more careful route to varietal development.

If the highly visible multinationals take a careful approach, the problem is not necessarily solved. There will always be the possibility that small, local companies could 'pirate' suboptimal *B. thuringiensis* genes by backcrossing genes from temperate varieties into parental lines for locally adapted hybrids. M. Goodman (North Carolina State University, 1993, personal communication) has pointed out that, in many tropical hybrids, one of the parental lines is often closely related to or is identical to parental lines used by temperate-zone maize breeders. This makes the use of pirated *B. thuringiensis* genes even easier.

In developing countries where hybrid varieties are produced and sold by organized seed companies or government programmes, it is possible that the hybrid status of the seed could increase the feasibility of using a seed-mixture strategy because farmers would not be in the position to save seed from only the *B. thuringiensis*-producing plants in their fields. Unfortunately, the seed distribution system in most countries is not well organized and controlled, so distributors could not be relied on to sell mixed seed.

Cotton

Although cotton is an inbred crop, on-farm seed production is difficult because of the need for delinting. Furthermore, there is an important maternal contribution to seed vigour that can be optimized by specialized seed producers. Because cotton is an important crop in developed countries, some companies are gambling on the fact that the difficulty with quality seed production and the potential for enforcement of patent rights will keep farmers in developed countries buying higher-priced seed.

The above strategy may work in developed countries but not in developing countries. The problem with local entrepreneurs in developing countries 'pirating' *B. thuringiensis* genes from cotton varieties in developed countries may be even more serious than it is with maize, given the inbred nature of the crop. Although some tropical annual cotton varieties are genetically distinct from temperate varieties, backcrossing is not difficult. In areas where perennial cotton is grown, backcrossing would be more difficult.

Rice

Rice seed is easy to harvest and farmers often save their own seed. This may be one of the reasons for there being less commercial effort put into the development of engineered rice than has been put into cotton and maize. Additionally, it is important to note that the temperate-zone rice varieties (*Japonica* rices) are, genetically, very different from the tropical rice varieties (*Indica* rices). If engineered rices were developed for Japan, they would not be agronomically acceptable in developing tropical countries, and backcrossing between *Japonica* and *Indica* varieties is not simple. Because the striped stem-borer is considered a major pest in Japan and the yellow stem-borer is not a problem there, Japanese breeders are not likely to incorporate genes into *Japonica* rices that are ideal candidates for sustainable control of yellow stem-borers, considered major pests of tropical rices.

A potential source of *B. thuringiensis* genes for tropical rice may come in the form of donations or purchases of genetic 'cassettes' of *B. thuringiensis* genes and promoters that were developed for insertion into other crops such as maize and cotton. These genes could be incorporated into rice varieties and could lead to high expression of the toxin. Hopefully, it would be an appropriate toxin. Getting a second-hand cassette with two appropriate toxins for a multiple toxin strategy seems unlikely, so concerted efforts by rice researchers would probably be needed to develop a rice variety with two useful genes.

Of the three crops discussed, it seems as if the initial 'flow' of *B. thuringiensis* genes into tropical varieties may be most controllable in rice, but once the genes are in tropical rices, the control would diminish greatly as farmers save and select their own seed. Any strategies that require control of expression, planting, etc. will have to be implemented early, or must not rely on farmers' choices (e.g. tissue-specific expression).

Contrasts and similarities

Although cotton and rice may be similar in terms of breeding practices, they are often dissimilar from a socioeconomic perspective. Tropical cotton is typically produced as a cash crop, and farms in many areas are run by the state or large landholders. Increasing yield is often viewed by the government as a way to deal with international debt and current trade balances.

In contrast, much of the maize and rice produced in developing countries is produced within a subsistence economy, where increased stability of yield is the major concern. There are certainly a growing number of large mechanized rice farms in Asia (e.g. Thailand, Malaysia) and export is becoming more important, but, in most areas, small farms are likely to remain an important part of a subsistence production system for the foreseeable future.

Most assessments of pest losses come as average yield losses for regions or average money spent on protection from pest damage. From a subsistence-farm perspective, the major issue is not average losses but variances in losses. The yellow stem-borer may only cause an average yield reduction of 1% in Asia, but it can cause enormous losses on single farms. When we are weighing the importance of a given pest for the resources of biotechnology, we must be careful not to use only estimates of overall crop losses as a guide.

The major problem for subsistence farmers is that when it rains it pours. Bottrell *et al.* (1992) pointed out that, in the Laguna province of the Philippines, yellow stem-borer usually causes minor losses to irrigated rice, but that they had seen a farm where the farmer was at the end of the 'irrigation line' and therefore did not get to flood his paddy and plant his crop until late. General lack of water led to his rice maturing later than the rice in the surrounding area. It appeared that all the yellow stem-borer females in the area congregated at his farm late in the season when his rice was still lush, while rice on the other farms was dry and ready for harvest. His crop was decimated.

This is just one dramatic example; however, it illustrates a more general problem. When crops have optimal levels of water and nutrients, they are generally more capable of withstanding insect damage. There is often a synergistic interaction between yield-reducing factors, so the poorer the land and the poorer the farmer, the more sensitive the remaining yield is to one more stress agent. Design and development of engineered crops should take this into account.

As an example, the farmer described above would certainly have benefited by having a variety in which 100% of the plants were producing a *B. thuringiensis* toxin, if the stem-borers had not already adapted to the toxin. The question must be posed whether the farmer would be better served with a mixture of rices in which 50–80% of the plants produced *B. thuringiensis* toxin. If research indicated that this mixture would remain effective longer, such a mixture could prevent unexpected calamities by ensuring protection of 50–80% of the crop. At least the farmer would have a chance of feeding his or her family.

Although cotton is a cash crop and governments are concerned with yields over large areas, there are still many individual farmers who must depend on stability of yield from single fields. In Brazil, for example, the southern region has many large mechanized cotton farms, but in the north small non-mechanized farms are the rule. Katre (Chapter 6, this volume, and personal communication) gives a description of Indian cotton farmers who committed suicide after their cotton crops were lost to insecticide-resistant insect pests. It is clear that any decision on the allocation of resources for crop protection by engineered plants will have to consider the farmers' long- and short-term needs.

Allocation of public resources

The public goals in engineering pesticidal plants will probably differ as much among geographic locations as they will among the three crops discussed. I hope it is apparent from the above discussion that any monolithic charge to genetic engineers interested in crop protection for developing countries would be mistaken. Instead, we must address the difficult issue of optimally allocating resources to deal with a heterogeneous set of circumstances. I shall offer a few examples to further illustrate this point.

As indicated above, a large part of the insect problems in tropical rice appear to be induced by broad-spectrum insecticides. Therefore, a major goal of any pest management strategy should be to decrease the use of these compounds. The question here is what are the best ways to use engineered crops for long-term reduction of pesticide use.

In rice-farming areas where there is no means to train farmers in IPM, one focus of engineering could be on removing evidence of leaf-folder damage. This would hopefully lead to a decrease in unnecessary use of insecticide and could decrease outbreaks of planthoppers. However, if a *B. thuringiensis* gene were used to control leaf-folders in a manner that did not give sustainable protection, farmers would probably go back to spraying insecticides once the leaf-folders adapted to the *B. thuringiensis* toxin. Therefore, to be defensible, any decision to target rice leaf-folders must have a solid base in a resistance-management programme that has been tested empirically, or the varietal deployment programme must be accompanied by a farmer training programme.

In areas where IPM training is currently practised or is on the way, it would not be appropriate to target leaf-folders in most localities because the educational system apparently offers farmers relief from this 'pest' (Kenmore, Chapter 4, this volume, and K.L. Heong personal communication). Instead, it would be more appropriate to use genetic engineering to decrease the chance of severe damage by stem-borers to reproductive-stage rice. (This type of damage is hard to control by any other approach, including insecticides.) This would certainly not require that 100% of the rice fields or 100% of the rice plants within a field be protected by *B. thuringiensis* toxin. Development of varieties that fit into a resistance-management programme would be important.

In Central America, maize farmers face some of the same issues as Asian rice farmers, except that a positive relationship between pesticide sprays and pest outbreaks is not as clear for some maize pests. In many areas of Africa, insecticide-induced outbreaks are not a problem because farmers cannot afford to use insecticides; they do, however, suffer crop losses.

Bacillus thuringiensis genes designed for temperate-zone maize problems will no doubt find their way into tropical maize germplasm without the help of public funds. Perhaps the most pressing and immediate need in these cases is to determine how these genes are likely to affect tropical maize pests. A simple programme to assess the sensitivity of these tropical pests to a variety of *B. thuringiensis* genes would be appropriate (our laboratory has standardized *B. thuringiensis* toxin formulations available for this purpose). This could help assess which important pests would be best and worst served by the 'standard' European corn-borer *B. thuringiensis* genes. If these genes seemed inappropriate for an important pest, such findings would direct the next efforts toward finding genes that were appropriate. Hopefully it would be possible to develop germplasm with the appropriate genes so that the 'flow' of inappropriate genes into tropical maize could be slowed or halted.

In the Central American situation, one might want to use *B. thuringiensis* maize as a temporary means of getting farmers off a pesticide treadmill by re-establishing natural enemies and by instituting training programmes. In Africa the major goal would be to avoid catastrophic crop losses for as long as possible. Therefore, resistance-management strategies that were straightforward in implementation would be useful in Africa. These strategies would not be constrained by the need to offer complete crop protection – they would just need to stabilize yields.

In Central America, the leafhopper *Dabulus maidis* has become a major problem because it transmits a pathogenic virus. If maize could be engineered to be resistant to the pathogen, the pest status of this leafhopper would drop drastically. However, if farmers are not educated about the change in pest status of the leafhopper, spraying might continue as it did before introduction of the virus-resistant plants. This case clearly illustrates the need to coordinate crop engineering and IPM training.

The cotton situation in some parts of India described by Katre (Chapter 6, this volume) offers a good example of engineered pesticidal crops that could help farmers get off the pesticide treadmill. According to her description, it is currently impossible for farmers to stop using insecticides, because there is a lack of natural enemies and unsprayed cotton is devastated by insect pests. If the Indian government could develop a programme that provided appropriate *B. thuringiensis*-expressing cultivars of cotton along with an IPM education programme and a publicity campaign warning (and demonstrations based on examples from other insects) that pests would soon adapt to the *B. thuringiensis* cotton, it might be possible to decrease long-term pesticide use. If wonderful new pesticidal plants became available by the time cotton pests adapted to the *B. thuringiensis* toxins, no harm would be done, and, if such wonderful new plants did not materialize, the educational programme might limit the return to misuse of insecticides.

None of the programmes outlined in this section can be accomplished without a concerted and sustained effort. The next section outlines one publicly funded programme in rice that offers an example of what can be done.

IRRI/Rockefeller rice project

The Rockefeller Foundation and IRRI recognized that engineering one excellent *B. thuringiensis* gene into a major crop plant does not equal successful, sustainable agriculture. They realized that there were many unresolved, complex issues, and many straightforward scientific questions that remained unanswered. In 1991, under the leadership of Dr D. Bottrell, a project was begun in an attempt to answer questions related to *B. thuringiensis* resistance management (Bottrell *et al.*, 1992). This project is now under the leadership of Dr M. Cohen.

The major motivation for undertaking this project, from an entomological perspective, was to learn as much as possible in a few years about the ecological and genetic interactions between *B. thuringiensis* toxins, rice, rice pests and natural enemies. It seemed clear that *B. thuringiensis*-expressing rice would sooner or later reach commercialization, and that the more understanding there was about the system, the more chance there was to implement the new technology in a sustainable manner. The project can be broken down into five interactive components, as follows:

1. search for useful *B. thuringiensis* genes;
2. assessing the potential for resistance in rice pests;
3. transformation technology and gene expression;
4. field tests of strategies to delay resistance; and
5. germplasm development and distribution.

Search for genes

The first part of the project was to search for unique and useful *B. thuringiensis* genes in a set of *B. thuringiensis* strains that had been collected and catalogued by IRRI and Plant Genetic Systems (Belgium). Because the collection consisted of nearly 4000 isolates and nearly 1000 potentially unique strains, this was a large undertaking. To streamline the programme it was important to prioritize the target rice pests and to set criteria for what constituted a unique, useful strain. After much debate, the decision was made to set pest priorities as follows: (i) yellow stem-borer; (ii) striped stem-borer; and (iii) leaf-folders. A *B. thuringiensis* strain was considered unique based on several criteria, including lack of genetic homology with toxins of standard strains that were active on the target pest(s), high efficacy at killing target pests and/or a unique mode of action as determined by biochemical studies of toxin binding to brush-border membranes of the pests' midgut epithelial cells. This first part of the project is now running smoothly, after difficult problems with bioassay development were addressed. Collaborative efforts with Ohio State University have made the biochemical assessments achievable.

Potential for resistance

The second part of the project, assessing the potential resistance in rice pests, has focused mainly on the two stem-borers. Because no *B. thuringiensis*-expressing rice plants were available, initial efforts in this work had to focus on more general questions. There were no quantitative studies on the genetic structure of populations of these two stem-borers, so the first studies examined this issue. Populations of the stem-borers from representative areas of the Philippines are being assessed for variability in ecological characteristics (e.g. host range) and for genetic differentiation at enzyme loci and in segments of mitochondrial DNA (in collaboration with the University of Maryland). As discussed in earlier sections of this chapter, an understanding of pest movement, which is reflected in population structure, is essential for logical development of resistance-management strategies. Philippine strains of stem-borers are now also being assessed for variation in response to selected *B. thuringiensis* toxins incorporated into artificial diets. The limitation to Philippine stem-borer strains is related to importation restrictions on pest organisms. A 'portable' bioassay kit that can be carried on an aeroplane has recently been developed so that stem-borer strains from other countries can be tested *in situ* and in parallel with those from the Philippines.

Selection experiments were begun with both stem-borer species, using *B. thuringiensis* toxins in artificial diet. Unfortunately, the yellow stem-borer has been difficult to maintain under these selection conditions, so the work is now being done solely with the striped stem-borer. Once *B. thuringiensis*-expressing rice is available, it will be possible to select yellow stem-borer strains for resistance.

Transformation and expression

Work on rice transformation and gene expression is being conducted in the Breeding Division at IRRI as well as in many other laboratories within developed and developing countries. Although entomologists have expressed interest in working with plants that have non-constitutive *B. thuringiensis* expression, development of such plants will probably come after the development of plants that have constitutive expression. From the entomological perspective, even having a plant with low constitutive expression would be valuable for selection experiments and for laboratory and field assessment of behavioural response of larvae to *B. thuringiensis* in the plant.

The technology of gene synthesis and plant transformation is still in a state of rapid change, so strategies that do not seem possible today may be possible 1 year from now. Any programme for developing *B. thuringiensis*-expressing rice will have to be flexible and incorporate new technology into strategy assessment.

Field tests

Field tests of resistance-management strategies have not begun at IRRI, but planning has. The first quantitative experiments will focus on: (i) larval movement and fitness in homogeneous plots of *B. thuringiensis*-expressing rice plants and in plots with mixtures of *B. thuringiensis* and non-*B. thuringiensis* rice plants (tests will also assess effects of the presence of weedy, wild-rice species within *B. thuringiensis* rice fields on fitness and movement); (ii) the interactions between *B. thuringiensis*-expressing rice and the efficacy of natural enemies at preying on and parasitizing pests that are and are not affected by the *B. thuringiensis* toxins; and (iii) general studies of the impact of *B. thuringiensis* plants on the community ecology of invertebrates in the rice ecosystem.

Later experiments will focus on comparing a number of engineered rice types that vary in *B. thuringiensis* genes and the expression of those genes. Comparisons will also be conducted regarding the farm-level feasibility of implementing various deployment strategies.

Development and distribution

Variety development and the distribution of *B. thuringiensis*-expressing rice should only occur after entomologists and breeders are fairly confident that they have developed rice plants that express *B. thuringiensis* toxins in a manner that could be used by farmers, without leading to rapid resistance development or disruption of the balance in other pest/natural-enemy systems. Getting to this point will require field tests of a number of strategies at many locations. This will be a major undertaking and will require cooperation between scientists in a number of countries. The rice IPM network may offer one mechanism for accomplishing this task.

There is a chance that this resistance-management project with *B. thuringiensis* rice will become somewhat of an academic exercise. It is clear that getting one semireasonable *B. thuringiensis* gene in a rice plant will not be too difficult, and that many laboratories in developed and developing countries will be able to accomplish this task in the next couple of years. Will the well-intentioned Rockefeller/IRRI project be left in the dust as biotechnology laboratories and rice breeders throughout Asia compete to be the first ones with any old *B. thuringiensis* gene in commercial use?

Even if this is the case, the Rockefeller/IRRI project is likely to have long-term benefits because agricultural scientists will need some of the same types of information when they start developing the replacement for *B. thuringiensis* rice. It would be much better, however, if a means could be found for the cooperative scientific development and deployment of *B. thuringiensis* rice.

Conclusions

We all like simple solutions to problems, but most of us have enough experience to know that we should not expect simple solutions to most real problems. I hope that this presentation has laid out some of the more important complexities involved in using pesticidal plants to deal with crop protection. I also hope that by presenting a lot of problems with the use of these plants I do not cause readers to throw up their hands in frustration. Yes, there are problems, but there are also opportunities.

There is little doubt that pesticidal plants will reach the market-place in developing countries. Without any intervention by the public sector, the plants that reach the market-place and the way these plants are used could be very inappropriate. Any well-thought-out approach to improving the situation could be beneficial. I have explained some of the problems that could be addressed, and no doubt there are more.

It has been argued that money spent on developing engineered plants could be better spent in other areas of pest management. Perhaps this is true, but this is not the issue. Engineered pesticidal plants will be developed. The question is about allocating resources to improve this development. A relatively small investment in developing appropriate use of pesticidal plants could avoid a large disaster caused by their misuse. Much of the investment in appropriate development and use of these plants is so closely tied to developing an understanding of the general ecology of pests and natural enemies that it is bound to improve our ability to develop other types of pest-management programmes.

Certainly the farmer will lose if engineered pesticidal plants are not developed and used appropriately. But the farmer will not be the only loser. One group of losers may be the corporations and non-profit groups whose *B. thuringiensis* genes are used. If pests adapt rapidly to *B. thuringiensis* plants and cause hardship, the finger will be pointed at crop engineering in general, and public funding of other useful engineering programmes may decrease. Large corporations have always been an easily identifiable culprit, so, if it turns out that genes that were originally engineered by company X turn up in the Indian cotton that has just been decimated by insect pests, we can guess where the buck will stop. Surely company X could spend money on public relations to alleviate the impact of such accusations on their image. However, they may be better served by getting involved, proactively, in facilitating more appropriate use of their technology.

One often hears that there is a mandate for researchers in agriculture to double the yield of food crops in the next 30 years. This may be an important goal, but along with this goal comes the responsibility to make sure that we do not also double the variation in yield. Crop protection from pests often increases yield but its most important function is to decrease the occurrence of catastrophic loss at the farm and regional level. If we look to engineered crop protection as a means of increasing yield without keeping an eye on the potential for the type of catastrophe that could occur when pests adapt to a resistant cultivar, we are the culprits.

Acknowledgements

This chapter has profited from discussions with D. Bottrell, M. Cohen, K.L. Heong, G. Khush, R. Aguda, C. Demayo, G. Kennedy and G. Roderick. I thank C. Satterwhite, R. Roush and M.T. Johnson for comments on the manuscript. This work has been sponsored in part by the Rockefeller Foundation.

New Diagnostics 17

Mark E. Whalon

Introduction

The revolution in molecular biology and its subsequent application across science has had far-reaching application in agriculture, education, industry and medicine. The forefathers of modern entomological science would scarcely recognize the laboratories or language of the modern molecular entomologist. Today the applications of molecular biology (biotechnology) in entomology are diverse and no longer require exacting bench-top capability. The development of various commercial nucleotides, enzymes, reagents and specialized equipment has opened the power and scope of molecular biology to virtually every practising entomologist with access to even rudimentary laboratory and, in some cases, field capability.

In the 1950s and 1960s, numerous advances in protein chemistry and gel electrophoresis allowed genetic polymorphism in proteins to be investigated. Entomologists studying insect populations soon realized that the ability to determine polymorphic proteins by electrophoresis could be an extremely useful tool. These techniques, which are based on the differential migration rates and distances of protein polymers with different charges associated with their amino acids, were so sensitive that a different charge on one amino acid could result in a mobility difference on a gel. Multiple forms of some proteins exist within single species, and some enzymes have several electrophoretically separable forms. If these polymorphisms followed the laws of genetic inheritance they became valuable genetic markers that could be used to determine genetic relationships. Many of these protein-electrophoretic techniques remain applicable today and, in some instances, may have advantages over biotechnology tools.

Proteins from different species are often immunochemically different due to amino acid differences resulting from mutations. Thus immunological tools were coupled with protein migration in gels to yield additional resolution for studying systematics and phylogeny. These tools were soon applied to population genetics and field (economic) entomology. By studying these protein and enzyme differences, entomologists were indirectly studying deoxyribonucleic acid (DNA) sequences. In molecular entomology today, many of the conventions, vocabulary, quantification statistics and applications have been carried over from these earlier techniques into the use of modern DNA diagnostics.

The total amount of information carried in an insect's genome is staggering, consisting of from over 0.5 to 2 billion (thousand million) nucleotides that comprise the chromosomes, which in turn are continuous strands of DNA ranging from 5 to 250 million nucleotides long. Much of this DNA is considered 'junk' because it does not code for any functional genes, and yet 'junk DNA' may have spacer or other unknown functions. Other nucleic acid (messenger ribonucleic acid (mRNA), ribosomal RNA (rRNA) and transfer RNA (tRNA)), organelles (mitochondrial DNA (mtDNA)), microorganism and, possibly, parasite genetic information may also be present. With this much information stored in each individual in a population, it is not surprising that considerable variation exists between individuals, populations and species.

The overall goal of this chapter is to present recent developments in the use of molecular tools for monitoring insects from the applied point of view. In so doing, I hope to demonstrate the scope of molecular biology techniques employed in monitoring insect populations, to discuss the recent advances in biotechnological diagnostics and to provide several examples of these applications from my own and other laboratories.

There are at least two approaches one could use to relate the use of molecular diagnostic tools in entomological population studies, techniques and applications. I shall attempt to organize this chapter along the lines of both molecular techniques used and their application arena. To facilitate the entomological discussion and not be overly encumbered by molecular methods, I shall introduce the techniques with a brief methodological description initially, and in all subsequent discussion make reference by common name, without any methodological explanation. This chapter is not intended to be an exhaustive review of the literature, nor shall I present a complete discussion of all the techniques or their applications. Readers seeking these developments will have to utilize the original research, some of which is cited here.

Molecular Methods

Restriction fragment length polymorphism (RFLP) analysis (see review by Caskey, 1987) has been commonly applied to various insect population studies, particularly in conjunction with Southern hybridization (Southern, 1975). These techniques are fairly straightforward, starting with the isolation of DNA, which is usually cleaved with restriction enzymes to produce different DNA fragment lengths. These are electrophoresed to separate them; the smaller fragments migrate faster than the larger pieces in the charged electrical field of the gel. After separation, the fragments are denatured chemically or with heat (100°C will cause double-stranded DNA to separate into single strands). The strands are then transferred from the gel to a solid matrix (usually a nitrocellulose or nylon filter), where they are immobilized with heat or ultraviolet (UV) radiation before the filter is submitted to prehybridization to prevent non-specific binding of the detection probe. A radioactive, biotin or similarly labelled probe (usually a biologically amplified piece of DNA, which is termed 'cloned'), a synthetic DNA or oligomer, a repetitive sequence (e.g. satellite DNA or a random amplified polymorphic DNA (RAPD), which may or may not have been previously characterized), is then hybridized to complementary sequences on the filter, and the annealed fragments are visualized with autoradiography, fluorescence or a colour reaction. The conditions of the reactions can be modified to make the annealing process more or less stringent, which provides for more or less specificity. This technique yields detection levels down to a single-copy gene, but requires considerable target DNA and time to accomplish.

Dot–blot hybridization is similar to RFLP analysis except that extracted, unrestricted and unelectrophoresed DNA or, in some instances, the whole insect is directly blotted on to the filter paper before fixation, prehybridization and hybridization. This technique has been widely used in studying insect vectors of human and plant disease. In general, it is faster and more economical and can be applied in near-field conditions, but usually yields less information than RFLP analysis.

Amplification of target DNA and RNA sequences represents one of the truly revolutionary steps forward in molecular biology, and it has subsequently been applied in insect population studies (Mullis and Faloona, 1987). In principle, a single copy of the target nucleic acid sequence can be amplified to yield 10^8 or 10^9 copies, thus vastly increasing the detection limits. In addition to detection, if the target sequence is polymorphic, meaningful population statistics can be determined. Two oligonucleotide primers, 7–18 base pairs (bp) long, are used which are complementary to the flanking regions upstream (5') and downstream (3') from the sequence that will be amplified. These primers define the ends of the target region. The double-stranded sample DNA is first denatured into two complementary single strands, using high temperature. The primers anneal to the flanking regions as the reaction cools, and the primers are then extended in the presence of nucleotides with DNA polymerase, using the target sequence as a template. Two copies of double-stranded DNA are thus formed. After the next cycle of heating and cooling, four double strands are made, then eight, then 16 ... until after 20 or 30 reaction cycles have produced sufficient target sequence to detect or to utilize in various other steps for further analysis. This technique is termed the polymerase chain reaction (PCR).

More recently, Williams *et al.* (1991) reported the development of a technique that uses a single 10 bp primer that hybridizes to and amplifies random regions of genomic DNA. This technique, RAPD, when coupled with PCR, RAPD-PCR, provides the most versatile and soon to be widely applied biotechnology technique in entomology. This process is as sensitive as PCR and produces amplified fragments from genomic regions that are repetitive and also regions that are unique. The disadvantage of this approach is that amplified regions are not well characterized, unlike most RFLP and dot–blot hybridization probes. However, the technique can often yield more than 30 polymorphisms from a single individual within an 18 h procedure. It does not require radioactivity or other labelling systems and the samples prior to the experiment are usually frozen, although this may not be necessary. Thus museum specimens or even ancient amber-embedded specimens can yield amplified fragments for study (Cano and Poinar, 1993).

In situ detection of single-copy genes is now possible with the application of PCR technology to fixed tissue on microscope slides (Gosden and Hanratty, 1993). This technique may prove useful in locating the expression sites of various genes (e.g. insecticide resistance genes), the tissues infected by symbionts and pathogens or the receptor sites for various peptides and regulator nucleotides. In addition, detection is becoming increasingly economical, non-isotopic, efficient, automated and even portable. Fluorescence can also accelerate 96-well RAPD-PCR experiments (Holland *et al.*, 1991), while specific detection is now possible with colour PCR (Kemp *et al.*, 1989). Automation is also commercially available for variously labelled PCR products (Mayrand *et al.*, 1991; DiCesare *et al.*, 1993), but perhaps the greatest step forward will be the field-portable PCR machines (MJ Research Inc.) with commercially prepared reaction kits.

Applications

Biotechnological tools have been applied broadly in insect population genetics, in studies addressing inheritance, introgression, genetic drift, heritability, diversity and gene linkage. All of the molecular biology techniques including dot–blot hybridization, RFLP, PCR, genomic oligonucleotide hybridization and RAPD-PCR have been applied in at least one of these arenas. These tools have also been applied to species separation, species detection and subspecies separation (biotypes, races, etc.). Some of these applications will have far-reaching implications for future systematic, phylogenetic, genetic and plant breeding studies. They may be particularly important in their application for identification of insecticide-resistant individuals and biotypes important to plant breeding programmes. Human and plant vector identification, association, detection and epidemiology have also benefited. In addition, biological control agents have been detected, and their spatial and temporal occurrence in host populations has been predicted on the basis of molecular diagnostic techniques.

A dramatic application of molecular biology in human health protection resulted from the geographical study of different populations of mosquitoes using RFLP. Raymond *et al.* (1991) were able to infer from their data that *Culex pipiens* amplified esterase genes, which account for organophosphate resistance, have spread from a single population in Africa or Asia to the Mediterranean, North America, Caribbean, India, Central Asia and South-East Asia. This remarkable event has occurred in the last 30 years and has resulted in significant resistance and, in some instances, loss of control and resultant economic and human impact. Another example of the application of biotechnology to human-health pest studies developed with various probes that distinguish closely related species of blackflies (Post and Crampton, 1988) and sandflies (Ready *et al.*, 1988). Various oligonucleotide probes have also been developed to the *Anopheles gambiae* complex (Gale and Crampton 1987; Hill *et al.*, 1991), and Hill *et al.* (1992) have prepared these and other probes for non-radioactive field identification. This latter work has eliminated the need for complex hybridization solutions, heat inactivation and posthybridization washes, while maintaining the same sensitivity as laboratory-based protocols. The techniques of RAPD-PCR have also been applied to mosquitoes for population, phylogeny and species identification (Kambhampati *et al.*, 1992a).

In many host-plant resistance (HPR) breeding programmes, there is often more genetic understanding of the plant species than of the target pest genetics. This situation can contribute to short-sighted HPR deployment strategies, and will be especially important with the emerging transgenic plant technology (see the summary in McGaughey and Whalon, 1992). An adequate genetic understanding of the pest population's evolution of biotypes that overcome HPR will be essential for wise deployment of both transgenic and conventionally selected plants in the future (Gould, 1988b). For example, the spatial and temporal distribution of aphids determined by intergenic, rRNA spacers demonstrated that sufficient clonal diversity was present on one sorghum leaf from a field population to evaluate sorghum HPR lines in conventional breeding programmes in central North America (Shufran *et al.*, 1991). Apparently these aphids arrive from sufficiently diverse clonal populations annually to assure breeders that small, replicated plots in one area are sufficient to evaluate HPR for a region. In other instances, the brown planthopper *Nilaparvata lugens* for example, populations within rice paddies may be much more genetically uniform (M.E. Whalon and R.A. Shufran, unpublished data). Thus molecular entomology can add critical information to plant breeding programmes, but much more needs to be done to ensure that other newly developed plant varieties will be durable when they are released. This is especially important in developing countries, where food reserves are marginal and where insufficient human, technological and economic capital exists to utilize conventional insecticides to avert major losses when pest-resistant crop varieties fail.

As a result of asexual reproduction, Aphiidae, Hymenoptera and Thysanoptera are often too isomorphic to find electrophoretically detectable allozyme polymorphisms; thus allozyme analysis has not been useful among these groups. However, RAPD-PCR techniques have produced discrete genetic markers in four species of aphids (Black *et al.*, 1992). Polymorphisms were detected among biotypes, populations, colour morphs and even individuals on single leaves. The amount of polymorphism also varied between species, perhaps indicating greater clonal diversity in some species. Thus RAPD-PCR may have far-reaching significance in the biology, ecology, phylogeny and management of many organisms.

The introduction of a new whitefly specics morphologically indistinguishable from *Bemisia tabaci* resulted in more than a $US0.5 billion loss in US agricultural production in 1991. The application of single-primer PCR techniques to these reported conspecifics has helped in the identification of two separate species and the clarification of the source of the economic losses (Perring *et al.*, 1993). Therefore, this technique may be useful to routinely monitor species and thus contribute to their control.

Until the development of suitable diagnostic probes (see the review by Lee and Davis, 1991), the epidemiology of deltacephaline leafhopper-borne mycoplasma-like organisms (MLO), a devastating yellows-syndrome disease of many economically important plants throughout the world, was often not possible to analyse. Kirkpatrick *et al.* (1987) first successfully cloned DNA fragments from the leafhopper vector *Colladonas montanus*, infected with the MLO that caused decline in *Prunus* trees. Through this biotechnological advance, my laboratory was able to develop oligonucleotides that definitively distinguished infected from non-infected leafhoppers. Laboratory and field application of these probes determined the stage and timing of vector infection by the MLO, the proportion of the vector population infected and the timing of transmission in the field (Rahardja *et al.*, 1992). We learned that the degree of disease transmission in the field was the result of the pathogenicity of the MLO for the vector, which was dramatically influenced by the orchard's ambient temperatures through the season (Garcia-Salazar *et al.*, 1992). Subsequently we have developed both rapid, non-isotopic, dot–blot hybridization and RAPD-PCR systems for monitoring this disease. Today, MLO disease transmission prediction in Michigan peach and cherry trees is possible, with routine monitoring of MLO-infected vectors and ambient temperatures.

Since most eukaryotic chromosomes are composed of repetitive sequences, – satellite DNA – the dispersion pattern of these sequences and their nature can be utilized in studying insect populations. Fortunately, the DNA of insects can be organized into three groups: single copy, intermediate and highly repetitive sequences. The organization of these three groups can be utilized to categorize different species (Table 17.1) into either short-period interspersion (SPI) or long-period interspersion (LPI) of single-copy structural genes and repetitive sequences (Crampton, 1992). In the SPI pattern 1–2 kb segments of single-copy genes alternate regularly with short 0.2–0.6 kb to 1.0–4.0 kb repetitive sequences, whereas the LPI pattern is characterized by 5.0–6.0 kb alternating with very long > 13 kb repetitive sequences. Thus SPI insects like *Aedes aegypti* will contain well-dispersed repetitive sequences, which can mask profiling or fingerprinting when using genomic DNA probes (Warren and Crampton, 1991). *Drosophila* and *Anopheles*, on the other hand, exhibit the LPI pattern, and cloned probes are more likely to be useful than single-copy or repetitive DNA probes.

Repetitive DNA sequences are thought to arise from sudden replications, followed by a mutational event or an unequal crossover, which may lead to a deletion and/or tandem duplication event (Smith, 1976). The pattern of repetitive sequences can be detected by minisatellite DNA and is sometimes used to develop unique fingerprints of individuals or populations (Jeffrys *et al.*, 1985). Hypervariable regions, usually in the chromosomes near the centromere, show multiallelic variation and correspond to or segregate with a trait. Thus they can be quite useful in insect population studies. The human minisatellite DNA probe 33.15 was used to characterize four Colorado potato beetle (*Leptinotarsa decemlineata*) populations. These polymorphisms could be used to segregate insecticide-resistant beetle populations selected with *Bacillus thuringiensis CryIIIA* toxin, pyrethroids and/or organophosphates (Whalon *et al.*, 1993a).

In addition, RAPD has been useful in segregating Colorado potato beetle (Whalon *et al.*, 1993b) and brown planthopper populations (K. Shufran and M. Whalon, unpublished data). The polymorphisms of RAPD, detected as DNA segments that amplify in one population at a higher frequency than in others, are inherited in a Mendelian fashion. A diagnostic Colorado potato beetle RAPD-PCR fragment pattern, using a 10 bp primer (Operon Inc.) has been developed. The diagnostic band has been shown to cosegregate with the *Bacillus thuringiensis CryIIIA* autosomally dominant, resistance gene in a field- and laboratory-selected strain (U. Rahardja and M. Whalon, unpublished data). We are currently assessing its potential

Table 17.1. Organization of single-copy and repetitive sequences in the genomes of insects. Short-period interspersion (SPI) and long-period interspersion (LPI) refer to the pattern of single-copy gene occurrence relative to repetitive sequences.

Species	Genome size	Gene pattern	Source
Aedes aegypti	Large	SPI	Cockburn and Mitchell, 1989
Aedes triseriatus		SPI	Black and Rai, 1988
Aedes albopictus		SPI	Black and Rai, 1988
Antherea pernyi		SPI	Crampton, 1992
Anopheles quadrimaculatus	Small	LPI	Black and Rai, 1988
Apis mellifera		LPI	Crampton, 1992
Bombyx mori		LPI	Crampton, 1992
Culex pipiens		SPI	Black and Rai, 1988
Drosophila melanogaster	Small	LPI	Crampton, 1992
Leptinotarsa decemlineata	Small	LPI	M.E. Whalon, unpublished data
Lucilia cuprina		SPI	Crampton, 1992
Musca domestica		SPI	Crampton, 1992
Nilaparvata lugens	Small	LPI	M.E. Whalon, unpublished data
Sarcophaga bullata		LPI	Crampton, 1992

utility for determination of gene frequency in the field. Such DNA probe technologies, together with other advances in DNA diagnostics, including automation and portable field equipment, will have a significant impact on integrated pest management and insect pest-resistance management in the near future.

Biotechnology advances

Molecular biological advances associated with *in situ* applications, portability, hybridization imaging, non-radioactive detection and automation are now being pioneered for medical diagnostics, but will soon find uses in entomology. Routine microscope-slide, *in situ* PCR amplification of genes involved in insect pathogen interaction, receptor–substrate complex formation, pesticide resistance and biotype evolution, adiopokinetic hormone expression and neuropeptide production are now possible (Arnheim *et al.*, 1990). Field-ready, portable PCR machines are now sold commercially (MJ Research Inc.). Both automated fingerprint reading (Ambis Inc.) and fluorescence revealed hybridization (Wallace Inc.) are available, and are being applied in a variety of medical and forensic diagnostics (Burke *et al.*, 1991; Kirby, 1992). Non-radioactive detection of hybridization reactions has also greatly increased the scope and range of DNA diagnostics. Various detection systems are available commercially, and any of these can greatly expand the number of laboratories and personnel capable of nucleic acid diagnostics. Many of the companies that market immunologically based diagnostic kits for application in medicine, industry, environment and agriculture are now marketing nucleic-acid-specific light- and heavy-chain (single-domain) antibodies haptenized to biotin for conventional binding with alkaline phosphatase or some other colorimetric system. These developments may mean that nucleic acid immunological probe kits will soon be available commercially in a wide array of diagnostic situations, including single-gene detection in the field.

Conclusions

Biotechnologically led developments have opened new opportunities for laboratory and field monitoring and detection of a broad array of genes and gene products in entomology. I have covered a range of applications of this technology in species identification, population monitoring, resistance detection, pest outbreak forecasting and vector– pathogen epidemiology. New developments in diagnostics occur almost weekly, and many will eventually find their use in entomology-related fields. In the short-term, entomology can expect reasonably priced, portable DNA amplification systems and commercially available DNA diagnostic kits. In the long-term, we can expect sweeping changes in pest-management consultancy, arthropod export–import diagnostic regulations, integrated pest management and biological control, fostered by the ability to detect genotypic features in individuals and populations. Taken as a whole, these developments will greatly expand our ability to monitor pests and to predict pest outbreaks.

Virus/Vector Control 18

MICHAEL E. IRWIN AND LOWELL R. NAULT

Introduction

Plant-pathogenic viruses cause considerable damage to crops by lowering yields and decreasing the value of harvested products. At times, plant viruses constitute the limiting factor to crop production. Arthropods, particularly insects of the order Homoptera, are responsible for spread of viruses within and between fields. Epidemics of many plant viruses, including most potyviruses, cannot be satisfactorily managed through vector-targeted chemical intervention. Some viruses, among them many that are aphid-transmitted, are difficult to control through conventional plant breeding, a practice that has traditionally focused on developing plant resistance to the virus. Thus, plant viruses are a major constraint on agricultural production, worldwide, and, as a whole, are exceptionally difficult to control. With the advent and expanding application of biotechnology, it behoves us to explore means through which these innovative technologies can be employed to understand and contain plant virus epidemics.

We are applied insect biologists with long-standing interests in formulating concepts of how viruses are transmitted by insects and the means by which plant viruses spread and thus can be contained. Much of our research has focused on understanding the behavioural ecology of vectors and their interrelationships with plant viruses. This has been done with the conviction that these aspects are fundamental and requisite to managing virus epidemics. We should like to highlight the fact that we are not molecular biologists and have little comprehension of the intricacies of that set of sciences. Nevertheless, we are convinced that biotechnology, and molecular biology in particular, will and should play important roles in helping to resolve perplexing questions surrounding the containment of virus epidemics. It is with this in mind that we contribute this chapter.

Pathosystems and Pathosystem Cycles

This chapter describes the dynamics of four fundamentally different pathosystems. The systems are distinct because they illustrate the four known types of relationships between plant-pathogenic viruses and their homopterous vectors:

1. Non-persistently transmitted, foregut-borne viruses are acquired and inoculated by vectors during brief probes to the plant epidermis. Virions reversibly attach to the maxillary food canal and lining of the foregut. Virus is retained by the vectors for a matter of minutes or hours. Soybean mosaic potyvirus (SMV), together with its aphid vectors, is the non-persistently transmitted, foregut-borne pathosystem that we feature.
2. Semipersistently transmitted, foregut-borne viruses are acquired and inoculated by vectors during prolonged probes to plant phloem. Virus attaches to the maxillary food canal and lining of the foregut. Virus is retained by the vectors for a matter of hours or days. Maize chlorotic dwarf waikavirus (MCDV), with its leafhopper vector, is the semipersistently transmitted, foregut-borne pathosystem that we feature.
3. Persistently transmitted, circulative viruses are acquired and inoculated by vectors during prolonged probes to the plant phloem. Virions ingested into the gut must pass into the haemocoele and then be taken up by the salivary glands before the vectors become viruliferous. The period of time from virus uptake to first transmission, the latent period, is 1–2 days for circulative viruses. Virus usually persists in the vector for the vector's lifetime but does not multiply in the vector. Barley yellow dwarf luteovirus (BYDV), with its aphid vectors, is the persistently transmitted, circulative pathosystem that we feature.
4. Persistently transmitted, propagative viruses are acquired and inoculated by the vector during prolonged probes in the plant phloem. Virions are ingested into the gut where they must first multiply in the gut wall before passage into the haemocoele. Virus can then circulate to other tissues and organs, including the salivary glands, where it multiplies and is incorporated into salivary secretions. The latent period lasts 1–4 weeks, and vectors retain infectivity for life. Some but not all propagative viruses are transovarially transmitted to their offspring. Maize rayado fino marafivirus (MRFV), with its leafhopper vector, is the persistently transmitted, propagative pathosystem that we feature.

There are many features common to the four pathosystems. Primary

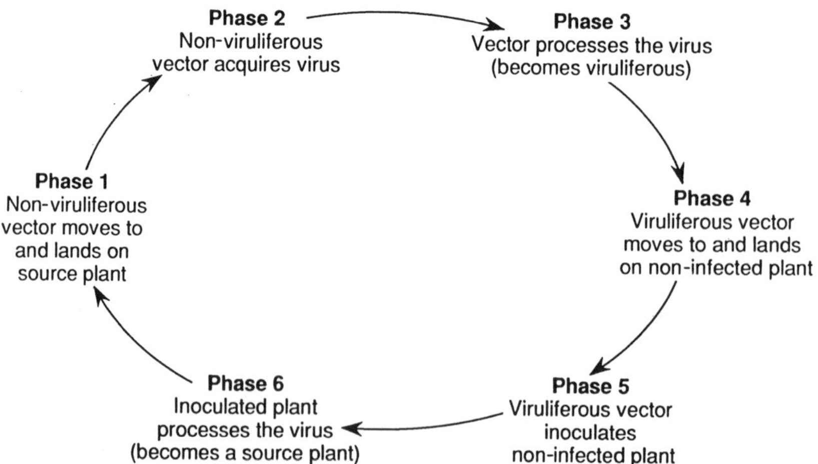

Fig. 18.1. Six shared phases in a pathosystem cycle.

among them is the cycle that must be completed for successful transmission and cycling of a plant-pathogenic virus. We identify six shared phases in a pathosystem's cycle (Fig. 18.1), which are described below, with a view to highlighting the junctures in each pathosystem's cycle through which biotechnology might best intercede in order to: (i) enhance our abilities to predict the intensity and timing of epidemics; and (ii) help contain such epidemics:

1. A non-viruliferous vector moves to and lands on a source plant, i.e. a plant that is infected and has processed the virus so that the virus can be acquired by a vector.
2. The non-viruliferous vector acquires the virus from the source plant.
3. The vector processes the virus so that it can be transmitted to other plants.
4. The viruliferous vector carries the virus to a plant.
5. The viruliferous vector inoculates the plant.
6. The plant processes the virus so that the virus can be acquired by a non-viruliferous vector. This completes the transmission cycle.

Viruses that are transmitted non-persistently most often have a natural entry point at the beginning of the season with phase 1, where a non-viruliferous vector moves to and lands on a source plant. In contrast to this, viruses that are persistently transmitted appear to have a natural entry point with phase 4, where the viruliferous vector carries the virus to a non-infected plant. We describe these four pathosystems and then explore the potential to use biotechnology to understand and help contain epidemics.

Featured Pathosystem Biologies

Soybean mosaic potyvirus (SMV) pathosystem

Soybean mosaic potyvirus (a non-persistently transmitted, foregut-borne pathosystem) is relatively specific to soybean, and is transmitted by more than 30 species of aphids (Irwin and Ruesink, 1986). The aphid species that are important vectors of SMV do not colonize soybean plants, at least not in the Americas, Europe or Africa (there is some question about the importance of colonizing aphids on the spread of SMV in Asia). Instead, they move to and land on soybean plants, probe, find the plants unsuitable for feeding and depart, only to land again nearby and repeat the process. As they fly short distances from plant to plant searching for a host, i.e. 'hop' through a soybean field, they are apt to land on a soybean plant infected with and which has processed SMV. If the encounter is early in the growing season, the likely source of the infected plant is a seedling from a seed that carried the virus from its infected parent (via vertical, seed-borne transmission) and, if the encounter is later in the season, by soybean plants inoculated by aphids (via horizontal transmission) (Irwin and Goodman, 1981). Seed-borne transmission, coupled with the fact that the seeds are transported around the world by humans, accounts for the global distribution of most strains of SMV (Irwin and Ruesink, 1986).

The virus is acquired during short, preingestion probes, as aphids penetrate epidermal cells with their stylets. The epidermal cell contents are transported to the aphid's precibarium where the aphid 'tastes' the fluid and determines if it should penetrate deeper into the leaf substrate, permitting the aphid to ingest. Short, preingestion probes into source plants bring the virus into the aphid's maxillary food canal and foregut. The sites in the aphid's food canal and cibarium where the virus attaches and the 'glue' that allows the virus particles to bind to those sites are important considerations and critical to the successful transmission of the virus. The helper component, produced only by virus-infected plants, is the putative 'glue', and is encoded by potyviral ribonucleic acid (RNA). Viral RNA encodes for structured protein (protein coat), helper component and other non-structural proteins. Without helper component, the virus will not bind to the appropriate sites, and the virus will not be properly processed, i.e. the aphid cannot retain and subsequently inject the virus into another plant. Once the virus particles are properly attached to the sites, the aphid is said to be viruliferous.

After having probed the plant and found the plant unsuitable for feeding and colonization, the aphid will leave the site and move (fly or walk) to another plant. This behaviour carries the virus to a new plant host. The aphid is now ready to inject the virus into a plant. After the viruliferous aphid has landed, it again initiates probing and preingestion feeding behaviours.

After the stylets are inserted into epidermal cells, the contents of the foregut may be expelled back to the plant by extravasation. Newly imbibed plant sap, mixed with previously acquired SMV virions that detach from the food canal and foregut, can thus be inoculated into epidermal cells of healthy plants. Once the virus particles have been introduced into the plant, they must multiply and be translocated if the cycle is to be completed.

During the processing of the virus by the plant, virus infection can be countered in several ways. These include the development of a hypersensitive reaction, which walls off the virus, producing local lesions; the channelling of the virus into non-vital organs, where it will be least disruptive to the reproductive efficiency of the plant; and, with regard to SMV and other potyviruses, walling off or cleansing the embryo of virus infection so that the virus does not become seedborne. Also, some cultivars are more resistant to, and others more tolerant of, infection by certain strains of the virus. A further reaction that reduces epidemics of potyviruses is not at first obvious. The plant can become necrotic and die shortly after becoming infected; that limits the time during which the virus is exposed for aphids to acquire it. Such a strategy can significantly decrease the rate at which epidemics progress.

Maize chlorotic dwarf waikavirus (MCDV) pathosystem

The maize chlorotic dwarf disease is caused by coinfections of at least two distinct strains of MCDV (a semipersistently transmitted, foregut-borne pathosystem) (Gingery and Nault, 1990). Maize plants in the eastern and south-eastern USA, where the disease is found, are also commonly coinfected with the maize dwarf mosaic potyvirus (MDMV). Johnsongrass, a perennial weed, is the overwintering host of both viruses. An important biological difference between MDMV and MCDV is that MDMV is transmitted by many aphid species, whereas the blackfaced leafhopper (*Graminella nigrifrons*) is the only known field vector of MCDV. The distribution of MCDV in the USA is delimited by the overlapping occurrences of Johnsongrass and the blackfaced leafhopper.

The leafhopper feeds and breeds on a number of annual and perennial grasses (Hunt and Nault, 1990). Early in the season, leafhoppers breed on winter grains or newly emerging leaves and shoots of perennials. Later, as these grains and grasses mature, leafhoppers invade annual grasses to complete additional generations. In the southern USA, leafhopper may breed year-round, but in northern states, such as Ohio, no more than three generations occur. The blackfaced leafhopper is a migrant, flying from southern to northern states in May and June, but the origin of adult leafhoppers that transmit MCDV is uncertain. Are migrants or local populations important in causing epidemics? Although the leafhopper also overwinters in northern states, cold harsh winters greatly reduce the numbers of overwintering adults or eggs that contribute to the cycle in the spring.

The MCDV pathosystem cycle begins when adult leafhoppers feed on infected Johnsongrass. When leafhoppers ingest from the phloem, they pick up individual virions or clusters of virions from inclusions (Ammar *et al.*, 1993). Virions are loosely bound by a matrix of what is probably viral RNA-encoded protein. The protein is probably the helper component required for leafhopper transmission (Hunt *et al.*, 1988). Virions with helper component attach to the lining of the upper region of the maxillary food canal and the precibarium, cibarium and pharynx (Ammar and Nault, 1991). Once leafhoppers have acquired virions, they are immediately capable of transmitting the virus but lose their capacity as vectors in a matter of hours or days. Retention of virus is temperature-dependent, with a half-life of 4–5 h at 30°C and 10–12 h at 20°C.

Leafhoppers dispersing from Johnsongrass apparently do not carry MCDV great distances (Rodriguez *et al.*, 1993). Spread of the virus to maize is always associated with the nearby presence of Johnsongrass. During the daytime, males employ a call–fly strategy, whereby they alternate between searching for virgin females and feeding (Hunt and Nault, 1991). Virgin females, in contrast to males, move very little during the day, and instead are perched high on plants, where they reply to acoustic signals from courting males. Subsequent to mating, females become even more quiescent, moving from plant to plant only to feed and lay eggs. Only at dusk do males and virgin females abandon the plant canopy and fly to more distant hosts.

After a viruliferous leafhopper alights on a maize seedling, it makes deep exploratory probes with its stylets. When it samples sap from vascular tissue, food is not always ingested into its midgut. Fluids taken into the foregut may be expelled back into the probed plant by extravasation. Virions and matrix detach from the food canal and foregut, mix with plant sap and are injected into the phloem of maize. In seedling maize, virus multiplies and symptoms appear in 7–10 days. These plants then serve as sources of inoculum for further spread of virus. Seedling maize, not older plants, are preferred by the blackfaced leafhopper as sites for feeding and breeding. When MCDV is transmitted during the seedling stage, there are serious economic losses to the crop. Maize, susceptible annuals or rhizomatous Johnsongrass infected by MCDV is the source of inoculum for seedling Johnsongrass, which in turn becomes the inoculum source the following year, thus completing the disease cycle.

Barley yellow dwarf luteovirus (BYDV) pathosystem

Barley yellow dwarf, a disease of more than 100 species of wild and cultivated grasses, including maize, sorghum, wheat, barley and oats, is caused by BYDV (a persistently transmitted, circulative pathosystem) – several closely related luteoviruses and luteovirus strains (Irwin and Thresh,

1990, and references therein). The disease causes severe yield losses in most of the small grains, worldwide. Although this complex of viruses is transmitted by over 20 species of aphids, *Rhopalosiphum padi*, *Rhopalosiphum maidis*, *Sitobion avenae*, *Schizaphis graminum* and *Metopolophium dirhodum*, species that feed on and colonize grasses, are primarily responsible for successful inoculation. This luteovirus is not seed-borne and, in temperate zones, is transmitted to crops from perennial, infected plant hosts or overseasoning viruliferous aphids; it cannot be transmitted vertically through either vector or plant hosts. In many parts of the world, movement to and colonization of grasses by viruliferous aphids occur relatively early in the growing season. As they move into cultivated grasses, aphids selectively carry with them various strains and types of viruses that are associated with the disease. Some of the isolates are specifically transmitted by certain aphid species, whereas others are non-specifically transmitted by several aphid species. It is the influx of and colonization by viruliferous aphids (phase 4 in the aforementioned phases in the cycling of a pathosystem) that produce the initial, scattered foci of BYDV infections early in the growing season in cultivated grasses (Irwin and Thresh, 1990).

A viruliferous aphid walks to or flies to and lands on a non-infected plant, probes it, and, once having found the crop to be a suitable host, penetrates the leaf tissue with its stylets, seeking out the phloem elements for long feeding sessions. During the inoculation process, the virions, along with the secretions from the accessory salivary and salivary glands, are injected into the phloem within the vascular bundles, where they are systemically transported to new plant tissue. The virus then establishes infection sites in susceptible meristematic tissue and begins to multiply. Processing the virus so that it is available for acquisition by an aphid takes the plant several days, and acquisition is most efficient when the virus titre is high, usually 1–2 weeks after virus processing by the plant. The virus is perpetuated by transfer from insect to plant and back to the insect.

A non-viruliferous aphid, if it lands on or walks to (phase 1 in the aforementioned phases in the cycling of a pathosystem), probes and, assuming the chemical and gustatory signals are correct, places its stylets into the phloem elements of the infected plant, will acquire the virus through ingestion. Virus passes to the midgut lumen, and then, depending on aphid species and virus strain, penetrates through the hindgut wall, invades the haemocoele, and circulates in the haemolymph. Matching the aphid species with the proper virus strain is not important for successful penetration of most virus strains through the gut wall and into the haemocoele; however, it is critical for some. If the virus has been ingested by the appropriate aphid species, virions eventually enter and concentrate in the accessory salivary glands, and are incorporated into the salivary secretions. Virions of specific isolates of BYDV not transmitted by some aphid species reach the haemocoele but are not taken up by the accessory salivary glands. This process is extremely selective. Thus, either the hindgut or salivary glands may be barriers to BYDV transmission (Gildow, 1990). Processing (i.e. the latent period) of this circulative virus takes from a few days to more than a week. Once an aphid becomes viruliferous, it remains so for life. When a viruliferous aphid moves to another plant, the cycle begins anew.

Maize rayado fino marafivirus (MRFV) pathosystem

Maize rayado fino marafivirus (a persistently-transmitted, propagative pathosystem) is transmitted primarily by the corn leafhopper *Dalbulus maidis*, throughout most of the neotropics (Gámez, 1980; Nault, 1990). At high elevations in Mexico, the Mexican corn leafhopper, *D. elimatus*, is also a vector. The virus and vector, for the most part, are restricted to the plant genus *Zea*, which includes maize and wild teosintes. The virus overseasons in the vector; there are no known perennial hosts for MRFV. In regions where maize is grown under irrigation during the dry season, vectors and virus perennate on maize year-round. In regions where maize is not grown during the dry season, the corn leafhoppers disperse from fields at harvest time to as yet unknown localities.

Leafhoppers appear on maize seedlings shortly after maize is planted at the beginning of the rainy season. Females alight on plant, using colour and odour cues from their hosts (Todd *et al.*, 1990). Females move to the maize whorl, where they begin feeding and egg-laying. Viruliferous females locate the phloem, where they secrete saliva and inoculate the plant with virus. The virus establishes infection sites in susceptible seedlings and systemically invades new plant tissues. Symptoms appear in 7–10 days, about the same time it takes for the first eggs laid by viruliferous females to hatch. The virus is not transovarially transmitted; thus MRFV is perpetuated solely by horizontal transfer from insect to plant and back to the insect. First-instar corn leafhoppers remain on natal leaves for several days, but later move upward on the plant, where they feed on virus-infected leaves. Virus is ingested when leafhoppers feed in the phloem of infected leaves. Virus passes to the midgut lumen, and then invades and multiplies in the gut tissues before passing into the haemocoele and circulating in the haemolymph. Eventually, virus enters the accessory salivary glands, multiplies and is incorporated into salivary secretions. Processing (i.e. latent period) of this propagative virus takes approximately 2–3 weeks, or about the time it takes for leafhoppers to mature from first-instar nymphs to adults. Once leafhoppers become viruliferous, they remain so for life.

Depending upon the condition of the natal host, young adult leafhoppers may mate and lay eggs and thus complete a second generation on the same plant (Todd *et al.*, 1991). Many leafhoppers, however, disperse to nearby plants or fly to other maize fields, preferably newly planted fields containing young maize plants. During daylight hours, leafhoppers disperse within the plant canopy, but at dusk they fly above the boundary layer and disperse to other maize fields. It is not known how far corn leafhoppers fly or whether they are capable of long-distance, migratory flight (Taylor *et al.*, 1993).

Comparing and Contrasting the Pathosystems

The four featured pathosystems (SMV, MCDV, BYDV and MRFV) contain several elements in common and several that are not. In each pathosystem, the virus, by definition, is the system's core, but these viruses differ greatly in morphology and in genome make-up. In one of the pathosystems (SMV), the virus multiplies in plant epidermis and invades all tissues, while, in the other three, the virus is transported in the phloem and multiplies primarily in phloem parenchyma and meristematic tissue. All four pathosystems are associated with homopterans (Insecta: Homoptera), two (SMV and BYDV) exclusively with several species of aphids (Homoptera: Aphididae) and two (MCDV and MRFV) each primarily with a single leafhopper species (Homoptera: Cicadellidae). Two of the viruses (SMV and MCDV) are foregut-borne, while the other two have more prolonged relationships with the vector, BYDV being circulative and MRFV being propagative. The single pathosystem featured here where the virus multiplies in the vector is MRFV, but, even in this complex system, the virus is not vertically transmitted transovarially. By passing through the seed, SMV can be transmitted vertically in the plant; it is thus the only virus in the four featured pathosystems that is seed-borne. All of the viruses can be transmitted horizontally by the vector, i.e. from plant to insect to plant within and between fields.

These pathosystems differ in other important ways as well. Host-plant diversity is one. Soybean mosaic potyvirus infects very few species of plants, the most prevalent being soybean; MCDV infects several grass species; BYDV infects well over 100 species of wild and domesticated grass species; and MRFV seems restricted to maize and its close relatives. Thus, viruses in two of the pathosystems (MCDV and BYDV) have wide plant host ranges whereas the other two (SMV and MRFV) have much narrower host ranges.

Where Biotechnology can Clarify the Process

This section advances concepts through which biotechnology might help to clarify various phases of a pathosystem's cycle. It is not meant to be all-inclusive; instead, it represents a short set of ideas that biotechnologists might wish to explore in their laboratories.

Phase 1: a non-viruliferous vector moves to and lands on a source plant

Both this phase and phase 4 encompass vector movement, an aspect of the pathosystem's cycle that is poorly understood. Virus–vector relationships are far less important to these phases than are the innate movement-related behaviours of the vectors and the interactions between vectors and the environment. Aphids and leafhoppers will be discussed collectively because both have species that tend to move long distances.

How and where vectors move is difficult to ascertain; thus it is nearly impossible to determine whether a given individual has travelled a great distance or has come from a neighbouring field or plant. Little is known about what factors stimulate these vectors to move and then land on plants. Labelling local populations, either through mechanical means or by describing their genetic characteristics, may enhance our ability to understand where individuals come from, and hence where they go, during long-distance flights (Irwin and Thresh, 1988).

The process of host selection by vectors is important, and yet far from fully understood. Some vector species apparently land on and colonize certain plant species but not others. Current wisdom suggests that vectors, at least aphids, land indiscriminately on plants, assuming that the colours of the plants are equally attractive, and select which hosts to feed on during the preingestion probing process. Information we have in hand (M.E. Irwin and G.E. Kampmeier, Illinois, 1991, unpublished data) counters that wisdom for aphids, suggesting that some additional selection mechanism may occur before landing takes place. Colour, texture, contrast and shape of plants or plant canopies appear to influence the landing of many aphid species. Odour, however, has never been shown to appreciably influence aphid landing.

In contrast, plant odours coupled with colour, but not odour alone, influence the attraction, landing and retention of leafhoppers on plants (Todd *et al.*, 1990). The yellow-green hues of the maize whorl are particularly attractive. Odours from non-hosts, such as sorghum, serve as deterrents to leafhopper retention and feeding. Much more needs to be learned about how colour, silhouette and size of the host plant interact with odour and other chemical stimuli to promote or retard aphid and leafhopper landing behaviours. Determining if aerial selection exists and understanding what is involved would be extremely useful in developing strategies for managing virus epidemics. We believe that biotechnology focused at this juncture would greatly enhance our capability to predict movement and landing probabilities for both aphids and leafhoppers.

Phase 2: the non-viruliferous vector acquires the virus from the source plant

Both the mechanism by which vectors process the virus (foregut, circulative or propagative) and the different plant tissues that are infected with virions influence this phase of the cycle. Many similarities exist in foregut-borne transmission of SMV and MCDV. The major difference is in the plant tissues that are affected by the virus, thus determining the site of virion acquisition. In the case of SMV, and potyviruses in general, the tissue involved is the epidermis. Pirone and colleagues have labelled a potyvirus related to SMV so that the process of acquisition could be followed closely (Berger and Pirone, 1986). With MCDV, phloem is the primary site of virion acquisition by the vectors. Knowledge of extravasation, which is critical to the transmission of foregut-borne viruses, and other feeding behaviours would benefit by microtechnology that could track the microscopic feeding activities of aphid and leafhopper vectors.

Many similarities also exist between circulative transmission of BYDV and propagative transmission of MRFV. In both pathosystems, phloem is the primary site of virion acquisition by the vectors. Aphid and leafhopper vectors insert their stylets and penetrate the sieve-tube elements of the host plant's phloem. Virus particles are siphoned from the phloem and through the stylet food canal by the sucking pump of the vector. Some evidence suggests that MRFV infection of maize may alter feeding behaviour of the corn leafhopper. Electronic feeding monitors have been employed to study this process (Backus and Bennett, 1992; Wayadande and Nault, 1993), but more research needs to be conducted, especially correlating electronic signals with specific feeding behaviours. Labelling of virus and vector saliva with molecular markers could further our basic understanding of this phase of virus transmission.

Phase 3: the vector processes the virus so that it can be transmitted to other plants

Once the virus has entered the vector, it must be processed before the vector is able to inoculate a plant. The processing differs greatly depending on the combination of virus and vector. Two major processing types are represented in the featured pathosystems, the foregut-borne types and the circulative/propagative types. Understanding this phase of foregut-borne interactions has depended on several aspects of biotechnology, including the labelling of virus particles, their antibodies or helper component so that virus-binding sites in the vector could be identified. This phase is being studied by Pirone and colleagues, using potyviruses related to SMV (Pirone, 1991). Virtually nothing is known about this phase of MCDV transmission. The putative helper component responsible for MCDV transmission (Hunt *et al.*, 1988) has not been isolated and identified. Techniques for isolation of potyviruses and caulimoviruses do not work for MCDV. The current approach is to sequence the MCDV genome and identify genes responsible for production of structural and non-structural proteins. By this approach, helper protein or protein fragments of RNA translation products might be produced. This will entail the use of polymerase chain reaction (PCR) and antibody technologies as aids in guiding the isolation and purification of helper component from MCDV-infected plants.

This phase of the cycle is rather complicated for BYDV and other circulative viruses. The dependence of certain virus isolates on specific vector species for successful transmission was established over 20 years ago (Rochow *et al.*, 1975). In doubly infected plants, the MAV isolate, for instance, is transmitted by both *S. avenae* and *R. padi*, whereas the RPV isolate is transmitted only by *R. padi* (Rochow, 1970). This type of dependent transmission has often been referred to as transcapsidation or genomic masking. Hu *et al.* (1988) studied the interaction of two related isolates of BYDV, the MAV isolate and the PAV isolate, in mixed infections. They found that, in mixed infections, virions consisted of coat proteins from both virus isolates. The newly assembled virus, which contained MAV RNA, was transmissible by *R. padi*. The phenotypic mixing of related capsid proteins within a single virion provides evidence that the entire capsid structure is not involved in virus recognition by aphids and that only specific epitopes of the capsid proteins are needed for virus transmission.

The virus passes through the oesophagus before entering the midgut, where it either is or is not capable of penetrating the wall of the hindgut. In most cases, failure of virions of most BYDV viruses to pass through the hindgut does not appear to be a major barrier to the completion of the processing of the virus within the aphid, but penetration of the accessory salivary glands is (Gildow, 1991). For some aphid virus combinations, the basal lamina of the accessory gland is a barrier (Gildow, 1993), whereas, in other cases, isolates of certain BYDV virions penetrate the basal lamina, but fail to initiate endocytosis at the basal plasmalemma, although in appropriate vector species this happens. Receptor-mediated endocytosis is the mechanism of virus-specific acquisition and transmission of BYDV (Gildow, 1991). In this model, receptor proteins embedded in the plasmalemma of the hindgut and accessory salivary gland must recognize specific receptor recognition sites on the surface of the virion capsid protein. Disruption of the recognition process is fertile ground for molecular biology research. Viruses that fail to be transmitted lack appropriate recognition sites, and thus do not attach to the membrane-bound receptors, and do not undergo endocytosis into, and are not transported through, the cell (Gildow, 1991).

Processing of MRFV by the corn leafhopper is not well understood. Electron microscopy (Kitajima and Gámez, 1983) and quantitative serology (Rivera and Gámez, 1986) have demonstrated that MRFV is propagative, but the dynamics of multiplication and transport remain unknown. Approaches such as those used for understanding BYDV transmission will help clarify the passage and multiplication of MRFV as it moves from the insect gut to the accessory salivary glands.

Phase 4: the viruliferous vector carries the virus to a plant

Two approaches to understanding vector movement seem plausible: (i) the genetic typing of the vector through such techniques as the identification of unique molecular sequences of specific populations (Lupoli *et al.*, 1990); and (ii) typing of virus isolates and the subsequent determination that the specific new infection was of a different genetic character from the one that was moving within the field. Because the dilution factor is so great, genetic tagging has more potential for success than does external physical marking in following aphid and leafhopper dispersal. The two approaches might be coupled to determine not only that a new virus isolate had entered an epidemic, but where that isolate originated (i.e. the source of the vector that carried the virus). Movement pattern and timing are extremely important to the magnitude of an epidemic. The use of genetic engineering and molecular biology in general to uniquely tag specific populations of vectors could well lead to the clarification of movement patterns.

The movement patterns and overseasoning strategies of the corn leafhopper are poorly understood. Does MRFV infection of the corn leafhopper enhance overseasoning ability, as does the corn stunt spiroplasma in this vector species? Whether the leafhopper is a migrant and what happens to adults during the maize-free period in tropical Latin America are questions that remain unanswered (Taylor *et al.*, 1993). Studies that deal with overseasoning ecology, biology and physiology are paramount to understanding this phase of the MRFV pathosystem.

Phase 5: the viruliferous vector inoculates the plant

Aspects of this phase of foregut-borne viruses are not well understood. The process of virus-particle detachment from the binding sites in the vector's foregut is being addressed by labelling and following the pathways of the virus and by understanding the nature of the binding to and detachment of the virus from the maxillary food canal and foregut of viruliferous vectors (Berger and Pirone, 1986). The detachment process is not fully understood, but may well be chemically mediated. Sloughing of the virus during extravasation may also play an important role in dislodging the virions from the binding sites. Until the helper component in the MCDV pathosystem can be isolated and identified, limited progress in understanding this phase of inoculation can be made in that pathosystem. The process of extravasation by the blackfaced leafhopper is another critical component that demands research. When does extravasation occur, what controls this behaviour, and how does it affect detachment and expulsion of virions and helper component during inoculation?

The more prolonged relationships exhibited by the BYDV and MRFV pathosystems may have similar inoculation modes, but, then again, they might not. It is assumed that when aphids and leafhoppers probe the phloem with their stylets, they occasionally secrete watery, enzyme-bearing saliva. This is the time when circulative and propagative viruses are thought to be introduced into vascular tissue. Secretion of watery saliva may occur when aphids and leafhoppers are involved in x-waveform behaviour, a complex electronically monitored waveform associated with phloem probing. The presumed function of this salivary secretion is to prevent plant phloem from sealing wounds made by the insect's penetrating stylets. The saliva may also predispose phloem to infection by plant viruses.

Phase 6: the plant becomes infected and processes the virus so that it can be acquired by a vector

Not only does the plant serve as a host to increase virus, but virus must be processed in a way to become available to vectors for transmission. The concept of viral replication within plant cells where the virus is first introduced by the vector is partially understood, but there are aspects that need further clarification. For instance, why does the virus replicate in cells of some plants but not in cells of others? Why are plant epidermal cells predisposed to initial virus replication when others are not? The MCDV inclusions, composed of large numbers of mature virions and matrix (Ammar *et al.*, 1993), the putative helper component, must have evolved as a response to facilitate transmission by leafhopper vectors. Fragments of inclusions have been detected in the maxillary food canal and foregut of vectors by electron microscopy (Ammar and Nault, 1991).

Knowledge of this phase of the cycle of the BYDV pathosystem is only beginning to be explored, and biotechnology is providing the tools. This work, currently being conducted by a CSIRO team (Gerlach *et al.*, 1990), has determined the genomic organization of an Australian isolate of BYDV and is exploring the stages of replication, effects of mutations on infection and nucleic acid molecules that interfere with the infection process. Molecular approaches can shed considerable light on this phase of the cycle.

Where Biotechnology can Help Contain the Epidemic

Each of the featured pathosystems consists of viruses, vectors, plant hosts (cultivated and, in some cases, wild) and the environment. Only the cultivated, plant hosts are easily manipulated in field situations by humans.

Thus, this section leans heavily on how the crop can be modified through biotechnology to help decrease epidemics. This section is not all-inclusive; instead, it represents a short set of ideas that biotechnologists might wish to explore in their laboratories and agriculturists might wish to examine in the field.

Phase 1: a non-viruliferous vector moves to and lands on a source plant

The modification of flight behaviours of wild populations of vector species would be extremely difficult to achieve using biotechnology. However, it seems entirely feasible, using a variety of biotechnologies and other means, to alter the attractiveness of cultivated plants to insect vectors. The cues for alighting response by aphids are similar (though not identical), no matter which aphid species are involved. Thus, the modes of intervention are similar for all aphid species. The physical aspects of plants that might be modified to alter alighting responses of aphids include canopy colour (higher landing rates occur in normal, dark green canopies than lighter green, chlorophyll-deficient canopies), background colour (contrasting backgrounds create a higher potential for landing), canopy density through plant spacing (the denser the foliage, the less likely an aphid will land) and foliage age (younger rather than older foliage seems to be preferred by some aphid species) (Irwin and Kampmeier, 1989). In addition, those aspects that pertain to plant attraction (or repulsion) by flying aphids might be considered, even though we do not yet understand what they are. There are some aphid species that frequently alight in small-grain systems but do not regularly alight on other canopies. Those species that tend to colonize grasses appear to alight less frequently in other canopies. Many of these aspects are properties of the crop plant and might lend themselves to modification, and genetic manipulation might be a logical tool. This could greatly alter the timing and degree of epidemics.

Because the most logical entry point of the MCDV cycle begins with the acquisition of virus from a perennial weed by the leafhopper, manipulation inspired by biotechnology is more difficult to attain. There is little question that elimination of Johnsongrass, the major perennial host plant of the virus, would effectively reduce or even eliminate MCDV epidemics. Currently, control of Johnsongrass is based on use of herbicides. Biological control has not been tried, but that may be important in the future should herbicides fail or government regulations restrict herbicide usage. The effectiveness of Johnsongrass-specific pathogens or insect herbivores could be enhanced through biotechnological advances. Little is known about how maize plants could be altered physically or chemically to prevent corn leafhoppers from landing or probing. Whether it is possible to alter the colour or odour of corn to discourage landing and probing is not known, but perhaps such resistance traits could be engineered by plant breeders using molecular techniques.

Phase 2: the vector acquires the virus from the source plant

For viruses that are acquired by vectors from the plant's epidermis, preingestion probes may not be dependent upon chemical stimuli. Instead, at least for aphids, they seem most dependent upon physical obstacles. For instance, the denser the pubescence on the probed surface, the less likely the aphid is to probe (Gunasinghe *et al.*, 1988).

At present no surface barrier is known that would prevent leafhoppers from initiating a probe. If, however, leafhoppers can be prevented from reaching the vascular tissue where phloem-limited viruses occur, virus acquisition and subsequent transmission could be reduced or prevented. Ingestion feeds by both aphids and leafhoppers appear to be dependent upon chemically induced taste stimuli. Engineering a plant so that vectors are not stimulated to feed would provide an obvious mechanism to curtail the cycles of circulative and propagative pathosystems. For persistently transmitted, phloem-limited viruses, this approach appears plausible and worth consideration. Thus, there may be ways to genetically alter the texture of leaf surfaces that are sites of probing to make aphids and leafhoppers less prone to probe. Oil sprays greatly reduce potyvirus acquisition by aphids and, to a lesser degree, MCDV by leafhoppers. Unlocking the mechanism of how oils reduce acquisition and incorporating these traits in the plant through genome manipulation could offer an exciting new means of control.

Phase 3: the vector processes the virus so that it can be transmitted to other plants

Certain viruses do not infect some plants because the viral genome cannot induce the production of helper component, which is key to the successful transmission of SMV and MCDV. Because helper component is a viral gene product, it is a good candidate for genetic manipulation. Virus coat protein, if deficient, can also obstruct vector transmission. If the plant's ability to react to the virus could be genetically altered, the cycle of the pathosystem might well be short-circuited. Raccah and colleagues have provided the first evidence that such an effect can be obtained for potyviruses by engineering the reverse of this situation, i.e. converting an aphid-non-transmissible virus strain into an aphid-transmissible strain (Gal-On *et al.*, 1991, 1992).

The ability to process virus is crucial to the completion of BYDV and MRFV cycles and is a juncture where biotechnology might well intercede to curb epidemics. In a real sense, this is more difficult than it might at first appear. Modifying viruses or vectors might alter the ability of a pathosystem's cycle to be completed, but neither the 'wild' viruses nor the 'wild' aphids are easily modified nor successfully introduced into the agroecosystem. The crop plant, which is introduced into the environment, is the part of the system that is most easily manipulated. Unlike non-persistently transmitted viruses, which require plant generated helper component, the plant appears not to be involved in this phase of the cycle in circulative and propagative, persistently transmitted pathosystems.

Phase 4: the vector carries the virus to a non-infected plant

This phase seems relatively devoid of junctures that permit biotechnological intervention in the process. Other than those posed in phase 1 above, it appears as though few opportunities exist within the plant itself for genetic manipulation that would influence this phase of the cycle. However, if, as we believe, initial movement from the colonized host is triggered by chemicals and other stress signals produced by the plant, it may be possible to alter the behavioural stimuli for colonization and initial dispersal by modifying those stimuli.

Phase 5: the vector inoculates the plant

Probing activity is directly dependent on the physical characteristics of the surface that is to be probed, as indicated in phase 2. Modifying that surface should alter probing behaviour. The biotechnological induction of specific oils or other chemicals by the plant could inhibit successful virus inoculation of plants; great success has been obtained for curbing potyvirus epidemics with regular applications of oils to plant surfaces. Any method that would arm the plant with thicker cuticle that might be more difficult to penetrate, with denser leaf pubescence that would reduce probing activity or with the ability to produce and coat leaf surfaces with oils would result in lowered inoculation rates, and therefore reduced transmission would, in effect, help curb epidemics.

An effective method to curb epidemics of circulative and propagative, persistently transmitted viruses during this phase is to reduce the propensity for aphids and leafhoppers to feed. As mentioned in phase 2 above, by developing ‘resistance’, vector feeding activity would cease. Chemicals sensed during preingestion probing by vectors elicit feeding responses. If the proper chemicals are absent, or if chemicals that deter the feeding response are present, vectors will not feed. This is an area that is ripe for exploration through biotechnology, and one that might prove extremely effective in curbing epidemics.

Phase 6: the plant becomes infected and processes the virus so that it can be acquired by a vector

This phase is the natural prime target for employing biotechnology to intercede in a pathosystem's cycle. By capturing and incorporating some of the defences plants have evolved to cope with viruses, epidemics in the field could readily be limited and delayed. Protein coats of viruses might be incorporated into the genome of a cultivated plant to protect it against related viruses. This strategy would work well for some virus groups, like potyviruses, but not others, such as luteoviruses. The coat-protein gene of SMV has been incorporated into the tobacco genome, and this seems to protect tobacco against infection by other potyviruses, including other strains of SMV (Stark and Beachy, 1989). Because several strains of BYDV can be found in a single, infected host plant, cross-protection may be an unsuitable target for luteoviruses and persistently transmitted viruses in general.

The process of infection may be key to developing novel ways to provide immunity and thus bring epidemics to a halt. Waterhouse and colleagues at CSIRO are paving the way in this area. This is an area that appears to hold great promise for reducing epidemics.

Summary

As has been repeatedly stated in this chapter, there are two major reasons for addressing pathosystem epidemics through biotechnological innovations. One is to better understand the process and the other is to have an impact on the process. Both are important, but the latter is far more restrictive in the targets available for modification. To have an impact on epidemics in the field, the biological target that is under the most genetic control of humans is the crop. Wild plant hosts and vectors are much less so. Within the physical setting, the temporal and spatial configuration of the crop and the cropping system is under human control, but the vagaries of the physical environment such as wind and rainfall clearly are not. Thus, to have an impact on epidemics, one might best concentrate on limited aspects of the physical layout of the crop in the field and on the genetic make-up of the crop. Biotechnology can greatly alter the genetic make-up of a plant. That, then, is a logical focus where the molecular biologist can affect virus epidemics in the field.

Two aspects of the plant genotype can be manipulated that will have an effect on epidemics. The first and most obvious is the modification of the plant–virus relationship. The second is the manipulation of the plant genotype to reduce the plant's interaction with the vector. It is the second conceptual process that needs further clarification. The plant and the vector have had a long, often convoluted, evolutionary relationship, although that relationship need not be entirely independent of the virus. The vector, the biological entity that moves relatively rapidly through the environment, is the active partner in causing epidemics. The vector is the biological component that governs the timing and extent of an epidemic. The evolutionary relationship between plant and vector has produced a number of built-in recognition signals in the vector. The types of signals for plant recognition differ among vector groups, but they appear to be somewhat consistent within a vector group. We believe that these strong relationships can be manipulated to reverse the alluring nature of the plant and replace it with aversion, thus reversing the vector's response to the plant. Instead of attraction, it might cause neutrality or even avoidance by the vector. The specific characteristics that might be manipulated in the plant are touched on in this chapter.

Acknowledgements

We wish to thank the Rockefeller Foundation and its Deputy Director for Agricultural Sciences, Dr Gary H. Toenniessen, and the World Bank and its Biotechnology Manager for Environmentally Sustainable Development, Dr Gabrielle Persley, the United Nations Development Programme, and the Australian Centre for International Agricultural Research, for organizing and sponsoring this workshop, and for allowing us to participate.

Insect Vector Control 19

ELIZABETH EVANS

Introduction

Many economically important phytopathogenic agents, plant viruses in particular, are transmitted by insect vectors, mainly aphids and leafhoppers (Matthews, 1991). There is no fully collated figure for worldwide losses to agriculture due to viruses, but it ranges in the tens of billions of dollars annually. On top of direct crop losses are the costs of insecticides used for vector control, and losses caused by other pathogens in plants weakened by virus disease (Hull and Davies, 1992). In some cases, the problems of endemic viral diseases have intensified as developing countries have adopted monoculture of ectopic, genetically homogeneous, high-yielding varieties. Some of the most serious outbreaks have been the mealy-bug-transmitted cocoa swollen-shoot virus (CSSV) in West Africa and the rice tungro virus transmitted by the green leafhopper (*Nephotettix virescens*) in South-East Asia (Bos, 1992).

There are no practical means yet to cure field-grown plants once they are infected, so efforts to control viral diseases have focused on: (i) use of healthy planting material and eradication of infected plants; (ii) cultural practices (e.g. time of planting); and (iii) interrupting transmission by depleting the insect vector population through elimination of breeding sites, by the release of sterile insects or, most often, with the use of organosynthetic insecticides (Hull and Davies, 1992). Temporary successes were achieved by the latter method in animal and human vector-borne disease transmission, but were not readily obtained for many plant viruses (mostly aphid-borne), because the insects acquire and inoculate the virus in too short a time for the chemical to exert toxicity.

Recently, rapid progress in insect molecular biology has helped to blur the dichotomy between basic and applied entomology, raising new options to deal with the old problem of vector control. Molecular studies of insect genomes are contributing to advances in diagnostics and the understanding of the population biology of vectors. The development of widely useful genetic transformation technology for insects is expected, and will provide powerful tools for genetic analysis, and the possibility of manipulating the genetic structure of vector populations in order to reduce their capacity to reproduce and/or transmit pathogens.

In seeking to survey the new ideas in insect vector control, this chapter draws on several examples from research into the control of vector-borne diseases of humans, using molecular biology. Although the differences in aetiology and entomology between medical and agricultural diseases and pests are many, some parallel goals and problems are shared in attempting to understand and limit disease transmission. With caveats in mind, strategies being developed for the control of medically important insect vectors can provide useful information, food for thought and, in some cases, the possibility of adapting these strategies to address agricultural problems.

Diagnostics for Insects and Insect-borne Pathogens

Effective application of any form of vector control measure must be based on an understanding of the frequency and distribution of insect vectors, as well as their taxonomy, behaviour and relationship to their target and the environment. Molecular tools such as restriction fragment length polymorphisms (RFLPs) and random amplified polymorphic deoxyribonucleic acid (DNA)–polymerase chain reaction (RAPD-PCR) are proving useful for monitoring populations, studying their genetic and ecological relationships and detecting nuclear and mitochondrial gene flow, which can be used to study such things as the migration of insects and natural enemies, the movement of genes conferring resistance to insecticides or the spread of insect-vectored plant diseases (Roderick, 1992; Whalon, Chapter 17, this volume). Understanding the population structure can be particularly important for vectors, which often exist within morphologically indistinguishable sibling species complexes that can vary significantly in the potential of individuals in a vector to actually transmit disease, having an impact on the control strategy (Meredith and James, 1990).

For example, RAPD-PCR has been used to discriminate between sibling and non-sibling populations of *Aedes aegypti* in studies of oviposition behaviour. Apostol *et al.* (1992) are determining the number of females that contribute eggs to a single oviposition container. By measuring the typical egg-batch size, they hope to estimate how many sites a female must visit in order to lay a full complement of eggs, an important factor in mosquito dispersal and transmission dynamics of arboviral disease. This type of study was extended to examine the genetic structure of mosquito populations in Puerto Rico. A spatial sampling of *A. aegypti* was derived from egg collections in oviposition traps from cities and neighbourhoods. The goal is to understand the seasonal and spatial variation in vector populations (Beaty, 1992). Similar probes have been developed for aphids (Black *et al.*, 1992).

Measuring the frequency of infected insect vectors can help in forecasting disease outbreaks and formulating a judicious programme of vector elimination. Hopp *et al.* (1991) used single-stranded DNA probes to detect potato virus X, potato virus Y, potato leafroll virus and potato spindle tuber viroid. A biotin/streptavidin alkaline phosphatase detection system revealed viral ribonucleic acid (RNA) in plant extracts and aphid (*Myzus persicae*) vectors with 20–50 times the sensitivity of enzyme-linked immunosorbent assays (ELISAs). Similarly, Nakashima *et al.* (1991) have developed complementary DNA (cDNA) probes that detect the causative agent (a mycoplasma-like organism) of rice yellow dwarf disease in plants and leafhopper vectors.

When laboratory facilities are available, PCR allows particularly sensitive detection of pathogens in hosts and vectors. Polymerase chain reaction primers have been developed to detect various flaviviruses, a group of mosquito-borne RNA viruses responsible for human encephalitis. Following reverse transcription of the viral RNA, amplification obtained with PCR allows low levels of virus to be detected. These assays are being developed to allow surveillance of the virus in mosquitoes in North America in order to prevent outbreaks, and may be extended to ecological studies to identify infected host-animal or insect reservoirs (Howe *et al.*, 1992).

Physical and Linkage Maps of Insect Genomes

Genetic analysis and manipulation of any species generally requires a collection of genetically mapped loci. Most genetic studies of insects have been limited to model organisms with morphological or biochemical markers amenable to classical genetic analysis, in particular the drosophilid flies (see Rodrigues and VijayRaghavan, Chapter 21, this volume). In contrast, most medically and agriculturally important vector species have been *terra incognita* to geneticists, making genctics a weak link in the development of biorational methods for control (Zraket *et al.*, 1990).

This situation is rapidly changing with the development of molecular markers and analytical techniques that enable the study of heretofore recalcitrant insect genomes. As a first step, the basic size, complexity and organization of insect genomes are being unravelled in order to understand the distribution of coding and repetitive sequences, necessary for the interpretation of mapping and hybridization studies. Secondly, physical and linkage maps are being generated using *ex* and *in situ* hybridization, RFLP and PCR-based methods. These studies should facilitate the understanding of the inheritance of such phenomena as insecticide resistance, the susceptibility to vector-borne pathogens, hybrid sterility, host-seeking behaviour, etc., which can have significance in insect control programmes.

Genome analysis

The DNA of all higher eukaryotes is subdivided into three components: highly repetitive, moderately repetitive and single-copy sequences. The organization of these components can be determined by the renaturation kinetics of complementary strands over time. Following heat denaturation, highly repetitive sequences will reassociate most quickly, whereas low- or single-copy sequences require a longer period to 'find' themselves. It must then be determined how these classes of sequences are distributed throughout the genome. These analyses on mosquito are revealing that different species vary significantly in their size and organization of repetitive sequences. These data are needed in order to make effective use of genomic and cDNA libraries for the isolation of DNA sequence. Dispersed repetitive sequences can mask the hybridization of DNA probes to single-copy genes (Crampton and Eggleston, 1992).

Linkage mapping with molecular markers

In 1991, the genetic map of *Drosophila melanogaster* has over 3700 loci. The next most detailed map was that of the silkworm *Bombyx mori*, with approximately 207 loci, while a few dozen other species were within the 10–150 range (Ashburner, 1991). This disparity can now be largely overcome by the use of RFLP, PCR or RAPD-PCR profiles as genetic markers. Restriction fragment length polymorphisms offer several advantages as genetic markers. First, unlike traditional mapping, they do not require the maintenance of large mutant strain collections (Ashburner, 1991). They are detected in any DNA-containing tissue at all life stages, are inherited as codominant, non-epistatic Mendelian loci and are widely and more or less randomly distributed in coding and non-coding regions. Furthermore, saturated RFLP maps can be used to locate the genes involved in quantitative trait loci (QTLs). Given a map of sufficient density surrounding the gene of interest, the nearest cloned probes can be used in chromosome walks or jumps through libraries (Romans *et al.*, 1991). The amount of work required to generate a useful map for a particular species will depend largely on genome size and the frequency of meiotic recombination (Ashburner, 1991).

Inherited as dominant Mendelian markers, RAPD-PCR markers are being used for linkage mapping and species identification in mosquito (Kambhampati *et al.*, 1992a). Another source of genetic markers takes advantage of the variation in the patterns of tandemly repeated nucleotide arrays interspersed throughout the genome, called satellite DNA sequences or micro-satellites (Blanchetot, 1991; Zheng *et al.*, 1991). Probes recognizing these short repetitive sequences on Southern blots define DNA fingerprint profiles for individual insects and are inherited in a Mendelian fashion in pedigree studies. Polymerase chain reaction amplification with primers that anneal to flanking unique sequences can detect variable numbers of simple sequence repeats (Besansky and Collins, 1992).

Physical maps

To correlate linkage markers with the physical map, genomic libraries can be screened with RFLP probes and RAPD primers to place the markers on a specific clone. Large genomic fragments can be carried in phage, cosmid or yeast artificial chromosome (YAC) libraries. In species that display polytene chromosomes, such as *Anopheles gambiae* and the medfly (and some other Diptera), individual chromosome bands can be dissected on a microscope and ligated to oligonucleotide linkers. The PCR primers homologous to the linkers are used to amplify the bands, which are cloned into libraries, representing a low-resolution physical map of the genome (Zheng *et al.*, 1991). Individual cloned genes and markers can then be mapped by dot–blot hybridization of filters representing the different bands. The microdissection strategy can also be adapted to metaphase chromosomes (Ashburner, 1991).

Once a marker is located in the genomic library, the clone can be mapped to the chromosome by *in situ* hybridization to polytene or mitotic (metaphase) chromosomes (Zacharopoulou *et al.*, 1992). Because mitotic analysis is difficult in some species (such as *B. mori*, where all the chromosomes assume an indistinguishable, spherical shape at metaphase), an improved pachytene (meiotic prophase) chromosome banding technique has been developed by staining early embryonic cells. Using this method, Tsuchida *et al.* (1992) localized cloned *B. mori* genes by fluorescence *in situ* hybridization on chromosomes prepared from testes.

As physical maps are developed, representative cloned fragments of the map may be sequenced to form so-called 'sequence tagged sites' (STS), which can be used to establish the physical relationship among clones from different sources and for selecting the same genomic region from new clone libraries. This approach applied to cDNA libraries yields 'expressed sequence tags' (Ashburner, 1991). A well-resolved map in one species may provide assistance in the analysis of closely related species or subspecies complexes by comparative study of gene distribution on the chromosomes (synteny analysis). In some cases, persistently linked gene clusters have been observed across wide phylogenetic distances. Once the synteny relationships are established, the regions of the genome likely to harbour genes of interest may be predicted (Zacharopoulou *et al.*, 1992).

Cloning Insect Genes

The starting-point for the molecular approaches to pest control is the mapping (see above) and cloning of genes. A representative (and by no means exhaustive) list of genes that have been cloned from a variety of insects is given in Table 19.1. Examples are included that may have a direct, immediate bearing on pest control, such as genes for insecticide resistance. Also listed are examples of genes whose cloning will lead to a more detailed understanding of insect physiology, which may, in turn, open up new pest-control options. Ultimately it is hoped that such complex processes as vector competence, sex determination and feeding behaviour will be understood at the molecular level. Recently, attention has been drawn to components of insect immune systems, reflected in the cloning of proteins such as the attacins and cecropins. Because these proteins may be related to ability to carry pathogens, Crampton *et al.* (1990) have retrieved the homologous genes from *A. aegypti* mosquitoes.

Probing for genes in insect libraries

Because many pest species are not amenable to classic mutational or mutant rescue analysis, the use of interspecific DNA hybridization and antibody probes is a boon to the identification and cloning of insect genes. A variety of strategies can be used to identify a particular gene of interest within a genomic or cDNA library. Despite the difference in the gross biology between *Drosophila* and other insects, in many (but not all) cases their evolutionary relatedness results in good conservation at the DNA level within coding sequences. Some of the scores of cloned *Drosophila* (and other) genes can now be used as probes for their homologues in insect genomic or cDNA libraries. With reduced stringency, even probes from different animal phyla can identify insect genes. For example, Palli *et al.* (1991) used human retinoic acid receptor cDNA clones to clone the juvenile hormone receptor from *Manduca sexta*.

Other types of probes include oligonucleotides synthesized on the basis of partial peptide sequence. Doyle and Knipple (1991) made use of PCR to generate probes for a sodium channel that is the target of pyrethroids and dichlorodiphenyltrichloroethane (DDT)-analogue insecticides and has been linked to resistance. They synthesized degenerate oligonucleotide primers to conserved regions of the *Drosophila* protein and amplified the corresponding region in different insect species. The amplification product was then used as a highly specific hybridization probe to pull out the corresponding genomic clone. Selected cDNA libraries can enrich for particular messenger RNAs (mRNAs) of interest expressed only in certain tissues (brain, midgut, ovary, salivary gland), only at certain times of development or only under certain conditions, such as heat shock or hormone treatment. In some cases, subtractive cDNA libraries (e.g. fed versus non-fed, vector versus non-vector) may be useful in analysing inducible and suppressible traits. Genes of interest have also been identified in cDNA expression libraries by probing with antibodies to purified proteins.

A method for isolating species-specific DNA sequences by subtraction has recently been applied to insects (Clapp *et al.*, 1993). Genomic DNA was prepared from two closely related *Drosophila* species. The DNA from the driver (non-target) species was biotinylated, while the target DNA was ligated to adapters. The DNAs were then allowed to hybridize so that homologous sequences common to the two species formed stable complexes. The driver/target hybrids and non- or rehybridized driver DNA was then removed by binding to streptavidin. Following this subtraction, the remaining target-derived DNA was amplified by PCR (using primers to the adapter ends) and cloned. This technique will be particularly useful in generating probes to differentiate among sibling species complexes and in analysing the differences between vector and non-vector relatives.

Applications of Transgenic Technology to Insect Vector Control

The ability to integrate defined genetic sequences into the cells of an organism is a powerful method of genetic analysis and can be an effective means of manipulating the biology of an organism (Handler and O'Brochta, 1991). For these reasons, various avenues for achieving transformation in non-drosophilids are being explored, and some are beginning to show promise (see Hoy, Chapter 9, this volume). Beyond the boon to basic molecular genetic research, several strategies to employ transgenic insects for insect vector control by modifying the genetic make-up of wild populations have been proposed and are outlined below. Substantial regulatory and safety issues surrounding the release of transgenic insects in the field can be anticipated but cannot be fully evaluated until transgenic technology is in place (Crampton and Eggleston, 1992; Hoy, Chapter 9, this volume).

Reducing vectoring capacity of insects for plant pathogens

Interrupting vector borne disease transmission by replacing insect vector populations with non-vector forms is an old idea that may become practicable with the use of molecular techniques and genetic engineering, which

Table 19.1. Representative list of cloned insect genes.

Gene/Protein	Insect	Function	Reference
Insecticide resistance			Hoy, Chapter 9, this volume
Hormones, neuropeptides and their receptors			
Ecdysteroid receptor	*Drosophila melanogaster*	Metamorphosis	Koelle *et al.*, 1991
Ecdysone response element	*D. melanogaster*	Receptor DNA binding site	Cherbas *et al.*, 1991
Eclosion hormone	*Bombyx mori*	Triggers ecdysis	Kamito *et al.*, 1992
Juvenile hormone	*Manduca sexta*	JH responsiveness	Palli *et al.*, 1991
Neuroparsin A	*Locusta*	Brain hormone	Lagueux *et al.*, 1992
Nicotinic acetylcholine receptor	*Schistocerca gregaria*	Neuropeptide receptor	Marshall *et al.*, 1990
PBAN	*Helicoverpa zea*	Pheromone biosynthesis activating neuropeptide	Davis *et al.*, 1992
Prothoracicotropic hormone (bombyxin)	*B. mori*	Triggers ecdysone	Kawakami *et al.*, 1990
Immune response proteins			
Attacins	*Hyalophora cecropia*	Humoral immunity	Sun *et al.*, 1990a
Cecropins	*H. cecropia*	Humoral immunity	Gudmundsson *et al.*, 1991
Diptericins	*D. melanogaster*	Antibacterial	Reichhart *et al.*, 1992
	Aedes aegypti		Crampton *et al.*, 1990

Miscellaneous genes			
Alcohol dehydrogenase	*Ceratitis capitata*	Detoxification/marker	Brogna *et al.*, 1993
Dihydrofolate reductase	*D. melanogaster*	Methotrexate resistance	Hao *et al.*, 1992
	A. aegypti	Nucleotide biosynthesis	Shotoski and Fallon, 1990
Homoeotic genes	*Tribolium castaneum*	Segmentation	Beeman *et al.*, 1990
	A. aegypti		Crampton *et al.*, 1990
Metallothionein	*D. melanogaster*	Heavy-metal detoxification	Theodore *et al.*, 1991
	A. aegypti	Marker	Fallon, 1991
Odorant binding proteins	Lepidoptera	Olfaction in antennae	Vogt, 1992
per	A. aegypti	Rhythmicity/timing	Crampton *et al.*, 1990
Red/white eye	*Lucilia cuprina*	Markers	Bedo and Howells, 1987
	Anopheles gambiae		Besansky *et al.*, 1993
Salivary gland genes	*A. aegypti*	Presumed involvement in pathogen transmission	James *et al.*, 1991
Segregation distorter	*D. melanogaster*	Meiotic drive	Powers and Ganetzky, 1991
Serine protease	*L. cuprina*		Elvin *et al.*, 1991
Sex peptides	*D. melanogaster*	Mating behaviour	Cited in Wyatt, 1991
Trypsin	*A. aegypti*	Digestion of blood	Barillas-Mury *et al.*, 1991

DNA, deoxyribonucleic acid; PBAN, pheromone biosynthesis activating neuropeptide.

could supply a population of selectively modified vectors. The goal would be to dilute or replace natural populations with the transformed individuals in order to reduce the reproductive capability of the vector population, or to interfere directly with the capacity of the vector to acquire and/or transmit the pathogen. Although hypothetical, apart from *Drosophila*, various advantages of the genetic engineering approach have been noted: (i) the potential to exploit genes across species barriers; (ii) the ability to perform site-specific modification of genes *in vitro* and reintroduce them to alter phenotypes; and (iii) the ability to introduce defined DNA sequences without the genomic disruptions and loss of fitness that accompanies conventional crosses or mutagenesis by irradiation, chemosterilization, etc. (Crampton *et al.*, 1990). Because vectoring a pathogen is not necessarily of benefit to the insect itself (and may be a burden), selectively removing the trait by targeted genetic manipulation would not necessarily compromise the fitness of the insect. Furthermore, targeted modification of a specific injurious function, in this case pathogen transmission, may prove easier and more ecologically sound than past attempts to eradicate whole insect populations, which have invited natural selection to overcome the intervention (Beard *et al.*, 1993a).

The refractory phenotype

In anticipation of transgenic technology, medical entomologists concerned with vector-borne diseases of humans are working to understand and isolate the molecules and genes responsible for vectorial capacity, with the aim of identifying candidate natural or novel genes that might be introduced into vector populations in order to block transmission of pathogens. To be useful, introduced genes must be dominant, completely penetrant and have few, if any, unlinked modifiers. Furthermore, these approaches must be integrated with population studies of the vectors in order to identify transmission conditions, since vectorial capacity can be influenced by ecological as well as genetic factors (Meredith and James, 1990). Studies into the genetic basis for naturally occurring resistance (neutralization of a pathogen by immune reaction within the vector) or refractoriness (incompatible interaction between the pathogen and vector) have been carried out in medically important vectors of parasitic and viral diseases. Although the mechanisms of transmission might be quite different from those for the various insect-borne plant pathogens, some studies from medicine are described below, providing examples of ways to identify key genes involved in refractoriness and suggesting biotechnological intervention to block vector transmission.

Genetic variation in mosquito susceptibility to species of *Plasmodium*, the malaria parasite, has been selected and studied (Feldmann and Ponnudurai, 1989; Vernick and Collins, 1989). Resistance is apparently due to an encapsulation reaction, which is an enhancement of the mosquito's normal defence against infection and may be bypassed by the parasite in susceptible strains. In some cases, the phenotype may be determined primarily by a single autosomal dominant gene, but, in others, complex genetic factors may be involved.

Genetic variation has been selected in *A. aegypti* for susceptibility or refractoriness to oral infection with the flavivirus that causes human yellow fever. Refractoriness can be manifest as three phenotypes: (i) blocked infection of the midgut; (ii) restriction of virus multiplication following infection; and (iii) inhibition of viral movement out of the midgut to other tissues, reducing titre. Miller and Mitchell (1991) selected flavivirus-susceptible and refractory lines from two divergent subspecies of *A. aegypti* in order to increase the likelihood for discovering RFLPs that differ between the selected lines, which will enable them to map the host genes responsible for restricting viral transmission. Crosses indicate that a single locus or a few major loci are involved. Similarly, *Culicoides variipennis* susceptibility to infection with bluetongue virus seems to be controlled by a single major locus and a second modifying locus (Tabachnick, 1991).

Wattam and Christensen (1992) took a molecular approach to identifying the gene products responsible for refractoriness of *A. aegypti* to *Dirofilaria immitis* and *Brugia malayi*. Using metabolic radiolabelling, they identified seven polypeptides induced in the thoracic tissue (the developmental site of the *B. malayi* parasite) by ingestion of the blood meal in *B. malayi*-refractory, but not susceptible, mosquito strains. The polypeptides were not induced in *D. immitis*-refractory strains, suggesting that the loci controlling susceptibility to the two parasites function independently.

Engineering novel forms of refractoriness

Because natural refractoriness may prove to be genetically unwieldy, biotechnology may provide novel ways to disrupt transmission by adding or removing specific insect functions that block or decoy host/pathogen interactions. For example, the interplay between a virus and its vector is mediated by specific molecular interactions, i.e. binding between viral and host proteins and/or nucleic acids, which occurs extra- and intracellularly as a virus is carried by the vector. The details of the particular viral life cycle (see Irwin and Nault, Chapter 18, this volume) will determine the nature of these interactions, and the possible mechanisms of refractoriness that occur naturally or might be designed. Some plant viruses may be carried passively by the insect mouthparts, food canal or gut (non-persistent viruses, including most aphid-borne viruses), while others persist and circulate through the haemolymph (circulative viruses, such as luteoviruses, transmitted by aphids, or gemini viruses, transmitted by leafhoppers) or replicate in the insect tissues (propagative viruses, such as Rhabdoviridae and reoviruses, transmitted by aphids and leafhoppers (Matthews, 1991).

The search for points of interaction that could be targeted to confer refractoriness has prompted investigation of vector organs and tissues where significant interactions with pathogens take place. In mosquitoes, studies of the insect immune system, salivary glands, midgut and reproductive system are turning up likely candidate genes that might be manipulated by recombinant DNA technology. Such genes might also provide a source of the regulatory elements that would permit abundant and targeted expression of hybrid genes in transgenic insects (Meredith and James, 1990). Because introduction of genes for refractoriness to pathogen transmission would place strong selection on the pathogen to evolve means of evading the effects of the introduced genes, two or more independent blocks might be introduced so that only a double mutant could be successfully transmitted. Thus multiple mechanisms of refractoriness may need to be developed (see Curtis, Chapter 20, this volume).

Strategies and targets to disrupt virus/host interaction

Viral coat proteins appear to be a key factor determining the ability of some viruses to be acquired and/or transmitted by vectors. In some cases, difference in coat protein can be the sole determinant of transmissibility of different viral strains (è.g. cucumovirus transmission by *Myzus persicae*), whereas, in other cases, a helper factor is involved (aphid potyvirus transmission (Lecoq *et al.*, 1991)). Thus an intervention that prevents the coat protein from interacting with docking receptors may be a way to block transmission. Alternatively, toxic or binding factors in the haemolymph might block the passage of circulative viruses. One group (Crampton and Eggleston, 1992) is exploring the use of mammalian antibody genes to disable the malarial parasite. In viruses that multiply in their insect vector, antisense RNAs that form duplexes with essential viral transcripts, thereby preventing their translation, could be used to block viral replication (Aldhouse, 1993).

Higgs *et al.* (1993) have explored inserting cDNA sequences of the La Crosse (LAC) virus genome into embryos of its mosquito vector in an effort to mimic the natural immunity to superinfection that is conferred by LAC virus replication in the salivary gland of *Aedes triseriatus*. Sindbis virus expression vectors, which infect mosquito cells, have been engineered to express LAC viral RNA sequences encoding the nucleocapsid protein or antisense sequences from the small viral RNA segment of LAC virus. These particular sequences were chosen because similar strategies for interference have been demonstrated in transgenic plants. These vectors appear to interfere with LAC replication in *Aedes* cell culture, and will ultimately be tested for their effects in whole animals. Although the cytoplasmic location of Sindbis replication precludes their use in germ-line transformation, these studies are enabling researchers to test new approaches to reducing vectorial capacity by genetic engineering. Similarly, Morris and James (1993) are testing the expression of a variety of potentially useful genes in a transient transfection assay of isolated mosquito salivary gland tissue. Effective genes identified in these ways may then be candidates for germ-line transformation, as the technology becomes available.

Transformation of insect symbionts

As an alternative to germ-line transformation, Beard *et al.* (1993a) are exploring the possibility of expressing antivectoring genes in the permanently associated symbiotic bacteria found in many vector species. Because insect vectors often have a single food source, such as blood or plant juice, many vectors carry symbiotic bacteria, which supplement their host's diet with amino acids, vitamins, nucleic acid bases, haem or select proteins. This symbiotic relationship is found in important medical and agricultural vector species, including tsetse flies, aphids, planthoppers and leafhoppers. The manipulation of genes in bacterial symbionts in order to obtain correct expression and targeting may prove easier than transformation of complex eukaryotic insect tissues.

In a demonstration system, Beard *et al.* (1993a) have shown that the midgut symbiont *Rodococcus rhodnii* of the Chagas' disease vector, *Rhodnius prolixus*, and the midgut rickettsia-like organism (RLO) of the tsetse fly, *Glossina morsitans* (vector for African trypanosomes, which cause sleeping sickness in humans and nagana in livestock), can be isolated, cultured and genetically transformed with a foreign marker gene on a plasmid. The altered symbionts can be stably reintroduced into their host insect. Various candidates for the antiparasitic or antiviral gene that might be introduced into insect endosymbionts are being explored. One possibility is to express single-chain antibodies raised against coat proteins, parasite attachment molecules, metabolite receptors and the like.

The presence of symbionts in agriculturally important vectors, such as aphids and leafhoppers, has piqued interest in adapting these strategies to block the transmission of plant viruses. To this end, the interactions between pea enation mosaic virus (PEMV) and its aphid (*Myzus persicae*) are being dissected and the possibility of expressing antiviral molecules in the symbionts of aphids is being explored. The virus is taken up in the aphid food canal and passes via the haemolymph to the salivary glands, where it is transmitted by subsequent feedings. The virus encodes a protein called the aphid transmission factor (ATF), which is necessary for binding to salivary gland cells. Treatment with anti-ATF antibodies, as well as certain mutations in the ATF gene, block transmission of PEMV by aphids (Demler and de Zoeten, 1991). Researchers speculate that single-chain anti-ATF antibodies might be engineered into the endosymbionts of the aphids. By supplying appropriate signal sequences, the antibodies might be secreted into the haemolymph (as other symbiont proteins are) and thereby bind the ATF of the circulating virus, preventing attachment to the salivary glands (F.F. Richards, Yale University, 1992, personal communication).

If antiviral genes are able to be expressed in this way, it then becomes necessary to spread these genes throughout the insect population. The favoured approach is to link the novel trait to the cytoplasmic incompatibility mechanism mediated by different symbiotic bacteria (*Wolbachia pipientis*) (Curtis and Graves, 1988; Curtis, Chapter 20, this volume).

Enhancing classic genetic control strategies

A range of genetic approaches to the control of insect pests and vectors have been attempted with varying degrees of success (for review, see Whitten, 1984). The aim of a genetic control programme is to manipulate the hereditary apparatus of individuals in a target population such that a high proportion of individuals in some ensuing generation will not survive. Genetic control generally involves the laboratory propagation (mass rearing) and release of genetically modified individuals which, by mating with residents of the target population, will serve to introduce the modified genetic material into the population. Some systems of genetic control involve a transporting mechanism, e.g. meiotic drive or negative heterosis (see Curtis, Chapter 20, this volume), which enable the genotype to spread through the target population, despite the genetic disability that it bestows on its carrier. The classic techniques used to achieve genetic modifications, such as sterility, including radiation, chemosterilization or chromosomal translocations, suffer from the introduction of gross genetic changes that may reduce fitness. Advances in the molecular biology of insects and the anticipation of transgenic technology maps provide the opportunity for making more precise genetic modification.

One method that may benefit from the application of molecular techniques is the sterile insect release method. This method has been used in control programmes against Mediterranean fruit fly (*Ceratitis capitata*) among other insects, but it can be costly or ineffective for many other pest and vector species. Generally, males and females are reared *en masse*, sterilized and released, although it is only the male that confers control. The presence of females adds costs to rearing and reduces the efficiency of sterile mating after release. To increase the efficiency of production, genetic sexing systems have been explored for the production of male-only strains. In essence, individuals are genetically marked in such a way that they can be screened or selected for in a laboratory population based on sex. Such systems have been developed by classic means for the Medfly, where a positive marker, such as resistance to a certain chemical (conferred by alcohol dehydrogenase or xanthine dehydrogenase), is relocated via a translocation on to the male-determining chromosome, thus providing a selection that is deleterious to females upon exposure (Robinson *et al.*, 1988).

The lack of good genetics of many species and the instability of strains obtained by classic means have hampered the use of these genetic sexing systems for mass-rearing, and have stimulated explorations into the use of molecular biology to provide greater understanding of underlying mechanisms and more precise genetic modifications. *Drosophila* research has identified mutants, such as sex-lethal, transformer, daughterless and double sex, which has led to the cloning of a cascade of genes and the understanding of the alternative, sex-specific, mRNA-splicing pathways, all involved in sex determination (Bownes, 1992). The sex-determining mechanisms vary widely in insects and it remains to be seen if homologous or analogous genes and mechanisms exist in other species.

As cloned genes are becoming available, it may be possible to transfer discrete, selectable loci to the male chromosome, such as alcohol dehydrogenase (*Adh*), which would allow male flies to survive in the presence of ethanol. Alternatively, stocks might be constructed that bear multiple conditional lethals on autosomes, masked by the wild-type allele, introduced on to the male chromosome. Males would thus be phenotypically wild-type and would be used as a vehicle for introducing a genetic load that would be manifest in subsequent generations of females (Cockburn *et al.*, 1989). These schemes, of course, only apply to species where the males are heterogametic. In insects such as Lepidoptera, where the female is XY, the selection scheme would need to be reversed. An alternative under consideration is to link a selectable product to a sex-, stage- or tissue-specific promoter. Genes for vitellogenin, the precursors of the yolk proteins, or the chorion genes, which specify the egg coat proteins, have been cloned and are expressed only in the female fat body and ovaria. Female-specific promoters might be fused to a conditional lethal gene, such as *Adh*, and then transformed into an *Adh*-minus strain. By altering the selection conditions, *Adh* can, in this case, provide selection against individuals expressing the gene. Chemical treatment with compounds that are metabolized by alcohol dehydrogenase into lethal products would eliminate all females (Shirk *et al.*, 1988). Female promoters might also be used to drive the expression of antisense RNA in order to selectively 'knock out' the expression of particular genes, effectively making a null mutant phenocopy in a sex-dependent manner (Robinson *et al.*, 1988).

Conclusions

Molecular techniques are making possible the genetic study of many insect species that were genetically unfathomable a few years ago. Application of these techniques to disease vectors ranges over studies of population structure, to genetic mapping, to monitoring of traits such as vector competence or insecticide resistance, all of which have immediate bearing on management programmes. Other techniques, such as genetic transformation of the germ line or endosymbionts, hold promise for innovative solutions to longstanding problems in vector control, although sizeable technical and regulatory problems remain to be solved. Fortunately, the history of biotechnology is filled with ideas that seemed improbable at first but grew to be feasible and even routine in a relatively short period.

Insect Vectors of Human Diseases

20

Christopher F. Curtis

Introduction

Most forms of pest control aim to reduce density of the pest population, but any such reduction tends to be nullified by density-dependent regulation due to relaxation of competition in the partially suppressed population. This relaxed competition benefits both locally born survivors of the control operations and immigrants or their progeny. To try to avoid population recovery, and hence to make pest or vector control more sustainable, the idea has developed of genetically rendering pest populations harmless to humans without reducing their survival, reproduction or population density (Curtis, 1968, 1979, 1992a; Curtis and Graves, 1988; World Health Organization, 1991).

In most parts of the tropics, malaria is far more important than any other obligately vector-borne disease (the qualification 'obligately' is inserted because of the evidence for a significant contribution of houseflies to transmission of diarrhoeal disease (Cohen *et al.*, 1991)). In Ghana, malaria ranks first of all diseases as a cause of loss of days of healthy life (Ghana Health Assessment Project Team, 1981), and no other obligately vector-borne disease ranks in the top 25; data indicating the seriousness of the malaria situation are summarized in Table 20.1 (partly based on Phillips-Howard and Doberstyn, 1990, and World Health Organization, 1993). In view of the overwhelming importance of malaria, this chapter will concentrate on this group of diseases and their *Anopheles* mosquito vectors.

Refractoriness to Malaria

With regard to mosquito species that are vectors of human disease, a list of examples of strains selected for non-susceptibility (refractoriness) was presented by Curtis (1979). To this list can be added an old example of natural refractoriness and new examples of selected refractoriness to human malarias in two of their vector species (Table 20.2).

The existing examples of refractoriness are not due to single genes, and this would make their use difficult or impossible to combine with most of the genetic systems discussed below. Furthermore, the existing examples do not show full dominance in hybrids, and this would reduce their effectiveness in the field unless and until fixation of the refractoriness was achieved.

Apart from selection of malaria refractoriness by old-fashioned animal breeding techniques, ideas have also been presented as to how refractoriness might be contrived by genetic engineering methods (World Health Organization, 1991; Crampton, 1994). These suggestions include the introduction of a gene coding for a transmission-blocking, single-chain antibody (Winger *et al.*, 1987), i.e. an antibody against the *Plasmodium* gametes, zygotes or oocysts that are produced in or on the stomach of the mosquito. A gene for

Table 20.1. The current status of malaria (*Plasmodium* spp.).

Prevalence of malaria infection in rural tropical Africa: about 50% of the population is infected at any time
Incidence of clinical attacks in Africa: approx. 250–450 million/year
No. of child deaths from malaria/year: 1–2 million (mostly in Africa)
Reported cases/year in South-East Asia: 0.5 million
Eradication from Europe, USA, Taiwan, most of Caribbean and former USSR
Resurgence in South Asia and tropical South America, where reported cases/year are:
India: 1.7 million
Brazilian Amazon: 0.5 million
Cases brought back by British travellers: 2000/year
Multidrug resistance in *P. falciparum* in South-East Asia
Chloroquine resistance widespread in *P. falciparum* in South America, East Africa, Papua New Guinea and India, appearing in West Africa, not yet in Central America
Chloroquine resistance appearing in *P. vivax* in Papua New Guinea
A partially effective synthetic vaccine against *P. falciparum* has been produced and field-tested in Colombia (Valero *et al.*, 1993) and Tanzania (Alonso *et al.*, 1994)

Table 20.2. Examples of *Anopheles* refractory to human malaria parasites. The first example refers to naturally existing refractoriness, the other two examples refer to naturally existing genes for refractoriness that can be selected to a high frequency.

Anopheles	*Plasmodium*	Mechanism	References
atroparvus	tropical *falciparum*	?	James *et al.*, 1932* Ramsdale and Coluzzi, 1975 Dashkova and Rasnicyn, 1982
gambiae	vivax, *malariae*, some *falciparum*	Encapsulation associated with esterases	Collins *et al.*, 1986 Vernick and Collins, 1989 Paskewitz *et al.*, 1989 S. Ahmad and C. Leake, personal communication
stephensi	*falciparum*	Abnormal digestion of haemoglobin	Feldmann and Ponnudurai, 1989 Feldmann *et al.*, 1990

*James *et al.* (1932) showed that *Anopheles atroparvus* was susceptible to European *P. falciparum* (which is now extinct).

Table 20.3. Possible methods of introducing malaria refractoriness into *Anopheles* populations.

Natural selection due to adverse effects of being infected
Ecological selection for a species that is refractory
'Dilution' by mass release without selection
Artificial selection for linked insecticide resistance genes
Meiotic drive and/or drive resistance
Homozygous translocations or hybrid sterility
Transposable elements
Wolbachia causing cytoplasmic incompatibility

an antibody that actively interferes with transmission may be expected to satisfy the above criterion of genetic dominance.

If a refractoriness gene were spread into a wild population that had previously been a major vector, there would be strong selection on the *Plasmodium* population for genes that conferred the ability to evade the effects of the refractoriness gene. In any serious attempt at vector replacement by a refractory strain, this should be equipped with two or more independently acting genes, each able to block *Plasmodium* development. By analogy with the use of mixtures of antibiotics or insecticides (Curtis *et al.*, 1993; Roush, 1993), the intention should be that double-mutant *Plasmodium* cells able to evade both refractoriness genes are initially non-existent, or at least much rarer than those that encounter residual susceptible mosquitoes, and they are therefore not selected for the evader genes. There is an obvious analogy to the introduction of mixtures of disease-resistant varieties of crop plant to delay the evolution of new virulent varieties of pathogen (Wolfe *et al.*, 1981).

Introduction of Refractoriness into Populations

The production of multiple forms of refractoriness seems within the grasp of available technology. However, development of practicable methods of spreading such genes in wild vector populations is much more of a challenge. Some suggestions as to how it might be done are shown in Table 20.3 (partly based on Curtis and Graves, 1988). It might be thought that one could rely on natural selection to drive refractoriness genes to fixation, because they avoid the damage that pathogens do to vectors. However, this damage is generally only slight (Klein *et al.*, 1982; Hogg *et al.*, 1995), especially in the case of well-established *Anopheles–Plasmodium* combinations. Furthermore, the proportion of infective individuals in wild vector populations is almost always low (usually less than 5%), so that any pressure of natural selection for refractoriness would be small. Were this not so, we might expect to find that all *Anopheles* populations had already been selected for naturally occurring refractory mutants.

The displacement of a vector species by introducing a non-vector species is a theoretical possibility that has been tried with respect to *Aedes* mosquitoes and dengue in certain Polynesian islands (Rosen *et al.*, 1976). This idea assumes that a naturally refractory species, such as *Anopheles atroparvus* with respect to tropical *P. falciparum* (Table 20.2), would be better adapted to tropical conditions than the local vectors, but this seems most unlikely. Also, in view of the sad history of accidental or deliberate introductions that went wrong, one must consider this a high-risk strategy that would be difficult to evaluate other than by an irrevocable release programme.

In principle, one could rely not on selection but on large and repeated releases of members of a refractory strain without any genetic mechanism to encourage spread of the refractoriness gene(s). One could describe this as 'dilution' of the susceptibility of the wild population. The method has the merit of simplicity and has been shown to work in cage experiments with *Anopheles* refractory to rodent malarias (Graves and Curtis, 1982; van der Kaay *et al.*, 1982). It is debatable whether it would be better to release non-biting males only or to release both sexes. The latter would increase the frequency of the refractoriness genes more efficiently, but would add to the biting nuisance, which is the aspect of mosquitoes that is usually of most immediate public concern. Also, if the refractoriness genes are recessive and if the population is not strongly regulated by density-dependent factors, female releases could end up being counterproductive by adding to the breeding potential of the population and producing large numbers of heterozygous, susceptible, mosquitoes (Curtis and Graves, 1988). These potential problems could be avoided by releasing males only and the dynamics would then be similar to the sterile-insect release method, except that the population density would not decline and one would therefore not expect an improving released : wild male ratio in successive generations.

Sterile-insect release has been successful in eradicating screw-worm flies from Mexico (Krafsur *et al.*, 1987) and Libya (Lindquist *et al.*, 1992). However, these countries have more resources than those in tropical Africa and South-East Asia that are worst afflicted with malaria. Also, screw-worm fly is a pest of a cash crop (cattle), protection of which has clearly identifiable economic benefits, which can be set against the cost of building and operating sterile-insect factories. In the present harsh politicoeconomic climate, protection of children's health in low-income countries is given low priority and governments are being 'advised' to restructure their economies to cut government expenditure on such items as health services, and to increase exports with which to pay bank interest. Therefore I am pessimistic about any malaria control method that requires large capital investment such as that needed for a mass-release programme. We should therefore contrive systems by which refractoriness genes will spread after a small release.

'Hitch-hiking' with Insecticide Resistance, Meiotic Drive or Transposable Elements

Because the spread of insecticide resistance genes is such a dramatic example of the operation of selection in wild populations, it is sometimes suggested that this process should be harnessed to cause the spread of desirable genes by the 'hitch-hiking' effect. There are, however, two problems: (i) there would have to be very close linkage between the insecticide resistance gene and the gene(s) which it was desired to spread; and (ii) one would have to commit oneself to 'using up' an effective insecticide by the deliberate release of resistance genes to it. It is hard to imagine that the producers of the compound and the vector-control authorities could be persuaded to use a new compound in this way; it would not be adequate to propose the use of an older insecticide to which resistance genes already exist in the wild population at an appreciable frequency, because the replacement process by the desired genes would then never go to completion.

Meiotic drive seems to have the potential to fix genes linked to the driven factor (Hamilton, 1967). However, several known examples of autosomal meiotic-drive factors are homozygous lethal (Whitten, 1979), but the only known examples in vector species involve a factor at or close to the male-determining locus in culicine mosquitoes (Hickey and Craig, 1966; Wood and Newton, 1991). With this, linked genes could only be driven into the non-biting sex. Drive-resistance factors are, however, common in wild populations and these are favoured by selection in the presence of a driving factor because the resistance is at no disadvantage to the drive factor, whereas the drive-susceptible factor is at a disadvantage to the driver. Wood *et al.* (1977) demonstrated, in an *Aedes aegypti* population cage, selection for a marker gene linked to drive resistance. It would be worthwhile to search more carefully for driving factors in *Anopheles* by crossing geographically remote populations, bearing in mind that, within any one population, drive-resistance factors are likely to be already masking the disruptive tendencies of any existing factors with meiotic-drive potential. Separation of drive from drive resistance by crossing might produce some surprising effects, but the many crosses made in the course of various genetic studies have not yet been reported to produce noticeable meiotic drive in *Anopheles*.

Negatively heterotic systems such as autosomal translocations (Curtis, 1968) or hybrid sterility due to interactions of the X chromosome (Curtis, 1982) would theoretically select for whichever genotype was in a majority after a release programme. However, the selective forces would be relatively weak and probably easily overcome by greater fitness of wild type genetic factors.

Transposons, such as P elements, have been proposed as a vehicle for driving malaria refractoriness genes into *Anopheles* populations (Curtis and Graves, 1988). Kidwell and Ribeiro (1992) used a very simple genetic model, in which heterozygotes for a transposon became homozygotes for it as a result of transposition (the 'jumping gene' phenomenon). On their model, transposition occurs equally in heterozygotes from either reciprocal cross, and, provided that it occurs in a large proportion of the heterozygotes, fixation of the transposon will occur even if it causes a marked reduction in fitness. In the case of P elements, standard accounts of the behaviour indicate that they are only free to cause transposition in heterozygotes from a P-carrying male mated to a non-P female, which would make them less inclined to invade a population than in the Kidwell and Ribeiro (1992) model. However, increase towards fixation of P in *Drosophila* has been shown to occur in cage experiments (Kidwell *et al.*, 1981; Meister and Grigliatti, 1993), and it has been inferred in wild populations worldwide since the 1950s (Kidwell, 1983). It is possible to 'load' transposons with extraneous genes, but this may affect their probability of transposition (Kiyasu and Kidwell, 1984) and there would seem to be a risk of deletion of the extraneous genes and selection in favour of the deleted version of the transposon. Once a population has become fixed for a P element, the population is no longer susceptible to further invasions of this element. It has been found that P elements in *Drosophila* have a narrow host range (Handler and O'Brochta, 1991), although it seems that not all elements share this host specificity. Mariner elements have been found in a wide variety of insects, including *Anopheles gambiae* (Robertson, 1993), where they are no longer active. Several research groups are trying to locate active transposons that would operate in *Anopheles* species.

Cytoplasmic Incompatibility due to *Wolbachia*

Sterility in certain crosses between strains of the *Culex pipiens* complex was first reported by Marshall (1938). Laven (1959) tabulated numerous examples of pairs of strains showing sterility either in both reciprocal crosses (bidirectional incompatibility) or only in one of the reciprocal crosses (unidirectional incompatibility). In such unidirectional cases it was possible

Table 20.4. Progeny from matings between an insect strain with uninfected cytoplasm and a strain infected with maternally inherited *Wolbachia*.

	Males	
Females	Uninfected	Infected
Uninfected	Uninfected	None
Infected	Infected	Infected

The uninfected females are at risk of failing to reproduce their cytoplasm if they cross-mate, but infected females are at no such risk.

to show by repeated backcrossing (Ghelelovitch, 1952; Laven, 1959) that the crossing type was strictly maternally inherited and the phenomenon was therefore called cytoplasmic incompatibility. Yen and Barr (1973) showed that, if larvae were reared in an appropriate concentration of tetracycline, one produced males which had been 'cured' of their incompatibility. However, the treated females and their descendants were only compatible with 'cured' males (Table 20.4). Electron microscopy showed that the 'curing' process was associated with disappearance of the symbiotic rickettsia-like bacterium *Wolbachia*, which had first been reported in *Culex* cytoplasm by Hertig and Wolbach (1924) and was described and named by Hertig (1936).

More recently, cytoplasmic incompatibility with all the above-described properties has been found in several other types of insect (Stevens and Wade, 1990), including *Drosophila* (Hoffmann *et al.*, 1986). The process of incorporation of *Wolbachia* into the germ line of the embryo and of disruption of early cleavage divisions after an incompatible fertilization has been described by O'Neill and Karr (1990). However, the process by which *Wolbachia* influences sperms and eggs to give incompatibility between wild populations as well as incompatibility between *Wolbachia*-infected males and uninfected females, but not in the reciprocal cross, has not been explained. *Wolbachia* has been observed to be eliminated during the process of spermatogenesis (Yen, 1975), but presumably the symbiont leaves behind an imprint that affects the sperm's chances of successfully fertilizing an egg of given *Wolbachia* infection status. Hurst (1991) has claimed the analogies between the 'rules' for incompatibility associated with *Wolbachia* and systems such as bacteriocins (reviewed by Maynard Smith, 1989), killer particles in *Paramecium* (Sonneborn, 1965), killer yeast (Somers and Bevan, 1969) and viruses of *Drosophila* sex-ratio spiroplasmas (Cohen *et al.*, 1987).

The mechanism of action of these systems has been partially worked out, but it is difficult to relate interactions between microbial cells to events in whole insects, and I still find it hard to imagine how: (i) a given strain of *Wolbachia* causes a specific incapacitating substance to be coated on to sperms; and (ii) the same strain of *Wolbachia*, when in an insect egg, has the effect of making a specific 'sperm-cleaning' enzyme.

Bidirectional Cytoplasmic Incompatibility Producing Negative Heterosis

The first attempt to employ cytoplasmic incompatibility for population replacement was in the filariasis vector *Culex quinquefasciatus*, using bidirectional incompatibility. This was expected to act in a mixed population as a powerful form of negative heterosis, since all cross-matings were expected to be sterile. The resulting genetic deaths should affect equal numbers of each strain and this would represent a larger proportion of whichever strain was initially rarer. Thus, if sufficient numbers of a strain with a foreign cytoplasm are released so as to constitute a majority in the mixed population, the indigenous strain is expected to be selectively eliminated (Table 20.5). Provided that there is complete sterility in all cross-matings, chromosomal factors will be 'transported' along with the process of cytoplasmic replacement (Laven and Aslamkhan, 1970). This type of selection could be so strong that it is expected to be able to replace a fully fit indigenous strain by a strain with a chromosomal factor causing considerable fitness reduction, but the 'equilibrium point' that has to be exceeded by the releases would then be in excess of 50% (Table 20.5).

Experiments were carried out in outdoor cages in India stocked with a local wild type. Releases were made of a strain with foreign cytoplasm bidirectionally incompatible with the wild type, and the release strain also had a semisterilizing male-linked translocation. The unexpected result was replacement of the local wild type by mosquitoes with foreign cytoplasm, but this was no longer associated with the translocation (Curtis, 1976). It was possible to explain this because, at about the time that the experiment was completed, several routes were discovered by which the rule of absolute sterility in cross-matings may break down. These include an effect of male age and polymorphism and segregation for cytoplasmic crossing type within wild populations (Subbarao *et al.*, 1974, 1977; Singh *et al.*, 1976). The disappointing result obtained in the outdoor cage underlined the point that extremely tight 'linkage' of a transporting system and a gene to be transported is required if the system is to work on a scale larger than on the bench-top.

Unidirectional Incompatibility

Caspari and Watson (1959) and Laven (1959) had perceived that, in cases of mixed populations of unidirectionally incompatible types, the strain

Table 20.5. Negative heterosis due to bidirectional cytoplasmic incompatibility producing frequency-dependent selection for chromosomal genes (homozygous refractory (RR)) even if they cause reduced fitness.

Matings			Progeny		
Female	Male	Mating frequency	Type	Relative no.	Frequency
(A)SS	(A)SS	0.25 × 0.25 = 0.0625	(A)SS	0.062	0.181*
(A)SS	(B)RR	0.25 × 0.75 = 0.1875	None (incompatibility)	–	
(B)RR	(A)SS	0.75 × 0.25 = 0.1875	None (incompatibility)	–	
(B)RR	(B)RR	0.75 × 0.75 = 0.5625	(B)RR	0.5625 × 0.5 = 0.281	0.819*
			Total	0.343	

It is assumed that a release produces a mixed population consisting of:
1 unit (25%) of wild type: cytoplasm: (A), homozygous susceptible genes: SS;
3 units (75%) of release strain: cytoplasm: (B), genes: RR (which cause refractoriness and reduce fitness to 50%).
* Note increase of frequency of (B)RR strain relative to initial conditions.

whose females were not sterilized would be at an advantage, even if it was initially present at a low frequency. Fine (1978) pointed out that this also applied to mixtures of *Wolbachia*-infected and uninfected strains, with the former being favoured by selection unless the infection caused seriously reduced viability. In fact, the unidirectional sterility in matings between infected and uninfected strains (Table 20.4) can be considered as a stratagem of *Wolbachia* to ensure their own transmission and to destroy uninfected members of the same insect species that compete with their hosts. The demonstration by O'Neill *et al.* (1992) of the similarity of 16S ribosomal ribonucleic acid (rRNA) sequences in the *Wolbachia* in several different types of insect suggests that *Wolbachia* from the same stock have successfully invaded and spread through populations of several host insect species and genera. Turelli and Hoffmann (1991) have observed *Wolbachia* spreading through wild *Drosophila* populations in California at the rate of about 100 km/year. By 1993 the population had reached equilibrium at the expected high frequency of the infected form (M. Turelli, personal communication).

The possibility has long been available of choosing an appropriate unidirectionally incompatible cytoplasmic type and using this to replace a wild-type cytoplasm. However, there seemed no point in doing this so long as all genes that one might want to drive into a population were located on chromosomes and could therefore be 'recombined' with cytoplasmic factors via the fully compatible cross-mating (Curtis, 1992b). However, with modern genetic engineering methods it would seem possible to introduce a gene into a maternally transmitted entity such as *Wolbachia* and then to introduce this into an uninfected insect and to arrange that the gene is expressed in its new location.

Beard *et al.* (1992) used a plasmid shuttle to transform an extracellular symbiont of *Rhodnius prolixus* (a vector of American trypanosomiasis), and successfully introduced the transformed symbiont into an uninfected strain of the insect by feeding. If the gene remains within the genome of the symbiont, there should be no chance of 'recombination' of the transporting system and the gene that one wished to transport. Furthermore, as reported by Beard *et al.* (1993b), it appears from polymerase chain reaction (PCR) tests that O'Neill has carried out on our various laboratory strains of *Anopheles* and on wild *Anopheles gambiae* from Tanzania that the vectors of human malaria are free of *Wolbachia*. Therefore they are apparently 'virgin territory' for artificial infection at the egg stage and for spreading the infected form after a relatively small release of it.

It is possible that *Anopheles gambiae* spp. are uninfected because they are not susceptible to *Wolbachia* infection. However, horizontal transfer events between species are probably extremely rare in nature, given that *Wolbachia* is obligately intracellular, and it is quite likely that the opportunity for invasion of *An. gambiae* by *Wolbachia* has never arisen. Introducing the bacterium directly into germ-line tissue is probably the only way to produce a new infection. Successful artificial interspecific transfers of *Wolbachia*, by microinjecting embryos with infected tissue or purified bacteria, have already been made between *Drosophila* species (Boyle *et al.*, 1993) and from *Aedes albopictus* into uninfected *Drosophila simulans* (Braig *et al.*, 1994). The latter represents a large phylogenetic distance, supporting the prediction that *Wolbachia* have a wide host range and fuelling optimism that infection of *Anopheles gambiae* may be achieved by similar means. Microinjection of *Anopheles* embryos with *Wolbachia* extracts is being attempted, using similar techniques to those described for DNA injection by Miller *et al.* (1987). Injection into *Anopheles* eggs is far more difficult than for *Drosophila*, and as yet the technical hurdles have not been overcome (S. O'Neill and S. Sinkins, personal communication). If an infected isofemale line can be produced, as determined by PCR using the *Wolbachia*-diagnostic primers of O'Neill *et al.* (1992), then infected males would be mated to uninfected females to examine whether cytoplasmic incompatibility is induced. Levels of incompatibility may be increased in subsequent generations by selection, as demonstrated in *Drosophila*, where such selection was associated with concomitant increase in *Wolbachia* density (Boyle *et al.*, 1993).

If attempts to infect *Anopheles gambiae* are successful, the next stage would be the production of a second infected stock that is unidirectionally incompatible with the first. The intention would be that, if the replacement of the wild symbiont-free *Anopheles* by an infected and genetically transformed strain was not fully successful in rendering the mosquito population unable to transmit malaria, a further replacement would be possible by a second strain whose males are incompatible with females of the first. Beard *et al.* (1993b) emphasize that the possibility of such successive *Wolbachia* replacements is a potential advantage of this method over the use of transposons. It has been shown in *Drosophila* that hosts infected with two bidirectionally incompatible strains of *Wolbachia* produce unidirectional incompatibility when crossed to stocks infected with only one of these bacterial strains (Sinkins *et al.*, 1995).

Driving Other Cytoplasmic Particles with *Wolbachia*

If a genetically transformed *Wolbachia* could be introduced into the ovaries of *Anopheles* and maternally transmitted, it is doubtful whether the product of an 'antimalaria' gene that had been introduced into the symbiont would diffuse to the gut or salivary gland, which are the sites occupied by malaria parasites. In tsetse, bacterial symbionts are well known to occur in gut cells (Nogge, 1978; Maudlin and Ellis, 1985), and there are *Wolbachia*-like organisms in the ovaries of some species (O'Neill *et al.*, 1993). The presence of the organisms in gut cells has been reported to be a necessary condition for susceptibility to trypanosome infection, this susceptibility being a maternally inherited character (Maudlin and Dukes, 1985; Maudlin and Ellis, 1985). Beard *et al.* (1993a) proposed transforming the gut symbionts with a plasmid carrying a gene that would interfere with trypanosome transmission, and the introduction of such symbionts into a strain 'cured' of its natural symbionts. It was further proposed that this transfected form should be caused to spread in a wild population by introduction of an appropriate *Wolbachia* into the ovaries, as discussed in the previous section.

No equivalent to the tsetse gut symbionts can be expected in *Anopheles* mosquitoes, which do not seem to be naturally dependent on symbionts in the way that tsetse are. We may consider whether it would be possible to introduce desired genes into some other reliably maternally inherited entity, such as a mitochondrion. This is expected to remain genetically 'linked' to a maternally inherited symbiont: Turelli *et al.* (1992) have shown that a mitochondrial variant is being carried along as *Wolbachia* invade wild Californian *Drosophila simulans*. Kambhampati *et al.* (1992b) found equivalent changes in *Aedes albopictus* (a dengue vector) in the laboratory. However, even if one could genetically transform some mitochondria, one could not have a mitochondrion-free insect into which transformed mitochondria could be introduced; therefore transformed mitochondria would be required to compete with pre-existing untransformed ones, and there is no reason to suppose that the former would win this competition. A better prospect may be to insert the desired genes into another maternally inherited factor not already in *Anopheles* but able to invade somatic cells, including gut and salivary gland cells. S.L. O'Neill (personal communication) has suggested use of a natural avirulent insect virus, which behaves similarly to the transovarially transmitted sigma virus of *Drosophila*. It may be possible to introduce such a factor into a mosquito stock, as well as introducing *Wolbachia* to provide selection for the new type of cytoplasm. The necessary 'linkage' of driving factor and genes to be driven would depend on high rates of transovarial transmission of both virus and *Wolbachia*.

The host range of *Wolbachia* seems to be so broad, and its peculiar effects on compatibility so widespread, that many of the above principles would seem to be directly applicable to agricultural pests. In the present politicoeconomic climate, it is more likely that finance will be provided to protect cash crops than to protect the lives and health of low-income people, and we may see the principles of population replacement applied first to a pest of a high-value exportable crop.

Acknowledgements

I am very grateful to Scott O'Neill and Steven Sinkins for their comments and for giving me permission to cite so far unpublished data.

Molecular Genetics 21

VERONICA RODRIGUES AND K. VIJAYRAGHAVAN

Introduction

Significant losses in food production occur because of insect pests. Insecticides are widely used to reduce this loss. Pesticides are usually not insect-specific and can affect other insects not harmful to plants. Their use can also result in contamination of food supplies and of the environment. Prolonged use of insecticides can result in the development of resistant strains of insects. Despite these drawbacks, few effective alternatives are currently available. The detailed study of the molecular mechanism of insect–plant interactions is thus of great importance. Such studies can potentially allow the rational development of drugs that can be used to block interactions between specific insects and their hosts. In addition, and of special significance, simply understanding these interactions opens up avenues for designing biological control methods that may have fewer disadvantages than many of the current methods of chemical intervention.

There are several events that precede insect feeding or deposition of eggs on a plant. First, the plant has to be identified in the environment. Several studies (see Brewer *et al.*, 1984, for examples) have shown that visual cues act over fairly long distances, although olfactory information may sometimes also be relevant. In the shorter range of metres to centimetres, olfactory cues are extremely important. Once the insect identifies a plant based on these cues, it examines the plant by tasting, smelling or touching it. At this stage, plants attractive by cues acting over long distances may be rejected if they 'fail' in the local gustatory, olfactory or tactile tests. If a plant is acceptable, the insect begins feeding to satiation. At this stage as well, chemosensory responses, especially taste, are important.

Insect–plant relationships are complex. Some insects are specific to a single plant, others to two or a few plants and yet others show no specificity at all (Brewer *et al.*, 1984). Thus the same or similar cues can and do result in quite different responses. A further level of complexity is added by the observation that many insects show a specificity at a particular stage of their development but do not require a specific host at other stages (Dethier, 1970). One explanation for the variation in the specificity or otherwise of insect–plant interactions may lie in the diversity of receptors for gustatory or olfactory stimuli (Buck and Axel, 1991), which allow the insect to discriminate between stimuli at a fine level. Receptors, however, merely report on the presence, absence or level of a stimulus. The response to this stimulus, a positive chemotaxis or a negative chemotaxis (avoidance) response (Baker, 1985), must emanate from the processing of information carried by peripheral neurons to the central nervous system (CNS). Some of the neural mechanisms underlying the processing of pheromone as well as other stimuli have been shown to occur in the antennal lobes, as well as the mushroom bodies in a number of insect species (Pinto *et al.*, 1988; Rodrigues, 1988; Homberg *et al.*, 1989). The antennal lobes are regions of the brain where the primary signal-detecting sensory neurons project. They can thus be called the 'secondary' level of information analysis. The mushroom bodies are groups of neurons located at more central regions of the brain, and there is evidence that they are involved in processing the input information at a higher ('tertiary') level.

The complexities of insect–plant interaction pose formidable barriers to analysis at the molecular level. However, general aspects of the process common to all insects allow a feasible approach to be designed, as the events in the analysis of sensory information, mentioned above, are essentially common to all insects. Insects recognize their environment in diverse ways and, as mentioned above, vision, smell, taste and touch all play important roles. Each specific stimulus usually elicits a highly stereotypic response in a given insect. The steps in detecting and responding to the environment can be split into the following units: detection of the signal, processing of sensory information and a motor response to the signal. At each of these levels of analysis, there are features shared by all insects at the anatomical, cellular and molecular level. Nerve cells in all animals share many common structural aspects. The roles of components of the nerve cell, such as the axon, the synapse, the cell body and the dendrites, have all been addressed extensively (Kuffler *et al.*, 1984). The anatomical and molecular features that give each nerve cell its characteristic physiology are conserved across all animals. There are many kinds of neurons, based on their shape, location and physiology, but nerve cells from all animals and within each animal share many common features. Invertebrate neurons differ from those of vertebrates in many ways, for example in the organization of their cell bodies in one region of the brain and also in their lack of myelination (Kandel, 1976), but, nevertheless, many aspects of nerve cell structure and function are conserved. Cellular mechanisms that operate, such as the steps in signal transduction, signal transmission and synaptic function, are also highly conserved between all animals. The development of the nervous system, including the mechanisms by which neuroblasts (NBs) segregate, the pattern of division of NBs and the specific cell lineages that give rise to identified neurons, are also highly conserved, even among insects that are not closely related (Thomas *et al.*, 1984; Zinn *et al.*, 1988; Patel *et al.*, 1989), and between insects and other arthropods. At the molecular level, most structural components of neurons, such as ion channels, neurotransmitter biosynthesis enzymes, neurotransmitter receptors and molecules involved in synaptic function, are not only conserved among insects but among all animals in general. In addition, analysis of nervous system development in the fruit fly, *Drosophila melanogaster*, has shown that many of the regulatory molecules involved in the development of the nervous system and many components used by the nervous system during development are highly conserved among insects.

Biology of *Drosophila* and Review of Molecular and Genetic Methods

The fertilized egg laid by the female *Drosophila melanogaster* takes about 10 days to develop into an adult fly. The egg hatches into a larva in about 24 h after egg-laying, and this larva undergoes two moults before beginning its pupal life to metamorphose into a fly. The moulting and pupation are controlled hormonally. The ease of maintenance of the fruit fly in the laboratory, its small genome and the cumulative efforts of *Drosophila* geneticists for over half a century have made this the organism of choice for studying animal development.

Adult males can be fed chemical mutagens, such as ethylmethanesulphonate (EMS), and their progeny examined for mutant phenotypes. After appropriate genetic crosses, flies (unless otherwise stated, by 'flies' and 'the fly' we mean *D. melanogaster*) that are homozygous for a mutagenized chromosome can be examined for lethal defects, for visually scorable phenotypes or for behavioural defects. Recessive lethal mutations can be maintained in the heterozygous condition over 'balancer' chromosomes. In other words, the classical genetics of the fly is facile and can be done under simple laboratory conditions. Powerful as these methods are, three major recent developments have changed the perception of the breadth of *Drosophila* genetics and its value to biology in general. The first is the recombinant deoxyribonucleic acid (DNA) revolution, which has allowed the cloning of DNA from any organism, the propagation of this cloned DNA in hosts and the ability to rapidly sequence this DNA and to predict the sequence of gene products. This advance is, of course, not unique to *Drosophila*. The second event resulted, interestingly, from the analysis of repetitive DNA in flies, combined with the study of the genetics of transposable elements. The pioneering work of Engels, Kidwell and Green, who made a detailed genetic study of a family of transposable elements in flies, called the P elements, and Rubin's work on repetitive elements and cloning of a repetitive element, *copia*, a copy of which was inserted in the *white* locus, are covered in Lawrence (1992). In cloning the *copia* element inserted in the *white* locus, Rubin and coworkers isolated DNA flanking the insert which corresponds to the *white* locus. The *white* gene was thus cloned. When the genetics of a particular *white* mutant hinted that it had a P element inserted at the locus, it was readily possible to clone the DNA from the mutant, because Rubin had isolated the wild-type *white* DNA earlier and could use this as a probe on a mutant genomic DNA library. The isolation of P element DNA changed the face of *Drosophila* molecular genetics. Rubin and Spradling demonstrated that P elements could be used as vectors to integrate cloned DNA into the germ-line chromosomes of flies (see Lawrence, 1992). This, combined with the powerful genetics of the fly, allowed the rapid analysis of gene function during development (see Lawrence, 1992, for a history of the above events). The third development was the analysis, at the molecular level, of homoeotic mutants in *Drosophila* and the comparative studies of the expression and function of these genes in higher animals. Nusslein-Volhard and Weischaus pioneered the isolation of mutants that affect the segmental body plan of the *Drosophila* larva, and Lewis had earlier pioneered the study of homoeotic genes, mutants which result in one segment or a part of a segment being transformed to another segment. The homoeotic genes of *Drosophila* contain a protein-coding motif called the homoeobox, conserved across several species. What is remarkable is that the expression pattern and chromosomal organization of these homoeobox-containing (or Hox) genes are conserved across several species. Recent results indicate that the mutant phenotype resulting from the loss of function or misexpression of certain mammalian Hox is remarkably similar to the corresponding mutations in *Drosophila*. Equally remarkable is that misexpression of specific homoeobox-containing genes from humans in *Drosophila* gives phenotypes that are similar to those resulting from misexpression of their *Drosophila* homologues. These results thus point to the conservation of both sequence and function over very distantly related animals, and this has a bearing on some of the arguments we shall present (McGinnis *et al.*, 1990; Zhao *et al.*, 1993).

The principal advantage that *Drosophila* offers is the rapid ability to understand the regulation and function of genes in a multicellular organism. While transgenics can be made in other animals, such as mice, their construction and analysis is significantly time-consuming. P-element-mediated germ-line transformation allows the transformation of genes from any organism into *Drosophila*. The function of these genes can then be studied in *Drosophila*. This method can be applied for those genes that have molecular homologue in flies, for which null mutations are available in flies and whose function the foreign gene product can complement when expressed in flies. Therefore the gene for, say, grasshopper hexokinase can be studied in flies. But there are better ways to study hexokinase and there are better uses of flies as a reaction vessel.

Development of Insect Nervous Systems

Studies on *Drosophila* and grasshopper in the past decade have greatly enhanced our understanding of how the nervous system develops. In *Drosophila* the first steps involve the definition of what is called the ventral neurogenic region of the egg. This is the region of the egg where NBs will form and then divide to form the components of the CNS. The definition of the ventral neurogenic region is achieved by the action of genes that are active in the mother (maternal genes) and by those that are active in the fertilized egg (zygotic genes). The maternal genes are involved in defining the axes of the embryo (the anterior–posterior, the dorsoventral and the termini). The zygotic genes are involved in the elaboration of these coordinates, and their action ultimately results in the segmental patterns seen in the epidermis, the nervous system and the muscles. In the ventral neurogenic region, what have been called 'proneural' genes are expressed in groups of cells. This expression declines in most cells, while one cell in each group continues to express the proneural gene. In the CNS, this cell will become an NB. The peripheral nervous systems (PNSs) of both the adult and larva develop, using remarkably similar early mechanisms. Analogous to the NBs are sensory mother cells (SMCs). These cells are chosen by the action of the proneural genes and subsequently divide to give rise to sensory neurons and the supporting cells of the peripheral sensillum. These events are illustrated in Fig. 21.1 (Doe and Smouse, 1990; Doe, 1992).

Once NBs or SMCs are determined, how do they go on to produce specific neurons? Let us take the case of the CNS first. Neuroblasts divide asymmetrically to produce an NB and a ganglion mother cell (GMC). The GMC divides once symmetrically to produce a pair of postmitotic neurons. In principle, each NB can be identified by its position in the developing CNS, its pattern of division, the genes it expresses and the genes that are expressed in its progeny and the pattern of neurons it produces. The progeny of an NB can derive its identity in several ways (Doe and Smouse, 1990). First, it is possible that environmental influences determine the fate of GMCs. This seems unlikely from experiments in grasshopper that laser-ablate NBs after they have produced GMCs (Doe and Goodman, 1985).

These experiments show that a new NB takes the place of an ablated one and this new cell divides to produce GMCs in a pattern as if the earlier generated GMCs did not exist. Thus the non-neuronal environment does not influence the sequence of specific GMCs being generated by the new cell. The second possibility is that GMC and NB interact to determine the specific GMC that will be produced. That this model is not valid is shown by *in vitro* culture experiments where the progeny neurons derived from dissociated NBs are examined. The neurotransmitter type and number of neurons produced is exactly the same as *in vivo*. These experiments show that contact between NB and GMCs are not necessary for the correct pattern of differentiating neurons. Finally, similar NB cell-culture experiments show that DNA replication but not cytokinesis is required for determining GMC fate (see Doe and Smouse, 1990, for a discussion). This result suggests that GMC determinants are generated in NBs and their activity is linked to DNA replication. In this model, GMC fate is controlled by the number of cycles of DNA replication undergone by the NB (Doe and Smouse, 1990). Thus, the position along the body axis and cell–cell interactions in the ventral ectoderm choose an NB from ectodermal cells (Fig. 21.1), and NB segmental identity is eventually determined by homoeotic selector genes and its position within a segment. Ganglion mother cell identity is controlled by the NB and linked to DNA replication. The fate of a neuron is determined by the fate of its parent GMC and by cell–cell interactions with its neighbouring neurons (see Yang, X. *et al.*, 1993). Ubiquitous expression of a transcription factor during NB segregation results in the progeny of a particular NB, NB4–2 (Fig. 21.2.), making two GMC-1 entities and a GMC-2, instead of one GMC-1 and one GMC-2 as would normally happen. This results in the duplication of the progeny of the GMC-1, namely the neuron RP1 (Fig. 21.2).

Work on the study of the CNS of grasshoppers and flies has come from several laboratories (see Thomas *et al.*, 1984; Doe, 1992, for reviews). While both grasshoppers and flies are insects, they differ in significant ways. Flies are holometabolous insects (undergo complete metamorphosis), whereas grasshoppers are hemimetabolous. The conservation of the cellular mechanisms that determine the manner in which the nervous system is laid out and defined is maintained over animals separated from each other for over hundreds of millions of years and many of the genes that mark NBs and GMCs do so in other arthropods in a similar or even identical pattern (Patel *et al.*, 1989).

The development of the PNS shares many features with that of the CNS. Here too the choice of a sensory mother cell (SMC) is controlled by cell–cell interactions in a region of 'proneural' gene activity. The SMC divides twice to give rise to three support cells and a peripheral neuron. Each sensillum on the fly has a specific position; it may often have more than one neuron, each with a different function, and each neuron must find its correct

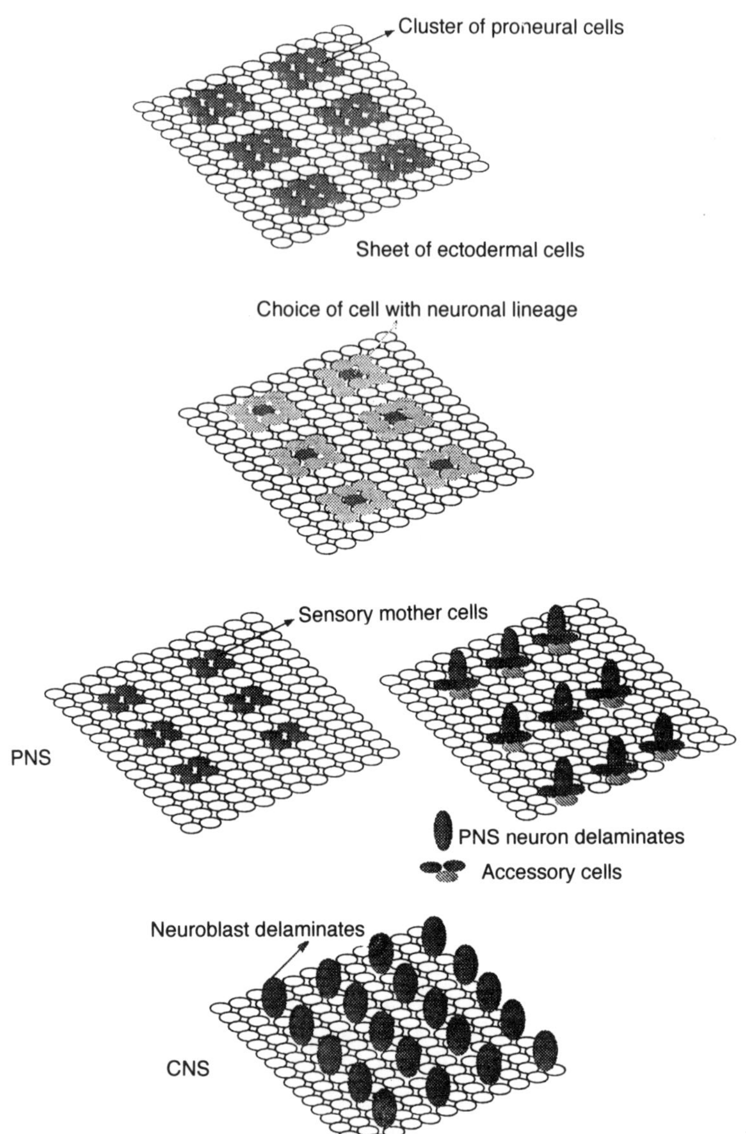

Fig. 21.1. Neuroblasts of the central nervous system (CNS) or sensory mother cells (SMC) of the peripheral nervous system (PNS) are chosen by cell interactions, form an epidermal sheet and delaminate. The identity of the progeny of neuroblasts or SMC depends on cell lineage and position. See text for details. Modified from Artavanis-Tsakonas and Simpson (1991).

target in the CNS. As in the CNS, both cell interactions and lineage play roles in deciding neuronal identity (Simpson, 1990; Ray *et al.*, 1992; Ray and Rodrigues, 1993, for the chemosensory pathway).

It is not only the cellular and molecular processes of NB segregation and determining neuronal identity that are conserved among insects. The mechanisms that neurons use to find their target also seem to be remarkably

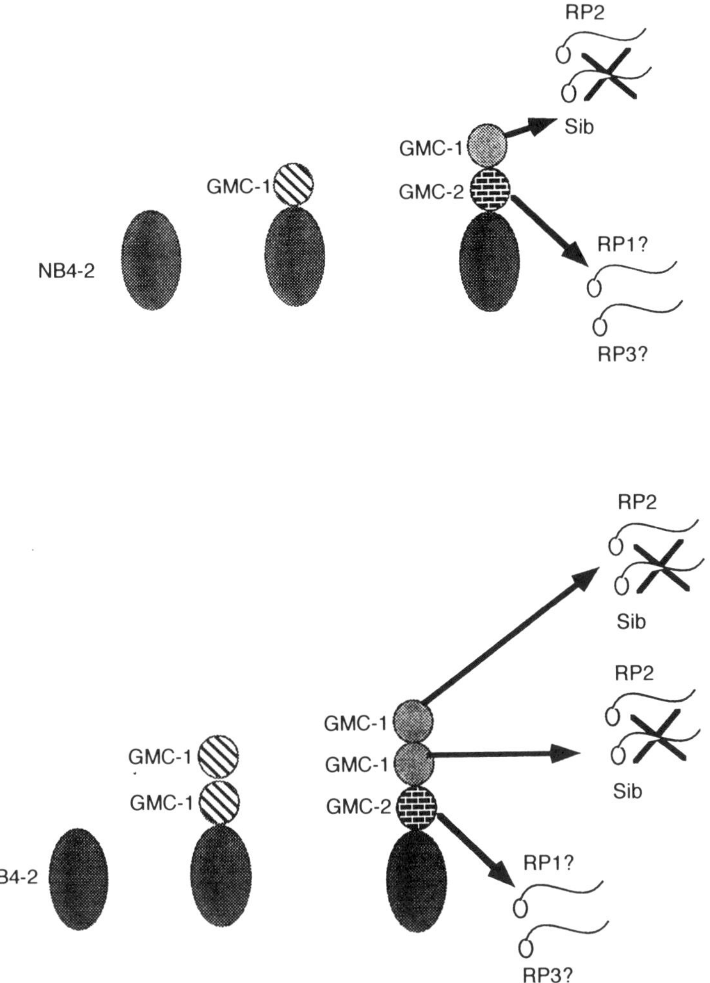

Fig. 21.2. The top panel shows the lineage of a neuroblast and the identity of its progeny. The misexpression of a transcription factor in the neuroblast 4-2 (NB4-2) lineage results in the duplication of ganglion mother cell 1 (GMC-1) and thereby two RP2 neurons are produced. The Sibling (Sib) of RP2 is speculated to die, in each case. Modified from Yang, X. *et al.* (1993).

conserved. The Goodman laboratory initiated a project to isolate cell-surface glycoproteins expressed on grasshopper axonal growth cones, glia and axon fascicles. They then purified the corresponding proteins from *Drosophila* and cloned the genes that encode them. Analysis of the expression pattern of the cell-surface molecules and genetic analysis of mutants are revealing about the way axons find their targets. These results suggest that pathways that growth cones recognize and prefer are labelled by homophilic or heterophilic adhesion molecules. Attractive and repellent cues guide the growing axon, which chooses between alternate pathways. It is probably a combinatorial expression of adhesion molecules that defines a path, and the redundancy present in the code makes it difficult, usually, to see a phenotype from a null mutation in a gene encoding these adhesion molecules (Goodman and Shatz, 1993).

Detection and Processing of Chemosensory Information

Once constructed, a neural circuit must function to precisely detect and respond to the environment. It is probable that chemosensory cues are recognized by specific molecular receptors. The recent cloning of odorant receptors from mammalian olfactory epithelia (Buck and Axel, 1991) and the diversity of receptors found allow the cloning of *Drosophila* homologues and suggest that much of the specificity of odorant recognition may lie at the most peripheral level. The distribution of receptors among the sensory neurons could result in a particular stimulus activating a set of neurons. Both the quantity and quality of the receptor distribution will determine the amplitude and frequency of the response of the peripheral neuron. There is thus a significant amount of processing of chemosensory cues at the most peripheral level. The signals from the peripheral neurons synapse on to the antennal lobe. The peripheral goes to both the ipsilateral and the contralateral lobes (Fig. 21.3). Functional anatomical studies using compounds that are preferentially taken up by active neurons show that specific odorants result in the activity of specific parts, or glomeruli, of the antennal lobe. Therefore, it seems that there is a combinatorial code that interprets and integrates the inputs received from the two antennae (Buchner and Rodrigues, 1984; Rodrigues and Buchner, 1984; Pinto *et al.*, 1988; Rodrigues, 1988). The signals from the antennal lobe then go to higher centres of the brain, where these are further integrated with other sensory inputs and with data resulting from experience and memory of earlier stimuli. This information then results in an appropriate motor output (Fig. 21.3).

Experiments on honey-bees, moths and larger flies all show that the processing of chemosensory and other stimuli, such as vision, is very highly advanced in insects. Insects are capable of detecting and responding rapidly to stimuli, and they are sensitive to quite small changes in the level of stimulus. However, the responses of insects to a specific concentration and quality of stimulus are stereotyped. Such responses can be used in behavioural screens for mutants that show altered responses to defined stimuli.

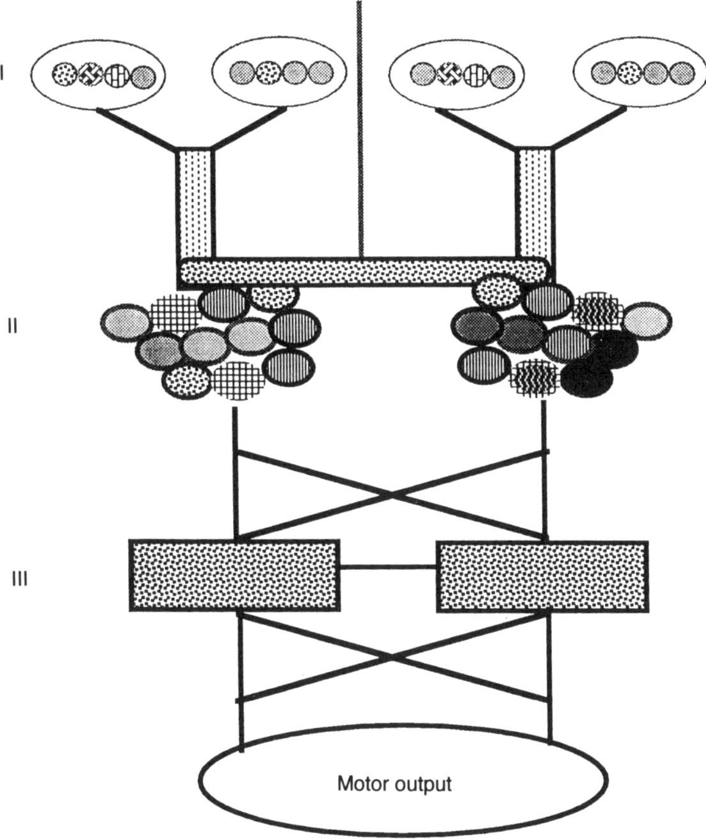

Fig. 21.3. Signal processing of chemosensory information: a model. I. Molecular receptors of differing specificities are expressed at different levels on the peripheral neurons. II. Peripheral signals are integrated in the antennal lobe or taste lobe (for gustatory inputs). III. Processing in the brain and comparing this information with past inputs and memory result in an appropriate motor output.

Genetics of the Chemosensory Pathway

The behavioural assays that are used to screen for and isolate mutants in chemosensory responses use either populations of flies or individuals. In the feeding preference test (Tanimura *et al.*, 1982), for example, alternate wells in a dish with 96 wells contained sugar added to an agar base while the remaining wells contained no sugar. The sugar-containing wells were uncoloured, whereas the other wells contained a food colouring additive. When starved flies were released on the plate, they ate from the sugar-containing wells after sensing them with the taste hair on their legs. Mutants that do not sense sugar will eat randomly and their abdomens will be coloured red, whereas the normal (wild-type) flies have uncoloured abdomens. Individual assays for taste behaviour involve the stimulation of the taste hairs on the legs of starved flies with a solution and testing for the extension of the proboscis in response. The olfactory population tests involve the use of a Y-shaped glass 'maze' (Rodrigues and Siddiqi, 1978; Ayyub *et al.*, 1990). Odorant flows through one arm of the Y while air flows through the other. Flies released at the bottom of the Y climb upwards and choose between the two arms of the Y according to their odorant preferences. Single-fly assays for olfactory responses have been developed by John Carlson and coworkers (McKenna *et al.*, 1989). In this assay, flies exposed to a puff of odorant jump if they can smell and fail to do so if they do not smell. Chemosensory assays for larval responses have also been devised in which larvae migrate to the region of an agar plate where a taste or smell cue is concentrated (Rodrigues, 1980).

These assays allowed the isolation, after EMS mutagenesis, of several mutants affected in their ability to smell and taste (for examples, see Siddiqi, 1987; Carlson, 1991; Rodrigues *et al.*, 1991). Among those with olfactory defects, *olfE* has been cloned (Hasan, 1989). The molecular analysis of several genes is in progress in different laboratories, and results from these studies are awaited. Behavioural abnormalities can result from mutations in genes that are not expressed in the nervous system. The 'enhancer-trap' approach (O'Kane and Gehring, 1987) and its rapid facility for isolating complementary DNAs (cDNAs) that represent messenger ribonucleic acid (mRNA) coding regions in the vicinity of the P insertion (Bellen *et al.*, 1989; Bier *et al.*, 1989; Wilson *et al.*, 1989) allow the *in situ* assay of gene activity by expression of the reporter gene and the verification that the mRNA pattern of expression is similar. We have used this method to isolate chemosensory mutants (Anand *et al.*, 1990). We briefly summarize results from two, namely *east* and *scalloped* (*sd*).

Mutants in the gene *east* show adult-specific defects in both olfaction and taste (VijayRaghavan *et al.*, 1992). Our initial study of mutant phenotypes examined olfactory responses, rapidly assayed in the 'jump test', in which flies jump when exposed to a pulse of odorant (McKenna *et al.*, 1989). Flies that are homozygous or hemizygous for the X chromosome bearing the P insertion in *east*, when examined in this test, show a clear mutant response to olfactory stimulation by several odours. The phenotypes were initially identified in odour-induced 'jump' assays, using benzaldehyde and isoamyl acetate as stimulants. These *east* flies are also mutant when tested with ethyl acetate, propionic acid and butanol. The chemosensory defects in *east* flies ate not restricted to olfaction. Males and females of the strain show specific gustatory defects. Wild-type adults are attracted to sugars and also to salt at low concentrations (Tanimura *et al.*, 1982; Arora *et al.*, 1987). These responses are normal in *east*. High salt concentrations normally repel flies (Arora *et al.*, 1987); *east* mutant flies, however, are more tolerant to high salt concentrations than the wild type.

Other sensory responses are normal in *east* mutants. The phototactic responses of the adult are normal and mutant flies also mate normally, are viable as homozygotes and exhibit no obvious motor abnormalities.

We have isolated several alleles of *east*, and none of the viable combination of alleles tested show any larval chemosensory phenotypes. We know that the gene is required for larval viability, since mutations with complete loss-of-function are lethal. However, we do not see, as yet, a role for the gene in larval chemoreception. All the viable allelic combinations show adult chemosensory defects. It is possible to examine the null alleles in clones of cells and test if a larval chemosensory phenotype can be seen (Xu and Rubin, 1993). In conjunction with a molecular analysis, chemosensory genetics can often provide clues to function of well-studied molecules in the cellular context of their domains of expression (Balakrishnan and Rodrigues, 1991) and can also identify new molecules (Hasan, 1989). The experiments with *east* are part of our efforts to dissect the functional roles of genes in diverse processes by examining their roles at levels from the molecular to behavioural. We are in the process of isolating full-length cDNA clones corresponding to mRNA from *east* gene and hope that the sequence combined with the expression pattern will be informative.

An analysis of the sequence and behaviour has been informative in our study of the (*sd*) locus (Campbell *et al.*, 1992; Inamdar *et al.*, 1993). The *sd* mutants are affected with regard to the responses of larvae and adults to several different taste stimuli. They require a higher threshold of the stimulus to elicit a behavioural response. In addition, an increased acceptance of low concentrations of sodium chloride as compared with the wild type is seen in one allele. Mutant phenotypes have been described before, and the lesion has been attributed to the more central levels of processing in the gustatory pathway (Rodrigues and Siddiqi, 1981): models of taste behaviour which propose that information from the sensory neurons is weighted and compared with results in behaviour are consistent with this (Dethier, 1976;

Balakrishnan and Rodrigues, 1991). Our electrophysiological results suggest that the reception of the stimuli at the receptor and the transduction processes are unaffected by the mutation (Inamdar *et al.*, 1993). The lesion at the central level of coding or processing of information could well arise by subtle defects during the development of the nervous system that allow survival in hypomorphic alleles. Indeed, the *sd* gene is expressed in the developing nervous system of the larva and the adult. Here again, an analysis of animals that are part mutant and part wild-type can determine the part of the brain that is responsible for the phenotype seen in *sd* mutants. Methods of analysing mosaic internal tissue at high cellular resolution are now available in *Drosophila* (Xu and Rubin, 1993), and we are using these methods in our experiments to dissect the function of different parts of the brain in the analysis of chemosensory information.

Conservation of Molecules

The sequence of *sd* cDNA clones shows a predicted protein homologous (68% identity) to a human transcription factor, TEF-1 (Campbell *et al.*, 1992). In a 72-amino acid region containing the conserved DNA-binding TEA domain, the conservation is extremely high (71 out of 72 amino acids identical). The high degree of conservation allows the dissection of the function of TEF-1 and of the *sd*-encoded protein rapidly. Studies on the *in vitro* transcription properties of TEF1 and *sd* are being undertaken in Strasburg (Xiao *et al.*, 1991).

We earlier remarked on the high conservation of cellular events in the development of the arthropod nervous system (Patel *et al.*, 1989). The high level of conservation of molecules such as *sd* the Hox genes and other regulatory genes, such as those of the *achete–scute* complex and those that encode components of ion channels, allows the cellular mechanisms to be understood in molecular terms. This conservation potentially allows the study of molecules from other insects in *Drosophila* and the understanding of their molecular function as well as their cellular role.

Studying Molecular and Cellular Function in *Drosophila*

The high degree of conservation of molecules expressed in the nervous system allows the study of genes isolated from one organism in another, such as *Drosophila*, where transformation of foreign DNA is readily possible. There are, however, some problems in doing this. First, these molecules should be studied in a genetic background where the homologous gene from

Drosophila is deleted, so that the phenotype of the introduced gene can be followed. This therefore necessitates a strong investment in the genetics and study of expression patterns of the *Drosophila* genes that are expressed in the nervous system (Palazzolo *et al.*, 1989; VijayRaghavan *et al.*, 1991). Such an investment, already widespread because of the large fly community, has gained a new thrust because of the progress made by the *Drosophila* genome projects in California and in Europe.

Very soon, a P-elements insertion will be available every 50 kb along the fly's genome, the entire genome will be organized in contigs of phage P1 DNA and several hundred cDNA clones will be mapped and sequenced. All of this means that it will become easier to map, clone and sequence genes, as well as to isolate alleles or mutants at a locus. Another problem in expressing foreign genes in flies is that the study of function may be hampered by the gene causing a dominant lethal phenotype. This would mean that a transformant expressing such a gene will not be viable even if the gene is present in one copy. The recent development of the GAL4 technique by Brand and Perrimon (1993) allows this problem to be circumvented. Thus, advances in the manipulative power in *Drosophila* allow the study of foreign genes with greater ease than earlier. Another advantage of the study of genes that have *Drosophila* homologues is that this allows the isolation of other interacting genes, using the genetics available in *Drosophila*. Several mammalian genes have been shown to function, and some even rescue phenotypes of *Drosophila* mutants (McGinnis *et al.*, 1990; Zhao *et al.*, 1993). It will be of interest to see, for example, if the cloning of specific chemoreceptors from other insects, when available, into flies allows the control of peripheral detection of specific cues.

Genetic Transformation of Other Insects

So far, efforts to transform insects other than *Drosophila* with foreign DNA have not had great success. This may change with the analysis of the *mariner* transposable element. Robertson (1993) has shown that this element is widespread in several insect species and that this spread occurred, most probably, by horizontal transfer. The *mariner* elements from *Drosophila mauritiana*, a very distant relative, can be transformed in *D. melanogaster*. This allows the detailed molecular analysis and functional testing of *mariner*, and opens the way for its use as a vector in the transformation of other insects. This provides exciting possibilities for studying the genetics of other insects. It must be emphasized, however, that the mere availability of transformation methods does not solve biological problems, but the availability will allow the reintroduction of genes that have been tested rapidly *in vitro* and in *Drosophila*.

Development of Biological Control Methods

What are the prospects for developing better biological control methods from the study of insects like *Drosophila*? Our assumption is that many basic cellular mechanisms have been conserved during evolution. This is clearly so at the molecular level too, as revealed by sequence similarity and functional substitution studies. Although these experiments have been crucial in understanding how the animal and its brain develops and has evolved, the prospects of the use of this information to intervene in the field are more distant.

Let us take a specific and quite hypothetical scenario. Assume that studies on *Drosophila* result in the identification and molecular cloning of an odorant receptor. The detailed study of such receptors and their ligands could result in the identification of a modified ligand that is highly attractive to some insects that are pests of rice. The spraying of areas near rice fields with such a modified ligand could result in attracting insects away from the crop. In a similar hypothetical vein, but more realistically, the detailed study of the recently characterized ecdysone receptors in *Drosophila* (Talbot *et al.*, 1993) allows the design of specific drugs that affect insect moulting in a specific way. *Drosophila* again affords a rapid system for selecting drugs that affect moulting and interact with the ecdysone receptor.

In our opinion, chemical approaches to pest control will continue to dominate the future, but such approaches will be quite different from the approaches currently being used. The use of model systems like *Drosophila*, the nematode *Caenorhabditis elegans* and the baker's yeast *Saccharomyces cerevisiae* allows the use of genetics and cell biology to design drugs that interfere with highly specific cellular events. The growing understanding of cellular processes in insect development, metamorphosis and function will allow the design and testing of drugs in the laboratory in a more rational manner and result in products that are environmentally benign. This requires the close interaction of geneticists with the applied researchers and field workers who will translate such dreams into reality.

Acknowledgements

We are grateful to Bill Chia and the Institute of Molecular and Cell Biology (IMCB), Singapore, for a wonderful intellectual atmosphere that encourages frequent visits from nearby India. In India our colleagues in our laboratory continue to help us in many ways.

A View from Industry 22

Ben Miflin

I have been asked to review industry's position with regard to linking biotechnology and integrated pest management (IPM), including options for strengthening such linkages. The term industry is broad and includes many enterprises. An enterprise such as Ciba is large and multifaceted, so it would take a considerable time for just one view to emerge. What follows is my personal view, written from within one company engaged in the agricultural business.

Industrial Motivation

In discussing industry's current and future role in IPM, it is important to consider some of the motivating forces that determine its behaviour. These are sketched out in Fig. 22.1 and discussed below.

People and their survival instinct

Companies are made up of individuals with many different backgrounds and motivations. However, many of them spend much of their working life in one company. These people wish to see their company survive and thus are strongly interested in the long-term perspective. Companies made up of this kind of employee are common in agriculture, and many have been in business for some time. For example, Ciba-Geigy arose from the merger of two companies that were founded in 1758 and 1884; it has had an agricultural division for over 50 years and is the leader in plant protection. It thus has a major investment in, and is committed to, having a sustainable business making sustainable products for a sustainable agriculture.

Culture and vision

All companies have a culture and vision, even if it is not stated. At the worst, a travelling door-to-door salesman might have a short-term vision, which might be stated as 'make a quick buck and move on'. In contrast, Ciba has a clearly stated long-term vision that states that by striking a balance between our economic, social and environmental responsibilities we want to ensure the prosperity of our enterprise beyond the year 2000.

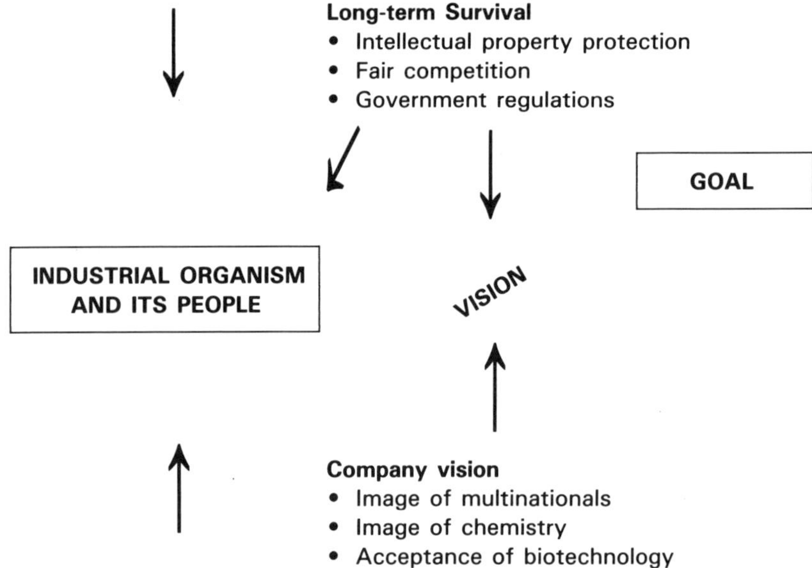

Fig. 22.1. The agricultural industry - driving forces.

Competitive position (market share)

Companies try to do better than the other companies in the industry; they do collaborate but only under conditions in which it is a 'win–win' situation. Any planned cooperation in IPM would have to meet these conditions.

Financial performance as reflected in the annual results

This is driven by the investment community who are managing the investments of banks, insurance companies, pension funds, etc. Much of this money ultimately belongs to private individuals, who demand the best returns for their savings, investments and pensions. Without the approval of the financial markets, companies are limited in their ability to make new investments. Companies only succeed financially if their products are accepted and bought over a long period of time.

Since IPM is important to the plant-protection industry, any role that industry plays in IPM has to be in tune with these forces if it is to succeed. Those people from the public sector who seek to influence the behaviour of industry in relation to a commitment to biotechnology and IPM will be more successful if they can work through these motivations. Stereotyped perceptions of the value of IPM solutions and of the players involved (Table 22.1) are not helpful.

Influence of Companies on Agriculture

Individual companies affect agriculture by the products that they develop and sell to solve farmers' problems. Such products or solutions may be complex and include active ingredients, delivery systems, advice and decision-making aids. Companies recognize that agriculture is becoming more complex and that society demands that the solutions proposed meet defined specifications. The competition between companies is also increasing, while the market size remains relatively stable. All of this places pressure on companies to devise flexible solutions tailored to specific users and environments. It is important to realize that a company such as Ciba is in the agriculture business and is no longer an 'agrochemical company' (Table 22.1). Its three agricultural businesses are plant protection, seeds and animal health. Within plant protection there are four sectors: insect control, weed control, disease control and seed treatment. These titles reflect the objectives of the businesses, which may or may not be served by a chemical remedy. This philosophy means that Ciba (like other companies) carries out research on multiple, competitive solutions within and between groups of its own business. At least part of the drive behind this approach is that IPM is important and will become more so in the future, and the company positively wishes to have products that can contribute to integrated crop management (Table 22.2).

An example of Ciba's multiple approach to a pest problem

In order to give substance to the thesis that companies are developing multiple solutions to agricultural problems I should like to review briefly some of the approaches used by Ciba that as of October 1993 were already on the market, or in the process of being developed, to combat lepidopteran pests.

Table 22.1. Integrated pest management (IPM) (and biotechnology and industry) should be value-, not emotion-driven.

System	Perceived value
Chemicals	Bad Unsafe Toxic
Biologicals	Good Safe Non-toxic
Natural	Good
Artificial	Bad
Biotech	Panacea *or* Evil
Industry	Bad
Multinationals	Very bad
Politicians	Misguided
Foundations/NGOs/UN	Good
Academics	Impractical
Entomologists	Sensitive

Chemicals

Currently , like its competitors, Ciba produces and sells many conventional agrochemicals that control a range of Lepidoptera (Table 22.3).

Pheromones

Ciba's SIRENE™, named after the sirens of Greek mythology and the Egyptian word for 'two secrets', is a pheromone-based product, currently undergoing trials in Egypt against the cotton pink bollworm. It is a complex viscous formulation of gossyplure, cypermethrin, carbon and an ultraviolet (UV) absorber to protect the double-bonded pheromone from the isomerizing effects of sunlight (Angst *et al.*, 1992). The latter component was critical to success and came out of collaborative work with other divisions of Ciba. The product is hand-applied as 50 μl droplets extruded from a mastic applicator gun. The males are attracted to the droplets and killed. In trials,

Table 22.2. The changing face of the agricultural industry.

From	To
Agrochemical	*Plant protection*
Herbicides	Weed control
Insecticides	Insect control
Fungicides	Disease control
Product-driven	*Problem solvers*
Simple approaches	*Multiple approaches*
Chemicals	Chemicals
	Microbial products
	Microorganisms
	Beneficials
	Seeds
Simple technology	*Many technologies*
Organic synthesis	Chemistry
Simple formulations	Biology
Spray and pray	Microbiology
	Genetics
	Biotechnology
	Multiple formulations
	Enhanced packaging
	Application technology
	Diagnostics

four applications of SIRENE proved significantly more effective than conventional applications, with up to ten times less insecticide applied per hectare. In spite of encouraging results, there are limitations: the product depends on manual application, requires considerable monitoring of pest build-up, is specific for only one of the insect pests of cotton and does not kill the females.

Macroorganisms for biological control

Recently, Ciba has set up a number of alliances and partnerships to develop macroorganisms for the control of insects. Two examples are relevant to the control of Lepidoptera. Firstly, *Trichogramma* spp. are naturally occurring parasitic wasps that can effectively control the eggs of lepidopteran pests. In conjunction with the Ontario government, a 5-year, $Can7 million project has been set up to investigate the biological control of spruce budworm. Secondly, Ciba has concluded a research and development (R & D) and marketing agreement with Biosys in the field of beneficial nematodes. One of the species concerned is *Steinernema carpocapsae*, which is effective against larvae that live in the soil and attack plant roots.

Table 22.3 Multiple, potentially competing solutions from one comapny for controlling Lepodoptera.

Chemicals
  Search for new modes of action

Bacillus thuringiensis
  AGREE, based on a strain produced by conjugation, went on sale in 1992

Pheromones
  SIRENE, a mixture of a pheromone, cypermethrin, carbon and a UV absorber applied as a droplet to control cotton bollworm in Egypt (six- to ten-fold reduction in insecticide and better control).

Macroorganisms for biological control
  Trichogramma – with Ontario government 5-year and $Can 7 million programme for control of spruce budwrom. Key questions: can the production problems and costs be improved?

Ciba–Biosys agreement – EXHIBIT based on the nematode *Steinernema carpocapsae*, effective against soil larvae

Transgenic crops
  Maize effective against *Ostrinia*

UV, ultraviolet

Products from microorganisms

In 1992 Ciba launched its first *Bacillus thuringiensis* product in the USA, under the trade name AGREE. The strain was constructed by the transfer through conjugation of the *cry1A(c)* gene from HD-191 to HD-135, which has *cry1A(a)*, *cry1C* and *cry1D* (Jarret and Burges, 1988). Current research involves looking for further improved strains and for genes for δ-endotoxins, in which biotechnology plays an important part. The product AGREE is also an example of how developments in packaging technology can contribute to providing user-friendly products. Unlike previous *B. thuringiensis* powders, this one has been formulated and adapted so that it can be sold in a water-soluble bag that can be dropped straight into the spray tank by the operator. The product thus needs minimum intervention by the user.

Transgenic crops

Ciba Seeds is primarily a maize company, and maize in all parts of the world is subject to attack by lepidopteran larvae that bore into the stem. The two major species are *Ostrinia nubialis* (European corn borer (ECB)) and *Diatrea saccharalis* (sugarcane borer), which between them cause an average annual estimated loss in yield of 5–7% worldwide, which may exceed 20% in certain environments. A truncated version of the *cry1A(b)* gene of *B. thuringiensis* has been completely synthesized, so that the deoxyribonucleic acid (DNA) codon usage reflects that used by maize (Koziel *et al.*, 1993b). This leads to expression of the protein in maize. This expression has been driven by different promoters, leading, in one case, to tissue-specific expression with only low amounts of protein in the seed (Koziel *et al.*, 1993a). Maize inbreds and hybrids have been field-tested in several locations and shown to be protected against attack by both ECB and sugarcane borers. In 1993, Ciba was granted an experimental use permit from the US Environmental Protection Agency to allow sufficient maize to be grown to give material for the necessary trials for the registration process.

The pest-resistant recombinant DNA products currently being tested are just the first of a series which will become more sophisticated with time. There are also many other insecticidal proteins that can be introduced. Further sophistication can be introduced by the use of different promoters, as research into the precise spatial, temporal and developmental control of gene expression in plants is expanding rapidly.

Summary

It is clear that not all these remedies will be applied at the same time or in the same environment; perhaps not all of them will make their way to the market. If they do, they will increase the spectrum of choice available to the farmer, the regulator and society, and increase the possibilities for IPM. Different approaches put considerable strain on a company to manage competing resources and call for Ciba researchers to master a range of technologies; biotechnology can clearly play a part in the R & D and the use of these different approaches but, in the end, it is only one tool among many others. The important thing for IPM in these approaches is the product and how it can be used; it is much less relevant to focus on the use or, conversely, the avoidance of any one tool. However, 'biotechnology' has attracted such a focus, and industry is expected to have a commitment for or against biotechnology. In the current political environment, it is important to address the use of biotechnology, but in any such discussion it is important to keep in mind the broad context in which biotechnology may be used – as one tool among many – in the development of effective strategies to combat pest and disease attack and to produce healthy plants.

Role of Biotechnology in Crop Protection

Within this context of diverse and competing solutions to agricultural problems, biotechnology may be important in several different ways, some of which are reviewed below.

Improvement of living products

Higher plants

The natural state of crop plants is resistance to or tolerance of pests and diseases. The ability of plants to look after themselves is the first weapon in pest management. Opportunities for plant-protection chemicals in insect and disease control only arise when the natural defences of the crop plants have been overcome. Plant breeders attempt to combat pests by introducing new sources of plant resistance, i.e. to reintroduce tolerance or resistance, but progress has been slow and the amount of variation available limited. The breeding rate and genetic skills of the pests have also ensured that they have often been able to overcome the introduced resistance by manipulation of their own composition. The advent of modern biotechnology has come about because we have devised techniques based on our knowledge of how microorganisms bring about genetic change. This technology gives some hope for more emphatically redressing the balance in favour of the crop plant. These technical developments affect both the speed and the accuracy with which genes can be moved within a given crop species, as well as increasing the range of variation that can be introduced into plants. The speeding up of conventional breeding depends on the enhancements brought by the use of genetic markers, such as restriction fragment length polymorphisms (RFLPs) (Prince and Tanksley, 1992) or random amplified polymorphic DNA markers (RAPDs) (Williams *et al.*, 1990). Marker technology is advancing rapidly and new opportunities for improvement arise regularly. These markers can be highly effective in introgressing resistance genes from wild relatives of crops, but they are also important in rapidly moving new traits, introduced by recombinant DNA techniques, into locally adapted varieties of the crop. Some of the traits that can be introduced into crop plants are described elsewhere in this volume.

It is my opinion that we have just scratched the surface of possibilities for pest and disease control, and that further research will reveal sophisticated and multiple mechanisms both for improving the yield of our crop plants and for making better use of the resources (soil, water, fertilizers, fossil fuel, etc.) needed for food production. This increasing improvement will undoubtedly be needed, because the pest and disease organisms will continue to evolve and change. These may be changes within organisms, such that they overcome the defence mechanisms (i.e. resistance develops), or changes in the pest populations within a given environment. For example, in Europe there has been a considerable change in the importance of the various disease organisms affecting cereals over the last 30 years.

Microorganisms

The major microbial inoculants sold in agriculture have been rhizobia to enhance nitrogen fixation. Although our knowledge of nitrogen fixation has been tremendously enhanced by studies on the molecular genetics of rhizobia, as far as I know none of the strains sold have been manipulated by recombinant DNA techniques. Nevertheless, improved strains depend on selection based on improved genetic knowledge. Microbial inoculants to control disease have often been promoted as a desirable technology, but so far they have suffered badly from unreliability in the field and are not widely used. Recently, the use of recombinant DNA technology has increased our knowledge of how these organisms bring about their effects. For example, recent studies have shown that *Pseudomonas* spp. produce a range of antimicrobial compounds, including inorganic chemicals (e.g. hydrocyanic acid (HCN)), organic chemicals (e.g. antibiotics like pyrrolnitrin) and proteins (e.g. chitinases). The production of this battery of active principles is directed by a group of genes under coordinated control, and the controlling elements have been identified (Gaffney *et al.*, 1992; Laville *et al.*, 1992). In the future, it will be possible to vary the components of the group of genes and of their control in such a way that more effective and stable control of disease can be obtained. Current research is also investigating the genetic aspects of root colonization and the relationships between the host root and the biological-control microorganism.

Improvement of the production of compounds by fermentation

The origins of biotechnology lie in fermentation processes, so it is no surprise that, as recombinant DNA technology has developed, it has been used in fermentation. In the pharmaceutical industry several products, chiefly proteins, are now on the market derived from fermentation of recombinant DNA organisms. This technology will have an impact on the agricultural industry, probably through the production of powders of strains of *B. thuringiensis* or through the production of the δ-endotoxins in other organisms. An example of the latter is Mycogen's M-cap technology, in which the proteins are produced in a non-pathogenic strain of *Pseudomonas fluorescens*. The bacterial cells are killed and chemically cross-linked and then used as a foliar application (Barnes and Cummings, 1987). Future uses of biotechnology will probably also include the production of secondary metabolites with biological activity.

Besides the use of biotechnology to improve the active principles used in pest and disease control, the technology will also improve the efficiency, effectiveness and quality of fermentation processes. In particular, universal host strains (e.g. of *B. thuringiensis*) are being developed that are optimized for fermentation and the genes for different active components transformed into them. The use of DNA-based probes and monoclonal antibodies will also be important in ensuring consistent quality in the fermentation of products.

Improvement in the selection and design of chemicals

Studies of structure–function relationships between chemicals and proteins are commonplace in biology and have been used to advance our knowledge of pharmaceuticals and how they act. So far, biotechnology has been little used in the design of agrochemicals. This is understandable, since the effects of target chemicals can be directly monitored by efficient *in vivo* screening systems. These systems effectively integrate the effects of uptake, transport and interaction with the target site into one test. Such screening has also been very effective in identifying a wide range of molecules with activity as crop-protection agents. In the face of such success, there are very few molecules that have been identified by 'rational' means. In cases where we know a great deal about the molecular nature of the target site, such as enolpyruvyl shikimate phosphate (EPSP) synthase, the target for glyphosate, or acetohydroxyacid synthase (AHAS), the target for the sulphonylureas, imidazolinones and other herbicides, there is no evidence that this information has yet led to the discovery of new active molecules. However, societal and environmental constraints are demanding ever more precise and selective remedies; the rate at which new lead structures are being discovered in some areas is slowing down, and the number of chemicals that must be screened to find a suitably active molecule is increasing. Increasing regulatory demands increase the number of selection criteria for chemicals, and thus fewer are identified which satisfy all the criteria. All of these factors create pressure to find new ways to identify active molecules or to extend the usefulness of existing ones. A clear example of the latter being approached by biotechnology is the development of herbicide-tolerant crops, particularly to compounds such as glyphosate and phosphinotricin.

Improvement in the interaction between higher plants and crop-protection chemicals

In some senses, the genetic make-up of crop plants is not part of IPM but is rather the substrate upon which IPM is practised. This is because plant genes are not readily subject to external influence and cannot be 'managed' in the way envisaged by IPM. However, this is no longer true in the conceptual sense, because of two developments.

Firstly, White (1979) discovered that the systemic acquired-resistance phenomenon in plants could be stimulated by aspirin (acetylsalicylic acid). This phenomenon at that time appeared to be mediated by the formation of new pathogenesis-related (PR) proteins, presumably due to gene action. This has subsequently been shown to be the case for salicylic acid and other chemicals (Ward *et al.*, 1991). Some of these chemicals are in development by Ciba for commercial use. This opens up the use of chemicals to trigger internal plant defences, alongside the use of chemicals to directly attack the pests, and thus increases the possible choices for pest management.

Secondly, it has proved possible to isolate the controlling elements from some of the PR genes and to link them upstream to other genes. These genes may then be turned on by the application of the external chemical. Williams *et al.* (1991) have shown that this may be done to control the expression of *B. thuringiensis* genes in tobacco. In principle, this adds another layer of potential control for IPM, in which genes introduced by recombinant DNA methods may be controlled. For example, it might be envisaged that the *B. thuringiensis* gene only becomes active in the plant when the plant is sprayed by a chemical inducer mixed with a second insecticidal active principle. This could have some advantages in resistance management. The system does have some potential constraints, in that the PR genes are induced by viruses etc. and are also developmentally regulated so that they are expressed at flowering, even in the absence of induction. There is also the need to have zero effective expression in the absence of inducer and a large response to the induction (for further discussion see Ward *et al.*, 1993). Nevertheless, existing and continued research will address these problems, and we may expect to see commercially useful developments that allow the expression of plant transgenes to be controlled in a manner consistent with a planned IPM strategy.

Improvement in the monitoring of pests

Intelligent pest management depends on a knowledge of which pests are present and of the dynamics of their build-up within the crop and its environment. Much pest control has, by necessity, to be either prophylactic or crisis management. Better knowledge of the build-up of pests and the most economic and effective timing of treatment, together with means to monitor the build-up in crop situations, would enhance the effectiveness of the use of crop-protection measures and probably minimize unwanted side-effects. Biotechnical methods are appropriate to the recognition of organisms by their DNA sequence and by using antibody-based technologies. This can be done over a wide range of specificities, according to need. The major constraints are the work and investment needed to devise methods sufficiently cheap and unsophisticated for field use. Where this is successful, the opportunities for more intelligent IPM are increased.

Summary

The scope for improving products, using the technology and knowledge arising from the rapid advances in molecular genetics (the new biotechnology), is immense. These products may be simple and relatively user-friendly – as in the case of crops improved by either direct addition of novel traits or more 'intelligent' breeding, or both. The use of improved seeds is likely to be equally applicable to small- and large-scale farmers of varying degrees of sophistication and to require no additional investment in machinery. These seeds are also applicable to IPM approaches as part of a package of measures. In addition, improved biological and chemical products may also be expected to arise, but there is no reason why these should be harder to use than existing products, although they are expected to be more appropriate to IPM. Finally, the advances in biotechnology may give rise to new, improved and more diverse strategies for pest monitoring and control, which will be more complex but also better and more sustainable approaches. The hope is that the technology will contribute to the increasing improvement of agricultural products. Ciba and other plant-protection companies have a commitment to biotechnology and are exploring and investigating some or all of the above opportunities. The question is not 'Will biotechnology have an impact on plant protection?' but 'When will this happen and will the commercial results justify the investment?' The latter concern is real and some of the factors that might work against the success of biotechnology are given below.

Table 22.4. Contrasting pressures on a manager in an insect-control company when choosing compounds for development.

IPM considerations	Commercial pressures
Selective	Broad market possibilities
Low-priced	Highly researched
No resistance problems	Long market life
Fast-acting	Not neurotoxic
Mobile in plants	Immobile in soil

IPM, integrated pest management.

Constraints on Investment in Biotechnology and IPM

As I pointed out earlier, industrial managers have to judge where to place their resources according to a range of criteria. In a free society, the manager responds as best he/she can to the motivations and restraints placed on his/her activities by the industrial system. If society as a whole, or a particular group within that society, wishes to change industrial behaviour, they can do this most effectively by understanding these motivations and restraints. Having earlier dealt with some of the motivating factors (Fig. 22.1 and Table 22.4), I should now like to cover some of the constraints.

Development costs

It currently takes about 10 years and many millions of dollars to bring a new chemical from time of first synthesis to widespread introduction on the market. It is, therefore, difficult for a company to invest in solutions for minor problems, which will only bring a small return. These problems have to rely on spin-off from major projects. This situation has reached such a state in the pharmaceutical field that the USA has introduced special conditions of market exclusivity to encourage commercial development of 'orphan drugs'. The reasons for the high cost of research are many, and include the extra research costs that go into improved screening (for primary and secondary effects of the chemical on the pest, the crop and the environment), into better formulations and application technology and into supporting a wider range of underlying technologies, including biotechnology. Development costs also increase because society requires more and more information on the products before their release; this is not to argue that the requirements are unfounded or unwise, but rather to state clearly that everything has a cost and this must be met through selling the products. If the size and value of the market are not sufficient to provide a return on the investment required for registration and development, then the product will not be developed for that market. Companies do not elect to invest in long-term R & D projects unless there is a reasonable expectation that the investments will be recovered, together with a profit equal to or better than that achieved from other opportunities. Investment in biotechnology is particularly difficult to plan because, as yet, the technology has not reached the market and the returns are very difficult to estimate.

The above situation works against many of the aims of IPM. Thus, the more selective a product is against a given pest, the narrower the range of products for which it is appropriate and the smaller the overall return that might be expected.

Acceptance of biotechnology

The use of modern biotechnology (recombinant DNA technology) in industry is under attack from critics in the industrial countries of North America and Europe. Surveys of public opinion indicate a mixed view, suggesting that people realize that biotechnology has a potential for benefits but that there is concern about risks and a desire that it be subject to regulation. However, several groups of critics have made it clear that it is their intention to stop the introduction of the products of biotechnology, irrespective of their benefits and whether or not they have met the rigorous standards set up by society. The activities of the critics and their campaigns have received wide coverage in the world's media. For example, it was reported that, in a series of radio, newspaper and TV advertisements, Jeremy Rifkin, a well known campaigner in the USA, said he was trying to halt commercialization of all genetically engineered food products (Kinsman, 1993). In Europe, there are small and large organizations that are taking direct action to try to prevent the field-testing of genetically modified organisms. DuBois (1992) wrote a detailed report of their activities in Europe, under the heading of 'Eco-terrorism'. These groups do not constrain their activities to legal and democratic political protest, but are willing to go outside the law and trespass on and/or physically destroy test crops that have been planted following all the agreed procedures and safeguards. Some of these critics are not so much against biotechnology *per se*, but against technology that they see worsening the exploitation of the Third World and supporting the existing agricultural structure in the European Community (EC). These groups are sophisticated enough to know and understand the motivations that govern industrial behaviour outlined above. Thus, not only do they seek to make the financial hurdles for the introduction of biotechnology as high as possible, but they also play upon the desire of the industrial manager and his workforce to be socially acceptable. Pressure is also indirectly put on investors and investment fund managers.

These remarks are not made by a defensive and embattled industrialist. Rather, they are made because I believe that the technology, properly researched, developed and applied, has much to contribute to the improvement of world agriculture. However, I also see many people who believe the same standing by silently. In the present climate surrounding the acceptance of biotechnology, it is important to realize that industry's (and perhaps even government's) continued investment in the improvement of plants and microorganisms by the introduction of new traits using recombinant DNA technology is at risk. This is one area in which progress has to be made if there is to be a strengthening of industry's commitment to the linkage of biotechnology and IPM. It does not matter that the first battle that is being fought concerns tomatoes for human consumption in the USA, with few implications for IPM. If the critics succeed in blocking the introduction of this product, then companies will clearly reconsider their investments in biotechnology.

In the face of such criticism what actions should be taken by those who desire to promote the use of biotechnology? A few suggestions follow (also see further discussions in Ingram, 1992; Miflin, 1992):

- Understand the motivations and reasons of the critics, differentiate between them and address them separately.
- Enter into a continued and involved dialogue with the critics and representatives of society.
- Explain the case for biotechnology with clear practical examples of how it can bring benefit.
- Ensure that these benefits are explained in terms that are relevant to the society at large, not just with reference to agricultural production.
- Be open about potential risks and drawbacks and counter these.
- Make non-technical people familiar with the products (not blinding them with the cleverness of the technology) and so enable them to feel they can participate in the debate.
- Realize who are the influencers in society that have to be convinced and ensure that they are addressed.

To give an example of explaining benefits, we have calculated the possible potential benefits of maize totally resistant to corn borers. We can explain this in terms of yield gain or resources saved. For example, total control of ECB in the USA and Europe gives 7–11 million more tonnes of grain, which is 66% of the production of France or the total production of Canada, or about twice the needs of France. Alternatively, if production in France is kept at the same level, total control of ECB in France saves over 200,000 ha of land, over 8 million litres of fossil fuel, 30,000 tonnes of fertilizer, $US12 million of pesticide and over 800 billion (thousand million) litres of irrigation water. Which is more relevant to the general public? The answer comes clearly from a recent survey in the UK, in which the researcher who carried out the survey, Joyce Tait, is quoted as saying, 'People said 'we don't need more milk, we have got enough already' . . . but they would look favourably on products which save resources or reduce consumption of

Table 22.5. Sustainable solutions – key questions.

Who pays for research?
What return do they expect and how can it be obtained?
Value of solutions – what is valuable gets looked after and used properly
Ownership by people in system (academics, industrialists, farmers, etc.)
Benefits and risks – regulations must be appropriate and consistent
Societal acceptance – society must be convinced that the benefits outweigh the risks; buy-in from society is essential
Trade barriers

fertilisers and pesticides' (Coglan, 1993). Whether ECB-resistant maize is used to produce more grain or to use fewer resources is dependent on the structure of the agriculture markets, which is in the realm of political and social decision-makers, not in the realm of industry and technology. However, unless we make all the benefits clear and explain the choices, the public and the decision-makers are not aware of the options (Table 22.5).

A further problem is who should communicate the benefits and risks to the general public. All surveys that I have seen suggest that people do not believe that industry is a reliable source of balanced information (e.g. Eurobarometer, 1991; Coglan, 1993). These surveys show that the groups most likely to be believed are the environmental and consumer pressure groups. Academics have a relatively comfortable position closer to the latter than the former. Industry has a lot to answer for in its past behaviour and has a heavy responsibility to become more open and transparent in what it does. It has an obligation to enter into serious dialogue on these issues with the public and with serious critics. However, those in academic and independent institutions who are informed about the issues and wish biotechnology to be of help in IPM and the developing world also have a duty to speak up and enter the debate with the critics, if they wish to see their ideas succeed.

Biotechnology regulations

Regulations governing the experimental and commercial release of recombinant DNA organisms are being proposed and implemented in the USA, Europe and elsewhere, even if this process appears to lag behind the pace at which products are coming forward. These regulations will involve industry in extra investments, which will be reflected in the cost of the products. However, if they are rationally based and meet the genuine concerns of the public, then they are unlikely to be a deterrent. In the context of IPM and developing countries, it is important that these countries develop regulations in parallel with the rest of the world. In this regard, the initiatives of several organizations such as the International Service for the Acquisition of Agri-biotech Application (ISAAA), various UN agencies and the World Bank in stimulating the introduction of recombinant DNA regulations in Latin America, Asia and Africa are of great importance. Some commentators have proposed that the regulations in developing countries are not as necessary, neither do they need to be as stringent, as in industrial countries (for example, see Hodgson (1990)). I do not agree, certainly not within the context of industrial involvement. If there are genuine reasons for the implementation of regulations in Europe and North America, then these should also hold in the rest of the world. Any other conclusion leads to double standards that widen the gaps between North and South, and would open industry to considerable criticism. Currently, Ciba breeds and sells conventional varieties of hybrid maize in Asia and Latin America. These conventional products will be supplemented by the development of appropriate recombinant DNA products (e.g. the current *B. thuringiensis* maize is capable of controlling sugarcane borers once the necessary locus has been transferred to locally adapted varieties), as soon as suitable regulations for the development and sale of products have been developed.

Intellectual-property protection

There is much current debate about intellectual-property protection (patents and plant-variety protection) with regard to biotechnology and biological control organisms. The debate surrounding the United Nations (UN) Convention on Biological Diversity and the Trade Related to Intellectual Property talks of the Uruguay Round of the General Agreement on Tariffs and Trade (GATT), and the various international dialogues (e.g. Keystone Center, 1991) have helped to identify some of the issues. Hopefully, the ratification of the Convention, and the establishment of detailed procedures that will follow, will ensure that biological resources worldwide are available to screen and develop. I believe intellectual-property protection requires a system that ensures that small or large companies, or private individuals, who invest in developing successful solutions retain sufficient of the benefits that these solutions bring, to give a fair return. This system should prevent companies and individuals from either, at one extreme, getting rich from the ideas and investments of others or, at the other, unfairly exerting monopoly power. The system should also publicize inventions so as to stimulate others to make improvements and develop alternative solutions. In the field of chemical plant-protection agents, the patent system has served to meet most of these needs reasonably well. Hopefully it can also be adapted to deal with biotechnology. Currently, this appears to be happening in the USA and Europe, although at a slow pace.

Return on investment for plant breeding

One important consulting firm in the seed industry calculated that, by 1990, companies had invested $US2.5 billion in plant biotechnology and were currently investing at the rate of $US250 million a year. Up to October 1993, not a single dollar has returned to the investors from the sale of recombinant DNA products. In many cases, the investors will never see a return because the technology was hyped up too soon and too greatly; this has resulted in many unwise investments. In many crops, it is still difficult to see how returns can accrue to the developer of improved varieties, because the nature of the farming industry does not provide sufficient returns for the breeders of inbred (self-replicating) crops. For example, in Europe breeders receive a return from royalties payable on seed; however, it was calculated in 1990 that in the UK the total royalty income for wheat and barley available was only about one-half of the costs of the breeding programmes in the country. If companies are to continue to do this, a system has to be developed to give them a fair return. In the USA many companies, including recently Pioneer, have withdrawn or reduced their wheat-breeding efforts. This might not have been so important in the past, when a large part of the investment in breeding in Europe and the USA was paid for by the taxpayer, but in recent years governments have considerably reduced or stopped their support for breeding. The situation becomes even more critical if these crops are to be improved by recombinant DNA means, which requires an even greater investment. Probably, the first test case will be the return that companies can obtain from the sale of recombinant DNA, insect-resistant cotton. Unless this is substantial and in some fair proportion to the savings the farmer makes on insecticides, it is unlikely that the investment will be forthcoming to support the research for the second-generation products that will undoubtedly be needed.

Conclusions

How do we strengthen industry's commitment to the linkage between biotechnology and IPM, and even to expand it? It is well recognized that biotechnology has a role to play in pest management and is well suited to IPM practices. Companies like Ciba are currently committed to making and selling solutions for agriculture in industrial and developing countries. They are already investing heavily in biotechnology, and undoubtedly further products suitable for IPM will be developed and released. However, if this investment and commitment are to be maintained in the long term, then the needs of the industry and their investors that I have outlined above have to be met.

One solution that has been proposed to the transfer of biotechnology to the Third World is to persuade industry to donate or to sell biotechnology to the developing world via the auspices of a foundation such as ISAAA (see Scott, 1993). While this is an attractive short-term measure, it is one that requires a lot of careful thought before being widely adopted. This is because it will only attack the symptoms and not address the long-term problems. Firstly, I think that the most useful long-term solutions are those that stimulate investment in, and returns from, technology in the developing countries, and these are greatly preferable to importing technology. Secondly, biotechnology is not a magic solution, and neither are the products produced from its application. Initially such products might appear to be magnificent remedies, but problems will re-emerge. For example, they may well invoke resistance in the pests unless carefully managed, they will suffer from the increase of other pest populations and they are unlikely to be permanent. In short, such products will require continued investment in R & D, tailored specifically to the problems of the region. It is not sure that this research will be forthcoming from industry under such a scheme. In fact, if a 'cheaper' solution 'donated' by one company competes unfairly with a solution being developed and sold in the region by another company, such that the second company does not get a fair return, the result may well be a withdrawal of investment by that company. To me, the situation is somewhat analogous to the importation of food to combat starvation. It is an understandable and necessary short-term remedy that may well obscure or endanger a longer-term approach to the problems.

Finally, I should like to raise the larger problem of world food production and to extend the question as to how to involve industry more in wider agricultural problems. The production of grain increased 2.6 times between 1950 and 1984, a growth rate of 3% per annum, greater than the growth rate of the world's population. Since agricultural areas increased much less, this means that the increases have been due to improvements in the yield of crops. Analyses for crops like maize, wheat and rice suggest that about half of this increase in yield has been due to genetic improvement and half to improvements in agricultural practices, including the use of irrigation, fertilizers and crop-protection chemicals. It is not easy to say how much of this progress came from advances in technology funded by the private sector and how much from public research. However, this progress has now stopped. Since 1984, the increase in grain yields has dropped to 1% per annum and is not predicted to keep up with the growth in population. Brown *et al.* (1993), from the Worldwatch Institute, produced an analysis of the situation which paints a very sombre picture for the future. Ironically, since 1984, it is very obvious that the investment in agricultural research by the public sector is diminishing rapidly. This has occurred in plant breeding and crop-production research, where American and European governments have reduced their investment dramatically. For example, in the UK, since 1983 the government has sold off the Plant Breeding Institute and reduced the number of scientists that it supports in agronomic research by about half. A parallel decline in industrial research investment does not appear to have occurred, and the relative importance in agricultural research of the industrial sector has increased. In addition, probably only industry has the know-how and resources to see a plant-protection agent through the current registration procedures. Since measures to limit population growth will take a long time to be effective, the growth in population predicted by Brown *et al.* (1993) will take place. This means that a vast increase in the world's food supply must also take place if population stabilization is not to come about through starvation. I cannot see how this increase can be achieved without industry playing a major part and its knowledge, skills, technologies and resources being utilized. The real challenge is thus how to involve industry, and under what conditions, in meeting this future need in the places where the need exists (i.e. in developing countries). This is a question that goes far beyond biotechnology, which can only play a minor, but important, part in solving the problems.

I do not think it appropriate for me, or for industry alone, to attempt to define in specific terms the conditions under which the commitment of industry can be increased. This has to be done by a wider community. I have, however, tried to outline some of the constraints and motivations that affect industry. I hope these are sufficient to stimulate the debate, within society at large, the agriculture establishment and the plant-protection industry, not only as to how to strengthen the commitment to the linkage between biotechnology and IPM, but also as to how the resources and technologies of industry can be used to meet the challenges of food production in the future.

Acknowledgements

I am grateful to Hilary Miflin, John Duesing and Elke Jarchow for several helpful discussions and suggestions during the preparation of this chapter.

Summary and Future Directions 23

Gabrielle J. Persley, Gary Toenniessen and Peter Dart

Introduction

This chapter provides our overview of the findings of the Bellagio Conference in terms of identifying key elements of IPM and the emerging new biotechnologies which may be able to contribute to sustainable pest management. It also identifies some international initiatives by international development agencies, through which these objectives may be pursued in the future. Crop loss through pests and pathogens is a highly complex and dynamic process. It is influenced by the genotypes of the crop population, the genetic composition and stability of pest and pathogen populations, the broader biological environment, including natural and introduced biocontrol agents, and the physical/geographical environment. There are interactions among all these factors, with existing populations influencing the evolution of future populations, all influenced by variable environments and the farmers' management practices. Biotechnology and new information technologies can help farmers and scientists to understand better this dynamic complexity and to deploy pest-management strategies more effectively.

Integrated pest management (IPM) is a system which utilizes all suitable methods, in a compatible manner, to maintain pest populations below levels causing economic injury. Pests include insects that attack crops, fungi, bacteria and nematodes that cause disease, and weeds. The choice of IPM action is defined by agroecological, socioeconomic and institutional factors. Effective choice depends on high-quality information about local crop ecosystems. Much research and development work is needed to develop techniques and approaches for gathering and interpreting timely and useful information, including the use of new technologies, such as geographical information systems (GIS), as a basis for decision-support systems for farmers and pest-control technicians. Once information is used to reach a decision, intervention can be based on a variety of specific techniques, either singly or in combination. Some are based on crop management techniques, including intercrops, rotations, planting times, mulching and flooding. Also, the development and deployment of resistant varieties, singly or in mixtures, is central to many IPM schemes. Information on the genetic and evolutionary variation within crop pest and natural enemy populations is essential for the management of varieties.

Integrated pest management programmes have been developed for many different food, cash-crop and forestry production systems. These programmes depend on farmers and pest-control technicians for continuing and careful assessments of the population of pests and their natural enemies, in order to decide if and when to intervene with alternative control measures. These may range from use of biological control agents and resistant varieties to carefully targeted application of chemical pesticides.

The key to successful implementation of IPM is the development by farmers and pest-control technicians of a practical and conceptually strong understanding, preferably based on their own local experiments and experience, of the ecology of their crops, their pests and their natural enemies. This understanding is then informed by complementary insights into the ecology and biology of production systems developed by agricultural research. Once combined, these kinds of knowledge can be translated into decision tools and practical control tactics to solve particular pest problems.

Biotechnology involves the use of living organisms, or parts of organisms, to make or modify specific products, to improve plant or animal production or to develop microorganisms for specific uses. Modern biotechnology encompasses an array of molecular techniques for manipulating or recognizing genetic variation, including recombinant deoxyribonucleic (DNA) techniques (the basis of genetic engineering), the use of monoclonal antibodies and new cell and tissue culture techniques. Much of the investment in modern biotechnology over the past 15 years has been directed at new technologies for pest and disease control, mainly in industrial countries. Several of these new biotechnologies are now at the stage of field evaluation and potential commercialization. The time is now opportune to consider their potential usefulness as part of an IPM approach.

Much of the development of appropriate, problem-orientated biotechnologies for IPM systems also requires an institutional agricultural research framework from which to develop a good understanding of the ecology and biology of the production system and to provide entomological, plant pathological and plant breeding support. The international agricultural research centres (IARCs), supported by the Consultative Group on

International Agricultural Research (CGIAR), have an important role to play here for many important crops, as also do advanced research institutes in several industrial countries with particular interests in IPM. The international IPM network established under the auspices of CGIAR offers an opportunity to strengthen IPM research, in support of IPM implementation.

Potential New Biotechnologies for IPM

New diagnostics

Several examples are available for using new diagnostic tools, based on monoclonal antibodies and molecular markers, to determine the evolutionary status and population dynamics of pests and pathogens, and also of the natural enemies of pests. The information generated, combined with molecular tags for host-plant resistance genes, is enabling breeders to develop new varieties, with a collection of resistance genes, selected to be effective against the specific pest populations present in a particular country or region. When these varieties are deployed in an IPM programme, it should be possible to use these same tools to monitor changes in populations of pests and of beneficial organisms. The deployment strategy can then evolve in ways that prolong the effectiveness of resistant varieties and maximize the benefits of other components of the IPM system.

In one elegant example, a new diagnostic tool is being used to monitor the local movement of a pest population. This had led to the identification of an essential secondary host, which could be eliminated by changing agronomic practices, thus solving the pest problem and eliminating the extensive use of pesticides. The movement of pests over broad geographical areas can similarly be monitored. In some cases it should be possible to predict a pest outbreak caused by new pests invading a particular region. Varieties resistant to these pests and other pest-management strategies could be deployed as a preventive measure.

In the short term, use will be made of new diagnostic techniques for identifying pests, detecting and measuring pesticides and identifying products produced by transgenic plants. These new diagnostics are developing rapidly, in part as a result of the developments in human health care. Some of these processes are rapid, inexpensive and robust enough for field use, such as the immunological assays based on the enzyme-linked immunosorb ent assay (ELISA). Nucleic acid-based methods are also being developed for field use. In general, the development costs for these diagnostics will need to be supported by public funding for developing countries. They will be useful: (i) for identifying pests; (ii) in enabling farmers to visualize the situation in the field, with respect to the presence of the pest or the disease;

(iii) in monitoring pest movement in time and space; (iv) in resistance breeding by simplifying or eliminating the need for disease bioassay; and (v) in monitoring pesticide contamination in the postharvest produce.

Novel products from genetic engineering

Using tools of genetic engineering, it is now possible to add specific genes (transgenes) to most microorganisms, many crop plants and some insects. This has greatly enhance the power of plant breeding. It should similarly enhance the prospects for and effectiveness of using biocontrol agents. It may even allow for the favourable modification of pest populations.

Novel plant virus resistance

Several examples are available on the development of transgenic plants for control of pests and diseases. Use of transgenes as a new source of resistance to plant viral infection is considered particularly promising and should help to reduce the use of insecticides for control of insect vectors of viral pathogens in many crops. Plants can be protected from infection by more than 20 viruses by genetically engineering them to develop virus coat protein-mediated protection (CPMP). Not all plant viruses are amenable to this process, and other systems of producing virus-resistant, transgenic plants are being sought, particularly for the cereals. Several systems, such as control of papaya ringspot virus and some potato viruses, for example, are in the field-test stage. Risks from deployment of such plants in IPM systems are considered to be small, in the light of field tests of several transgenic crops modified for their resistance to selected viruses, in several countries.

Novel insect resistance

Transgenic plants that have some resistance to certain lepidopteran insects are now being field-tested, for crops such as cotton, maize and potato, in several countries. The deployment of transgenic plants will need to be carefully managed, in an IPM context, to maintain natural enemies and prevent the rapid development of insect resistance. Risk/benefit analyses are important aspects of any biotechnological intervention

Use of transgenes to give crop plants resistance to insect pests is controversial, particularly in the case of transgenes for toxins, such as those which naturally occur in the microbial insecticide *Bacillus thuringiensis*. The concern is that these transgenes will be no more durable than other major genes for insect resistance, and that, if the host-plant resistance breaks down, the usefulness of naturally occurring or genetically engineered *B. thuringiensis* as a control agent may also be lost. In recognition of this potential problem, research is being conducted on transgene design options and transgenic crop deployment strategies that can prolong the usefulness of these genes within IPM systems. Inducible transgenes that are expressed only upon predation, tissue-specific expression, use of mixed cultivars, trap crops and other types of refuges are all strategies being explored for the use of novel *B. thuringiensis* genes. These strategies will require farmer participation in order to be successful. The basic idea is to maintain a low level of the natural pest population, which causes no or minimal economic loss, generates no selective pressure to overcome resistance and maintains the natural enemies. If this works for *B. thuringiensis* genes, it may also provide strategies for more effective deployment of naturally occurring major genes for resistance. Effective gene deployment will be needed to maximize the usefulness of novel pest-resistant transgenic plants in IPM systems.

In the future, transgenes for pest resistance are likely to become more sophisticated than the various genes for toxins being employed today. Rather than killing pests, they may reduce the ability of pests to recognize and/or colonize the crop, or they may modify the damage-causing behaviour of the pests. For example, transgenes are being developed that produce a protein which inactivates the binding site in insect vectors where plant viral pathogens nominally bind. There is research under way aimed at introducing similar antivectoring transgenes into the insects themselves, either directly or indirectly via symbiotic bacteria, and then using genetics to force the spread of these genes through the insect population.

New biological control agents

Another type of new technology now available is the use of biologically produced pesticides (biopesticides), such as *B. thuringiensis*, pheromones as attractants to traps or as mating disruptants, and biological control agents. Several different approaches are being developed for biological control. For insects, pathogenic viruses, bacteria, fungi and nematodes, parasitoid wasps and other natural enemies are in current use or near to deployment in the field in many countries. For instance, entomopathogenic baculoviruses are being extensively used in Brazil on about 7 million ha to control insect pests on soybeans.

Research on improved production systems for biopesticides and improved strains through selection and genetic engineering is occurring in several industrial-country laboratories in both the public and private sectors. Multiplication and formulation of these biocontrol agents to be robust enough for field use are difficult and often labour-intensive. It is likely that novel biopesticides will be of most commercial interest to small companies. Bacterial and fungal inoculants have also been developed and released commercially, on a small scale, as biological control agents for plant fungal and nematode diseases.

Biocontrol agents are being used increasingly as alternatives to chemical pesticides. They have proven to be particularly effective in dealing with exotic pests causing havoc in areas where their natural enemies do not exist. Examples of classic biological control are where the introduction of one or two carefully selected natural enemies has brought a devastating pest problem under control on a region-wide basis (e.g. cassava mealy bug in Africa). The ability of most biocontrol agents to self-renew is critical to their success in developing countries.

Biotechnology can help make these 'biopesticides' an even more effective component of IPM systems. Production of such agents could be developed as a small-scale biotechnology industry in some developing countries. Biotechnology may have a particular role in making production less expensive and formulation and storage more effective. Transgenic biopesticides may prove faster-acting and broader-spectrum than conventional products, but care must be taken not to sacrifice the valuable, self-renewing nature of biological control agents in developing more biopesticide-like products, which require repeated applications.

Resistant plant varieties

A key and continuing need is for ecological, social and systems analysis of problems in production systems, to identify the technologies that may be of value to farmers, including new plant varieties. The breeding of resistant plants is an important component in IPM and will be assisted by use of molecular-marker technologies to 'tag' resistance genes during the process and to enable development of durable, multigenic resistance.

Future Directions

Effective IPM programmes combine various methods of pest control into an integrated management system. Emphasis is placed on farmers assessing and monitoring pest populations and on bringing stability to the system by avoiding overdependence on any single method, especially chemical pesticides. However, this does not mean that farmers using IPM should not have access to the most robust new pest-control methods becoming available through advances in biotechnology. Rather, it means that these new tools need to be designed and deployed such that they strengthen IPM and give farmers greater management control over their production systems. To accomplish this, advocates and practitioners of IPM must work with biotechnologists in the design phase to identify the priority problems in the field, and must assume the major responsibility for evaluating and implementing deployment strategies of potentially useful novel products of biotechnology, within an integrated approach to pest management.

International IPM facility

An important follow-up to the Bellagio Conference was the establishment in 1995 of an international IPM facility, cosponsored by the World Bank and three United Nations agencies: the Food and Agriculture Organization (FAO), the United Nations Development Programme (UNDP) and the United Nations Environment Programme (UNEP).

The proposal arose from the 1994 report of an interagency task force on IPM, composed of representatives of CAB INTERNATIONAL, FAO, UNDP, UNEP and the World Bank. The task force reported that, although a number of effective programmes are under way, for example on rice in Asia, the realization of IPM at the farmer level has proceeded too slowly relative to the number of opportunities available. The actual implementation of IPM activities has proved to be complex, requiring specialized technical expertise, detailed analysis of policy, social and economic issues at the national and local level, political commitment and careful attention to detail in project design and implementation. Also, resources have not always been available in sufficient quantities and at the appropriate times.

These problems suggested that a mechanism was needed to draw upon local, national and international expertise, knowledge and resources to facilitate the process of project identification, design and implementation. The creation of such a mechanism would provide for more rapid and effective response by international agencies to the needs of farmers and national IPM programmes, and greater efficiency of resource use by governments and development agencies.

The facility is being established under the auspices of the current FAO/World Bank cooperative programme. It is attached to FAO for administrative purposes and CAB INTERNATIONAL is assisting FAO in the implementation of its programme. The aims of the facility are: to identify, with national partners, areas of high priority for IPM implementation; provide technical assistance in the design and monitoring of IPM projects, including those supported by World Bank loans and credits; facilitate funding of new projects; assist in the preparation of national IPM programmes for national and external support; and constitute a service of advice to national IPM programmes on all aspects of IPM policy development and implementation.

The functions of the IPM facility are, specifically, to:

1. catalyse and facilitate consultations and collaboration among national policy-makers, development agencies and non-government organizations on the planning and implementation of IPM activities;
2. stimulate the development of improved IPM concepts and practices through scientific research and other strategies that increase the participation of farmers, extension agents and on-farm researchers;

3. identify, assemble, and assist with the preparation of promising, high-priority pilot and large-scale projects for investment by national, bilateral and multilateral sources;
4. facilitate and promote the implementation of a small set of pilot projects leading to larger national programmes, through the provision of technical and financial support and training;
5. document, analyse and evaluate IPM pilot projects and other experiences in order to provide best practices, policy and management options to improve the quality of IPM initiatives and to accelerate and expand the uptake of IPM;
6. advise and assist national programmes on the design, implementation and evaluation of IPM programmes; and
7. identify specific scientific, technical, social and political constraints affecting IPM implementation, and propose means of removing such constraints.

The facility will also provide a means for wider adoption of IPM technologies emerging from the research conducted by advanced research institutes, including the IARCs. There needs to be effective means by which emerging technologies can be taken up and used widely by national governments and international development agencies. Integrated pest management has been indentified by a recent CGIAR Task Force on Sustainable Development as one of the priority areas for sustaining agricultural production in the next decades.

The facility came into operation in July 1995. Its cosponsors are actively seeking the participation of all the key players in IPM and biotechnology, in both the public and private sector and the non-governmental community, to facilitate the wider adoption of IPM. The facility will also provide one mechanism for the field evaluation of some of the emerging biotechnologies for pest management described in this volume.

References

Abdul-Sattar, A.A. and Watson, T.F. (1982) Survival of tobacco bud worm (Lepidoptera: Noctuidae) larvae after short term feeding periods on cotton treated with *Bacillus thuringiensis*. *Journal of Economic Entomology* 75, 630–632.

Abot, A.R. (1993) Avaliação da resistência de *Anticarsia gemmatalis* Hubner, 1818 (Lepidoptera: Noctuidae) ao seu virus de poliedrose nuclear, *Baculovirus anticarsia*. MSc thesis. Universidade Federal do Parana, Curitiba, Brazil.

Abou-Jawdeh, Y. (1992) Pests of vegetables in Africa: pest control practices and prospects for integrated pest management. FAO Background Paper distributed at: TCP/RAF/2256(T) – Regional Workshop on Development and Application of IPM for Vegetable Production in Africa, 23–30 November 1992, Dakar, Senegal, 36 pp.

Adang, M.J., Brody, M.S., Cardineau, G., Eagan, N., Roush, R.T., Shewmaker, C.K., Jones, A., Oakes, J.V. and McBride, K.E. (1993) The reconstruction and suppression of a *Bacillus thuringiensis cryIIIA* gene in protoplasts and transgenic potato plants. *Plant Molecular Biology* 21, 1131–1145.

Adhikari, T.B., Vera Cruz, C.M., Zhang, Q., Nelson, R.J., Mew, T.W. and Leach, J.E. (1995) Genetic diversity of *Xanthomonas oryzae* pv. *oryzae* in Asia. *Applied Environmental Microbiology* 61(3), 966–971.

Agarwal, S. and Kumar, K. (1992) Science, technology and women for sustainable environment and rural development. *International Journal of Sustainable Development* 1, 56–60.

Aggarwal, P.C. (1989) How natural farming is successful. *International Agricultural Development* 9, 14–15.

Aggarwal, P.C. (1991) Natural farming works: an Indian success story. *Third World Resurgence* 13, 39–40.

AGROW (1989) Facts and figures. *World Agrochemical News* September, 125–131.

AGROW (1992) Two faces of the Indian pesticide industry. *World Agrochemical News* 173, 21–23.

Agyen-Sampong, M. (1978) Pests of cowpea and their control in Ghana. In: Singh, S.R., Van Emden, H.F. and Taylor, T.A. (eds) *Pests of Grain Legumes: Ecology and Control.* Academic Press, London, pp. 85–92.

Akhurst, R.J. (1982) Antibiotic activity of *Xenorhabdus* spp., bacteria symbiotically associated with insect pathogenic nematodes of the families Heterorhabditidae and Steinernematidae. *Journal of General Microbiology* 128, 3061–3065.

Aldhouse, P. (1993) Malaria: focus on mosquito genes. *Science* 261, 546–548.

Altman, M. (1992) 'Biopesticides' turning into new pests? *Trends in Evolution and Ecology* 7, 65.

Alonso, P.L., Smith, T., Armstrong Schellenberg, J.R.M. *et al.* (1994) Randomised trial of efficacy of SPf66 vaccine against *Plasmodium falciparum* malaria in children in southern Tanzania. *Lancet* 344, 1175–1181.

Amanor, K.S. (1994) *The New Frontier. Farmer Responses to Land Degradation: a West African Study.* Zed Books, London.

Amanor, K.S. (1995) The folk farmer and the global monocrop economy: responses to degrading environment in Ghana. Presented at 13th International Plant Protection Congress, The Hague, 7 July 1995.

Amante-Bordeos, A., Sitch, L.A., Nelson, R., Dalmacio, R., Oliva, N.P., Aswidinnoor, H. and Leung, H. (1992) Transfer of bacterial blight and blast resistance from the tetraploid wild rice *Oryza minuta* to cultivated rice, *Oryza sativa. Theoretical Applied Genetics* 84, 345–354.

Amarasinghe, L.D. (1993) Entomopathogenic nematodes for control of the tea termite, *Postelectrotermes militaris* in Sri Lanka. PhD thesis, University of London.

Ammar, E.D. and Nault, L.R. (1991) Maize chlorotic dwarf viruslike particles associated with the foregut in vector and nonvector leafhopper species. *Phytopathology* 81, 444–448.

Ammar, E.D., Gingery, R.E. and Nault, L.R. (1993) Cytopathology and ultrastructure of mild and severe strains of maize chlorotic dwarf virus in maize and Johnsongrass. *Canadian Journal of Botany* 71, 718–724.

Anand, A., Fernandes, J., Arunan, M.C., Bhosekar, S., Chopra, A., Dedhai, N., Sequiera, K., Hasan, G., Palazzolo, M.J., VijayRaghavan, K. and Rodrigues, V. (1990) *Drosophila* 'enhancer-trap' transposants: gene expression in chemosensory and motor pathways and identification of mutants affected in smell and taste ability. *Journal of Genetics* 69, 151–168.

Ananthakrishnan, T.N. (1973) *Thrips: Biology and Control.* MacMillan India, Delhi, 120 pp.

Ananthakrishnan, T.N. (ed.) (1992) *Biotechnological Approaches to the Biological Control of Phytophagous Insects.* Entomology Research Institute, Madras, India.

Andersen, W.R. and Fairbanks, D.J. (1990) Molecular markers: important tools for plant genetic resource characterization. *Diversity* 6, 51–53.

Anderson, R.M. and May, R.M. (1981) The population dynamics of microparasites and their invertebrate hosts. *Philosophical Transactions of the Royal Society of London B* 291, 451–524.

Andrews, J.H. (1992) Biological control in the phyllosphere. *Annual Review of Phytopathology* 30, 603–635.

Andrews, K.L., Sanchez, R. and Cave, R.D. (1992) Management of diamondback moth in Central America. In: Talekar, N.S. (ed.) *Diamondback Moth and Other Crucifer Pests: Proceedings of the Second International Workshop, Tainan, Taiwan, 10–14 December 1990.* AVRDC Publication No. 92-368, pp. 487–498.

Angelichio, M.L., Beck, J.A., Johansen, H. and Ivey-Hoyle, M. (1991) Comparison of several promoters and polyadenylation signals for use in heterologous gene expression in cultured *Drosophila* cells. *Nucleic Acids Research* 19, 5037–5073.

Angst, M., Gugumus, F., Rist, G., Vogt, M. and Rody, J. (1992) European Patent EPA 376880.

Annis, P.C. (1990) Sealed storage of bag stacks: status of the technology. In: Champ, B.R., Highley, E. and Banks, H.J. (eds) *Fumigation and Controlled Atmosphere Storage of Grain: Proceedings of an International Conference.* ACIAR Proceedings No. 25, pp. 180–187.

Anon. (1990) Another look at viral inputs. *Agrichemical Age* October, 26–27.

Anon. (1991) Demonstration of IPM of diamondback moth on farmers' fields in the lowlands. In: *Progress Report.* Asian Vegetable Research and Development Center, Shanhua, Taiwan.

Anon. (1992) Call to save *Bt. Biocontrol News and Information* 13(2), 19N.

Anon. (1993a) B.t. management working group (BTWG). *Resistant Pest Management* 5, 2.

Anon. (1993b) *International Institute of Tropical Agriculture, Plant Health Division 1992 Annual Report.* IITA, Benin Station, Republic of Benin, 149 pp.

Anon. (1994) *Proceedings of the East/Central/Southern African Integrated Pest Management Implementation Workshop, 19–24 April 1993.* Harare, Zimbabwe, 159 pp. Natural Resources Institute, Chatham.

Apostol, B.A., Reiter, P., Black IV, W.C., Miller, B.R. and Beaty, B.J. (1992) Studies of oviposition behavior of *Aedes aegypti* using RAPD-PCR markers. Abstract. *American Journal of Tropical Medicine and Hygiene* 47(4) (Supplement).

Aquino, G. and Heinrichs, E.A. (1979) Brown planthopper populations on resistant varieties treated with a resurgence-causing insecticide. *International Rice Research Newsletter* 5, 12.

Ardates, E.Y., Leung, H., Vera Cruz, C.M., Leach, J.E., Mew, T.W. and Nelson, R.J. (1994) Hierarchical analysis of spatial variation of the rice bacterial blight pathogen across agroecosystems in the Philippines. *Phytopathology*, in press.

Arnheim, N., White, T. and Rainey, W.E. (1990) Application of PCR: organismal and population biology. *BioScience* 40(3), 174–182.

Aronson, A.I., Beckman, W. and Dunn, P. (1986) *Bacillus thuringiensis* and related insect pathogens. *Microbiological Review* 50, 1–24.

Arora, K., Rodrigues, V., Joshi, S. and Siddiqi, O. (1987) A gene affecting the specificity of chemosensory neurons of *Drosophila. Nature* 330, 62–63.

Artavanis-Tsakonas, S. and Simpson, P. (1991) Choosing a cell fate: a view from the *Notch* locus. *Trends in Genetics* 7, 403–408.

Arumugam, V. (1992) *Victims Without Voice: A Study of Women Pesticide Workers in Malaysia.* Tenaganita and PANS, Penang, Malaysia. 192 pp.

Ashburner, M. (1992) Mapping insect genomes. In: Crampton, J.M. and Eggleston, P. (eds) *Insect Molecular Science.* Academic Press, London, pp. 51–75.

Atkinson, P.W., Hines, E.R., Beaton, S., Matthaei, K.I., Reed, K.C. and Bradley, M.P. (1991) Association of exogenous DNA with cattle and insect spermatozoa *in vitro. Molecular Reproduction and Development* 23, 1–5.

Ayyub, C., Paranjape, J., Rodrigues, V. and Siddiqi, O. (1990) Genetics of olfactory behavior in *Drosophila melanogaster. Journal of Neurogenetics* 6, 243–262.

Backus, E.A. and Bennett, W.H. (1992) New AC electronic feeding monitor for time-structure analysis of waveforms. *Annals of the Entomological Society of America* 85, 437–444.

Baker, G.L. (1986) The ecology of mermithid nematode parasites of grasshoppers and locusts in south eastern Australia. In: Samson, A.R., Vlak, J.M. and Peters, D. (eds) *Fundamental and Applied Aspects of Invertebrate Pathology*. Foundation of the 4th International Colloquium of Invertebrate Pathology, Wageningen, The Netherlands, pp. 277–280.

Baker, T.C. (1985) Chemical control of behaviour. In: Kerkut, G.A. and Gilbert, L.I. (eds) *Comprehensive Insect Physiology, Biochemistry and Pharmacology*, Vol. 9, *Behaviour*. Pergamon, Oxford, UK, pp. 621–672.

Bakke, A. (1982) Utilization of aggregation pheromones for control of the spruce bark beetle. In: Leonhardt, B.A. and Beroza, M. (eds) *Insect Pheromone Technology: Chemistry and Applications*. American Chemical Society, Washington, D.C., pp. 219–227.

Balakrishnan, R. and Rodrigues, V. (1991). The *Shaker* and *shaking-B* genes specify elements in the processing of gustatory information in *Drosophila melanogaster. Journal of Experimental Biology* 157, 161–181.

Balderelli, R.M. and Lengyel, J.A. (1990) Transient expression of DNA after ballistic introduction into *Drosophila* embryos. *Nucleic Acids Research* 18, 5903–5904.

Ballinger, D.G. and Benzer, S. (1989) Targeted gene mutations in *Drosophila. Proceedings of the National Academy of Science USA* 86, 9402–9406.

Ballinger-Crabtree, M.E., Black, W.C. and Miller, B.R. (1992) Use of genetic polymorphisms detected by the random amplified polymorphic DNA polymerase chain reaction (RAPD-PCR) for differentiation and identification of *Aedes aegypti* subspecies and populations. *American Journal of Tropical Medicine and Hygiene* 47(6), 893–901.

Barillas-Mury, C., Graf, R., Hagedorn, H.H. and Wells, M.A. (1991) cDNA and deduced amino acid sequence of a blood meal-induced trypsin from the mosquito, *Aedes aegypti. Insect Biochemistry* 21, 825–831.

Barnes, A.C. and Cummings, S.G. (1987) Cellular encapsulation of pesticides produced by the expression of heterologous genes. US Patent.

Bateman, M.A. (1982) Chemical methods for suppression or eradication of fruit fly populations. In: Drew, R.A.I., Hopper, G.H.S. and Bateman, M.A. (eds) *Economic Fruit Flies of the South Pacific Region*. Queensland Department of Primary Industries, Brisbane, pp. 115–128.

Baumgärtner, J. and Gutierrez, A.P. (1989) Simulation techniques applied to crops and pest models. In: Cavalloro, R. and Delucchi, V. (eds) *PARASITIS 88. Proceedings of a Scientific Congress*, Barcelona, 25–28 October 1988. *Boletin de Sanidad Vegetal, Fuera de Sanidad Vegetal, Fuera de Serie* 17, 175–214.

Beard, C.B., Mason, P.W., Aksoy, S., Tesh, R.B. and Richards, F.F. (1992) Transformation of an insect symbiont and expression of a foreign gene in the

Chagas' disease vector *Rhodnius prolixus*. *American Journal of Tropical Medicine and Hygiene* 46, 195–200.

Beard, C.B., O'Neill, S.L., Mason, P., Mandelco, L., Woese, C.R., Tesh, R.B., Richards, F.F. and Aksoy, S. (1993a) Genetic transformation and phylogeny of bacterial symbionts from tsetse. *Insect Molecular Biology* 1, 123–131.

Beard, C.B., O'Neill, S.L., Tesh, R.B., Richards, F.F. and Aksoy, S. (1993b) Modification of arthropod vector competence via symbiotic bacteria. *Parasitology Today* 9, 179–183.

Beaty, B.J. (1992) *Annual Report to the MacArthur Foundation*. Network on the Biology of Disease Vectors (communicated by Denis Prager).

Beck, D.L., Van Dolleweerd, C.J., Lough, T.J., Balmore, E., Voot, D.M., Andersen, M.T., O'Brien, I.E.W. and Forster, R.L.S. (1994) Disruption of virus movement confers broad-spectrum resistance against systemic infection by plant viruses with a triple gene block. *Proceedings of the National Academy of Sciences of the United States of America* 91, 10310–10314.

Beck, E., Ludwig, G., Auerswald, E.A., Reiss, B. and Schaller, H. (1982) Nucleotide sequence and exact localization of the *Neomycin phosphotransferase* gene from transposon Tn5. *Gene* 19, 327–336.

Beckendorf, S.K. and Hoy, M.A. (1985) Genetic improvement of arthropod natural enemies through selection, hybridization or genetic engineering techniques. In: Hoy, M.A. and Herzog, D.C. (eds) *Biological Control in Agricultural IPM Systems*. Academic Press, Orlando, pp. 167–187.

Bedard, W.D. and Wood, D.L. (1974) Programs utilizing pheromones in survey and control. Bark beetles – the western pine beetle. In: Birch, M.D. (ed.) *Pheromones*. North-Holland, Amsterdam, pp. 441–449.

Bedding, R.A. (1972) Biology of *Deladenus siricidicola* (Neotylenchidae), an entomophagous nematode parasitic in siricid woodwasps. *Nematologica* 18, 482–493.

Bedding, R.A. (1984a) Nematode parasites of Hymenoptera. In: Nickle, W.R. (ed.) *Plant and Insect Parasitic Nematodes*. Marcel Dekker, New York, pp. 755–795.

Bedding, R.A. (1984b) Large-scale production, storage and transport of the insect parasitic nematodes *Neoaplectana* spp. and *Heterorhabditis* spp. *Annals of Applied Biology* 104, 117–120.

Bedding, R.A. (1988) Apparatus and method for rearing and harvesting nematodes. Australian Patent Application No. PJ 0630/88.

Bedding, R.A. (1990) Logistics and strategies for introducing entomopathogenic nematode technology into developing countries. In: *Entomopathogenic Nematodes in Biological Control*. CRC Press, Boca Raton, Florida, pp. 233–246.

Bedding, R.A. and Molyneux, A.S. (1982) Penetration of insect cuticle by infective juveniles of *Heterorhabditis* spp. (Heterorhabditidae: Nematoda). *Nematologica* 28, 354–359.

Bedding, R.A., Stanfield, M.S. and Crompton, G.W. (1991) Apparatus and method for rearing nematodes, fungi, tissue cultures and the like, and for harvesting nematodes. International Patent Application No. PCT/AU91/00136.

Bedding, R.A., Akhurst, R.J. and Kaya, H.K. (1993) Future prospects for entomogenous and entomopathogenic nematodes. In: Bedding, R.A., Akhurst, R.J. and Kaya, H.K. (eds) *Nematodes and the Biological Control of Insect Pests*. CSIRO Publications, Canberra.

Bedo, D.G. and Howells, A.J. (1987) Chromosomal localization of the white gene of *Lucilia cuprina* (Diptera: Calliphoridae) by *in situ* hybridization. *Genome* 29, 72–75.

Beeman, R.W., Brown, S.J., Stuart, J.J. and Denell, R.E. (1990) Homoeotic genes of the red flour beetle, *Tribolium castaneum*. In: Hagedorn, H.H. *et al.* (eds) *Molecular Insect Science*. Plenum Press, New York, pp. 21–29.

Begley, J.W. (1990) Efficacy against insects in habitats other than soil. In: Gaugler, R. and Kaya, H.K. (eds) *Entomopathogenic Nematodes in Biological Control*. CRC Press, Boca Raton, Florida, pp. 215–227.

Bellen, H.J., O'Kane, C.J., Wilson, C., Grossniklaus, U., Pearson, R.K. and Gehring, W.J. (1989) P-element mediated enhancer detection: a versatile method to study development in *Drosophila*. *Genes and Development* 3, 1288–1300.

Benbow, R.M., Zhao, J. and Larson, D.D. (1992) On the nature of origins of DNA replication in eukaryotes. *BioEssays* 14, 661–670.

Benbrook, C.M. and Marquart, D.J. (1993) *Challenge and Change. A Progressive Approach to Pesticide Regulation in California*. Sacramento, California.

Benedict, J.H., Sachs, E.S., Attman, D.W., Ring, D.R., Stone, T.B. and Sims, S.R. (1993) Impact of delta-endotoxin-producing transgenic cotton on insect–plant interactions with *Heliothis virescens* and *Helicoverpa zea* (Lepidoptera: Noctuidae). *Environmental Entomology* 22, 1–9.

Bennett, F.D. and Yaseen, M. (1980) Investigations on the natural enemies of cassava mealybugs (*Phenacoccus* spp.) in the Neotropics. CIBC report, CAB International, Wallingford, UK, mimeograph, 19 pp.

Bentley, J. (1994) Stimulating farmer experiments in non-chemical pest control in Central America. In: Scoones, I. and Thompson, J. (eds) *Beyond Farmer First: Rural People's Knowledge, Agricultural Research, and Extension Practice*. IT Publications, London.

Berg, D.E. and Howe, M.N. (eds) (1989) *Mobile DNA*. American Society of Microbiologists, Washington, DC.

Berger, P.H. and Pirone, T.P. (1986) The effect of helper component on the uptake and localization of potyviruses in *Myzus persicae*. *Virology* 153, 256–261.

Bernardo, M.A., Naqvi, N., Leung, H., Zeigler, R.S. and Nelson, R.J. (1993) A rapid method for DNA fingerprinting of the rice blast fungus *Pyricularia grisea*. *International Rice Research Newsletter* 18, 48–50.

Besansky, N.J. (1990) A retrotransposable element from the mosquito *Anopheles gambiae*. *Molecular and Cellular Biology* 10, 863–871.

Besansky, N.J. and Collins, F.H. (1992) The mosquito genome: organization, evolution and manipulation. *Parasitology Today* 8, 186–192.

Besansky, N.J., Bedell, J.A. and Collins, F.J. (1993) Cloning and characterization of the white gene of *Anopheles gambiae*. Abstract. In: 2nd International Symposium on Molecular Insect Science. Flagstaff, Arizona, USA.

Bhumannavar, P.S. *et al.* (1988) Evaluation of citrus germplasm for resistance to oriental red mite, *Eutetranychus orientalis* (Klein) under tropical humid South India conditions. *Tropical Pest Management* 34, 193–198.

Biassangama, A., le Rü, B., Iziquel, Y., Kiyindou, A. and Bimangou, A.S. (1989) L'entomocénose inféodée à la cochenille du manioc, *Phenacoccus manihoti* (Homoptera: Pseudococcidae), au Congo, cinq ans après l'introduction d'*Epidinocarsis lopezi* (Hymenoptera: Encyrtidae). *Annales de la Société Entomologique de France (NS)* 25, 315–320.

Bier, E., Vaessin, H., Shepherd, S., Lee, K., McCall, K., Barbel, S., Ackerman, L., Caretto, R., Uemura, T., Grell, E., Jan, L.Y. and Jan, Y.N. (1989) Searching for pattern and mutation in the *Drosophila* genome with a P-*lacZ* vector. *Genes and Development* 3, 1273–1287.

Billimoria, S.L. (1986) Taxonomy and identification of baculoviruses. In: Granados, R.R. and Federici, B.A. (eds) *The Biology of Baculoviruses*, Vol. I. *Biological Properties and Molecular Biology*. CRC Press, Boca Raton, Florida, pp. 37–59.

Bird, T.J. (1987) Fighting African cassava pests from the air. *Aerogram* 4, 6–7.

Bishop, D.H.I. (1986) UK release of genetically marked virus. *Nature* 323, 496.

Bishop, D.H.I., Entwistle, P.F., Cameron, I.R., Allen, C.J. and Possee, R.D. (1988) Field trials of genetically engineered baculovirus insecticides. In: Sussman, M., Collins, C.H., Skinner, F.A. and Stewart-Tull, D.E. (eds) *The Release of Genetically Engineered Micro-organisms*. Academic Press, New York, pp. 143–179.

Black, W.C. and Rai, K.S. (1988) *Cell* 25, 693–704.

Black, W.C. IV, DuTeau, N.M., Puterka, G.J., Nechols, J.R. and Pettorini, J.M. (1992) Use of the random amplified polymorphic DNA polymerase chain reaction (RAPD-PCR) to detect DNA polymorphisms in aphids (Homoptera: Aphididae). *Bulletin of Entomological Research* 82, 151–159.

Blanchetot, A. (1991) A *Musca domestica* satellite sequence detects individual polymorphic regions in insect genome. *Nucleic Acids Research* 19, 929–932.

Blissard, G.W. and Rohrmann, G.F. (1990) Baculovirus diversity and molecular biology. *Annual Review of Entomology* 35, 127–155.

Bonato, E.R. and Bonato, A.L.V. (1987) *A soja no Brasil: História e estatística*. Documentos 21, EMBRAPA-CNPSo, Londrina, Brasil, 61 pp.

Bonman, J.M. and Mackill, D.J. (1988) Durable resistance to rice blast disease. *Oryza* 25, 103–110.

Bonman, J.M., Khush, G.S. and Nelson, R.J. (1992) Breeding rice for resistance to pests. *Annual Review of Phytopathology* 30, 507–528.

Bonning, B.C. and Hammock, B.D. (1992) Development and potential of genetically engineered viral insecticides. *Biotechnology and Genetic Engineering Reviews* 10, 455–489.

Booker, R.H. (1965) Pests of cowpea and their control in northern Nigeria. *Bulletin of Entomological Research* 55, 663–672.

Borromeo, E.S. (1990) Molecular characterization of *Pyricularia oryzae* cv. populations from rice and other hosts. Ph.D Thesis, University of the Philippines at Los Baños. 124 pp.

Borromeo, E.S., Nelson, R.J., Bonman, J.M. and Leung, H. (1993) Genetic differentiation among isolates of *Pyricularia grisea* infecting rice and weed hosts. *Phytopathology* 83, 393–399.

Bos, L. (1992) New plant virus problems in developing countries: a corollary of agricultural modernization. *Advances in Virus Research* 41, 349–407.

Bosque-Pérez, N.A. and Buddenhagen, I.W. (1992) The development of host-plant resistance to insect pests: outlook for the tropics. In: Menken, S.B.J., Visser, J.H. and Harrewijn, P. (eds) *Proceedings of the 8th International Symposium on Insect-Plant Relationships*. Kluwer Academic Publishers, Dordrecht, The Netherlands.

Bottrell, D.G., Aguda, R.M., Gould, F.L., Theunis, W., Demayo, C.G. and Magalit, V.F. (1992) Potential strategies for prolonging the usefulness of *Bacillus thuringiensis* in engineered rice. *Korean Journal of Applied Entomology* 31, 247–255.

Boulter, D. (1992) Biotechnology and plant protection against insect pests. In: Sasson, A. and Costarini, V. (eds) *Plant Biotechnologies for Developing Countries*. Technical Centre for Agricultural and Rural Cooperation (CTA), Wageningen, The Netherlands, p. 195.

Boussienguet, J. (1986) Le complexe entomophage de la cochenille du manioc, *Phenacoccus manihoti* (Hom. Coccoidea Pseudococcidae) au Gabon. I. Inventaire faunistique et relations trophiques. *Annales de la Société Entomologique de France (NS)* 22, 35–44.

Bowen, A.T. (1991) Innovative IPM application technology. *Journal of Arboculture* 17, 138–140.

Bownes, M. (1992) Molecular aspects of sex determination in insects. In: Crampton, J.M. and Eggleston, P. (eds) *Insect Molecular Science*. Academic Press, London, pp. 76–100.

Boyle, L., O'Neill, S.L., Robertson, H.M. and Karr, T.L. (1993) Interspecific and intraspecific horizontal transfer of *Wolbachia* and *Drosophila*. *Science* 260, 1796–1799.

Boza-Barducci, T. (1972) Ecological consequences of pesticides used for the control of cotton insects in Canete Valley, Peru. In: Farva, M.T. and Milton, J.P. (eds) *The Careless Technology*. Natural History Press, New York, USA, pp. 423–438.

Bradley, J.R. Jr (1993) Influence of habitat on the pest status and management of *Heliothis* species on cotton in the southern United States. In: Kim, K.C. and McPherson, D.A. (eds) *Evolution of Insect Pests: Pattern of Variation*. Wiley, New York, pp. 375–392.

Braig, H.R., Guzman, H., Tesh, R.B. and O'Neill, S.L. (1994) Replacement of the natural *Wolbachia* symbiont of *Drosophila simulans* with a mosquito counterpart. *Nature* 367, 453–455.

Brand, A. and Perrimon, N. (1993) Targeted gene expression as a means of altering cell fates and generating dominant phenotypes. *Development* 118, 401–415.

Brewer, J.W., Capinera, J.L. and Cates, R.J. (1984) The relationship of plants and insects. In: Evans, H.E. (ed.) *Insect Biology, a Textbook of Entomology*. Addison-Wesley, London, pp. 214–251.

Broglie, R., Broglie, K., Roby, D. and Chet, I. (1993) Production of transgenic plants with enhanced resistance to microbial pathogens. In: Lynch, J. (ed.) *Benefits and Risks of introducing Biological Control Agents. Proceedings of a Workshop, 3–6 April 1992. Saariselka, Finland.* OECD, Paris.

Brogna, S., Bourtzis, K. and Savakis, C. (1993) The evolution and expression of the genes encoding alcohol dehydrogenase in *Ceritata capitata*. Abstract. In: *2nd International Symposium on Molecular Insect Science*. Flagstaff, Arizona, USA.

Brown, I. (1993) Molecular detection assays for plant pathogenic fungi. *Agbiotech News and Information* 5, 219–222.

Brown, L.R., Kane, H. and Ayres, E. (1993) *Vital Signs 1993: The Trends that are Shaping our Future*. W.W. Norton and Worldwatch Institute, USA.

Broza, M., Sneh, B., Yawetz, A., Oron, U. and Honigman, A. (1984) Commercial application of *Bacillus thuringiensis* var. *entomocidus* to cotton fields for the control of *Spodoptera littoralis* Boisduval (Lepidoptera: Noctuidae). *Journal of Economic Entomology* 77, 1530–1533.

Buchner, E. and Rodrigues, V. (1984) Autoradiographic localization of ^{3}H-choline uptake in the brain of *Drosophila melanogaster. Neuroscience Letters* 44, 25–31.

Buck, L. and Axel, R. (1991) A novel multigene family may encode odorant receptors: a molecular basis for odor recognition. *Cell* 65, 175–187.

Buddenhagen, I.W. (1977) Resistance and vulnerability of tropical crops in relation to their evolution and breeding. *Annals of the New York Academy of Sciences* 287, 309–326.

Buddenhagen, I.W. (1983a) Crop improvement in relation to virus diseases and their epidemiology. In: Plumb, R.T. and Thresh, J.W. (eds) *Plant Virus Epidemiology*. Blackwell Scientific Publications, Oxford, UK.

Buddenhagen, I.W. (1983b) Disease resistance in rice. In: Lamberti, F., Waller, J.M. and Van Der Graaff, N.A. (eds) *Durable Resistance in Crops*. Plenum, New York, USA.

Buddenhagen, I.W. (1983c) Agroecosystems, disease resistance and crop improvement. In: Thor, K. and Williams, P.H. (eds) *Challenging Problems in Plant Health*. American Phytopathological Society, St Paul, USA.

Buddenhagen, I.W. (1983d) Breeding strategies for stress and disease resistance in developing countries. *Annual Review of Phytopathology* 21, 385–409.

Buddenhagen, I.W. (1987) Disease susceptibility and genetics in relation to breeding of bananas and plantains. In: Persley, G.J. and de Langhe, E.A. (eds) *Banana and Plantain Breeding Strategies*. ACIAR Proceedings No. 21, Australian Centre for International Agricultural Research, Canberra, Australia.

Buddenhagen, I.W. (1991) Better cultivars for resource-poor farmers. In: *Proceedings of a Seminar on Crop Protection for Resource-Poor Farmers, Isle of Thorns Conference Centre, East Sussex, UK, 4–8 November 1991*. Natural Resources Institute, Chatham, UK, pp. 83–94.

Buddenhagen, I.W. (1992) *Prospects and Challenges for African Agricultural Systems: an Evolutionary Approach*. Prepared for Carter Lecture Series on Sustainability in Africa: Integrating Concepts, 9–11 April 1992, University of Florida, Gainesville, Florida, USA.

Burdon, J.J. and Roelfs, A.P. (1985) The effect of sexual and asexual reproduction on the isozyme structure of population of *Puccinia graminis. Phytopathology* 75, 1068–1073.

Burke, T., Dolf, G., Jeffrys, A.J. and Wolff, R. (1991) *DNA Fingerprinting: Approaches and Applications*. Birkhauser Verlag, Switzerland.

Caetano-Anolles, G., Bassam, B.J. and Gressoff, P.M. (1991) DNA amplification fingerprinting using very short arbitrary oligonucleotide primers. *Biotechnology* 9, 553–557.

Cagauan, A.G. (1990) Fish toxicity, degradation period and residues of selected pesticides in rice-fish culture. Paper presented at the Workshop on the Environmental and Health Impacts of Pesticide Use in Rice Culture. IRRI, Los Baños, Philippines, 3 pp.

Campbell, S., Inamdar, M., Rodrigues, V., VijayRaghavan, K., Palazzolo, M.J. and Chovnick, A. (1992) The *scalloped* gene encodes a novel evolutionarily conserved transcription factor required for sensory organ differentiation in *Drosophila. Gene and Development* 6, 367–379.

Campion, D.G. and McVeigh, E.M. (1984) Controlling insect pests with pheromones. *Span* 27, 100–102.

Campion, D.G. and Nesbitt, B.F. (1981) Lepidopteran sex pheromones and pest management in developing countries. *Tropical Pest Management* 27, 53–61.

Cano, R.J. and Poinar, H.N. (1993) Rapid isolation of DNA from fossil and museum specimens suitable for PCR. *BioTechniq* 15, 432–435.

Caprio, M.A. and Tabashnik, B.E. (1992) Gene flow accelerates local adaptation among finite populations: simulating the evolution of insecticide resistance. *Journal of Economic Entomology* 85, 611–620.

Caprio, M.A., Hoy, M.A. and Tabashnik, B.E. (1991) A model for implementing a genetically-improved strain of the parasitoid *Trioxys pallidus* Haliday (Hymenoptera: Aphididae). *American Entomologist* 34, 232–239.

Carde, R.T. (1986) Epilogue: behavioural mechanisms. In: Payne, T.L., Birch, M.C. and Kennedy, C.E.J. (eds) *Mechanisms in Insect Olfaction.* Clarendon, Oxford, pp. 175–186.

Cariño, F.O., Dyck, V.A. and Kenmore, P.E. (1982) *Role of Natural Enemies in Population Suppression and Pest Management of Green Leafhoppers.* IRRI Saturday Seminar, International Rice Research Institute, Los Baños, Philippines, 29 pp.

Carlson, J. (1991) Olfaction in *Drosophila*: genetic and molecular analysis. *Trends in Neuroscience* 14, 520–524.

Carminati, J., Johnston, C.G. and Orr-Weaver, T.L. (1992) The *Drosophila ACE3* chorion element autonomously induces amplification. *Molecular and Cellular Biology* 12, 2444–2453.

Carozzi, N.B., Warren, G.W., Desai, N., Jayne, S.M., Lotstein, R., Rice, D.A., Evola, S. and Koziel, M.G. (1992) Expression of a *chimerie CaMV 355 Bacillus thuringiensis* insecticidal protein gene in transgenic tobacco. *Plant Molecular Biology* 20, 539–548.

Carr, J.P., Marsh, L.E., Lomonossoff, G.P., Sekiya, M.E. and Zaitlan, M. (1992) Resistance to tobacco mosaic virus induced by the 54-kDa gene sequence requires expression of the 54-kDa protein. *Molecular Plant Microbe Interactions* 5, 397–404.

Caskey, C.T. (1987) Disease diagnosis by recombinant DNA methods. *Science* 236, 1223–1229.

Caspari, E. and Watson, G.S. (1959) On the evolutionary importance of cytoplasmic sterility in mosquitoes. *Evolution* 13, 568–570.

Castillo, M.R. (1995) IPM – institutional constraints and opportunity in the Philippines. Presented at 13th International Plant Protection Congress, The Hague, 6 July 1995.

Castro, L.A.B. (1992) Plantas transgênicas resistentes a insetos, perspectivas e limitações. *Pesquisa Agropecuaria Brasileira* 27, 319–324.

Cate, J.R. (1990) Biological control of pests and diseases: integrating a diverse heritage. In: Baker, R.R. and Dunn, P.E. (eds) *New Directions in Biological Control.* Alan Liss, New York, pp. 23–43.

Causse, M.A., Fulton, T.M., Cho, Y.G., Ahn, S.N., Chunwongse, J., Wu, K., Xiao, J., Yu, Z., Ronald, P.C., Harrington, S.E., Second, G., McCouch, S.R. and Tanksley, S.D. (1994) Saturated molecular map of the rice genome based on an interspecific backcross population. *Genetics* 138, 1251–1274.

Chambers, R. (1991) Problems of paradigms. In: Swaminathan, M.S. (ed.) *Biotechnology in Agriculture: A Dialogue.* Macmillan India Ltd., Madras, pp. 245–255.

Champ, B.R., Highley, E. and Banks, H.J. (eds) (1990) *Fumigation and Controlled Atmosphere Storage of Grain. Proceedings of an International Conference, Singapore, 14–18 February 1989.* ACIAR Proceedings No. 25, Australian Centre for International Agricultural Research, Canberra, Australia, 301 pp.

Chang, N.T. (1990) *Ceranisus menes* (Walker) (Eulophidae: Hymenoptera), a new parasite of bean flower thrips, *Megalurothrips usitatus* (Bagnall) (Thripidae: Thysanoptera). *Plant Protection Bulletin* 32, 237–238.

Chapco, W., Ashton, N.W., Martel, R.K.B. and Antonishyn, N. (1992) A feasibility study of the use of random amplified polymorphic DNA in the population genetics and systematics of grasshoppers. *Genome* 35, 569–574.

Charudattan, R. (1991) The mycoherbicide approach with plant pathogens. In: Tebeest O.D. (ed.) *Microbial Control of Weeds.* Chapman and Hall, New York, pp. 24–57.

Chemical Weekly (1992a) India likely to get US genetic engineering know-how. *Chemical Weekly* 13 October, 77.

Chemical Weekly (1992b) Sandoz to market biological pesticide. *Chemical Weekly* 20 October, 81.

Chemical Weekly (1992c) TAC to produce two biocides. *Chemical Weekly* 17 November, 89.

Chen, D.H., Zeigler, R.S., Leung, H. and Nelson, R.J. 1995. Population structure of *Pyricularia grisea* at two screening sites in the Philippines. Phytopathology 85, 1011–1020.

Cherbas, L., Lee, K. and Cherbas, P. (1991) Identification of ecdysone response elements by analysis of the *Drosophila* Eip 28/29 gene. *Genes and Development* 5, 120–131.

Chin, K.M. (1985) Virulence analysis as a tool in disease management. In: *Progress in Upland Rice Research. Proceedings of the 1985 Jakarta Conference.* International Rice Research Institute, Los Baños, Philippines, pp. 392–401.

Chunwongse, J., Martin, G.B. and Tanksley, S.D. (1993) Pre-germination genotypic screening using PCR amplification of half-seeds. *Theoretical and Applied Genetics* 86, 694–698.

Clapp, J.P., McKee, R.A., Allen-Williams, L., Jopley, J.G. and Slater, R.J. (1993) Genomic subtractive hybridization to isolate species specific DNA sequences in insects. *Insect Molecular Biology* 1, 133–138.

Claridge, M.F., den Hollander, J. and Morgan, J.C. (1982) Variation within and between populations of the brown planthopper *Nilaparvata lugens* (Stål). In: Knight, W.J., Patt, N.C., Robertson, T.S. and Wilson, M.R. (eds) *First International Workshop on Leafhoppers and Planthoppers of Economic Importance.* Commonwealth Institute of Entomology, London, pp. 306–318.

Cockburn, A.F. and Mitchell, S.F. (1989) Repetitive DNA interspersion patterns in Diptera. *Archives of Insect Biochemistry and Physiology* 10, 105–113.

Cockburn, A.F., Howells, A.J. and Whitten, M.J. (1989) Recombinant DNA technology and genetic control of pest insects. In: Russel, G.E. (ed.) *Management and Control of Invertebrate Crop Pests.* Intercept, Andover, Hampshire, pp. 211–241.

Coglan, A. (1993) Gene industry fails to win hearts and minds. *New Scientist* 19 June, 4.

Cohen, A.J., Williamson, D.C. and Oishi, K. (1987) SPV3 viruses of *Drosophila* spiroplasma. *Israeli Journal of Medical Science* 23, 429–433.

Cohen, D., Green, M., Block, C., Slepon, R., Ambar, R., Wasserman, S. and Levine, M. (1991) Reduction of transmission of shigellosis by control of houseflies *Musca domestica. Lancet* 337, 993–997.

Collinge, D.B. and Slusarenko, A.J. (1987) Plant gene expression in response to pathogens. *Plant Molecular Biology* 9, 389–410.

Collins, F.H., Sakai, R.K., Vernick, K.D., Paskewitz, S., Seeley, D.C., Miller, L.H., Collins, W.E., Campbell, C.C. and Gwadz, R.W. (1986) Genetic selection of a *Plasmodium* refractory strain of the malaria vector *Anopheles gambiae. Science* 234, 607–610.

Cook, A.G. and Perfect, T.J. (1985) The influence of immigration on population development of *Nilaparvata lugens* and *Sogatella furcifera* and its interaction with immigration by predators. *Crop Protection* 4, 423–433.

Cook, A.G. and Perfect, T.J. (1989a) Population dynamics of three leafhopper vectors of rice tungro viruses *Nephotettix virescens* (Distant), *N. nigropictas* (Stål) and *Recilia dorsalis* (Motschuleky) (Hemiptera: Cicadellidae), in farmers' fields in the Philippines. *Bulletin of Entomological Research* 79, 437–451.

Cook, A.G. and Perfect, T.J. (1989b) Population characteristics of brown planthopper *Nilaparvata lugens* in the Philippines. *Ecological Entomology* 14 1–9.

Cook, R.J. (1990) Twenty-five years of progress towards biological control. In: Hornby, D. (ed.) *Biological Control of Soil-Borne Plant Pathogens.* CAB International, Wallingford, Oxon., UK.

Cooley, R.N., van Gorcom, R.F.M., van den Hondel, C.A.M.J.J. and Caten, C.E. (1991) Isolation of a benomyl resistant allele of the β-tubulin gene from *Septoria nodorum* and its use as a dominant selectable marker. *Journal of General Microbiology* 137, 2085–2091.

Cooper, B., Lapidot, M., Heick, J.A. and Beachy, R.N. (1995) Multi virus resistance in transgenic tobacco plants expressing a dysfunctional movement protein of tobacco mosaic virus. *Virology* 206, 307–313.

Corey, J.S. (1991) Releases of genetically modified viruses. *Medical Virology* 1, 79–88.

Cornelissen, B.J. and Melchers, L.S. (1993) Strategies for control of fungal diseases with transgenic plants. *Plant Pathology* 101, 709–712.

Correa, B.S., Panizzi, A.R., Newman, G.G. and Turnipseed, S.G. (1977) Distribuição geográfica e abundância estacional dos principais insetos-pragas da soja e seus predadores. *Anais da Sociedade Entomologica do Brasil* 6, 40–50.

Correa, F.J. and Zeigler, R.S. (1991) Stable resistance and pathogenic variability in the rice – *Pyricularia oryzae* complex. In: Cuevas-Perez (ed.) *Rice in Latin America: Improvement, Management and Marketing.* CIAT, Cali, Colombia, p. 240.

Correa, F.J. and Zeigler, R.S. (1993) Pathogenic variability in *Pyricularia grisea* at a rice-blast 'hot spot' breeding site in Eastern Colombia. *Plant Disease* 7, 1029–1035.

Correa-Ferreira, B.S. (1980) Controle biológico de pragas da soja. In: Ramiro, Z.A., Grazia, J. and Lara, F.M. (eds) *Anais do XI Congresso Brasileiro de Entomologia*. Fundação Cargill, Campinas, pp. 277–301.

Correa-Ferreira, B.S. (1985) *Criação massal do percevejo verde, Nezara viridula*. Documentos 11, EMBRAPA-CNPSo, Londrina, Brasil, 16 pp.

Correa-Ferreira, B.S. (1986) Ocorrencia natural do complexo de parasitoides de ovos de percevejos da soja no Paraná. *Anais da Sociedade Entomologica do Brasil* 15, 189–199.

Correa-Ferreira, B.S. (1991) Parasitoides de ovos de percevejos: incidência natural, biologia e efeito sobre a população de percevejos da soja. PhD thesis, Universidade Federal do Paraná, Curitiba, Brazil.

Corso, I.C. (1990) *Uso do sal de cozinha na redução da dose de inseticida para controle de percevejos da soja*. Comunicado Tecnico 45, CNPSo-EMBRAPA, Londrina, 7 pp.

Cousins, Y.L., Lyon, B.R. and Llewellyn, D.J. (1991) Transformation of an Australian cotton cultivar: prospects for cotton improvement through genetic engineering. *Australian Journal of Plant Physiology* 18, 481–494.

Cox, J. and Williams, D.J. (1981) An account of cassava mealybug (Hemiptera: Pseudococcidae) with a description of a new species. *Bulletin of Entomological Research* 71, 247–258.

Crampton, J.M. (1992) Potential application of molecular biology in entomology. In: Crampton, J.M. and Eggleston, P. (eds) *Insect Molecular Science*. Academic Press, London, pp. 3–18.

Crampton, J.M. (1994) Approaches to vector control: new and trusted, 3. Prospects for genetic manipulation of insect vectors. *Transactions of the Royal Society of Tropical Medicine and Hygiene* 88, 141–143.

Crampton, J.M. and Eggleston, P. (1992) Biotechnology and the control of mosquitoes. In: Yong, W.K. (ed.) *Animal Parasite Control Using Biotechnology*. CRC Press, Boca Raton, Florida, pp. 333–350.

Crampton, J.M., Morris, A., Lycett, G., Warren, A. and Eggleston, P. (1990) Transgenic mosquitoes: a future vector control strategy? *Parasitology Today* 6, 31–36.

Crawley, M.J., Hails, R.S., Rees, M., Kohn, D. and Buxton, J. (1993) Ecology of transgenic oilseed rape in natural habitats. *Nature* 363, 620–623.

Critchley, B.R., Campion, D.G., McVeigh, L.J., Cavanagh, G.C., Hosny, M.M., Nasr, R.S.A., Khidr, A.A. and Naguib, M. (1985) Control of pink bollworm *Pectinophora gossypiella* (Saunders) (Lepidoptera: Gelechiidae), in Egypt by mating disruption using hollow-fibre, laminate-flake and microencapsulated formulations of synthetic pheromone. *Bulletin of Entomological Research* 75, 329–345.

Crowhurst, R.N., Hawthorne, B.T., Rikkerink, E.H.A. and Templeton, M.D. (1991) Differentiation of *Fusarium solani* f.sp. *cucurbitae* races 1 and 2 by random amplification of polymorphic DNA. *Current Genetics* 20, 391–396.

Cunningham, J.C. (1988) Baculoviruses: their status compared to *Bacillus thuringiensis* as microbial insecticides. *Outlook on Agriculture* 17, 10–17.

Curtis, C.F. (1968) Possible use of translocations to fix desirable genes in insect pest populations. *Nature* 218, 368–369.

Curtis, C.F. (1976) Population replacement in *Culex fatigans* by means of cytoplasmic incompatibility. 2. Field cage experiments with overlapping generations. *Bulletin of the World Health Organization* 53, 107–119.

Curtis, C.F. (1979) Translocations, hybrid sterility, and the introduction into pest populations of genes favourable to man. In: Hoy, M.A. and McKelvey, J.J. Jr (eds) *Genetics in Relation to Insect Management.* Rockefeller Foundation, New York, pp. 19–30.

Curtis, C.F. (1982) The mechanism of hybrid male sterility from crosses in the *Anopheles gambiae* and *Glossina morsitans* complexes. In: Steiner, W.W.M., Tabachnick, W.J., Rai, K.S. and Narang, S. (eds.) *Recent Developments in the Genetics of Insect Disease Vectors.* Stipes Publishing Company, Champaign, Illinois, USA, pp. 290–312.

Curtis, C.F. (1985) Theoretical models of the use of insecticide mixtures for the management of resistance. *Bulletin of Entomological Research* 75, 259–265.

Curtis, C.F. (1991) *Control of Disease Vectors in the Community.* Wolfe Publishing, London, 233 pp.

Curtis, C.F. (1992a) Selfish genes in mosquitoes. *Nature* 357, 450.

Curtis, C.F. (1992b) Making mosquitoes harmless. *Parasitology Today* 8, 305.

Curtis, C.F. and Graves, P.M. (1988) Methods for replacement of malaria vector populations. *Journal of Tropical Medicine and Hygiene* 91, 43–48.

Curtis, C.F., Hill, N. and Kasim, S.H. (1993) Are there effective resistance management strategies for vectors of human disease? *Biological Journal of the Linnean Society* 48, 3–18.

Dai, K., Zhang, L., Ma, Z., Zhong, L., Zhang, Q., Cao, A., Xu, K., Li, Q. and Gao, Y. (1988) Research and utilization of artificial host egg for propagation of parasitoid *Trichogramma* and other egg parasites. *Les Colloques de L'INRA* 43, 311–418.

Daitota, I. (1989) Indian farmer practices organic farming successfully. *Third World Network Features* 384, 2–5.

Daly, J.C. and McKenzie, J.A. (1986) Resistance management strategies in Australia: the *Heliothis* and 'Wormkill' programmes. In: *Proceedings of British Crop Protection Conference on Pests and Diseases.* BCPC, Brighton, UK, pp. 951–959.

Dantharayana, W. and Vitarana, S.I. (1987) Control of the live-wood termite *Glyptotermes dilatatus* using *Heterorhabditis* sp. (Nemat.). *Agriculture, Ecosystems and Environment* 19, 332–342.

Dashkova, N.G. and Rasnicyn, S.P. (1982) Review of data on susceptibility of mosquitoes in the USSR to imported strains of malaria parasites. *Bulletin of the World Health Organization* 60, 893–897.

Davidson, E.W. (1992) Development of insect resistance to biopesticides. *Pesquisa Agropecuária Brasileira* 27, 47–57.

Davies, W.P. (1992) Prospects for pest resistance to pesticides. In: Aziz, A., Kadir, S.A. and Barlow, H.S. (eds) *Pest Management and the Environment in 2000.* CAB International, Wallingford, UK, pp. 95–110.

Davis, M.T.B., Vakharia, V.N., Henry, J., Kempe, T.G. and Raina, A.K. (1992) Molecular cloning of the pheromone biosynthesis activating neuropeptide in

Helicoverpa zea. *Proceedings of the National Academy of Sciences, USA* 89, 142–146.

De Hoogh, J. (1987a) Agricultural policies in industrial countries and their effects on the Third World. A critical view on the comparative static analysis of a dynamic process. *Tijdschrift voor Sociaalwetenschappelijk Onderzoek in de Landbouw* 2, 68–81.

De Hoogh, J. (1987b) International division of labour in world agriculture. Reply to Alan Mathews, Rod Tyers and Kym Anderson. *Tijdschrift voor Sociaalwetenschappelijk Onderzoek in de Landbouw* 2, 330–334.

Demler, S.A. and de Zoeten, G.A. (1991) The nucleotide sequence and luteovirus-like nature of RNA-1 of an aphid nontransmissible strain of pea enation mosaic virus. *Journal of General Virology* 72, 1819–1834.

Denno, R.F. and G. Roderick (1970) Population biology of planthopper. *Annual Review of Entomology* 37, 489–520.

Dent, D. (1991) *Insect Pest Management*. CAB International, Wallingford, UK.

DePamphilis, M.L. (1993) Origins of DNA replication in metazoan chromosomes. *Journal of Biological Chemistry* 268, 1–4.

Derksen, A.C.G. and Granados, R.R. (1988) Alteration of lepidopteran peritrophic membrane by baculovirus and enhancement of viral infectivity. *Virology* 167, 242–250.

Desowitz, R.S. (1993) *The Malaria Capers. Tales of Parasites and People*. W.W. Norton, New York, USA, 288 pp.

Dethier, V. (1970) Chemical interactions between plants and insects. In: Sondheimer, E. and Simeone, J.B. (eds) *Chemical Ecology*. Academic Press, New York, USA, pp. 83–102.

Dethier, V. (1976) *The Hungry Fly*. Harvard University Press, Cambridge, Massachusetts, USA.

de Wit, P.J.G.M. (1992) Molecular characterization of gene-for-gene systems in plant-fungus. *Annual Review of Phytopathology* 30, 391–418.

Dey, K. and Pande, Y.D. (1987) Evaluation of certain non-insecticidal methods of reducing infestation of the mango nut weevil, *Sternochetus gravis* (F) in India. *Tropical Pest Management* 33, 27–28.

Dhavle, S. (1990) Natural farming has come to stay. *Third World Network Features* 485, 2–4.

DiCesare, B., Grossman, E., Katz, E., Picozza, E., Ragusa, R. and Woudenberg, T. (1993) A high-sensitivity electrochemiluminescence-based detection system for automated PCR product quantitation. *BioTech* 15, 152–157.

Dimock, M., Beach, R.M. and Carison, P.S. (1988) Endophytic bacteria for delivery of crop protection agents. In: Roberts, D.W. and Granados, R.R. (eds) *Proceedings of Biotechnology, Biological Pesticides and Novel Plant-Pest Resistance for Insect Pest Management*. Ithaca, New York, USA, pp. 88–93.

Dinham, B. (1993) Pesticides in India. *Pesticide News* 19, 12–14.

Doe, C.Q. (1992) The generation of neuronal diversity in the *Drosophila* central nervous system. In: *Determinants of Neuronal Identity*. Academic Press Inc., New York. pp. 119–154.

Doe, C.Q. and Goodman, C.S. (1985) Early events in insect neurogenesis II. The role of cell interactions and cell lineages in the development of neural precursor cells. *Development Biology* 111, 206–219.

Doe, C.Q. and Smouse, D.T. (1990) The origins of cell diversity in the *Drosophila* central nervous system. *Seminars in Cell Biology* 1, 211–218.

Doeleman, J.A. (1990) Benefits and costs of entomopathogenic nematodes: two biological control applications in China. *ACIAR Economic Assessment Series* 4, 15.

Dowling, D.N., Boesten, B., O'Sullivan, D.J., Stephens, P., Morris, J. and O'Gara, F. (1993) Genetically-engineered fluorescent pseudomonads for improved bio-control of plant pathogens. In: Lynch, J. (ed.) *Benefits and Risks of Introducing Biological Control Agents. Proceedings of a Workshop, 3–6 April 1992, Saariselka, Finland.* OECD, Paris.

Doyle, J.J. and Persley, G.J. (1996) Enabling the safe use of biotechnology: Principles and Practice. Environmentally Sustainable Development Studies and Monographs Series No. 10. The World Bank, Washington, D.C. USA. 140 pp.

Doyle, K.E. and Knipple, D.C. (1991) PCR based phylogenetic walking: isolation of para-homologous sodium channel gene sequences from seven insect species and an arachnid. *Insect Biochemistry* 21, 689–696.

Dreze, J. and Sen, A. (1989) *Hunger and Public Action.* Clarendon Press, Oxford.

DuBois, M. (1992) Eco-terrorism. *Wall Street Journal Europe* 7 September, 1.

Dumas, D.P., Wild, J.R. and Rauschel, F.M. (1990) Expression of *Pseudomonas* phosphotriesterase activity in the fall armyworm confers resistance to insecticides. *Experientia* 46, 729–734.

Edwards, C.R. and Ford, R.E. (1992) Integrated pest management in the corn/soyabean agroecosystem. In: Zalom, F.G. and Fry, W.E. (eds) *Food Crop Pests and the Environment, the Need and Potential for Biologically Intensive Integrated Pest Management.* APS Press, St Paul, Minnesota, USA, pp. 13–57.

Edwards, J.W. and Coruzzi, G.M. (1990) Cell-specific gene expression in plants. *Annual Review of Genetics* 24, 275–303.

Edwards, O.R. and Hoy, M.A. (1993) Polymorphism in two parasitoids detected using random amplified polymorphic DNA (RAPD) PCR. *Biological Control Theory and Application in Pest Management* 3, 243–257.

Ehler, L.E. (1990) Environmental impact of introduced biological control agents: implications for agricultural biotechnology. In: Marois, J.J. and Bruyening, G. (eds) *Risk Assessment in Agricultural Biotechnology.* Publ. No. 1928, University of California, Division of Agriculture and Natural Resources, Berkeley, pp. 85–96.

Ehler, R-U. and Peters, A. (1993) Entomopathogenic nematodes in biological control: feasibility, perspectives and possible risks. In: Lynch, J. (ed.) *Benefits and Risks of Introducing Biological Control Agents. Proceedings of a Workshop, 3–6 April 1992, Saariselka, Finland.* OECD, Paris. p. xx.

Ehtesham, N.Z., Bentur, J.S. and Bennett, J. (1993) DNA based probe for distinguishing different Indian biotypes of Asian rice gall midge *Orseolia oryzae.* Paper presented at the Sixth Annual Meeting of the International Program on Rice Biotechnology, 1–5 February 1993, Chiang Mai, Thailand.

Elliott, H.J., Bashford, R., Greener, A. and Candy, S.G. (1993) Integrated pest management of the Tasmanian eucalyptus leaf beetle, *Chrysophtharta bimaculata* (Olivier) (Coleoptera: Chrysomelidae). *Forest Ecology Management* 53, 29–38.

El Titi, A. (1987) Environmental manipulation detrimental to pests. In: Delucchi, V. (ed.) *Integrated Pest Management Quo Vadis?* Parasitis, Geneva, pp. 105–121.

Elvin, C.M., Riddles, P.W., Whan, V. and Vuocolo, T. (1991) Analysis of multigene family expression: serine protease gene from ectoparasitic flies (sheep blowfly and buffalo fly). Abstract. In: *XIX International Congress of Entomology*. Beijing, China, p. 129.

Eurobarometer (1991) *Opinions of Europeans on Biotechnology in 1991*. Eurobarometer 35.1, Report of the Commission of the European Communities. Brussels, 76 pp.

Ezueh, M.I. (1981) Nature and significance of preflowering damage by thrips to cowpea. *Entomologia Experimentalis et Applicata* 29, 305–312.

Ezuka, A. (1979) Breeding for and genetics of blast resistance in Japan. In: *Proceedings of the Rice Blast Workshop*. International Rice Research Institute, Los Baños, The Philippines, pp. 27–48.

Ezuka, A. and Sakaguchi, S. (1978) Host parasite relationship in bacterial leaf blight of rice caused by *Xanthomonas oryzae*. *Review of Plant Protection Research* 11, 93–118.

Fabres, G. (1981) Première quantification du phénomène de gradation des populations de *Phenacoccus manihoti* Matile-Ferrero (Hom. Pseudococcidae) en République Populaire du Congo. *Agronomie* 1, 483–486.

Fabres, G. (1982) Bioécologie de la cochenille du manioc *Phenacoccus manihoti* (Hom. Pseudococcidae) en République Populaire du Congo II. Variations d'abondance et facteurs de régulation. *Agronomie Tropicale* 36, 369–377.

Fabres, G. and Boussienguet, J. (1981) Bioécologie de la cochenille du manioc *Phenacoccus manihoti* (Hom. Pseudococcidae) en République Populaire du Congo. *Agronomie Tropicale* 36, 82–89.

Fabres, G. and Le Rü, B. (1986) Etude des relations plante–insecte pour la mise au point de méthodes de régulation des populations de la cochenille du manioc. In: *La Cochenille du Manioc et sa Biocoenose au Congo 1979– 1984. Travaux de l'Équipe Franco-Congolaise.* ORSTOM-DGRS, Brazzaville, pp. 57–71.

Fabres, G. and Matile-Ferrero, D. (1980) Les entomophages inféodés à la cochenille du manioc, *Phenacoccus manihoti* (Hom. Coccoidea Pseudococcidae) en République Populaire du Congo. I. Les composantes de l'entomocoenose et leurs inter-relations. *Annales de la Société d'Entomologie de France (NS)* 16, 509–515.

Fairbanks, D.J., Waldridges, A., Ruas, C.F., Ruas, M.P., Maughan, P.J., Robinson, R.L., Andersen, W.R., Riede, C.R., Pauley, C.S., Caetano, L.G., Arantes, O.M.N., Fungaro, M.H.G.P., Vidotto, M.C. and Jankevicius, S.E. (1993) Efficient characterization of biological diversity DNA extraction and random amplified polymorphic DNA markers. *Revista Brasileira de Genetica* 16(1), 11–22.

Fakih, M. (1993) The IPM field school action research process: a case study in Indonesian National IPM Program (1993). pp. 7–28.

Fallon, A. (1991) DNA-mediated gene transfer: applications to mosquitoes. *Nature* 352, 828–829.

FAO (1991) *Mid-term Review of FAO Inter-country Program for the Development and Application of Integrated Pest Control in Rice in South and South East Asia. Mission Report.* FAO, Rome, 181 pp.

FAO (1993) *Global IPM: a report on the Global IPM Field Study Tour and Meeting August–September 1993.* FAO, Manila.

FAO (1994) Sustainable agriculture through integrated pest management. Presented at the 22nd Regional Conference for Asia and the Pacific, Manila, 3–7 October.

Fassuliotis, G. and Creighton, C.S. (1982) *In vitro* cultivation of the entomogenous nematode *Fillipjevmermis leipsandra. Journal of Nematology* 14(1), 126–131.

Faure, J.C. (1960) Thysanoptera of Africa, 3: *Taeniothrips sjostedti* (Trybom) 1908. *Journal of the Entomological Society of South Africa* 23, 34–44.

Federici, B.A. (1990) Bright horizons for invertebrate pathology. In: *Proceedings of the Vth International Colloquium on Invertebrate Pathology and Microbial Control.* Society for Invertebrate Pathology, Adelaide, Australia, pp. v–ix.

Feitelson, J.S., Payne, J. and Kim, L. (1992) *Bacillus thuringiensis*: insects and beyond. *Bio/Technology* 10, 271–275.

Feldmann, A.M. and Ponnudurai, T. (1989) Selection of *Anopheles stephensi* for refractoriness and susceptibility to *Plasmodium falciparum. Medical and Veterinary Entomology* 3, 41–52.

Feldmann, A.M., Billingsley, P.F. and Savelkoul, E. (1990) Bloodmeal digestion by strains of *Anopheles stephensi* Liston (Diptera: Culicidae) of differing susceptibility to *Plasmodium falciparum. Parasitology* 101, 193–200.

Felton, W. (1991) Selective sprayer ushers in new era in weed control. *Aust. Grain* 1(3), 12–14.

Ferre, J., Real, M.D., van Rie, J., Jansens, S. and Peferoen, M. (1991) Resistance to the *Bacillus thuringiensis* bioinsecticide in a field population of *Plutella xylostella* is due to a change in a midgut membrane receptor. *Proceedings of the National Academy of Sciences USA* 88, 5119–5123.

Ferro, D.N. (1993a) Potential for resistance to *Bacillus thuringiensis*: Colorado potato beetle (Coleoptera: Chrysomelidae): a model system. *American Entomologist* 39(1), 38–44.

Ferro, D.N. (1993b) Integrated pest management in vegetables in Massachusetts. In: Leslie, A.R. and Cuperus, G.W. (eds) *Successful Implementation of Integrated Pest Management for Agricultural Crops.* Lewis Publications, Boca Raton, Florida, USA, pp. 95–107.

Ferro, D.N. and Gelernter, W.D. (1989) Toxicity of a new strain of *Bacillus thuringiensis* to Colorado potato beetle (Coleoptera: Chrysomelidae). *Journal of Economic Entomology* 82, 750–755.

Ferro, D.N. and Lyon, S.M. (1991) Colorado potato beetle (Coleoptera: Chrysomelidae) larval mortality: operative effects of *Bacillus thuringiensis* subsp. *san diego. Journal of Economic Entomology* 84, 806–809.

Ferro, D.N., Morzuch, B.S. and Margolies, D. (1983) Crop loss assessment of the Colorado potato beetle (Coleoptera: Chrysomelidae) on potatoes in western Massachusetts. *Journal of Economic Entomology* 76, 349–356.

Ffrench-Constant, R.H., Mortlock, D.P., Shaffer, C.D., MacIntyre, R.J. and Roush, R.T. (1991) Molecular cloning and transformation of cyclodiene resistance in *Drosophila*: an invertebrate γ-aminobutyric acid subtype A receptor locus. *Proceedings of the National Academy of Sciences USA* 88, 7209–7213.

Ffrench-Constant, R.H., Aronstein, K. and Roush, R.T. (1993) A single-amino acid substitution in a γ-aminobutryic acid subtype A receptor locus is associated with cyclodiene insecticide resistance in *Drosophila* populations. *Proceedings of the National Academy of Sciences USA* 90, 1957–1961.

Filippov, N.A. (1990) Biological methods in integrated plant protection from agricultural pests in the USSR. In: FAO/UNEP/USSR Workshop on Integrated Pest Management, 12–15 June 1990, Kishinev, Moldavia, USSR (unpublished).

Finardi, C.E. and Souza, G.L. (1980) *Acão da extensão rural no manejo integrado de pragas.* Associação de Crédito e Assistência Rural do Paranã (ACARPA), Curitiba, Brasil, 16 pp.

Fine, P.E.M. (1978) On the dynamics of symbiote-dependent cytoplasmic incompatibility in culicine mosquitoes. *Journal of Invertebrate Pathology* 30, 10–18.

Fischhoff, D.A. (1992) Management of lepidopteran pests with insect resistant cotton: recommended approaches. In: Herber, D.J. and Richter, D.A. (eds.) *Proceedings–Beltwide Cotton Conferences.* National Cotton Council of America, Memphis, Tennessee, pp. 751–753.

Fisher, R.A. (1930) *The Genetical Theory of Natural Selection.* Clarendon Press, Oxford, UK.

Fitchen, J.H. and Beachy, R.N. (1993) Genetically engineered protection against viruses in transgenic plants. *Annual Review of Microbiology* 47, 739–763.

Fitt, G.P. (1989) The ecology of *Heliothis* species in relation to agroecosystems. *Annual Review of Entomology* 34, 17–52.

Fitt, G.P. (1994) Perspectives on cotton pest management in Australia. *Annual Review of Entomology.* 39, 543–562.

Flexner, J.L., Lighthart, B. and Croft, B.A. (1986) The effects of microbial pesticides on non-target, beneficial arthropods. *Agriculture, Ecosystems and Environment* 16, 203–254.

Follett, P.A., Kennedy, G.G. and Gould, F. (1993) REPO: a simulation model that explores Colorado potato beetle (Coleoptera: Chrysomelidae) adaptation to insecticides. *Environmental Entomology* 22, 283–296.

Foo, G.S. (1989) Sri Lanka: pesticide use. In: Fernando, R. (ed.) *Pesticides in Sri Lanka.* Friedrich-Ebert-Stiftung, Colombo, Sri Lanka, pp. 50–54.

Fournier, D., Karch, F., Bride, J., Hall, L.M.C., Berge, J.B. and Spierer, P. (1989) *Drosophila melanogaster* acetylcholinesterase gene structure, evolution and mutations. *Journal of Molecular Biology* 210, 15–22.

Fournier, D., Bridge, J.M., Poirie, M., Berge, J. and Plapp, F.W. Jr (1992) Insect glutathione S-transferases, biochemical characteristics of the major forms from houseflies susceptible and resistant to insecticides. *Journal of Biological Chemistry* 267, 1840–1845.

Fox, M.W. (1992) Superpigs and Wondercorn. *Lyons and Buford* 21, 8.

Friedman, J.J. (1990) Commercialization and application technology. In: Gougler, R. and Kaya, H.K. (eds) *Entomopathogenic Nematodes in Biological Control.* CRC Press, Boca Raton, Florida, USA, pp. 153–170.

Friedman, J.J., Langston, S.L. and Pollitt, S. (1989) Mass production in liquid culture of insect-killing nematodes. International Patent No. WO89/04602.

Frisbie, R.E., Kamal, M., El-Zik, and Wilson L. Ted. (1988) *Integrated Pest Management Systems and Cotton Production.* John Wiley & Sons, New York.

Fullaway, D.T. and Dobroscky, I.D. (1934) A new *Thripoctenus* parasite from the Philippines. *Proceedings of the 5th Pacific Scientific Congress* (Canada, 1933) 5, 3439–3444.

Futuyma, D.J. (1979) *Evolutionary Biology*. Sinauer, Sunderland, Massachusetts, USA.

Fuxa, J.R. (1990) New directions for insect control with baculoviruses. In: Baker, R.R. and Dunn, P.E. (eds) *New Directions in Biological Control*. UCLA Symposia 112, Alan Liss, New York, pp. 97–113.

Gaffney, T.D., Lam, S.T., Ligon, J., Gates, K., Frazelle, A., Di Maio, J., Hill, D.S., Goodwin, S., Torkewitz, N. and Becker, J.O. (1992) Global regulation of expression of antifungal factors by a *Pseudomonas fluorescens* biological control strain. In: *Abstracts of 6th International Symposium on Molecular Plant–Microbe Interactions, 11–16 July, Seattle, USA*, abstr. 230.

Gahan, A.B. (1932) Miscellaneous descriptions and notes on parasitic Hymenoptera. *Annals of the Entomological Society of America* 25, 736–757.

Gale, K.R. and Crampton, J.M. (1987) DNA probes for species identification of mosquitoes in the *Anopheles gambiae* complex. *Medical and Veterinary Entomology* 1, 127–136.

Gallagher, K.D. (1992) IPM development in the Indonesian national IPM program. In: Aziz, A., Kadir, S.A. and Barlow, H.S. (eds) *Pest Management and the Environment in 2000*. CAB International, Wallingford, UK, pp. 82–89.

Gallagher, K.D., Kenmore, P.E. and Sogawa, K. (1994) Judicial use of insecticide deters planthopper outbreaks and extend the role of resistant varieties on Southeast Asian rice. In: Denno, R.F. and Perfect, T.J. (eds) *Planthoppers, Their Ecology and Management*. Chapman and Hall, London, pp. 599–614.

Gal-On, A., Antignus, Y., Rosner, A. and Raccah, B. (1991) Infectious *in vitro* RNA transcripts derived from cloned cDNA of the cucurbit potyvirus, zucchini yellow mosaic virus. *Journal of General Virology* 72, 2639–2643.

Gal-On, A., Antignus, Y., Rosner, A. and Raccah, B. (1992) A zucchini yellow mosaic virus coat protein gene mutation restores aphid transmissibility but has no effect on multiplication. *Journal of General Virology* 73, 2183–2187.

Gámez, R. (1980) Rayado fino virus disease in the American tropics. *Tropical Pest Management* 26, 26–33.

Gandhimathi, O.Q. (1992) Networking for LEISA development. *ILEIA Newsletter* 8, 10–12.

Garcia-Salazar, C., Whalon, M.E. and Rahardja, U. (1992) Temperature-dependent pathogenicity of the X-disease mycoplasma-like organism to its vector, *Paraphlepsius irroratus*. *Environment* 20, 179–184.

Garro, J.E., de la Cruz, R., Shannon, P.J. and de la Cruz, R. (1991) Propanil resistance in *Echinochloa colona* populations with different herbicide use histories. *Proceedings of the Brighton Crop Protection Conference, Weeds* 3, 1079–1083.

Gasser, C.S. and Fraley, R.T. (1989) Genetically engineered plants for crop improvement. *Science* 244, 1293–1299.

Gazzoni, D.L. and Oliveira, E.B. (1979) Distribuição estacional de *Epinotia aporema* e seu efeito sobre o rendimento e seus componentes, e caracteristicas agronômicas da soja cv. UFV-1, semeada em diversas épocas. In: *Anais do I Seminário Nacional de Pesquisa de Soja*, Vol. II. EMBRAPA-CNPSo, Londrina, Brasil, pp.

94–105.
Gazzoni, D.L. and Oliveira, E.B. (1984) Soyabean insect pest management in Brazil: I Research effort; II Programme implementation. In: Matteson, P.C. (ed.) *Proceedings of International Workshop on Integrated Pest Control of Grain Legumes*. EMBRAPA, Goiania, Brasil, pp. 312–325.
Gazzoni, D.L., Oliveira, E.B., Corso, I.C., Correa-Ferreira, B.S., Villas Boas, G.L., Moscardi, F. and Panizzi, A.R. (1981) *Maenjo de pragas da soja*. Circular Tecnica 5, EMBRAPA-CNPSo, Londrina, Brasil, 44 pp.
Gelernter, W.D. (1990) MVP™ bioinsecticide: a bioengineered, bioencapsulated product for control of lepidopteran larvae. In: *Proceedings of the Vth International Colloquium on Invertebrate Pathology and Microbial Control*. Society for Invertebrate Pathology, Adelaide, Australia, p. 14.
Gelernter, W.D., Toscano, N.C., Kido, K. and Federici, B.A. (1986) Comparison of a nuclear polyhedrosis virus and chemical insecticides for control of the beet armyworm (Lepidoptera: Noctuidae) on head lettuce. *Journal of Economic Entomology* 79, 714–717.
Georghiou, G.P. (1990) Overview of insecticide resistance. In: Green, M.B., LeBaron, H.M. and Moberg, W.K. (eds) *Managing Resistance to Agrochemicals*. American Chemical Society, Washington DC, USA, pp. 18–41.
Gerlach, W.L., Miller, W.A., Cheng, Z.G. and Waterhouse, P.M. (1990) Molecular biology of barley yellow dwarf virus. In: Burnett, P.A. (ed.) *World Perspectives on Barley Yellow Dwarf*. CIMMYT, Mexico, DF, Mexico, pp. 105–110.
Ghana Health Assessment Project Team (1981) A quantitative method of assessing the health impact of different diseases in less developed countries. *International Journal of Epidemiology* 10, 73–92.
Ghauri, M.S.K. (1980) Notes on Anthocoridae (Heteroptera) with description of two new species of economic importance from Africa. *Bulletin of Entomological Research* 70, 287–291.
Ghelelovitch, M.S. (1952) Sur le déterminisme génétique de la stérilité dans les croisements entre différentes sourches de *Culex autogenicus* Roubaud. *Comptes Rendus de l'Académie des Sciences* 234, 2386–2388.
Gibbons, A. (1991) Moths take the field against biopesticide. *Science* 254, 646.
Gildow, F.E. (1990) Barley yellow dwarf virus–aphid vector interactions associated with virus transmission and vector specificity. In: Burnett, P.A. (ed.) *World Perspectives on Barley Yellow Dwarf*. CIMMYT, Mexico, DF, Mexico, pp. 111–122.
Gildow, F.E. (1991) Barley yellow dwarf virus transport through aphids. In: Peters, D.C., Webster, J.A. and Chlouber, C.S. (eds) *Proceedings, Aphid–Plant Interactions: Populations to Molecules*. Oklahoma State University, Stillwater, Oklahoma, USA, pp. 165–177.
Gildow, F.E. (1993) Evidence for receptor-mediated endocytosis regulating luteovirus acquisition by aphids. *Phytopathology* 83, 270–277.
Gingery, R.E. and Nault, L.R. (1990) Severe maize chlorotic dwarf disease caused by double infection with mild virus strains. *Phytopathology* 80, 687–691.
Gips, T. (1990) *Breaking the Pesticide Habit: Alternatives to 12 Hazardous Pesticides*. International Organization of Consumer Unions, Penang, Malaysia.
Glass, E.H. (1992) Constraints to the implementation and adoption of IPM. In: Zalom, F.G. and Fry, W.E. (eds) *Food, Crop Pests and the Environment, the Need*

and Potential for Biologically Intensive Integrated Pest Management. APS Press, St Paul, Minnesota, USA, pp. 167–175.

Goettel, M.S., St Leger, R.J., Bhairi, S., Jung, M.K., Oakley, B.R., Roberts, D.W. and Staples, R.C. (1990) Pathogenicity and growth of *Metarhizium anisopliae* stably transformed to benomyl resistance. *Current Genetics* 17, 129–132.

Goodman, C.S. and Shatz, C.J. (1993) Developmental mechanisms that generate precise patterns of neuronal connectivity. *Cell* 72/*Neuron* 10 (Suppl.), 77–98.

Goodwin, P.H. and Annis, S.L. (1991) Rapid identification of genetic variation and pathotype of *Leptosphaeria maculans* by random amplified polymorphic DNA assay. *Applied and Environmental Microbiology* 57, 2482– 2486.

Goodwin, S.B., Drenth, A. and Fry, W.E. (1992) Cloning and genetic analyses of two highly polymorphic, moderately repetitive nuclear DNAs from *Phytophthora infestans*. *Current Genetics* 22, 1107–1115.

Gonsalves, D., Chee, P., Provvidenti, R., Seem, R. and Slightom, J.L. (1992) Comparison of coat protein-mediated and genetically-derived resistance in cucumbers to infection by cucumber mosaic virus under field conditions with natural challenge inoculations by vectors. *Biotechnology* 10, 1562–1570.

Gosden, J. and Hanratty, D. (1993) PCR *in situ*: a rapid alternative to *in situ* hybridization for mapping short, low copy number sequences without isotopes. *BioTech* 15, 78–80.

Gotz, B. (1940) Sexualduftstoffe als Lockmittel in der Schadlingsbekamfung. *Umschau* 44, 794–796.

Gould, F. (1983) Genetics of plant–herbivore systems: interactions between applied and basic study. In: Denno, R.F. and McClure, M.S. (eds.) *Variable Plants and Herbivores in Natural and Managed Systems*. Academic Press, New York pp. 599–653.

Gould, F. (1986a) Simulation models for predicting durability of insect-resistant germplasm: a deterministic diploid, two locus model. *Environmental Entomology* 15, 1–10.

Gould, F. (1986b) Simulation models for predicting durability of insect-resistant germplasm: Hessian fly (Diptera: Cecidomyiidae) resistant winter wheat. *Environmental Entomology* 15, 11–23.

Gould, F. (1988a) Evolutionary biology and the design of genetically engineered crops. *Bioscience* 38, 26–32.

Gould, F. (1988b) Genetic engineering, integrated pest management and the evolution of pests. *TIBTECH* 6(4), 15–18.

Gould, F. (1991) The evolutionary potential of crop pests. *American Scientist* 79, 496–507.

Gould, F. (1994) Potential and problems with high-dose strategies for pesticidal engineered crops. *Biocontrol Science and Technology* 4, 451–461.

Gould, F. and Anderson, A. (1991) Effects of *Bacillus thuringiensis* and HD-73 delta-endotoxin on growth, behaviour, and fitness of susceptible and toxin-adapted strains of *Heliothis virescens* (Lepidoptera: Noctuidae). *Environmental Entomology* 20, 30–38.

Gould, F., Kennedy, G.G. and Johnson, M.T. (1991a) Effects of natural enemies on the rate of herbivore adaptation to resistant host plants. *Entomologia Experimentalis et Applicata* 58, 1–14.

Gould, F., Anderson, A., Landis, D. and van Mellacri, H. (1991b) Feeding behaviour and growth of *Heliothis vitrescens* larvae on diets containing *Bacillus thuringiensis* formulations or endotoxins. *Entomologia Experimentalis et Applicata* 58, 199–210.

Gould, F, Wilhoit, L. and Via, S. (1991c) The use of ecological genetics in developing and deploying aphid-resistant crop cultivars. In: Peters, D.C., Webster, J.A. and Chlowber, C.S. (eds) *Proceedings Aphid–Plant Interactions: Populations to Molecules.* MP-132, USDA/ARS, Oklahoma State University, Oklahoma, USA, pp. 71–85.

Gould, F., Martinez-Ramirez, A., Anderson, A., Ferre, J., Silva, F.J. and Moar, W.J. (1992) Broad-spectrum resistance to *Bacillus thuringiensis* toxins in *Heliothis virescens. Proceedings of the National Academy of Sciences USA* 89, 7986–7990.

Gould, F., Follett, P., Nault, B. and Kennedy, G.G. (1993) Resistance management strategies for transgenic potato plants. In: Zehnder, G., Jansson, R.K. Powelson, M.L., Raman, K. and Raman, K.V. (eds) *Potato Pest Management: A Global Perspective.* APS Press, St Paul, Minnesota, USA. p. xx.

Gould, F., Anderson, A., Reynolds, A., Bumgarner, L. and Moar, W. (1995) Selection and genetic analysis of *Heliothis virescens* (Lepidoptera, Noctuidae) strain with high levels of resistance to *Bacillus thuringiensis* toxins. *Journal of Economic Entomology* 88, 209–223.

Graves, P.M. and Curtis, C.F. (1982) A cage replacement experiment involving introduction of genes for refractoriness to *Plasmodium yoelii nigeriensis* into a population of *Anopheles gambiae. Journal of Medical Entomology* 19, 127–132.

Greathead, D.J. (1990) Prospects for the use of natural enemies in combinations with pesticides. In: *The Use of Natural Enemies to Control Agricultural Pests.* FFTC Book Series No. 40, FFTC, Taiwan, pp. 1–7.

Greathead, D.J. and Waage, J.F. (1983) *Opportunities for Biological Control of Agricultural Pests in Developing Countries.* World Bank Technical Paper No. 11, World Bank, Washington, DC, USA.

Green, G.J. and Campbell, A.B. (1979) Wheat cultivars resistant to *Puccinia graminis tritici* in western Canada: their development, performance and economic value. *Canadian Journal of Plant Pathology* 1, 3–11.

Griffin, C.T., Joyce, S.A., Dix, I., Burnell, A.M. and Downes, M.J. (1993) Characterization of the entomopathogenic nematode *Heterorhabditis* (Nematoda: Heterorhabditidae) from Ireland and Britain by molecular and cross-breeding techniques, and the occurrence of the genus in these islands. *Fundamental and Applied Nematology* 17, 245–253.

Griffiths, W.T. (1984) A review of the development of cotton pest problems in the Sudan Gezira. MSc thesis, Imperial College, University of London, UK.

Groeters, F.R., Tabashnik, B.E., Finson, N. and Johnson, M.W. (1993) Effects of resistance to *Bacillus thuringiensis* on mating success of the diamondback moth (Lepidoptera: Plutellidae). *Journal of Economic Entomology* 86, 1035–1039.

Gudmundsson, G.H., Lidholm, D.A., Asling, B., Gan, R. and Boman, H.G. (1991) The eecropin locus: cloning and expression of a gene cluster encoding three antibacterial peptides in *Hyalophora eecropia. Journal of Biological Chemistry* 266, 11510–11517.

Gunasinghe, U.B., Irwin, M.E. and Kampmeier, G.E. (1988) Soyabean leaf pubescence affects aphid vector transmission and field spread of soyabean mosaic virus. *Annals of Applied Biology* 112, 259–272.

Gunning, R.V., Easton, L.R., Greenup, L.R. and Edge, V.E. (1984) Pyrethroid resistance in *Heliothis armigera* (Hubner) (Lepidoptera: Noctuidae) in Australia. *Journal of Economic Entomology* 77, 1283–1287.

Gutierrez, A.P., Wermelinger, B., Schulthess, F., Ellis, C.K., Baumgärtner, J.U. and Yaninek, J.S. (1988a) An analysis of the biological control of cassava pests in Africa: I Simulation of carbon, nitrogen and water dynamics in cassava. *Journal of Applied Ecology* 25, 901–920.

Gutierrez, A.P., Neuenschwander, P., Schulthess, F., Herren, H.R., Baumgärtner, J.U., Wermelinger, B., Löhr, B. and Ellis, C.K. (1988b) An analysis of the biological control of cassava pests in Africa: II. The biological control of the cassava mealybug (*Phenacoccus manihoti* Mat. Fer.). *Journal of Applied Ecology* 25, 921–940.

Hadrys, H., Balick, M. and Schierwater, B. (1992) Applications of random amplified polymorphic DNA (RAPD) in molecular ecology. *Molecular Ecology* 1, 55–63.

Hahn, S.K. and Williams, R.J. (1973) Enquête sur le manioc en République du Zaïre. Report to the Minister of Agriculture of the Republic of Zaire, March, 12 pp. (mimeograph).

Halford, W.T. (1987) The use of selective microbial insecticides for control of specific field generations of *Heliothis* spp. on cotton and economic injury levels and damage levels for specific field generations of *Heliothis* spp. on cotton. PhD thesis, Mississippi State University, USA.

Hall, L.M.C. and Malcolm, C.A. (1991) The acetycholinesterase gene of *Anopheles stephensi*. *Cellular and Molecular Neurobiology* 11, 131–141.

Hall, L.M.C. and Spierer, P. (1986) The *ACE* locus on *Drosophila melanogaster* structural gene for acetylcholinesterase with an unusual 5' leader. *European Molecular Biology Organization Journal* 5, 2949–2954.

Halward, T.M., Stalker, H.T., LaRue, E.A. and Kochert, G. (1991) Genetic variation detectable with molecular markers among unadapted germplasm resources of cultivated peanut and related wild species. *Genome* 34, 1013–1020.

Hamer, J.E. (1991) Molecular probes for rice blast disease. *Science* 252, 632–633.

Hamer, J.E., Farrall, L., Orbach, M., Valent, B. and Chumley, F.G. (1989) Host species-specific conservation of a family of repeated DNA sequences in the genome of a fungal plant pathogen. *Proceedings of National Academy of Sciences USA* 86, 9981–9985.

Hamilton, W.D. (1967) Extraordinary sex ratios. *Science* 156, 477–481.

Hammond, W.N.O., Neuenschwander, P. and Herren, H.R. (1987) Impact of the exotic parasitoid *Epidinocarsis lopezi* on cassava mealybug (*Phenacoccus manihoti*) populations. *Insect Science Application* 8, 887–891.

Handler, A.M. and O'Brochta, D.A. (1991) Prospects for gene transformation in insects. *Annual Review of Entomology* 36, 159–183.

Hansen, M. (1987) *Escape from the Pesticide Treadmill: Alternatives to Pesticides in Developing Countries*. International Organization of Consumers Unions, Penang, Malaysia, 185 pp.

Hanzlik, T.N., Dorrian, S.J., Gordon, K.H.J. and Christian, P.D. (1993) A novel small RNA virus isolated from the cotton bollworm, *Helicoverpa armigera*. *Journal of General Virology* 74, 1805–1810.

Hao, H., Tyschenko, G.M., Rancour, S.L. and Walker, V.K. (1992) Dihydrofolate reductase and methotrexate resistance in *Drosophila*. Abstract. In: *XIX International Congress of Entomology*. Beijing, China, p. 142.

Hara, A.H., Gaugler, R., Kaya, H.K. and Lebeck, L.M. (1991) Natural populations of entomopathogenic nematodes (Rhabditida: Heterorhabditidae, Steinernematidae) from the Hawaiian Islands. *Environmental Entomology* 20, 211–216.

Harris, M.K. (1980) *Biology and Breeding for Resistance to Arthropods and Pathogens in Agricultural Plants*. Texas Agricultural Experiment Station Publication MP-1451.

Harris, P. (1985) Biocontrol and the law. *Bulletin of the Entomological Society of Canada* 17, 1.

Harrison, B.D. (1992) Advances and prospects in potato virology with special reference to virus resistance. *Netherlands Journal of Plant Pathology* 98 (Supplement 2), 1–12.

Hasan, G. (1989) Molecular cloning of an olfactory gene from *Drosophila melanogaster*. *Proceedings of National Academy of Sciences USA* 86, 2908–2912.

Haugen, D.A., Bedding, R.A., Underdown, M.G. and Neumann, F.G. (1990) National strategy for control of *Sirex noctilio* in Australia. *Australian Forest Grower* 13(2) (lift-out).

Hawksworth, D.L. (1991) *The Biodiversity of Microorganisms and Invertebrates: Its Role in Sustainable Agriculture*. CAB International, Wallingford, UK.

Hawksworth, D.L. (ed.) (1994) *The Identification and Characterization of Pest Organisms*. CAB International, Wallingford, UK.

Headley, J.C. and Hoy, M.A. (1987) Benefit/cost analysis of an integrated mite management program for almonds. *Journal of Economic Entomology* 80, 555–559.

Hearn, A.B. and Fitt, G.P. (1992) Cotton cropping systems. In: Pearson, C. (ed.) *Field Crop Ecosystems of the World*. Elsevier Press, pp. 85–142.

Hedin, P.A. (1983) *Plant Resistance to Insects*. ACS Symposium Series 208, American Chemical Society, Washington DC, USA.

Heinrichs, E.A. (1979) Chemical control of the brown planthopper. In: *Brown Planthopper: Threat to Rice Production in Asia*. International Rice Research Institute, Los Baños, Philippines, pp. 145–170.

Heinrichs, E.A. and Mochida, O. (1984) From secondary to major pest status: the case of insecticide-induced rice brown planthoppers, *Nilaparvata lugens*, resurgence. *Protection Ecology* 7, 201–218.

Heinrichs, E.A., Ressig, W.H., Valencis, S.L. and Chelliah, S. (1982) Rates and effect of resurgence-including insecticides on populations of *Nilaparvata lugens* (Hemiptera: Delphscides) and its predators. *Environmental Entomology* 11, 1269–1273.

Heinrichs, E.A., Medrano, F.G. and Rapusas, H.R. (1985) *Genetic Evaluation for Insect Resistance in Rice*. International Rice Research Institute, Los Baños, Philippines, 356 pp.

Helgeson, J.P. (1992) New genes for disease resistance through somatic hybridization. *Netherlands Journal of Plant Pathology* 98 (Supplement 2), 223–229.

Herren, H.R. and Neuenschwander, P. (1991) Biological control of cassava pests in Africa. *Annual Review of Entomology* 36, 257–283.

Herren, H.R., Bird, T.J. and Nadel, D.J. (1987a) Technology for automated aerial release of natural enemies of the cassava mealybug and cassava green mite. *Insect Science Applications* 8, 883–885.

Herren, H.R., Neuenschwander, P., Hennessey, R.D. and Hammond, W.N.O. (1987b) Introduction and dispersal of *Epidinocarsis lopezi* (Hym. Encyrtidae), an exotic parasitoid of the cassava mealybug *Phenacoccus manihoti* (Hom. Pseudococcidae) in Africa. *Agricultural Ecosystems and Environments* 19, 131–141.

Hertig, M. (1936) The rickettsia, *Wolbachia pipientis* (gen. et sp.n.) and associated inclusions of the mosquito, *Culex pipiens*. *Parasitology* 28, 453–490.

Hertig, M. and Wolbach, S.B. (1924) Studies on rickettsia-like micro-organisms in insects. *Journal of Medical Research* 44, 329–374.

Hickey, W.A. and Craig, G.B. (1966) Genetic distortion of sex ratio in the mosquito *Aedes aegypti*. *Genetics* 53, 1177–1183.

Hidaka, K. (1993) Farming systems for rice cultivation which pronounce the regulation of pest populations by natural enemies: planthopper management in traditional, intensive farming, and LISA rice cultivation in Japan. *ASPAC Food and Fertilizer Technology Center Extension Bulletin (Taipei)* No. 374, December.

Higgs, S., Powers, A.H. and Olson, K.E. (1993) *Alphavirus* expression systems: applications to mosquito sector studies. *Parasitology Today* 9, 444–452.

Hilder, V.A., Gatehouse, A.M.R., Sheerman, S.E., Barker, R.F. and Boulter, D. (1987) A novel mechanism of insect resistance engineered into tobacco. *Nature* 330, 160–163.

Hilder, V.A., Gatehouse, A.M.R. and Boulter, D. (1993) Transgenic plants conferring insect tolerance: protease inhibitor approach. In: Kung, S.D. and Wu, R. (ed.) *Transgenic Plants*. Academic Press, San Diego, California, USA, pp. 317–338.

Hill, S.B. (1990) Pest control in sustainable agriculture. *Proceedings of Entomological Society of Ontario* 121, 5–12.

Hill, S.M., Urwin, R., Knapp, T.F. and Crampton, J.M. (1991) Synthetic DNA probes for the identification of sibling species in the *Anopheles gambiae* complex. *Medical and Veterinary Entomology* 5, 455–463.

Hill, S.M., Urwin, R. and Crampton, J.M. (1992) A simplified, non-radioactive DNA probe protocol for the field identification of insect vector specimens. *Transactions of the Royal Society for Tropical Medicine and Hygiene* 86, 213–215.

Hinchee, M.A.W., Padgette, S.R., Kishore, G.M., Delanney, X. and Fraley, R.T. (1993) Herbicide-tolerant plants. In: Kung, S.D. and Wu, R. (eds) *Transgenic Plants*. Academic Press, pp. 243–263.

Hodgson, J. (1990) Appropriate biotech for Africa. *Bio/Technology* 8, 793.

Hoffmann, A.A., Turelli, M. and Simmon, G.M. (1986) Unidirectional incompatibility between populations of *Drosophila simulans*. *Evolution* 40, 692–701.

Hoffmann, F., Fournier, D. and Spierer, P. (1992) Minigene rescue acetylcholinesterase lethal mutations in *Drosophila melanogaster*. *Journal of Molecular Biology* 223, 17–22.

Hoffmann-Campo, C.B. (1989) Tamandua da soja: aspectos biologicos, danos e comportamento. EMBRAPA-CNPSo, Londrina, Brasil (folder to farmers).

Hoffmann-Campo, C.B., Oliveira, E.B., Mazzarin, R.M. and Oliveira, M.C.N. (1990) Niveis de infestaçia nos rendimentos e caracteristicas agronômicas da soja. *Pesquisa Agropecuaria Brasileira* 25(2), 221–227.

Hofte, H., and Whiteley, H. (1989) Insecticidal crystal proteins of *Bacillus thuringiensis*. *Microbiology Review* 53, 242–255.

Hogg, J.C., Carwardine, S.L. and Hurd, H. (1995) Low *Plasmodium* oocyst burdens are capable of reducing anopheline mosquito fecundity. *Annals of Tropical Medicine and Parasitology* 89, 185.

Hokkanen, H. (1985) Exploiter–victim relationships of major plant diseases: implications for biological weed control. *Agriculture, Ecosystems and Environment* 14, 63–76.

Holland, P.M., Abramson, R.D., Watson, R. and Gelfand, D.H. (1991) Detection and specific polymerase chain reaction product by utilizing the 5'–3' exonuclease activity of *Thermus aquaticus* DNA polymerase. *Proceedings of National Academy of Sciences USA* 88, 7276–7280.

Holmes, R. (1993) Can sustainable farming win the battle of the bottom line? *Science* 260, 1893–1895.

Homberg, U., Christensen, T.A. and Hildebrand, J.G. (1989) Structure and function of the deutocerebrum in insects. *Annual Review of Entomology* 34, 477–501.

Hopkins, A.R., Moore, R.F. and James, W. (1982) Economic injury level for *Heliothis* spp. larvae on cotton plants in the four tree-leaf and pinhead square stage. *Journal of Economic Entomology* 75, 328–332.

Hopkins, C.M., White, F.F., Choi, S.H., Guo, A. and Leach, J.E. (1992) Identification of a family of avirulence genes from *Xanthomonas oryzae* pv *oryzae*. *Molecular Plant–Microbe Interactions* 5, 451–459.

Hopp, H.E., Hain, L., Bravo Almonacid, F., Tozzini, A.C., Orman, B., Arese, A.I., Ceriani, M.F., Saladrigas, M.V., Celnik, R., del Vas, M. and Mentaberry, A.N. (1991) Development and application of a nonradioactive nucleic acid hybridization system for simultaneous detection of four potato pathogens. *Journal of Virology Methods* 31, 11–30.

Hopper, K.R. and Roush, R.T. (1993) Mate-finding, dispersal, number released and the success of biological control introductions. *Ecological Entomology* 18, 321–331.

Hopper, K.R., Roush, R.T. and Powell, W. (1993) Management of genetics of biological control introductions. *Annual Review of Entomology* 38, 27–51.

Houck, M.A., Clark, J.B., Peterson, K.R. and Kidwell, M.G. (1991) Possible horizontal transfer of *Drosophila* genes by the mite *Proctolaelaps regalis*. *Science* 253, 1125–1129.

Howarth, F.G. (1991) Environmental impacts of classical biological control. *Annual Review of Entomology* 36, 485–509.

Howe, D.K., Vodkin, M.H., Novak, R.J., Shope, R.E. and McLaughlin, G.L. (1992) Use of the polymerase chain reaction for the sensitive detection of St. Louis encephalitis viral RNA. *Journal of Virology Methods* 36, 101–110.

Hoy, M.A. (1976) Genetic improvement of insects: fact or fantasy. *Environmental Entomology* 5, 833–839.

Hoy, M.A. (1990a) Genetic improvement of arthropod natural enemies: becoming a conventional tactic? In: Baker, R.R. and Dunn, P.E. (eds) *New Directions in Biological Control: Alternative for Suppressing Agricultural Pests and Diseases*. UCLA Symposium of Molecular and Cellular Biology, New Series Vol. 112, Alan R. Liss, New York, USA, pp. 405–417.

Hoy, M.A. (1990b) Pesticide resistance in arthropod natural enemies: variability and selection responses. In: Roush, R.T. and Tabashnik, B.E. (eds) *Pesticide Resistance in Arthropods*. Chapman and Hall, New York, USA, pp. 203–236.

Hoy, M.A. (1992a) Criteria for release of genetically-improved phytoseiids: an examination of the risks associated with release of biological control agents. *Experimental and Applied Acarology* 14, 393–416.

Hoy, M.A. (1992b) Biological control of arthropods: genetic engineering and environmental risks. *Biological Control* 2, 166–170.

Hoy, M.A. (1994) *Insect Molecular Genetics*, Academic Press, San Diego, 452 pp.

Hoy, M.A. (1995) Impact of risk analyses on pest-management programs employing transgenic arthropods. *Parasitology Today* 11, 229–232.

Hoy, M.A. and Cave, F.E. (1989) Toxicity of pesticides used on walnuts to a wild and azimphosmethyl resistant strain of *Trioxys pallidus* (Hymenoptera: Aphididae). *Journal of Economic Entomology* 82, 1585–1592.

Hoy, M.A., Castro, D. and Cahn, D. (1982) Two methods for large scale production of pesticide resistant strains of the spider mite predator *Metaseiulus occidentalis* (Nesbitt) (Acarina, Phytoseiidae). *Zeitschrift für Angewandte Entomologie* 94, 1–9.

Hu, J.S., Rochow, W.F., Palukaitis, P. and Dietert, R.R. (1988) Phenotypic mixing: mechanism of dependent transmission for two related isolates of barley yellow dwarf virus. *Phytopathology* 78, 1326–1330.

Huffaker, C.B. (1979) *New Technologies of Pest Control*. Wiley, New York, USA.

Hull, R. (1990) Non-conventional resistance to viruses in plants – concepts and risks. In: Gustafsen, J.P. (ed.) *Gene Manipulation in Plant Improvement II*. Plenum Press, London, UK, pp. 289–304.

Hull, R. and Davies, J.W. (1992) Approaches to nonconventional control of plant virus diseases. *Critical Reviews in Plant Sciences* 11, 17–33.

Humber, R.A. (1990) Systematic and taxonomic approaches to entomophthoralean species. In: *Proceedings of the Vth International Colloquium on Invertebrate Pathology and Microbial Control*. Society for Invertebrate Pathology, Adelaide, Australia, pp. 133–137.

Hunt, G.J. and Page, R.E. Jr (1992) Patterns of inheritance with RAPD molecular markers reveal novel types of polymorphism in the honey bee. *Theoretical and Applied Genetics* 85, 15–20.

Hunt, R.E. and Nault, L.R. (1990) Influence of life history of grasses and maize chlorotic dwarf virus on the biotic potential of the leafhopper *Graminella nigrifrons* (Homoptera: Cicadellidae). *Environmental Entomology* 19, 76–84.

Hunt, R.E. and Nault, L.R. (1991) Roles of interplant movement, acoustic communication and phototaxis in mate-location behaviour of the leafhopper *Graminella nigrifrons*. *Behavioural Ecology and Sociobiology* 28, 315–320.

Hunt, R.E., Nault, L.R. and Gingery, R.E. (1988) Evidence for infectivity of maize chlorotic dwarf virus and for a helper component in its leafhopper transmission. *Phytopathology* 78, 499–504.

Hurst, L.D. (1991) The evolution of cytoplasmic incompatibility, or when spite can be successful. *Journal of Theoretical Biology* 148, 269–277.

Huynh, N. (1993) Genetic variation in yellow stemborer, *Scirpophaga incertulas* (Walker), populations and biological basis of resistance of the *Oryza sativa* (IR56) × *Oryza brachyantha* cross to the pest. Unpublished PhD thesis, University of the Philippines at Los Baños, Philippines.

Hyong, K.K. (1987) Science, technology and mankind. In: *Creating the Future of Mankind.* Yoko Civilization Research Institute, Tokyo, Japan, pp. 103–107.

Iatrou, K. and Meidinger, R.G. (1990) Tissue-specific expression of silkmoth chorion genes *in vivo* using *Bombyx mori* nuclear polyhedrosis virus as a transducing vector. *Proceedings of the National Academy of Sciences USA*, 87, 3650–3654.

Ignoffo, C.M. (1992) Environmental factors affecting persistence of entomopathogens. *Florida Entomologist* 75(4), 516–525.

Ignoffo, C.M. and Garcia, C. (1992) Combinations of environmental factors and simulated sunlight affecting activity of inclusion bodies of the *Heliothis* (Lepidoptera: Noctuidae) nucleopolyhedrosis virus. *Environmental Entomology* 21, 210–213.

Ikeda, R., Khush, G.S. and Tabien, R.E. (1990) A new resistance gene to bacterial blight derived from *O. longistaminata. Japanese Journal of Breeding* 40, 280–281.

Ikehashi, I. and Khush, G.S. (1979) Breeding for blast resistance at IRRI. In: *Proceedings of the Rice Blast Workshop.* International Rice Research Institute, Los Baños, Philippines, p. 220.

Iman, M., Soekarna, D., Situmorang, J., Adiputra, M.G. and Manti, I. (1986) Effects of insecticides on various field strains of diamondback moth and its parasitoids in Indonesia. In: Talekar, N.S. and Griggs, T.D. (eds) *Diamondback Moth Management.* Asian Vegetable Research and Development Center, Shanhua, Taiwan, pp. 313–323.

Inamdar, M., VijayRaghavan, K. and Rodrigues, V. (1993) The function of the *scalloped* locus is essential for normal taste behaviour in *Drosophila melanogaster. Journal of Neurogenetics* 9, 123–139.

Indonesian National IPM Programme (1993) *IPM Farmer Training: the Indonesian Case.* Indonesian National IPM Programme, Yogjakarta.

Indonesian National IPM Programme (1994) *IPM Consolidation: Developing a Farmer-led IPM Program.* Indonesian National IPM Programme, Yogjakarta.

Ingram, D.S. (1992) Towards an informed public. *Proceedings of the Royal Society of Edinburgh* 99B(3/4), 121–134.

Inscoe, M.N., Leonhardt, B.A. and Ridgway, R.L. (1990) Commercial availability of insect pheromones and other attractants. In: Ridgway, R.L., Silverstein, R.M. and Inscoe, M.N. (eds) *Behaviour-modifying Chemicals for Insect Pest Management.* Marcel Dekker, New York, USA, pp. 631–715.

Irwin, M.E. and Goodman, R.M. (1981) Ecology and control of soyabean mosaic virus. In: Maramorosch, K. and Harris, K.F. (eds) *Plant Diseases and Their Vectors: Ecology and Epidemiology.* Academic Press, New York, USA, pp. 181–220.

Irwin, M.E. and Kampmeier, G.E. (1989) Vector behaviour, environmental stimuli and the dynamics of plant virus epidemics. In: Jeger, M.J. (ed.) *Spatial Components of Plant Disease Epidemics.* Prentice-Hall, Englewood Cliffs, New Jersey, USA, pp. 14–39.

Irwin, M.E. and Ruesink, W.G. (1986) Vector intensity: a product of propensity and activity. In: McLean, G.D., Garrett, R.G. and Ruesink, W.G. (eds) *Plant Virus*

Epidemics: Monitoring, Modelling and Predicting Outbreaks. Academic Press, Sydney, Australia, pp. 13–33.

Irwin, M.E. and Thresh, J.M. (1988) Long-range dispersal of cereal aphids as virus vectors in North America. In: Anderson, R.M. and Thresh, J.M. (eds) *The Epidemiology and Ecology of Infectious Diseases Agents. Proceedings of the Royal Society of London* pp. 95–120.

Irwin, M.E. and Thresh, J.M. (1990) Epidemiology of barley yellow dwarf: a study in ecological complexity. *Annual Review of Phytopathology* 28, 393–424.

Jackai, L.E.N. and Daoust, R.A. (1986) Insect pests of cowpeas. *Annual Review of Entomology* 31, 95–119.

Jackai, L.E.N. and Rawlston, J.R. (1988) Rearing the legume pod borer, *Maruca testulalis* Geyer (Lepidoptera: Pyralidae) on artificial diet. *Tropical Pest Management* 34, 168–172.

James, A.A., Blackmer, K., Marinotti, O. Ghosn, C.R. and Racioppi, J.V. (1991) Isolation and characterisation of the gene expressing the major salivary gland protein of the female mosquito *Aedes aegypti. Molecular Biochemistry and Parasitology* 44, 245–253.

James, S.P., Nicol, W.D. and Shut, P.G. (1932) A study of induced malignant tertian malaria. *Proceedings of the Royal Society of Medicine* 25, 1153–1186.

Jaques, R.P. and Laing, D.R. (1989) Effectiveness of microbial and chemical insecticides in control of the Colorado potato beetle (Coleoptera: Chrysomelidae) on potatoes and tomatoes. *Canadian Entomology* 121, 1123–1131.

Jarret, P. and Burges, D.H. (1988) Preparation of strains of *Bacillus thuringiensis* having improved activity against certain lepidopterous pests and novel strains produced thereby. UK Patent GB 2165261. Patent Office, London, UK.

Jayappa, B.G. and Lingappa, S. (1988) Screening cowpea germplasm for resistance to *Aphis cracciwora* Koch, in India. *Tropical Pest Management* 34, 62–64.

Jayaraj, S. (1989) Advances in biological means of pest control. *Hindu Survey of Indian Agriculture* 181–187.

Jeffrys, A., Wilson, V. and Thein, S. (1985) Hypervariable 'minisatellite' regions in human DNA. *Nature* 314, 67–73.

Jena, K.K. and Khush, G.S. (1990) Introgression of genes from *Oryza officinalis* Well Ex. Watt to cultivated rice, *O. sativa* L. *Theoretical and Applied Genetics* 80, 737–745.

Jenns, A.E. and Leonard, K.J. (1985) Reliability of statistical analyses for estimating relative specificity in quantitative resistance in a model host–pathogen system. *Theoretical and Applied Genetics* 69, 503–513.

Johnson, M.T. and Gould, F. (1992) Interaction of genetically engineered host plant resistance and natural enemies of *Heliothis virescens* (Lepidoptera: Noctuidae) in tobacco. *Environmental Entomology* 21, 586–597.

Johnson, R. (1981) Durable resistance: definition of genetic control and attainment in plant breeding. *Phytopathology* 71, 567–568.

Jones, D.A. and Kerr, A. (1989) The efficacy of *Agrobacterium radiobacter* strain K1026, a genetically engineered derivative of strain K84, in the biological control of crown gall. *Plant Disease* 73, 15–18.

Jones, T. (1994) BioNET: a global network for biosystematics of arthropods and microorganisms. In: Hawksworth, D.L. (ed.) *Identification and Characterisation of Pest Organisms.* CAB International, Wallingford, UK, pp. 81–91.

Justin, C.G.L., Rabindra, R.J. and Jayaraj, S. (1990) *Bacillus thuringiensis* Berliner and some insecticides against DBM (*Plutella xylostella*) on cauliflower. *Journal of Biological Control* 4, 40–43.

Kainoh, Y. (1990) Chemicals controlling the behaviour of parasitoids and predators. In: *The Use of Natural Enemies to Control Agricultural Pests.* FFTC Book Series No. 40, FFTC, Taiwan, pp. 212–221.

Kajiwara, T. (1992) Application of plant biotechnology to plant protection in developing countries. In: Sasson, A. and Costarini, V. (eds) *Plant Biotechnologies for Developing Countries. Proceedings of an International Symposium. 26–30 June 1989.* CTA, Luxemburg, pp. 181–189.

Kalshoven, L.G.E. and Van Der Vecht, J. (1950) *De plagen van de culturgewassen in Indonesia.* Van Hoeve, The Hague, 512 pp.

Kambhampati, S., Black, W.C. and Rai, K.S. (1992a) Random amplified polymorphic DNA of mosquito species and populations (Diptera: Culicidae): techniques, statistical analysis and applications. *Journal of Medical Entomology* 29, 939–945.

Kambhampati, S., Rai, K.S. and Verleye, D.M. (1992b) Frequencies of mitochondrial DNA haplotypes in laboratory cage populations of the mosquito *Aedes albopictus. Genetics* 132, 205–209.

Kamdar, P., von Allmen, G. and Finnerty, V. (1992) Transient expression of DNAT in *Drosophila* via electroporation. *Nucleic Acids Research* 20, 3526.

Kamito, T., Tanaka, H., Sato, B., Nagasawa, H. and Suzuki, A. (1992) Nucleotide sequence of cDNA for the eclosion hormone of the silkworm, *Bombyx mori*, and expression in the brain. *Biochemistry and Biophysics Research Communications* 182, 514–519.

Kandel, E.R. (1976) *Cellular Basis of Behaviour.* W.H. Freeman, San Fancisco, California, USA.

Karieva, P. (1993) Transgenic plants on trial. *Nature* 363, 580–581.

Karl, S.A. and Avise, J.C. (1993) PCR-based assays of Mendelian polymorphisms from anonymous single-copy nuclear DNA: techniques and applications for population genetics. *Molecular Biology and Evolution* 10(2), 342–361.

Katre, N.V. (1990) *Biological Pest Control and Sustainable Agriculture in Less-industrialised Nations.* American Association for Advancement of Science Publication 90–59S, Washington DC, USA.

Katre, N.V. (1993) The conjugation of proteins with polyethylene glycol and other polymers. *Advanced Drug Delivery Reviews* 10, 91–114.

Kawakami, A., Kataoka, H., Oka, T., Mizoguchi, A., Kimura-Kawakami, M., Adachi, T., Iwami, M., Nagasawa, H., Suzuki, A. and Ishizaki, H. (1990) Molecular cloning of the *Bombyx mori* prothoracicotrophic hormone. *Science* 247, 1333–1335.

Kaya, H.K. and Gaugler, R. (1993) Entomopathogenic nematodes. *Annual Review of Entomology* 38, 181–206.

Keller, S. (1992) The *Beauveria–Melolontha* project: experiences with regard to locust and grasshopper control. In: Lomes, C.J. and Prior, C. (eds) *Biological Control of Locusts and Grasshoppers. Proceedings of a Workshop held at the International Institute of Tropical Agriculture, Cotonou, Republic of Benin.* CAB International, Wallingford, UK, pp. 279–286.

Kemp, D.J., Smith, D.B., Foote, S.J., Samaras, N. and Peterson, M.G. (1989) Colorimetric detection of specific DNA segments amplified by polymerase chain reaction. *Proceedings of the National Academy of Sciences USA* 86, 2423–2427.

Kenmore, P.E. (1980) Ecology and outbreaks of a tropical insect pest of the green revolution. The rice brown planthopper, *Nilaparvata lugens* (Stål.) PhD dissertation, University of California, Berkeley, USA.

Kenmore, P.E. (1991) *Indonesia's Integrated Pest Management – A Model for Asia.* FAO Inter-country Programme for Integrated Pest Control in Rice in South and Southeast Asia. FAO, Manila, Philippines, 56 pp.

Kenmore, P.E., Cariño, F.O., Perez, C.A., Dyck, V.A. and Guierrez, A.P. (1984) Population regulation of the rice brown planthopper (*Nilaparvata lugens* Stål) within rice fields in the Philippines. *Journal of Plant Protection in the Tropics* 1, 19–37.

Kennedy, J.W. (1981) Practical application of pheromones in regulatory pest management programs. In: Mitchell, E.R. (ed.) *Management of Insect Pests with Semiochemicals.* Plenum Press, New York, USA, pp. 1–11.

Keystone Center (1991) *Final Consensus Report: Global Initiative for the Security and Sustainable Use of Plant Genetic Resources.* Keystone Center, Colorado, USA.

Khush, G.S. (1977) Disease and insect resistance in rice. *Advances in Agronomy* 29, 265–341.

Kidwell, M.G. (1983) Evolution of hybrid dysgenesis determinants in *Drosophila melanogaster. Proceedings of the National Academy of Sciences of the USA* 80, 1655–1659.

Kidwell, M.G. (1992) Horizontal transfer. *Current Opinion in Genetics and Development* 2, 868–873.

Kidwell, M.G. (1993) Voyage of an ancient *mariner. Nature* 362, 202.

Kidwell, M.G. and Ribeiro, J.M.C. (1992) Can transposable elements be used to drive disease refractoriness genes into vector populations? *Parasitology Today* 8, 325–329.

Kidwell, M.G., Novy, J.B. and Feeley, S.M.J. (1981) Rapid unidirectional change of hybrid dysgenesis potential in *Drosophila. Journal of Heredity* 72, 32–38.

Kinsman, J. (1993) Rifkin's consumer alert on BST catching headlines. *Biotech Reporter* March 1993, 7.

Kirby, L.T. (1992) *DNA Fingerprinting.* W.H. Freeman, Oxford, UK.

Kiritani, K. (1979) Pest management in rice. *Annual Review of Entomology* 24, 279–312.

Kirkpatrick, B.C., Stenger, D.C., Morris, T.J. and Purcell, A.H. (1987) Cloning and selection of DNA from a nonculturable plant pathogenic mycoplasmalike organism. *Science* 238, 199–200.

Kirschbaum, J.B. (1985) Potential implication of genetic engineering and other biotechnologies to insect control. *Annual Review of Entomology* 30, 51–70.

Kishi, M., Hirschhom, N., Djajadisastra, M., Satteriee, L., Strowinan, S. and Dilta, R. (1995) Relationship of pesticide spraying to signs and symptoms in Indonesian farmers. *Scandinavian Journal of Work Environment and Health* 21, 124–133.

Kiss, A. and Meerman, F. (1991) *Integrated Pest Management and African Agriculture.* World Bank Technical Paper No. 142, World Bank, Washington DC, USA.

Kistler, H.C., Momol, E.A. and Benny, U. (1991) Repetitive genomic sequences for determining relatedness among strains of *Fusarium oxysporum*. *Phytopathology* 81, 331–336.

Kitajima, E.W. and Gámez, R. (1983) Electron microscopy of maize rayado fino virus in the internal organs of its leafhopper vector. *Intervirology* 19, 129–134.

Kiyasu, P.K. and Kidwell, M.G. (1984) Hybrid dysgenesis in *Drosophila melanogaster*: the evolution of mixed P and M populations maintained at high temperature. *Genetical Research* 44, 251–259.

Kiyosawa, S. (1981) Gene analysis for blast resistance. *Oryza* 18, 196–203.

Kiyosawa, S. (1982) Genetic and epidemiological modeling of breakdown of plant disease resistance. *Annual Review of Phytopathology* 20, 93–117.

Klassen, W., Ridgway, R.L. and Inscoe, M. (1982) Chemical attractants in integrated pest management programs. In: Kidonilus, A.F. and Beroza, M. (eds) *Insect Suppression with Controlled Release Pheromone Systems*, Vol. 1. CRC Press, Boca Raton, Florida, USA, pp. 13–130.

Klein, T.A., Harrison, B.A., Andre, R.G., Whitmore, R.E. and Inlaá, I. (1982) Detrimental effects of *Plasmodium cynomolgi* infections on the longevity of *Anopheles dirus*. *Mosquito News* 42, 265–271.

Knight, A.L. and Hull, L.A. (1989) Use of sex pheromone traps to monitor azinphosmethyl resistance in tufted apple bud moth (Lepidoptera: Tortricidae). *Journal of Economic Entomology* 82(4), 1019–1026.

Koch, M.F. and Parlevleit, J.E. (1991) Genetic analysis of, and selection for, factors affecting quantitative resistance to *Xanthomonas campestris* pv. *oryzae* in rice. *Euphytica* 53, 235–245.

Koelle, M., Talbot, W.S., Segraves, W.A., Bender, M.T., Cherbas, P. and Hogness, D.H. (1991) The *Drosophila* EcR gene encodes an ecdysone receptor, a new member of the steroid receptor superfamily. *Cell* 67, 59–77.

Kogan, M. (1982) Plant resistance in pest management. In: Metcalf, R.L. and Luckman, W.H. (eds) *Introduction to Insect Pest Management*. John Wiley and Sons, New York, USA, pp. 93–134.

Kogan, M. (1988) Integrated pest management theory and practice. *Journal of Applied and Experimental Entomology* 49, 59–70.

Kogan, M. (1989) Plant resistance in soyabean insect control. In: Pascale, A.J. (ed.) *Proceedings of the World Soyabean Research Conference IV, Buenos Aires, Argentina, 1989*. Association Argentina de la Soja, Buenos Aires, pp. 1519–1525.

Kogan, M. and Turnipseed, S.G. (1987) Ecology and management for soyabean arthropods. *Annual Review of Entomology* 32, 507–538.

Kogan, M., Turnipseed, S.G., Shepard, M., Oliveira, E.B. and Borgo, A. (1977) Pilot insect pest management program for soyabean in southern Brazil. *Journal of Economic Entomology* 70, 659–663.

Koziel, M.G., Beland, G., Bowman, C., Carozzi, N.B., Crenshaw, R., Crossland, L., Dawson, J., Desai, N., Hill, M., Kadwell, S., Launis, K., Lewis, K., Maddox, D., McPherson, K., Meghji, M., Merlin, E., Rhodes, R., Warren, G.W., Wright, M. and Evola, S. (1993a) Field performance of elite transgenic maize plants expressing an insecticidal protein derived from *Bacillus thuringiensis*. *Bio/Technology* 11, 194–200.

Koziel, M.G., Desai, N.M., Lewis, K.S., Kramer, V.C., Warren, G.W., Evola, S.V., Crossland, L.D., Wright, M.S., Merlin, E.J., Launis, K.L. and Rothstein, S.J.

(1993b) Synthetic DNA sequence having enhanced insecticidal activity in maize. PCT Patent No. WO 93/07278.

Krafsur, E.S., Whitten, C.J. and Novy J.E. (1987) Screwworm eradication in North and Central America. *Parasitology Today* 3, 131–137.

Kritalugsana, S. (1988) Pesticide poisoning studies and data collection in Thailand. In: Teng, P.S. and Heong, K.L. (eds) *Pesticide Management and Integrated Pest Management in Southeast Asia.* Consortium for International Crop Protection, College Park, Maryland, USA, pp. 321–324.

Kuffler, S.W., Nicholls, J.G. and Martin, R.W. (1984) *From Neuron to Brain,* 2nd edn. Sinauer, Sunderland, Massachusetts, USA.

Kumar, N. (1992) Biological pest control becomes established technique in Asia. *Biotechnology and Development Monitor* 13, 9.

Kurata, N., Nagamura, Y., Yamamoto, K., Harushima, Y., Sue, N., Wu, J., Antonio, B.A., Shomura, A., Shimizu, T., Lin, S.-Y., Fukuda, A., Shimano, T., Kuboki, Y., Toyama, T., Miyamoto, Y., Kirihara, T., Hayasaka, K., Miyao, A., Monna, L., Zhong, H.S., Tamura, Y., Wang, Z.-X., Momma, T., Umehana, Y., Yano, M., Sasaki, T. and Minobe, Y. (1994) A 300 kilobase interval genetic map of rice including 883 expressed sequences. *Nature Genetics* 8, 365–372.

Kurodo, Y. (1993) Challenging Japan's pesticide aid. *Global Pesticide Campaigner PAN* (San Francisco, USA) 3(2), 1, 11– 13.

Lagueux, M., Kromer, E. and Girardie, J. (1992) Cloning of a *Locusta* cDNA encoding neuroparsin A. *Insect Biochemistry and Molecular Biology* 22, 511–516.

Lamb, C.J., Lawton, M.A., Dron, M. and Dixon, R.A. (1989) Signals and transduction mechanisms for activation of plant defences against microbial attack. *Cell* 56, 215– 224.

Lander, E.S. and Botstein, D. (1989) Mapping Mendelian factors underlying quantitative traits using RFLP linkage maps. *Genetics* 121, 185–199.

Langeweg, Ir. F. (1989) *Concern for Tomorrow: A National Environmental Survey 1985–2010.* National Institute of Public Health and Environmental Protection, Bilthoven, The Netherlands, 350 pp.

Lanier, G.N. (1990) Principles of attraction–annihilation. In: Ridgway, R.L., Silverstein, R.M. and Inscoe, M.N. (eds) *Behaviour-modifying Chemicals for Insect Pest Management.* Marcel Dekker, New York, USA, pp. 25–45.

Lapidot, M., Gafny, R., Ding, B., Wolf, S., Lucas, W.J. and Beachy, R.N. (1993) A dysfunctional movement protein of tobacco mosaic virus that partially modifies the plasmodesmata and limits virus spread in transgenic plants. *Plant Journal,* 4, 959–970.

Laven, H. (1959) Speciation by cytoplasmic isolation in the *Culex pipiens* complex. *Cold Spring Harbor Symposium* 24, 166–173.

Laven, H. and Aslamkhan, M. (1970) Control of *Culex pipiens* and *C.p. fatigans* with integrated genetical systems. *Pakistan Journal of Science* 22, 303–312.

Laville, J., Voisard, C., Keel, C., Maurhofer, M., Defago, G. and Haas, D. (1992) Global control in *Pseudomonas fluorescens* mediating antibiotic synthesis and suppression of black root rot of tobacco. *Proceedings of the National Academy of Sciences USA* 89, 1562–1566.

Lawrence, P.A. (1992) *The Making of a Fly. The Genetics of Animal Design.* Blackwell Scientific Publications, London, UK.

Leach, J.E., White, F.W., Rhoads, M.L. and Leung, H. (1990) A repetitive DNA sequence differentiates *Xanthomonas campestris* pv. *oryzae* from other pathovars of *Xanthomonas campestris. Molecular Plant–Microbe Interactions* 3, 238–246.

Leach, J.E., Rhodes, M.L., Vera Cruz, C.M., White, F.F., Mew, T.W. and Leung, H. (1992) Assessment of genetic diversity and population structure of *Xanthomonas oryzae* pv. *oryzae* with a repetitive DNA element. *Applied and Environmental Microbiology* 58, 2188–2195.

Lecoq, H., Bourdin, D., Raccah, B., Hiebert, E. and Percifull, D.E. (1991) Characterization of a zucchini yellow mosaic virus isolate with a deficient helper component. *Phytopathology* 81, 1087–1091.

Lee, E.J., Zhang, Q. and Mew, T.W. (1989) *Progress in Irrigated Rice Research.* International Rice Research Institute, Los Baños, Philippines.

Lee, I. and Davis, R.E. (1991) Mycoplasmas which infect plants and insects. In: *Bergey's Manual of Systematic Bacteriology,* Vol. 2. Williams and Wilkins, Baltimore, Maryland, USA.

Lenski, R.E. and Nguyen, T.T. (1988) Stability of recombinant DNA and its effects on fitness. *Trends in Ecology and Evolution* 6, 18–20.

Leonard, K.J. and Mundt, C.C. (1984) Methods for estimating epidemiological effects of quantitative resistance to plant diseases. *Theoretical and Applied Genetics* 67, 219– 230.

Le Rü, B. and Papierok, B. (1987) Taux intrinsèque d'accroissement naturel de la cochenille du manioc, *Phenacoccus manihoti* Matile-Ferrero (Homoptera, Pseudococcidae). Intérêt d'une méthode simplifiée d'estimation de r_m. *Acta Oecologica, Oecologia Applicata.* 8, 3–14.

Leslie, A.R. and Cuperus, G.W. (1993) *Successful Implementation of Integrated Pest Management for Agricultural Crops.* Lewis Publications, Boca Raton, Florida, USA.

Leung, H. and Williams, P.H. (1986) Enzyme polymorphism and genetic differentiation among geographic isolates of the rice blast fungus. *Phytopathology* 76, 778–783.

Leung, H., Nelson, R.J. and Leach, J.E. (1993a) Population structure of plant pathogenic fungi and bacteria. *Advances in Plant Pathology* 10, 157–205.

Leung, H., Shi, Z., Nelson, R., Bonman, J.M., Estrada, B., Chen, D., Bernardo, M., Scott, R., Zeigler, R.S., Correa, F., Bustaman, M., Syahril, D., Mukelar, A., Shahjahan, A.K.M. and Sy, A. (1993b) Analysis of populations of the rice blast fungus at upland rice screening sites. Paper presented at the Sixth Annual Meeting of the International Program on Rice Biotechnology, Chiang Mai, Thailand, 1–5 February 1993. Rockefeller Foundation, New York, USA.

Levy, M., Romao, J., Marchetti, M.A. and Hamer, J.E. (1991) DNA fingerprinting with a dispersed repeated sequence resolves pathotype diversity in the rice blast fungus. *Plant Cell* 3, 95–102.

Levy, M., Correa-Victoria, F.J., Zeigler, R.S., Shen, Y., Shajahan, A.K.M., Gnanmanickam, S., Nelson, R.J., Manry, J. and Hamer, J.E. (1993) International atlas of genetic diversity in the rice blast fungus: 1992–1993. Paper presented at the Annual Meeting of the Rockefeller Foundation's International Program on Rice Biotechnology, Chiang Mai, Thailand, 2–5 February 1993. Rockefeller Foundation, New York, USA.

Lewis, T. (1973) *Thrips: Their Biology, Ecology, and Economic Importance.* Academic Press, London, UK, 349 pp.

Li, L.Y., Liv, W.H., Chen, C.S., Han, S.T. and Shin, J.C. (1988) *In vitro* rearing of *Trichogramma* sp. and *Anastatus* sp. in artificial 'eggs' and the methods of mass production of *Trichogramma* and other egg parasitoids. *Les Colloques de l'INRA* 43, 339–352.

Lidholm, D.A., Gudmundsson, G.H. and Boman, H.G. (1992) A highly repetitive, *mariner*-like element in the genome of *Hyalophora cecropia. Journal of Biological Chemistry* 266, 11518–11521.

Lim, G.S. (1990) Overview of vegetable IPM in Asia. *FAO Plant Protection Bulletin* 38, 73–87.

Lim, G.S., Ong, S.H. and Cheah, U.B. (1983) Environmental hazards associated with pesticide use on vegetables in Malaysia. In: *Proceedings of the International Conference on Environmental Hazards by Agrochemicals,* Vol. 1. Alexandria University, Alexandria, pp. 372–392.

Lindquist, D.A., Abusowa, M. and Hall, J.R. (1992) The New World screwworm fly in Libya: a review of its introduction and eradication. *Medical and Veterinary Entomology* 6, 2–8.

Lisansky, S.G. (1993) Crop protection without chemicals: the present and future of biopesticides. In: Cartwright, A. (ed.) *World Agriculture 1993.* Sterling Publications, London, UK, pp. 39–71.

Litsinger, J.A., Quirino, C.B., Lumaban, M.D. and Bandong, J.P. (1978) Grain legume pest complex in rice-based cropping systems at three locations in the Philippines. In: Singh, S.R., Van Emden, H.F. and Taylor, T.A. (eds) *Pests of Grain Legumes: Ecology and Control.* Academic Press, London, UK, pp. 309–320.

Loevinsohn, M.E. (1987) Insecticide use and increased mortality in rural central Luzon, Philippines. *Lancet* 13 June, 1359–1362.

Longstaff, B.C. (1993) Role of expert systems in the management of stored grain pests. In: Corey, S.A., Dall, D.J. and Milne, W.M. (eds) *Pest Control and Sustainable Agriculture.* CSIRO, Australia, pp. 475–478.

Longstaff, B.C. and Cornish, P. (1994) Pest man – a decision support system for pest management in the Australian grain industry. *Al Applications* 8, 13–23.

Loudon, P.T. and Roy, P. (1991) Assembly of five bluetongue virus proteins expressed by recombinant baculoviruses: inclusion of the largest protein VP1 in the core and virus-like particles. *Virology* 180, 798–802.

Love, J.M., Knight, A.M., McAleer, M.A. and Todd, J.A. (1990) Towards construction of a high-resolution map of the mouse genome using PCR-analyzed microsatellites. *Nucleic Acid Research* 18, 4123–4130.

Luckow, V.A. and Summers, M.D. (1988) Trends in the development of baculovirus expression vectors. *Bio/Technology* 6, 47–55.

Lupoli, R., Irwin, M.E. and Vossbrinck, C.R. (1990) A ribosomal DNA probe to distinguish populations of *Rhopalosiphum maidis* (Homoptera: Aphididae). *Annals of Applied Biology* 117, 3–8.

McCouch, S.R., Kochert, G., Yu, Z.H., Wang, Z.Y., Khush, G.S., Coffman, W.R. and Tanksley, S.D. (1988) Molecular mapping of rice chromosomes. *Theoretical and Applied Genetics* 76, 815–829.

McCouch, S.R., Abenes, M.L., Angeles, R., Khush, G.S. and Tanksley, S.D.

(1991) Molecular tagging of a recessive gene, *xa-5*, for resistance to bacterial blight of rice. *Rice Genetics Newsletter* 8, 143–145.

McDonald, B.A. and Martinez, J.P. (1990) DNA restriction fragment length polymorphisms among *Mycosphaerella graminicola* (anamorph *Septoria tritici*) isolates collected from a single wheat field. *Phytopathology* 80, 1368–1373.

McDonald, B.A., McDermott, J.M., Goodwin, S.B. and Allard, R.W. (1989) The population biology of host–pathogen interactions. *Annual Review of Phytopathology* 27, 77–94.

McGaughey, W.M. (1985) Insect resistance to the biological insecticide *Bacillus thuringiensis*. *Science* 229, 193–195.

McGaughey, W.H. and Beeman, R.W. (1988) Resistance to *Bacillus thuringiensis* in colonies of Indianmeal moth and almond moth (Lepidoptera: Pyralidae). *Journal of Economic Entomology* 81, 28–33.

McGaughey, W.H. and Johnson, D.E. (1987) Toxicity of different serotypes and toxins of *Bacillus thuringiensis* to resistant and susceptible Indianmeal moth (Lepidoptera: Pyralidae). *Journal of Economic Entomology* 80, 1122– 1126.

McGaughey, W.H. and Whalon, M.E. (1992) Managing insect resistance to *Bacillus thuringiensis* toxins. *Science* 258, 1451–1455.

McGinnis, N., Kuziora, M.A. and McGinnis (1990) Human Hox-4.2 and *Drosophila Deformed* encode similar regulatory specificities in *Drosophila* embryos and larvae. *Cell* 63, 969–976.

McGrane, V., Carlson, J.O., Miller, B.R. and Beaty, B.J. (1988) Microinjection of DNA into *Aedes triseriatus* ova and detection of integration. *American Journal of Tropical Medicine and Hygiene* 39, 502–510.

McInnis, D.O., Haymer, D.S., Tam, S.Y.T. and Thanaphum, S. (1990) *Ceratitis capitata* (Diptera: Tephritidae): transient expression of a heterologous gene for resistance to the antibiotic geneticin. *Annals of the Entomological Society of America* 83, 982–986.

McKenna, M., Monte, P., Helfand, S.L. and Carlson, J.R. (1989) A simple chemosensory response and the isolation of *acj* mutants in which it is affected. *Proceedings of the National Academy of Sciences USA* 86, 8118–8122.

Mackill, D.J. and Bonman, J.M. (1992) Inheritance of blast resistance in near-isogenic lines of rice. *Phytopathology* 82, 746–749.

McMurtry, J.A. (1983) Phytoseiid predators in orchard systems: a classical biological control success story. In: Hoy, M.A., Cunningham, G.L. and Knutson, L. (eds) *Biological Control of Pests by Mites*. University of California Special Publication 3304, Berkeley, California, USA, pp. 21–26.

McPherson, M.J., Quirke, P. and Taylor, G.R. (1992) *PCR, A Practical Approach*. IRL Press, Oxford, UK.

McVeigh, L.J., Campion, D.G. and Critchley, B.R. (1990) The use of pheromones for the control of cotton bollworms and *Spodoptera* spp. in Africa and Asia. In: Ridgway, R.L., Silverstein, R.M. and Inscoe, M.N. (eds) *Behaviour-modifying Chemicals for Insect Pest Management*. Marcel Dekker, New York, USA, pp. 407–416.

Maeda, S. (1989) Expression of foreign genes in insects using baculovirus vectors. *Annual Review of Entomology* 34, 351–372.

Magallona, E.D. (1986) Developments in diamondback moth management in the Philippines. In: Talekar, N.S. and Griggs, T.D. (eds) *Diamondback Moth*

Management. Asian Vegetable Research and Development Center, Shanhua, Taiwan, pp. 423–435.

Magallona, E.D., Verzola, E.A., Reyes, T.M. and Dajet, C.M. (1982) Insecticide management with particular relevance to Baguio-La Trinidad area. In: *Proceedings of the 19th Conference of the National Pest Control Council of the Philippines*. Pest Control Council of the Philippines, Baguio City, Philippines, pp. 43–48.

Mallet, J. and Porter, P. (1992) Preventing insect adaptation to insect resistant crop seed mixtures or refugia, the best strategy. *Proceedings of the Royal Society of London Series B* 250, 165–169.

Marshall, J.F. (1938) *The British Mosquitoes*. British Museum (Natural History), London, UK.

Marshall, J., Buckingham, S.D., Shingai, R., Lunt, G.G., Goosey, M.W., Darlison, M.G., Sattelle, D.B. and Barnard, E.A. (1990) Sequence and functional expression of a single alpha subunit of an insect nicotinic acetylcholine receptor. *EMBO Journal* 9, 4391–4398.

Martin, G.B., Williams, J.G.K. and Tanksley, S. (1991) Rapid identification of markers linked to a *Pseudomonas* resistance gene in tomato by using random primers and near-isogenic lines. *Proceedings of the National Academy of Sciences USA* 88, 2336–2340.

Maruniak, J.E. (1992) Contribution of molecular biology to the improvement of insect viruses as biological control products. *Pesquisa Agropecuaria Brasileira* 27, 143–150.

Maruyama, K. and Hartl, D.L. (1991) Evidence for interspecific transfer of the transposable element *mariner* between *Drosophila* and *Zaprionus*. *Journal of Molecular Evolution* 33, 514–524.

Matile-Ferrero, D. (1978) Cassava mealybug in the People's Republic of Congo. In: Nwanze, K.F.and Leuschner, K. (eds) *Proceedings of an International Workshop on the Cassava Mealybug* Phenacoccus manihoti *Mat.Ferr. (Pseudococcidae), M'vuazi, Zaïre, June 1977*. pp. 29–46.

Matthews, R.E.F. (1991) *Plant Virology*. Academic Press, San Diego, California, USA, pp. 520–553.

Maudlin, I. and Dukes, P. (1985) Extrachromosomal inheritance of susceptibility to trypanosome infection in tsetse flies. I. Selection of susceptible and refractory lines of *Glossina morsitans morsitans*. *Annals of Tropical Medicine and Parasitology* 79, 317–324.

Maudlin, I. and Ellis, D.S. (1985) Association between intracellular rickettsial-like infections of midgut cells and susceptibility to trypanosome infection in *Glossina* spp. *Zeitschrift für Parasitenkunde* 71, 683–687.

Maxwell, F.G. (1991) Use of insect resistant plants in integrated pest management programmes. *FAO Plant Protection Bulletin* 39, 139–146.

Maynard Smith, J. (1989) *Evolutionary Genetics*. Oxford University Press, Oxford, UK.

Mayrand, P.E., Robertson, J., Ziegle, J., Hoff, L.B., McBridge, L.J., Chamberlain, J.S. and Kronick, M.N. (1991) Automated genetic analysis. *Annales de Biologie Clinique* 49, 1–7.

Mazur, P., Cole, K.W., Hall, J.W., Schreuders, P.D. and Mahowald, A.P. (1992) Cryobiological preservation of *Drosophila* embryos. *Science* 258, 1932–1935.

Mazurier, S., van de Giessen, A., Heuvelman, K. and Wernars, K. (1992) RAPD analysis of *Campylobacter* isolates: DNA fingerprinting without the need to purify DNA. *Letters in Applied Microbiology* 14, 260–262.

M'boob, S. and Ketelaar, J.W. (1995) Pilot rice farmers' field schools in Africa: first season's results. Presented at 13th International Plant Protection Congress, The Hague, 6 July, 1995.

Meadows, M. (1993) *Bacillus thuringiensis* in the environment: ecology and risk assessment. In: Entwistle, P.F., Cory, J.S., Bailey, M.J. and Higgs, S. (eds) *Bacillus thuringiensis, an Environmental Biopesticide: Theory and Practice*. John Wiley and Sons, Chichester, UK, pp. 193–220.

Medford, J.I., Elmer, J.S. and Klee, H.J. (1991) Molecular cloning and characterisation of genes expressed in shoot apical meristems. *Plant Cell* 3, 359–370.

Medina, J.R., Calumpang, S.M.F. and Barrodo-Medina, M.J.V. (1991) Insecticide residues in well water. Paper presented at the Workshop on the Environmental and Health Impacts of Pesticide Use in Rice Culture, International Rice Research Institute, Los Baños, Philippines, 6 pp.

Meister, G.A. and Grigliatti, T.A. (1993) Rapid spread of a P element/Adh construct through experimental populations of *Drosophila melanogaster*. *Genome* 36, 1169–1175.

Meon, S. (1986) Behaviour of a benomyl resistant isolate of *Colletotrichum capsici*. In: *Proceedings of the 2nd International Conference of Plant Protection in the Tropics, Extended Abstracts*. Malaysian Plant Protection Society, Kuala Lumpur, Malaysia, pp. 360–370.

Meredith, S.E.E.O. and James, A.A. (1990) Biotechnology as applied to vectors and vector control. *Annual Parasitology Hum. Comp.* 65 (Supplement I) (1), 113–118.

Mew, T.W., Vera Cruz, C.M. and Medalla, E.S. (1992) Changes in race frequency of *Xanthomonas oryzae* pv. *oryzae* in response to rice cultivars planted in the Philippines. *Plant Disease* 76, 1029–1032.

Michaille, J.J., Mathavan, S., Gaillard, J. and Garel, A. (1990) The complete sequence of *mag*, a new retrotransposon in *Bombyx mori*. *Nucleic Acids Research* 18, 674.

Michelmore, R.W., Paran, I. and Kesseli, K.V. (1991) Identification of markers linked to disease-resistance genes by bulked segregant analysis: a rapid method to detect markers in specific genomic regions by using segregating populations. *Proceedings of the National Academy of Sciences USA* 88, 9828–9832.

Miflin, B.J. (1992) Plant biotechnology: aspects of its application to industry. *Proceedings of the Royal Society of Edinburgh* 99B(3/4), 153–163.

Miller, B.R. and Mitchell, C.J. (1991) Genetic selection of a flavivirus-refractory strain of the yellow fever mosquito *Aedes aegypti*. *American Journal of Tropical Medicine and Hygiene* 45, 399–407.

Miller, D.W. (1988) Genetically engineered viral insecticides: practical considerations. In: Hedin, P.A., Hollingworth, R.M. and Mann, J.J. (eds) *Biotechnology for Crop Protection*. American Chemical Society, Washington DC, USA, pp. 405–421.

Miller, L.H. (1992) The challenge of malaria. *Science* 257, 36–37.

Miller, L.H., Sakai, R.K., Romans, P., Gwadz, W., Kantoff, P. and Coon, H.G. (1987) Stable integration and expression of a bacterial gene in the mosquito *Anopheles gambiae*. *Science* 237, 779–781.

Miller, L.K. (1988) Baculoviruses as gene expression vectors. *Annual Review of Microbiology* 42, 177–199.

Milne, C.P., Jr, Phillips, J.P. and Krell, P.J. (1988) Microinjection of early honeybee embryos. *Journal of Apicultural Research* 27, 84–89.

Milner, R.J. and Prior, C. (1994) Susceptibility of the Australian plague locust, *Chortoicetes terminifera*, and the wingless grasshopper, *Phaulacridium vittatum*, to the fungi *Metarhizium* spp. *Biological Control* 4, 132–137.

Milner, R.J., Hartley, T.F., Lutton, G.G. and Prior, C. (1994) Control of *Phaulacridium vittatum* (Sjostedt) (Orthoptera: Acrididae), in field cages using an oil-based spray of *Metarhizium flavoviride* Gams and Rozsypal (Deuteromycetina: Hyphomycetes). *Journal of Australian Entomological Society*.

Miranda, M.A.C., Rosseto, C.J., Rosseto, D., Braga, N.R., Mascarenhas, H.A.A., Teixeira, J.P.F. and Massariol, A. (1979) Resistencia de soja a *Nezara viridula* e *Piezodorus guildinii* em condicoes de campo. *Bragantia* 38, 181–188.

Mitsuda, H. and Yamamato, A. (1980) Advances in grain storage in a CO_2 atmosphere in Japan. In: Shejbal, J. (ed.) *Controlled Atmosphere Storage of Grains*. Elsevier, Amsterdam, The Netherlands, pp. 235–246.

Miyata, T., Sinchaisri, J., Sayampol, B., Rushtaparkornchai, W. and Vattanatangum, A. (1988) Toxicological experiments: (a) Insecticide resistance patterns. In: *Report of Meeting of the Joint Research Project 'Insect Toxicological Studies on Resistance to Insecticides and Integrated Control of the Diamondback Moth'*. Department of Agriculture, Bangkok, Thailand, pp. 9–23.

Moar, W.J. and Trumble, J.T. (1987) Biologically derived insecticides for use against beet armyworm. *California Agriculture* November–December, 13–15.

Moffat, A.S. (1993) New chemicals seek to outwit insect pests. *Science* 261, 550–551.

Mohan, K.S., Jayapal, S.P. and Pillai, G.B. (1989) Biological suppression of coconut rhinoceros beetle, *Oryctes rhinoceros*, by *O. baculovirus* – impact on pest population and damage. *Journal of Plantation Crops* 16, 163–170.

Moore, J. (1991) Insecticide program stresses B.t. products. *American Vegetable Grower* January, 32–34.

Moraes, R.R., Loeck, A.E. and Belarmino, L.C. (1991) Inimigos naturais de *Rachiplusia nu* (Guenee, 1852) e de *Pseudoplusia includens* (Walker, 1857) (Lepidoptera: Noctuidae) em soja no Rio Grande do Sul. *Pesquisa Agropecuaria Brasileira* 26(1), 57–64.

Morales, L. (1991) Potencial de uso do Baculovirus de *Autographa californica* (Speyer) no controle de *Pseudoplusia includens* (Walker) e *Anticarsia gemmatalis* (Huber) e *Rachiplusia nu* (Guenee) (Lep.: Noctuidae). MSc thesis, Universidade Estadual Paulista – Faculdade de Ciencias Agrarias e Veterinarias, Jaboticabal, Brasil.

Morales, L., Moscardi, F. and Gravena, S. (1993) Potencial do Baculovirus de *Autographa californica* (Speyer) no controle de *Chrysodeixis includens* (Walker) e *Anticarsia gemmatalis* (Lep.: Noctuidae). *Pesquisa Agropecuaria Brasileira* 28(2), 237–243.

Moran, V.C. (1983) The phytophagous mites and insects of cultivated plants in South Africa – patterns and pest status. *Journal of Applied Ecology* 20, 439–450.

Morell, V. (1993) Australian pest control virus causes concern. *Science* 261,

683–684.
Morley, A.W., van Rijswijk, G. and Reis, R.G. (1989) Is there life after DDT? In: *Proceedings of the 5th Australian Conference on Grassland Invertebrate Ecology*, pp. 102–108.
Morris, A.C. and James, A.A. (1993) Transfection of salivary glands from the mosquito *Aedes aegypti*. Abstract. In: *2nd International Symposium on Molecular Insect Science*. Flagstaff, Arizona, USA.
Morris, A.C., Eggleston, P. and Crampton, J.M. (1989) Genetic transformation of the mosquito *Aedes aegypti* by micro-injection of DNA. *Medical and Veterinary Entomology* 3, 1–7.
Morris, A.C., Schaub, T.L. and James, A.A. (1991) FLP-mediated recombination in the vector mosquito, *Aedes aegypti. Nucleic Acids Research* 19, 5895–5900.
Moscardi, F. (1984) Microbial control of insect pests in grain legume crops. In: Matteson, P.C. (ed.) *Proceedings of the International Workshop on Integrated Pest Control in Grain Legumes, Goiania, Goias, Brasil, 1983*. EMBRAPA, Brasilia, Brazil, pp. 189–223.
Moscardi, F. (1989) Use of viruses for pest control in Brazil: the case of the nuclear polyhedrosis virus of the soyabean caterpillar, *Anticarsia gemmatalis. Memorias do Instituto Oswaldo Cruz, Rio de Janeiro* 84(3), 51–56.
Moscardi, F. (1990) Uso de entomopatogenos no manejo de pragas da soja no Brasil. In: Fernandes, O.A., Correia, A.C.B. and Bortoli, S.A. (eds) *Manejo Integrado de Pragas e Nematoides*. FUNEP–UNESP, Jaboticabal, pp. 207–220.
Moscardi, F. and Correa-Ferreira, B.S. (1985) Biological control of soybean caterpillars. In: Shibles, R. (ed.) *World Soybean Research Conference III: Proceedings*. Westview Press, Boulder, Colorado, USA, pp. 703–711.
Moscardi, F. and Sosa-Gómez, D.R. (1992) Use of viruses against soybean caterpillars in Brazil. In: Copping, L.G., Green, M.B. and Rees, R.T. (eds) *Pest Management in Soybean*. Elsevier Applied Science, London, UK, pp. 98–109.
Moscardi, F., Kastelic, J. and Sosa-Gómez, D.R. (1992) Suscetibilidade de tres especies de lepidopteros, associados a soja, a tres isolados do fungo *Nomuraea rileyi* (Farlow) Samson. *Anais da Sociedade Entomologica do Brasil* 21, 93–100.
Mouches, C., Pasteur, N., Berge, J.B., Hyrien, O., Raymond, M., de Saint Vincent, B.R., de Silvestri, M. and Georghiou, G.P. (1986) Amplification of an esterase gene is responsible for insecticide resistance in a California *Culex* mosquito. *Science* 233, 778–780.
Mouches, C., Pauplin, Y., Agarwal, M., Lemieux, L., Herzog, M., Abadon, M., Beyssat-Arnaouty, V., Hyrien, O., de Saint Vincent, B.R., Georghiou, G.P. and Pasteur, N. (1990) Characterizatin of amplification core and esterase B1 gene responsible for insecticide resistance in *Culex. Proceedings of the National Academy of Sciences USA* 98, 2574–2578.
Mueller, E., Gilbert, J., Davenport, G., Brigneti, G. and Baulcombe, D. (1995) Homology dependent resistance: transgenic virus resistance in plants related to homology-dependent gene silencing. *The Plant Journal* 7, 1001–1013.
Mulbry, W.W. and Karns, J.S. (1989) Parathion hydrolase specified by the *Flavobacterium opd* gene: relationship between the gene and protein. *Journal of Bacteriology* 171, 6740–6746.
Mullis, K.B. and Faloona, F.A. (1987) Specific synthesis of DNA *in vitro* via a polymerase-catalysed chain reaction. *Methods in Enzymology* 155, 335–350.

Mundt, C.C. (1990) Probability of mutation to multiple virulence and durability of resistance gene pyramids. *Phytopathology* 80, 221–223.

Mundt, C.C. (1991) Probability of mutation to multiple virulence and durability of resistance gene pyramids; further comments. *Phytopathology* 81, 240–242.

Murdoch, L.L. (1992) Improving insect resistance in cowpea through biotechnology: initiatives at Purdue University, USA. In: Thottappilly, G., Monti, L.M., Mohan Raj, D.R. and Moore, A.W. (eds) *Biotechnology: Enhancing Research on Tropical Crops in Africa*, p. 364.

Nagarajan, S. (1992) Safety aspects gain recognition: pest management. *Hindu Survey of Indian Agriculture* 139–145.

Nakashima, K., Kato, S., Iwanami, S. and Murata, N. (1991) Cloning and detection of chromosomal and extrachromosomal DNA from mycoplasmalike organisms that cause yellow dwarf disease of rice. *Applied Environmental Microbiology* 57, 3570–3575.

National Research Council (1986) *Pesticide Resistance: Strategies and Tactics for Management.* National Academy Press, Washington DC, USA.

National Research Council (1989) *Alternative Agriculture.* National Academy Press, Washington DC, USA, 448 pp.

Nault, L.R. (1990) Evolution of an insect pest: maize and the corn leafhopper, a case study. *Maydica* 35, 165–175.

Nault, L.R. (1993) Arthropod transmission of plant viruses. In: Madden, L.V., Raccah, B. and Thresh, J.M. (eds) *Epidemiology and Management of Plant Virus Diseases.* Springer-Verlag, New York, USA.

Nei, M. (1987) *Molecular Evolutionary Genetics.* Columbia University Press, New York, USA.

Nelson, R.R. (1978) Genetics of horizontal resistance to plant diseases. *Annual Review of Phytopathology* 16, 359–378.

Nelson, R., Ardales, E., Vera Cruz, C., Mew, T.W. and Leung, H. (1993) Microgeographic differentiation of *Xanthomonas oryzae* pv. *oryzae* in the Philippines. Paper presented at the Sixth Annual Meeting of the International Program on Rice Biotechnology, 1–5 February 1993, Chiang Mai, Thailand.

Nelson, R.J. Baraoidan, M.R., Vera Cruz, C.M., Yap, I.V., Leach, J.E., Mew, T.W. and Leung, H. (1994) Relationship between phylogeny and pathotype for the bacterial blight pathogen of rice. *Applied Environmental Microbiology* 60, 3275–3283.

Nelson, R.S., McCormick, S.M., Delannay, X., Dube, P., Layton, J., Anderson, E.J., Kaniewska, M., Proksch, R.K., Horsch, R.B., Rogers, S.G., Fraley, R.T. and Beachy, R.N. (1988) Virus tolerance, plant growth, and field performance transgenic tomato plants expressing coat protein from tobacco mosaic virus. *Biotechnology* 6, 403–409.

Neuenschwander, P. and Hammond, W.N.O. (1988) Natural enemy activity following the introduction of *Epidinocarsis lopezi* (Hymenoptera, Encyrtidae) against the cassava mealybug, *Phenacoccus manihoti* (Homoptera, Pseudococcidae), in southwestern Nigeria. *Environmental Entomology* 17, 894–902.

Neuenschwander, P. and Haug, T. (1990) New technologies for rearing *Epidinocarsis lopezi*, a biological control agent against the cassava mealybug. In: Anderson, T.A. and Leppla, N. (eds) *Advances and Applications in Insect Rearing.* Westview Press, Boulder, Colorado, USA, pp. 353–377.

Neuenschwander, P., Hennessey, R.D. and Herren, H.R. (1987) Food web of insects associated with the cassava mealybug, *Phenacoccus manihoti* Matile-Ferrero (Hemiptera: Pseudococcidae), and its introduced parasitoid *Epidinocarsis lopezi* (De Santis) (Hymenoptera: Encyrtidae), in Africa. *Bulletin of Entomological Research* 77, 177–189.

Neuenschwander, P., Hammond, W.N.O., Gutierrez, A.P., Cudjoe, A.R. and Baumgärtner, J.U. (1989) Impact assessment of the biological control of the cassava mealybug, *Phenacoccus manihoti* Matile-Ferrer (Hemiptera: Pseudococcidae) by the introduced parasitoid *Epidinocarsis lopezi* (De Santis) (Hymenoptera: Encyrtidae). *Bulletin of Entomological Research* 79, 579–594.

Neuenschwander, P., Hammond, W.N.O., Ajuonu, O., Gado, A. and Echendu, N. (1990) Biological control of the cassava mealybug, *Phenacoccus manihoti* (Hom., Pseudococcidae) by *Epidinocarsis lopezi* (Hym., Encyrtidae) in West Africa, as influenced by climate and soil. *Agricultural Ecosystems Environment* 32, 39–55.

Ng, N.Q. and Marechal, R. (1985) Cowpea taxonomy, origin and germ plasm. In: Singh, S.R. and Rachie, K.O. (eds) *Cowpea Research, Production and Utilisation.* John Wiley & Sons, New York, USA, pp. 11–22.

Nogge, G. (1978) Aposymbiotic tsetse flies, *Glossina morsitans morsitans* obtained by feeding on rabbit immunized specifically with symbionts. *Journal of Insect Physiology* 24, 299–304.

Norgaard, R.B. (1988) The biological control of cassava mealybug in Africa. *American Journal of Agricultural Economics* 70, 366–371.

Notteghem, J.L. (1985) Definition of a strategy for the use of resistance through genetic analysis of the host–pathogen relationship: the case of the rice *Pyricularia oryzae* relation. *Agronomie Tropicale* 40, 129–147.

Novozhilov, K.V. (1990) Integrated pest control on plants in the USSR. FAO/ UNEP/ USSR Workshop on Integrated Pest Management, 12–15 June 1990, Kishinev, Moldavia, USSR (unpublished).

NRI (1992) *A Synopsis of Integrated Pest Management in Developing Countries in the Tropics.* Natural Resources Institute, Chatham, 20 pp.

Nsiama She, H.D. (1985) The bioecology of the predator *Hyperaspis jucunda* Muls. (Coleoptera: Coccinellidae) and the temperature responses of its prey, the cassava mealybug *Phenacoccus manihoti* Mat. Ferr. PhD Thesis, University of Ibadan, Nigeria, 300 pp.

Nwanze, K.F. (1982) Relationships between cassava root yields and infestations by the mealybug, *Phenacoccus manihoti. Tropical Pest Management* 28, 27–32.

Nyiira, Z.M. (1971) The status of insect pests on cowpea, *Vigna unguiculata* (L.) Walp. in Uganda and their control. *PANS* 17, 194–197.

Nyiira, Z.M. (1973) Pest status of thrips and lepidopterous species on vegetables in Uganda. *East African Agricultural Forestry Journal* 39, 131–135.

Ocampo, P.P., Varca, L.M. and Tejada, A.W. (1991) Pesticide residue on vertebrate and invertebrate organisms in rice agroecosystem. Paper presented at the Workshop on the Environmental and Health Impacts of Pesticide Use in Rice Culture, International Rice Research Institute, Los Baños, Philippines, 14 pp.

Ogawa, T., Tabien, R.E., Yamamoto, T., Busto, G.A. and Ikeda, R. (1990) Breeding for near-isogenic lines for resistance to bacterial blight in rice. *Rice Genetics Newsletter* 7, 10.

Oka, I.N. (1990) The Indonesia National Integrated Pest Management Program: success and challenges. Paper presented at the UNDP Workshop on Integrated Pest Management, New York, USA, 14 pp.

O'Kane, G.J.P. and Gehring, W.J. (1987) Detection *in situ* of genomic regulatory elements in *Drosophila. Proceedings of the National Academy of Sciences USA* 84, 9123–9127.

Okwakpam, B.A. (1967) Three species of thrips in cowpea flowers in the dry season at Badeggi, Nigeria. *Nigerian Entomologist Magazine* 3, 45–46.

Oliveira, E.B., Gazzoni, D.L., Corso, I.C., Villas Boas, G.L. and Hoffmann-Campo, C.B. (1988) *Pesquisa com inseticidas em soja: sumario dos resultados alcancados entre 1975 e 1987.* Documentos No. 30, EMBRAPA-CNPSo, Londrina, PR, Brasil, 260 pp.

O'Neill, S.L. and Karr, T.L. (1990) Bidirectional incompatibility between conspecific populations of *Drosophila simulans. Nature* 348, 178–180.

O'Neill, S.L. Giordano, R., Colbert, A.M.E., Karr, T.L. and Robertson, H.M. (1992) 16S rRNA phylogenetic analysis of the bacterial endosymbionts associated with cytoplasmic incompatibility in insects. *Proceedings of the National Academy of Sciences USA* 89, 2699–2702.

O'Neill, S.L., Gooding, R.H. and Aksoy, S. (1993) Phylogenetically distant symbiotic microorganisms reside in *Glossina* midgut and ovary tissue. *Medical and Veterinary Entomology* 7, 377–383.

Ooi, P.A.C. (1988) Ecology and surveillance of *Nilaparvata lugens* (Stål.): implications for its management in Malaysia. Unpublished PhD thesis, University of Malaya, Kuala Lumpur, Malaysia.

Ooi, P.A.C. and Shepard, B.M. (1994) Predators and parasitoids of rice insect pests. In: Heinrichs, E.A. (ed) *Biology and Management of Rice Insect Pests.* Wiley Eastern, New Delhi, pp. 583–612.

Ooi, P.A.C. and Sudderuddin, K.I. (1978) Control of diamondback moth in Cameron Highlands, Malaysia. In: Amin, L.L., Abdul Aziz, S.A.K., Lim, G.S., Singh, K.C., Tan, A.M. and Varghese, G. (eds) *Proceedings Plant Protection Conference 1978.* Rubber Research Institute, Kuala Lumpur, Malaysia, pp. 193–227.

Ooi, P.A.C. and Waage, J.K. (1994) Biological control in rice applications and research needs. In: Teng, P.S. *et al.* (eds) *Rice Pest Science and Management.* IRRI, Los Baños, The Philippines. pp. 209–216.

Ooi, P.A.C., Lim, G.S., Ho, T.H., Manalo, P.L. and Waage, J.K. (1992) *Integrated Pest Management in the Asia–Pacific Region. Proceedings of a Conference on Integrated Pest Management in the Asia–Pacific Region, 23–27 September 1991, Kuala Lumpur.* CAB International, Wallingford, Oxon., UK.

Orbach, M.J., Porro, E.B. and Yanofsky, C. (1986) Cloning and characterization of the gene for β-tubulin from a benomyl-resistant mutant of *Neurospora crassa* and its use as a dominant selectable marker. *Molecular and Cellular Biology* 6, 2452–2461.

Orr, W.C. and Sohal, R.S. (1992) The effects of catalase gene overexpression on life span and resistance to oxidative stress in transgenic *Drosophila melanogaster. Archives of Biochemistry and Biophysics* 297, 35–41.

Otvos, I.S., Cunningham, J.C. and Knapp, W.J. (1989) Aerial application of two baculoviruses against the western spruce budworm, *Choristoneura occidentalis*

Freeman (Lepidoptera: Tortricidae) in British Columbia. *Canadian Entomology* 121, 209–217.

Ou, S.H. (1980) Pathogen variability and host resistance in rice blast disease. *Annual Review of Phytopathology* 18, 167–187.

Ou, S.H. (1985) *Rice Diseases*, 2nd edn. Commonwealth Mycological Institute, Kew, Surrey, UK.

Pace, G.W., Grote, W., Pitt, D.E. and Pitt, J.M. (1986) Liquid culture of nematodes. International Patent No. WO 86/10174.

Packer, L. and Owen, R.E. (1992) Variable enzyme systems in the Hymenoptera. *Biochemical Systematics and Ecology* 20, 1–7.

Padidam, M. (1991) Rational deployment of B.t. strains for control of insect pests in India. *Current Science* 60, 464–465.

Palazzolo, M.J., Hyde, D., VijayRaghavan, K., Mechlenburg, K., Benzer, S. and Meyerowitz, E.M. (1989) Use of a new strategy to isolate and characterize 436 *Drosophila* cDNA clones corresponding to RNAs detected in adult heads but not in early embryos. *Neuron* 3, 527–539.

Palittapongarnpim, P., Chomyc, S., Fanning, A. and Kunimoto, D. (1993) DNA fingerprinting of *Mycobacterium tuberculosis* isolates by ligation-mediated polymerase chain reaction. *Nucleic Acids Research* 21, 761–762.

Palli, S.R., Riddiford, L.M. and Hiruma, K. (1991) Juvenile hormone and retinoic acid receptors in *Manduca epidermis. Insect Biochemistry* 21, 7–15.

Panizzi, A.R. (1985) Dynamics of phytophagous pentatomids associated to soybean in Brazil. In: Shibles, R. (ed.) *World Soybean Research Conference III: Proceedings.* Westview Press, Boulder, Colorado, USA, pp. 674–680.

Panizzi, A.R., Correa, B.S., Gazzoni, D.L., Oliveira, E.B., Newman, G.G. and Turnipseed, S.G. (1977a) *Insetos da soja no Brasil.* Boletim Tecnico 1, CNPSo-EMBRAPA, Londrina, Brasil, 20 pp.

Panizzi, A.R., Correa, B.S., Newman, G.G. and Turnipseed, S.G. (1977b) Efeito de inseticidas na populacao das principais pragas da soja. *Anais da Sociedade Entomologica do Brasil* 6, 264–275.

Paran, I., Kesseli, R. and Michelmore, R. (1991) Identification of RFLP and RAPD markers linked to downy mildew resistance genes in lettuce using near-isogenic lines. *Genome* 34, 1021–1027.

Parlevleit, J.E. (1979) Components of resistance that reduce the rate of epidemic development. *Annual Review of Phytopathology* 17, 203–222.

Parlevleit, J.E. (1981) Stabilizing selection in crop pathosystems: an empty concept or a reality? *Euphytica* 30, 259–269.

Parlevleit, J.E. (1983) Models explaining the specificity and durability of host resistance derived from observations on the barley–*Puccinia hordei* system. In: Lamberti, F., Waller, J.M. and Van der Graaff, N.A. (eds) *Durable Resistance in Crops*. Plenum, London, UK, pp. 57–80.

Parlevleit, J.E. (1988) Identification and evaluation of quantitative resistance. In: Leonard, K.J. and Fry, W.E. (eds) *Plant Disease Epidemiology, Genetics, Resistance and Management*. McGraw-Hill, New York, USA, pp. 215–248.

Parlevleit, J.E. and Zadoks, J.C. (1977) The integrated concept of disease resistance: a view including horizontal and vertical resistance in plants. *Euphytica* 26, 5–21.

Paskewitz, S.M., Brown, M.R., Collins, F.H. and Lea, A.O. (1989) Ultrastructural localization of phenoloxidase in the midgut of refractory *Anopheles gambiae* and association of the enzyme with encapsulated *Plasmodium cynomolgi*. *Journal of Parasitology* 75, 594–600.

Patel, N.H., Martin-Blanco, E., Coleman, K.G., Poole, S.J., Ellis, M.C., Kornberg, T.B. and Goodman, C.S. (1989) Expression of engrailed protein in arthropods, annelids and chordates. *Cell* 58, 955–968.

Paterson, A.H., Lander, E.S., Hewitt, J.D., Peterson, S., Lincoln, S.E. and Tanksley, S.D. (1988) Resolution of quantitative traits into Mendelian factors, using a complete linkage of restriction fragment length polymorphisms. *Nature* 335, 721–726.

Paterson, A.H., Damon, S., Hewitt, J.D., Zamir, D., Rabinowitch, H.D., Lincoln, S.E., Lander, E.S. and Tanksley, S.D. (1991a) Mendelian factors underlying quantitative traits in tomato: comparison across species, generations and environments. *Genetics* 127, 181–197.

Paterson, A.H., Tanksley, S.D. and Sorrells, M.E. (1991b) DNA markers in plant improvement. *Advances in Agronomy* 46, 39–90.

Pathak, M.D. and Dyck, V.A. (1974) Developing an integrated method of rice insect pest control. *PANS* 19, 534–544.

Paul, A.V.N. and Agarwal, R.A. (1990) Persistent toxicity of some insecticides to egg parasitoid, *Trichogramma brasiliensis* Ashmead. *Indian Journal of Entomology* 51, 273–277.

Pavan, O.H.O. and Ribeiro, H.C.T. (1989) Selection of a baculovirus strain with a bivalent insecticidal activity. *Memorias do Instituto Oswaldo Cruz* 84(3), 63–65.

Pawar, C.S., Sithanantham, S., Bhatnagar, V.S., Srivastava, C.P. and Reed, W. (1988) The development of sex pheromone trapping of *Heliothis armigera* at ICRISAT, India. *Tropical Pest Management* 34, 39–43.

Pedersen, W.L. and Leath, S. (1988) Pyramiding major genes for resistance to maintain residual effects. *Annual Review of Phytopathology* 26, 369–378.

Perkins, H.D. and Howells, A.J. (1992) Genomic sequences with homology to the P element of *Drosophila melanogaster* occur in the blowfly *Lucilia cuprina*. *Proceedings of the National Academy of Sciences USA* 89, 10753–10757.

Perkins, J.H. (1982) *Insects, Experts and the Insecticide Crisis*. Plenum Press, London, UK.

Perlak, F.J. and Fischhoff, D.A. (1990) Expression of *Bacillus thuringiensis* insect control proteins in genetically modified plants. In: Osmond, G. (ed.) *Proceedings of the Vth International Colloquium on Invertebrate Pathology and Microbial Control*. Society for Invertebrate Pathology Adelaide, Australia, pp. 461–465.

Perlak, F.J., Deaton, R.W., Armstrong, T.A., Fuchs, R.L., Sims, S.R., Greenplate, J.T. and Fischhoff, D.A. (1990) Insect resistant cotton plants. *Bio/Technology* 8, 939–943.

Perlak, F.J., Fuchs, R.L., Dean, D.A., McPherson, S.L. and Fischhoff, D.A. (1991) Modifications of the coding sequence enhances plant expression of insect control protein genes. *Proceedings of the National Academy of Sciences USA* 88, 3324–3328.

Perlak, F.J., Stone, T.B., Muskopf, Y.M., Petersen, L.J., Parker, G.B., McPherson, S.A., Wyman, J., Love, S., Reed, G., Biever, D. and Fischhoff, D.A. (1993) Genetically improved potatoes: protection from damage by Colorado potato beetles. *Plant Molecular Biology* 22, 313–321.

Perring, T.M., Cooper, A.D., Rodriguez, R.J., Farrar, C.A. and Bellows, T.S. (1993) Identification of a whitefly species by genomic and behavioural studies. *Science* 259, 74–76.

Perry, J.N. and Wall, C. (1986) The effect of habitat on the flight of moths orienting to pheromone sources. In: Payne, T.L., Birch, M.C. and Kennedy, C.E.J. (eds) *Mechanisms in Insect Olfaction.* Clarendon, Oxford, UK, pp. 91–96.

Persley, G.J. (1990) *Beyond Mendel's Garden: Biotechnology in the Service of World Agriculture.* CAB International, Wallingford, Oxon., UK, 155 pp.

Persley, G.J. (ed). (1991) *Agricultural Biotechnology: Opportunities for International Development.* CAB International, Wallingford, Oxon., UK.

Phillips, J.P., Xin, J.H., Kirby, K., Milne, C.P. Jr, Krell, P. and Wild, J.R. (1990) Transfer and expression of an organophosphate insecticide-degrading gene from *Pseudomonas* in *Drosophila melanogaster. Proceedings of the National Academy of Sciences USA* 87, 8155–8159.

Phillips-Howard, P.A. and Doberstyn, E.B. (1990) *Malaria, Present Situation of the Disease.* WHO document, CTD/CP12/90.6, Geneva.

Pingali, P.L. (1992) Impact of pesticides on the lowland paddy eco-system: results from a multi-disciplinary study in the Philippines. Paper presented at the Workshop on Measuring the Health and Environmental Effects of Pesticides, Bellagio, Italy, 43 pp.

Pingali, P.L., Moya, P.P. and Velasco, L.E. (1990) Prospects for rice yield improvement in the post-green revolution Philippines. *Philippine Review of Economics and Business* 27(2).

Pinheiro, M.L.S., Castro, M.E.B., Sihler, W., Irineu, B.P. and Moscardi, F. (1990) Analysis of DNA and proteins of four geographical isolates of *Anticarsia gemmatalis* multiple nuclear polyhedrosis virus. In: *Proceedings of the Vth International Colloquium on Invertebrate Pathology and Microbial Control.* Society for Invertebrate Pathology, Adelaide, Australia, pp. 264.

Pinto, L., Stocker, R.F. and Rodrigues, V. (1988) Anatomical and neurochemical classification of the antenna glomeruli in *Drosophila melanogaster* Meigen (Diptera: Drosophilidae). *International Journal of Insect Morphology and Embryology* 17, 335–344.

PIRG (1993) *Farmers against Dunkel. Action Alert.* Public Interest Research Group, New Delhi, India.

Pirone, T.P. (1991) Viral genes and gene products that determine insect transmissibility. *Seminars in Virology* 2, 81–87.

Podgwaite, J.D., Reardon, R.C., Kolodny-Hirsch, D.M. and Walton, G.S. (1991) Efficacy of ground application of the gypsy moth (Lepidoptera: Lymantriidae) nucleopolyhedrosis virus product, Gypchek. *Journal of Economic Entomology* 84, 440–444.

Poinar, G.O. (1979) *Nematodes for Control of Insect Pests.* CRC Press, Boca Raton, Florida, USA, 277 pp.

Poinar, G.O. and Thomas, G.M. (1966) Significance of *Achromobacter nematophilus* Poinar and Thomas (Achromobacteraceae: Eubacteriales) in the development of the nematode, DD136 (*Neoaplectana* sp. Steinernematidae). *Parasitology* 56, 385–390.

Post, R.J. and Crampton, J.M. (1988) The taxonomic use of variation in repetitive DNA sequences in the *Simulium damnasum* complex. In: Service, M.W. (ed.) *Biosystematics of Haemotophagous Insects*. Oxford University Press, Oxford, UK, pp. 245–256.

Powell, P.A., Nelson, R.S., De, B., Hoffmann, N., Rogers, S.G. *et al.* (1986) Delay of disease development in transgenic plants that express the tobacco mosaic virus coat protein gene. *Science* 232, 738–743.

Powell, P.A., Stark, D., Sanders, P.R. and Beachy, R.N. (1989) Protection against tobacco mosaic virus in transgenic plants that express tobacco mosaic virus antisense RNA. *Proceedings of the National Academy of Sciences USA* 86, 6949–6952.

Powers, P.A. and Ganetzky, B. (1991) On the components of segregation distortion in *Drosophila melanogaster* V. Molecular analysis of the Sd locus. *Genetics* 129, 131–144.

Presnail, J.K. and Hoy, M.A. (1992) Stable genetic transformation of a beneficial arthropod by microinjection. *Proceedings of the National Academy of Sciences USA* 89, 7732–7736.

Prince, J.P. and Tanksley, S.D. (1992) Restriction fragment length polymorphisms in plant breeding and genetics. *Proceedings of the Royal Society of Edinburgh* 99B(3/4), 23–29.

Prior, C. (1989) Biological pesticides for low external-input agriculture. *Biocontrol News and Information* 10, 17–22.

Qin, X., Kao, R., Yang, and Zhang, G. (1988) Study on application of entomopathogenic nematodes of *Steinernema bibionis* and *Steinernema feltiae* to control wood borers. *Forest Research* 1, 179–185.

Rafalski, J.A. and Tingey, S.V. (1993) Genetic diagnostics in plant breeding: RAPDs, microsatellites and machines. *Trends in Genetics* 9, 275–280.

Rahardja, U., Whalon, M.E., Garcia-Salazar, C. and Yan, Y.T. (1992) Field detection of x-disease mycoplasma-like organism in *Paraphlepsius irroratus* (Say) using a DNA probe. *Environmental Entomology* 21, 81–88.

Ram, S. and Gupta, M.P. (1990) Integrated pest management in lucerne (*Medicago sativa* L.) and its economics in India. *Tropical Pest Management* 36, 258–262.

Ramalho, F.S., McCarty, J.C., Jenkins, J.N. and Parrott, W.L. (1984) Distribution of tobacco budworm (Lepidoptera: Noctuidae) larvae within cotton plants. *Journal of Economic Entomology* 77, 591–594.

Ramsdale, C.D. and Coluzzi, M. (1975) Studies on the infectivity of tropical African strains of *Plasmodium falciparum* to some southern European vectors of malaria. *Parasitologia* 17, 39–48.

Rancourt, D.E., Peters, I.D., Walker, V.K. and Davies, P.L. (1990) Wolffish antifreeze protein from transgenic *Drosphila*. *Bio/Technology* 8, 453–457.

Rancourt, D.E., Davies, P.L. and Walker, V.K. (1992) Differential translatability of antifreeze protein mRNAs in a transgenic host. *Biochimica Biophysica Acta* 1129, 188–194.

Ray, K. and Rodrigues, V. (1993) Spatiotemporal appearance of the labellar taste bristles of *Drosophila melanogaster. Roux's Archives of Development Biology* 203, 340–350.

Ray, K., Hartenstein, V. and Rodrigues, V. (1992) Development of the labellar taste bristles in *Drosophila. Development Biology* 155, 26–37.

Raymond, M., Callaghan, A., Fort, P. and Pasteur, N. (1991) Worldwide migration of amplified insecticide resistance genes in mosquitoes. *Nature* 350, 151–153.

Ready, P.D., Smith, D.F. and Killick-Kendrick, R. (1988) DNA hybridizations on squash blotted sandflies to identify both *Phlebotomus papatasi* and infecting *Leishmania major*. *Medical and Veterinary Entomology* 2, 109–116.

Reichhart, J., Meister, M., Dumarcq, J., Zachary, D., Hoffman, D., Ruiz, C., Richards, G. and Hoffman, J.A. (1992) Insect immunity: developmental and inducible activity of the *Drosophila* diptericin promotor. *EMBO Journal* 11, 1469–1477.

Reimers, P.J. and Leach J.E. (1991) Race specific resistance to *Xanthomonas oryzae* pv. *oryzae* conferred by bacterial blight resistance gene *Xa-10* in rice (*Oryza sativa*) involves accumulation of a lignin-like substance in host tissues. *Physiological and Molecular Plant Pathology* 38, 39–55.

Reimers, P.J., Guo, A. and Leach, J.E. (1992) Increased activity of a cationic peroxidase associated with an incompatible interaction between *Xanthomonas oryzae* pv *oryzae* and rice (*Oryza sativa*). *Plant Physiology* 99, 1044–1050.

Reimers, P.J., Consignado, B. and Nelson, R.J. (1993) Wild species of *Oryza* with resistance to rice blast (Bl). *International Rice Research Notes* 18, 5.

Reimers, P.J., Amante-Bordeos, A., Calvero, A.C., Estrada, B.A., Mauleon, R., Nelson, R.J., Nahar, N.S., Shahjahan, A.K.M., Darwin, S. and Correa, F. (1994) Resistance to rice blast in a line derived from *Oryza minuta*. *International Rice Research Notes*, 19(2), 9–10.

Renvoize, B.S. (1973) The area of origin of *Manihot esculenta* as a crop plant – a review of the evidence. *Economic Botany* 26, 352–360.

Repetto, R. (1985) *Paying the Price: Pesticide Subsidies in Developing countries*. Research Report 2, World Resources Institute, Washington DC, USA.

Richards, P. (1995) Farmers' science and its application: the human habitat for IPM. Presented at 13th International Plant Protection Congress, The Hague, 7 July 1995.

Rigden, J. and Coutts, R. (1988) Pathogenesis-related proteins in plants. *Trends in Genetics* 4, 87–89.

Ripp, B.E., Banks, H.J., Bend, E.J., Calverley, P.J., Gay, J.E. and Navarro, S. (1984) *Controlled Atmospheres and Fumigation in Grain Storages*. Elsevier, Amsterdam, The Netherlands, 798 pp.

Rivera, C. and Gámez, R. (1986) Multiplication of maize rayado fino virus in the leaf-hopper vector, *Dalbulus maidis*. *Intervirology* 25, 76–82.

Robertson, H.M. (1993) The *mariner* transposable element is widespread in insects. *Nature* 362, 241–245.

Robinson, A.S., Savakis, C. and Louis, C. (1988) Status of molecular genetic studies in the medfly, *Ceratitis capitata*, in relation to genetic sexing. In: *Modern Insect Control: Nuclear Techniques and Biotechnology*. pp. 241–261.

Robinson, R.A. (1976) *Plant Pathosystems*. Springer-Verlag, Berlin, 184 pp.

Robinson, R.A. (1991) The genetic controversy concerning vertical and horizontal resistance. *Revista Mexicana de Fitopatologia* 9, 57–63.

Rochow, W.F. (1970) Barley yellow dwarf virus: phenotypic mixing and vector specificity. *Science* 167, 875–878.

Rochow, W.F., Foxe, M.J. and Muller, I. (1975) A mechanism of vector specificity for circulative aphid-transmitted plant viruses. *Annals of the New York Academy of Sciences* 266, 293–301.

Roderick, G.K. (1992) Estimating gene flow among insect populations in managed and natural systems. Abstract. In: *XIX International Congress of Entomology*. Beijing, China, p. 132.

Rodrigues, V. (1980) Olfactory behaviour of *Drosophila melanogaster*. In: Siddiqi, O., Babu, P., Hall, L. and Hall, J. (eds) *Development and Neurobiology of Drosophila*. Plenum, New York, USA. pp. 361–371.

Rodrigues, V. (1988) Spatial coding of olfactory information in the antennal lobe of *Drosophila melanogaster*. *Brain Research* 453, 299–307.

Rodrigues, V. and Buchner, E. (1984) 3H–2 deoxyglucose mapping of odour-induced activity in the antennal lobes of *Drosophila melanogaster*. *Brain Research* 324, 372–378.

Rodrigues, V. and Siddiqi, O. (1978) Genetic analysis of chemosensory pathway. *Proceedings of the Indian Academy of Sciences* 87, 147–160.

Rodrigues, V. and Siddiqi, O. (1981) A gustatory mutant of *Drosophila* defective in pyranose receptors. *Molecular and General Genetics* 181, 406–408.

Rodrigues, V., Sathe, S., Pinto, L., Balakrishnan, R. and Siddiqi, O. (1991) Closely linked regions of the X-chromosome define central and peripheral steps in gustatory processing. *Molecular and General Genetics* 226, 265–276.

Rodriguez, C.M., Madden, L.V., Nault, L.R. and Louie, R. (1993) Spread of maize chlorotic dwarf virus from infected corn and Johnsongrass by *Graminella nigrifrons*. *Plant Disease* 77, 55–60.

Rola, A. and Pingali, P.L. (1993) *Pesticides, Rice Productivity, and Farmers' Health – an Economic Assessment*. IRRI, Philippines.

Romans, P., Seeley, D.C., Kew, Y. and Gwadz, R.W. (1991) Use of a restriction fragment length polymorphism (RFLP) as a genetic marker in crosses of *Anopheles gambiae* (Diptera: Culicidae): independent assortment of a diphenol oxidase RFLP and an esterase locus. *Journal of Medical Entomology* 18, 147–151.

Romao, J. and Hamer, J.E. (1992) Genetic organization of a repeated DNA sequence family in the rice blast fungus. *Proceedings of the National Academy of Sciences USA* 89, 5316–5330.

Rombäch, M.C. and Gallagher, K.D. (1994) The brown planthopper: promises, problems and prospects. In Heinrichs, E.A. (ed.) *Biology and Management of Rice Insects*. John Wiley Eastern, New Delhi.

Ronald, P.C. and Tanksley, S.D. (1991) Genetic and physical mapping of the bacterial blight resistance gene *Xa–21*. *Rice Genetics Newsletter* 8, 142.

Rosen, L., Rozeboom, L.E., Reeves, W.C., Sangrain, J. and Gubler, D.R. (1976) A field trial of competitive displacement of *Aedes albopictus* on a Pacific atoll. *American Journal of Tropical Medicine and Hygiene* 25, 906.

Rosseto, C.J. (1989) Breeding for resistance to stink bugs. In: Pascale, A.J. (ed.) *Proceedings of World Soybean Research Conference IV, Buenos Aires, Argentina*. Asociacion Argentina de la Soja, Buenos Aires, pp. 2060–2064.

Rosseto, C.J., Lourencao, A.L., Igue, T. and Miranda, M.A.C. (1981) Picadas de alimentacao de *Nezara viridula* em cultivares e linhagens de soja de diferentes graus de suscetibilidade. *Bragantia* 40, 109–114.

Roush, R.T. (1989) Designing resistance management programs: how can you choose? *Pesticide Science* 26, 423–441.

Roush, R.T. (1993) Occurrence, genetics and management of insecticide resistance. *Parasitology Today* 9, 174–179.

Roush, R.T. and Daly, J.C. (1990) The role of population genetics in resistance research and management. In: Roush, R.T. and Tabashnik, B.E. (eds) *Pesticide Resistance in Arthropods*, Chapman and Hall, New York, USA, pp. 97–152.

Roush, R.T. and McKenzie, J.A.(1987) Ecological genetics of insecticide and acaricide resistance. *Annual Review of Entomology* 32, 361–380.

Roush, R.T. and Tabashnik, B.E. (1990) *Pesticide Resistance in Arthropods*. Chapman and Hall, New York, USA, 343 pp.

Roush, R.T. and Tingey, W.M. (1992) Evolution and management of resistance in the Colorado potato beetle, *Leptinotarsa decemlineata*. In: Denholm, I., Devonshire, A.I. and Holloman, D.W. (eds) *Resistance '91 – Achievements and Developments in Combating Pesticide Resistance*. Elsevier Applied Science, Harlow, Essex, UK, pp. 61–74.

Rubia, E.C., Shepard, B.M. Yambao, E.B., Ingram, K.T., Arida, G.S. and Penning de Vries, F. (1989) Stem borer damage and grain yield of flooded rice. *Journal of Protection in the Tropics* 6, 205–211.

Rubia, E.G. (1990) Simulation of rice yield reduction caused by stem borer. *International Rice Research Newsletter* 15, 34–35.

Rubin, G.M. and Spradling, A.C. (1982) Genetic transformation of *Drosophila* with transposable element vectors. *Science* 218, 348–353.

Rushtapakornchai, W. and Vattanatangum, A. (1986) Present status of insecticidal control of diamondback moth in Thailand. In: Talekar, N.S. and Griggs, T.D. (eds) *Diamondback Moth Management*. Asian Vegetable Research and Development Center, Shanhua, Taiwan, pp. 307–312.

Ryder, M.H. and Jones, D.A. (1990) Biological control of crown gall. In: Hornby, D. (ed.) *Biological Control of Soil-borne Plant Pathogens*. CAB International, Wallingford, UK, pp. 45–64.

Saiki, R.K., Gelfand, D.H., Stoffel, S., Scharf, S.J., Higuchi, R., Horn, G.T., Mullis, K.B. and Erlich, H.A. (1988) Primer-directed enzymatic amplification of DNA with a thermostable DNA polymerase. *Science* 239, 487–491.

Sakai, R.K. and Miller, L.H. (1992) Effects of heat shock on the survival of transgenic *Anopheles gambiae* (Diptera: Culicidae) under antibiotic selection. *Journal of Medical Entomology* 29, 374–375.

Salifu, A.B. (1986) Studies on aspects of the biology of the flower thrips, *Megalurothrips sjostedti* (Trybom) with particular reference to resistance in host cowpea, *Vigna unguiculata* (L.) Walp. PhD thesis, Wye College, University of London, UK.

Santamaria, P. (1986) Injecting eggs. In: Roberts, D.B. (ed.) *Drosophila, A Practical Approach*. IRL Press, Oxford, UK, pp. 159–173.

Santos, M.A., dos Ferraz, S. and Muchojev, J.J. (1992) Evaluation of 20 species of fungi from Brazil for biocontrol of *Meloidogyne incognita* race 3. *Nematropica* 22, 183–192.

Saxena, H. (1991) Microbial control of insect pests. *Farmer and Parliament* July, 7–11.

Schafer, J.F. and Roelfs, A.P. (1985) Estimated relation between numbers of urediniospores of *Puccinia graminis* f. sp. *tritici* and rates of occurrence of virulence. *Phytopathology* 75, 749–750.

Schedl, A., Beermann, F., Thies, F., Montoliu, L., Kelsey, G. and Schutz, G. (1992) Transgenic mice generated by pronuclear injection of a yeast artificial chromosome. *Nucleic Acids Research* 20, 3073–3077.

Schneider, J.C., Roush, R.T., Kitten, W.F. and Laster, M.L. (1989) Movement of *Heliothis virescens* (F.) (Lepidoptera: Noctuidae) in Mississippi in the spring: implications for area wide management. *Environmental Entomology* 18, 438–446.

Schots, A., De Boer, J., Schouten, A, Roosien, J., Zilverentant, J.F., Pomp, H., Bouwman-Smits, L., Overmars, H., Gommers, J.F., Visser, B., Stiekema, W.J. and Bakker, J. (1992) Plantibodies: a flexible approach to design resistance against pathogens. *Netherlands Journal of Plant Pathology* 98 (Supplement 2), 183–191.

Schulthess, F. (1987) The interactions between cassava mealybug (*Phenacoccus manihoti* Mat.-Ferr) populations and cassava *Manihot esculenta* (Crantz) as influenced by weather. PhD thesis. Swiss Federal Institute of Technology Zurich, Switzerland, 136 pp.

Scott, M.P. and Williams, S.M. (1993) Comparative reproductive success of communally breeding burying beetles as assessed by PCR with randomly amplified polymorphic DNA. *Proceedings of the National Academy of Sciences USA* 90, 2242–2245.

Scott, M.P., Haymes, K.M. and Williams, S.M. (1992) Parentage analysis using RAPD PCR. *Nucleic Acids Research* 20(20), 5493.

Scott, P.R. (1993) Biotechnology R&D trends: science policy for development. *AgBiotech News and Information* 5(1), 29N–31N.

Sears, M.K., Jaques, R.P. and Laing, J.E. (1983) Utilization of action thresholds for microbial and chemical control of lepidopterous pests (Lepidoptera: Noctuidae, Pieridae) on cabbage. *Journal of Economic Entomology* 76, 368– 374.

Sedlmeier, R. and Altenbuchner, J. (1992) Cloning and DNA sequence analysis of the mercury resistance genes of *Streptomyces lividans*. *Molecular and General Genetics* 236, 76–85.

Sen, A. (1981) *Poverty and Famines: an essay on Entitlement and Deprivation*. Clarendon Press, Oxford.

Sentry, J.W. and Kaiser, K. (1992) P element transposition and targeted manipulation of the *Drosophila* genome. *Trends in Genetics* 8, 329–331.

Serdar, C.M., Murdock, D.C. and Rohde, M.F. (1989) Parathion hydrolase gene from *Pseudomonas diminuta*: subcloning, complete nucleotide sequence, and expression of the mature portion of the enzyme in *Escherichia coli*. *Bio/Technology* 7, 1151–1155.

Serikawa, T., Kuramoto, T., Hilbert, P., Mori, M., Yamada, J., Dubay, C.J., Lindpainter, K., Ganten, D., Guenet, J.L., Lathrop, G.M. and Beckman, J.S. (1992) Rat gene mapping using PCR-analyzed microsatellites. *Genetics* 131, 701–721.

Sharma, D. (1991) India battles to eradicate major crop pests. *New Scientist* 10, 15.

Shirk, P., O'Brochta, D.A., Roberts, E. and Handler, A.M. (1988) Sex-specific

selection using chimeric genes. In: Hedlin, P.A. (ed.) *Biotechnology for Crop Protection*. American Chemical Society, pp. 135–145.
Shotoski, F.A. and Fallon, A.M. (1990) An amplified dihydrofolate reductase gene contains a single intron. *European Journal of Biochemistry* 201, 157–160.
Showers, W.B. (1993) Diversity and variation of European corn borer populations. In: Kim, K.C. and McPheron, B.A. (eds) *Evolution of Insect Pests: Patterns of Variation*. John Wiley and Sons, New York, USA, pp. 287–309.
Shufran, K.A., Black, W.C. and Margolies, D.C. (1991) DNA fingerprinting to study spatial and temporal distributions of an aphid, *Schizaphis graminum*. *Bulletin of Entomological Research* 81, 303–313.
Siddiqi, O. (1987) Neurogenetics of olfaction in *Drosophila melanogaster*. *Trends in Genetics* 3, 137–142.
Sikora, R.A. (1992) Management of the antagonistic potential in agricultural ecosystems for the biological control of plant parasitic nematodes. *Annual Review of Phytopathology* 30, 245–270.
Silva, M.T.B., Corso, I.C., Belarmino, L.C., Link, D., Tonet, G.L., Gomez, S.A. and Santos, B. (1988) Avaliacao de inseticidas sobre predadores de pragas da soja, em dez anos agricolas, no Brasil. *Trigo e Soja* 96, 3–16.
Simmonds, N.W. (1991) Genetics of horizontal resistance to diseases of crops. *Biological Reviews* 66, 189–241.
Simpson, P. (1990) Lateral inhibition and the development of the sensory bristles of the adult peripheral nervous system of *Drosophila*. *Development* 109, 509–519.
Singh, K.R.P., Curtis, C.F. and Krishnamurthy, B.S. (1976) Partial loss of cytoplasmic incompatibility with age in males of *Culex fatigans*. *Annals of Tropical Medicine and Parasitology* 70, 463–466.
Singh, S.P. (1991) Highlights of the Biological Control Centre. *Biocontrol Newsletter* 1(1), 4.
Singh, S.P. (1992) Natural enemies of crop pests and weeds. Paper presented at CABI–ICAR Workshop, November 1992, Bangalore, India.
Singh, S.R. and Taylor, T.A. (1978) Pests of grain legumes and their control in Nigeria. In: Singh, S.R., Van Emden, H.F. and Taylor, T.A. (eds) *Pests of Grain Legumes: Ecology and Control*. Academic Press, London, UK, pp. 99–111.
Singh, S.R., Jackai, L.E.N., Dos Santos, J.H.R. and Adalla, C.B. (1990) Insect pests of cowpea. In: Singh, S.R. (ed.) *Insect Pests of Tropical Food Legumes*. John Wiley & Sons, Chichester, UK, pp. 43–89.
Sinkins, S.P., Braig, H.R. and O'Neill, S.L. (1995) *Wolbachia* superinfections and the expression of cytoplasmic incompatibility. *Proceedings of the Royal Society of London*, 261, 325–330.
Sitch, L.A. (1990) Incompatibility barriers operating in crosses of *Oryza sativa* with related species and genera. In: Gustafson, J.P. (ed.) *Gene Manipulation in Plant Improvement II*. Plenum Press, New York, USA, pp. 77– 94.
Sitch, L.A., Amante, A.D., Dalmacio, R.D. and Leung, H. (1989) *Oryza minuta*, a source of disease resistance for rice improvement. In: Mujeeb-Kazi, A. and Sitch, L.A. (eds) *Review of Advances in Plant Biotechnology 1985–1988. Proceedings of 2nd International Symposium on Genetic Manipulation in Crops*. International Maize and Wheat Improvement Centre, Lisbon, Mexico DF, Mexico, and International Rice Research Institute, Los Baños, Philippines, pp. 315–322.

Skinner, D., Leung, H. and Leong, S.A. (1990) Genetic map of the blast fungus *Magnaporthe grisea*. In: O'Brien, S.J. (ed.) *Genetic Maps, Locus Maps of Complex Genomes*. Cold Spring Harbor Laboratory, Cold Spring Harbor, New York, USA, pp. 382–383.

Slee, R. and Bownes, M. (1990) Sex determination in *Drosophila melanogaster*. *Quarterly Review of Biology* 65, 175–204.

Smith, G.P. (1976) Evolution of repeated DNA sequences by unequal crossover. *Science* 191, 528–535.

Smith, R.A. and Couche, G.A. (1991) The phylloplane as a source of *Bacillus thuringiensis* variants. *Applied Environmental Microbiology* 57, 311–315.

Smits, P.H., van Velden, M.C., van de Vrie, M. and Vlak, J.M. (1987a) Feeding and dispersion of *Spodoptera exigua* larvae and its relevance for control with a nuclear polyhedrosis virus. *Entomologia Experimentalis et Applicata* 43, 67–72.

Smits, P.H., van de Vrie, M. and Vlak, J.M. (1987b) Nuclear polyhedrosis virus for control of *Spodoptera exigua* larvae on glasshouse crops. *Entomologia Experimentalis et Applicata*. 43, 73–80.

Smits, P.H., Rietstra, I.P. and Vlak, J.M. (1988) Influence of application techniques on the control of beet armyworm larvae (Lepidoptera: Noctuidae) with nuclear polyhedrosis virus. *Journal of Economic Entomology* 81, 470–475.

Soares, G.G. and Quick, T.C. (1992) MVP a novel bioinsecticide for the control of diamondback moth. In: Talekar, N.S. (ed.) *Diamondback Moth and Other Crucifer Pests: Proceedings of the Second International Workshop*, 10–14 December 1990. Asian Vegetable Research and Development Center, Tainan, Taiwan, pp. 129–138.

Sobral, B.W.S. and Honeycutt, R.J. (1993) High output genetic mapping of polyploids using PCR-generated markers. *Theoretical and Applied Genetics 86*, 105–112.

Soekardi, M. (1988) Pesticide residue control and monitoring in Indonesia. In: Teng, P.S. and Heong, K.L. (eds) *Pesticide Management and Integrated Pest Management in Southeast Asia*. Consortium for International Crop Protection, College Park, Maryland, USA, pp. 373–378.

Somers, J.M. and Bevan, E.A. (1969) The inheritance of the killer character in yeast. *Genetical Research* 13, 71– 83.

Sonneborn, T.M. (1965) The metagon: RNA and cytoplasmic inheritance. *American Naturalist* 99, 279–307.

Sosa-Gómez, D.R. (1990) Caracterizacao de isolados de *Beauveria* sp. e determinacao das exigencias termicas e hidricas de *Beauveria bassiana* (Bals.) Vuill. PhD thesis, Escola Superior 'Luiz de Queiroz', Universidade de Sao Paulo, Piracicaba, Brasil.

Southern, E.M. (1975) Detection of specific sequences among DNA fragments separated by gel electrophoresis. *Journal of Molecular Biology* 98, 503–517.

Southwood, T.R.E. (1966) *Ecological Methods*. Chapman and Hall, London, UK.

Spradling, A.C. and Rubin, G.M. (1982) Transposition of cloned P elements into *Drosophila* germline chromosomes. *Science* 218, 341–347.

Srinivas, K.R. (1992) Private investment in biotechnology promoted in India. *Biotech and Development Monitor* 11, 10–11.

Srinivas, K. and Krishna Moorthy, P.N. (1991) Indian mustard as a trap crop for management of major lepidopterous pests on cabbage. *Tropical Pest Manage-*

ment 37, 26–32.

Stanley, J., Frischmuth, T. and Ellwood, S. (1990) Defective viral DNA ameliorates symptoms of geminivirus infection in transgenic plants. *Proceedings of the National Academy of Sciences USA* 86, 8949–8952.

Stark, D.M. and Beachy, R.N. (1989) Protection against potyvirus infection in transgenic plants: evidence for broad spectrum resistance. *Bio/Technology* 7, 1257–1262.

Sternlicht, M., Barzakay, I. and Tamim, M. (1990) Management of *Prays citri* in lemon orchards by mass trapping of males. *Entomologia Experimentalis et Applicata* 55, 59–67.

Stevens, L. and Wade, M.J. (1990) Cytoplasmically inherited reproductive incompatibility in *Tribolium* flour beetles and effect on population size. *Genetics* 124, 367–372.

Stone, T.B., Simms, S.R. and Marrone, P.G. (1989) Selection of tobacco budworm for resistance to genetically engineered *Pseudomonas fluorescens* containing the delta endotoxin from *Bacillus thuringiensis* subsp. *kurstaki*. *Journal of Invertebrate Pathology* 53, 228–234.

Stouthamer, R., Luck, R.F. and Werren, J.H. (1992) Genetics of sex determination and the improvement of biological control using parasitoids. *Environmental Entomology* 21(3), 427–435.

Strong, D.R., Lawton, J.H. and Southwood, T.R.E. (1984) *Insects on Plants: Community Patterns and Mechanisms*. Blackwell Scientific Publications, Oxford, UK.

Subbarao, S.K., Curtis, C.F., Singh, K.R.P. and Krishnamurthy, B.S. (1974) Variation in cytoplasmic crossing type in populations of *Culex pipiens fatigans* Wied. from the Delhi area. *Journal of Communicable Disease* 6, 80–83.

Subbarao, S.K., Krishnamurthy, B.S., Curtis, C.F., Adak, T. and Chandrahas, R.K. (1977) Segregation of cytoplasmic incompatibility properties of *Culex pipiens fatigans*. *Genetics* 87, 381–390.

Suckling, D.M., Rodgers, D.J., Shaw, P.W., Wearing, C.H., Penman, D.R. and Chapman, R.B. (1987) Monitoring azinphosmethyl resistance in the light brown apple moth (Lepidoptera: Tortricidae) in New Zealand. *Journal of Economic Entomology* 80(4), 733–738.

Sun, S., Lindstrom, I. and Faye, I. (1990) The attacins, their structure, function and gene organization. Abstract. In: Hagedorn, H.H. *et al.* (eds) *Molecular Insect Science*. Plenum Press, New York, USA, pp. 367.

Suri, S., Nagarajan, S. and Pawar, A.D. (1992) State of art of IPM research and adoption in India. In: Ooi, P.A.C., Lim, G.S., Ho, T.H., Manalo, P.L. and Waage, J.K. (eds) *Integrated Pest Management in the Asia–Pacific Region*. CAB International, Wallingford, UK, pp. 95–111.

Sutherland, J.W. (1986) Grain aeration in Australia. In: Champ, B.R. and Highley, E. (eds) *Preserving Grain Quality by Aeration and In-store Drying*. ACIAR Proceedings 15, pp. 206–218.

Sutherland, J.W. (1988) The drying of grain with heated air. *Australian Refrigeration, Air Conditioning and Heating* January, 35–38.

Sutherland, J.W., Evans, D.E., Fane, A.G. and Thorpe, G.R. (1987) Disinfestation of grain with heated air. In: *Proceedings of 4th International Working Conference on Stored-products, September 1986*, Tel Aviv, Israel, pp. 261–274.

Swaminathan, M.S. (1991) *Biotechnology in Agriculture*. MacMillan, India.

Sylvestre, P. (1973) Aspects agronomiques de la production du manioc à la ferme d'état de Mantsumba (Rep. Pop. Congo). Inst. Rech. Agron. Trop. Paris, France, 35 pp. (mimeograph).

Szentasi, A. (1985) Behavioural aspects of female guarding and inter-male conflict in the Colorado potato beetle. In: Ferro, D.N. and Voss, R.M. (eds) *Proceedings of the Symposium on Colorado Potato Beetle. XVIIth International Congress on Entomology Research Bulletin* 704, pp. 127–137.

Tabachnick, W.J. (1991) Genetic control of oral susceptibility to infection of *Culicoides variipennis* with bluetongue virus. *American Journal of Tropical Medicine and Hygiene* 45, 666–671.

Tabashnik, B.E. (1993) Delaying insect adaptation to transgenic crops: seed mixtures and refugia reconsidered. *Proceedings of the Royal Society of London, Series B*.

Tabashnik, B.E. (1994) Evolution of resistance to *Bacillus thuringiensis*. *Annual Review of Entomology* 39, 47–79.

Tabashnik, B.E. and Croft, B.A. (1982) Managing pesticide resistance in crop–arthropod complexes: interactions between biological and operational factors. *Environmental Entomology* 11, 1137–1144.

Tabashnik, B.E., Finson, N., Schwartz, J.M., Caprio, M.A. and Johnson, M.W. (1992a) Diamondback moth resistance to *Bacillus thuringiensis* in Hawaii. In: Talekar, N.S. (ed.) *Diamondback Moth and Other Crucifer Pests: Proceedings of an International Workshop*, 10–14 December 1990, Shanhua, Taiwan. AVRDC Publication No. 92–368, pp. 175–204.

Tabashnik, B.E., Finson, N., Schwartz, J.M. and Johnson, M.W. (1992b) Inheritance of resistance to *Bacillus thuringiensis* in diamondback moth (Lepidoptera: Plutellidae). *Journal of Economic Entomology* 85, 1046–1055.

Taborsky, V. (1992) Small-scale processing of microbial pesticides. *FAO Agricultural Services Bulletin* 96.

Takasu, K. and Yagi, S. (1992) *In vitro* rearing of the egg parasitoid *Ooencyrtus nezarae ishii* (Hymenoptera: Encyrtidae). *Applied Entomology and Zoology* 27(1), 171–173.

Talbot, W.S., Swyryd, E.A. and Hogness, D.S. (1993) *Drosophila* tissues with different metamorphic responses to ecdysone express different ecdysone receptor isoforms. *Cell* 73, 1323–1337.

Talekar, N.S. (1992) Integrated management of diamondback moth: a collaborative approach in Southeast Asia. In: Ooi, P.A.C., Lim, G.S., Ho, T.H., Manalo, P.L. and Waage, J.K. (eds) *Integrated Pest Management in the Asia–Pacific Region*. CAB International, Wallingford, Oxon. UK, pp. 37–50.

Tamò, M. and Baumgärtner, J. (1993) Analysis of the cowpea agro-ecosystem in West Africa: I. A demographic model for carbon acquisition and allocation in cowpea, *Vigna unguiculata* (L.) Walp. *Ecological Modelling* 65, 95–121.

Tamò, M., Baumgärtner, J., Delucchi, V. and Herren, H.R. (1993a) Assessment of key factors responsible for the pest status of the bean flower thrips *Megalurothrips sjostedti* (Thysanoptera: Thripidae) in West Africa. *Bulletin of Entomological Research* 83, 251–258.

Tamò, M., Baumgärtner, J. and Gutierrez, A.P. (1993b) Analysis of the cowpea agro-ecosystem in West Africa: II. Modelling the interactions between cowpea and the bean flower thrips *Megalurothrips sjostedti* (Trybom) (Thysanoptera: Thripidae). *Ecological Modelling* 70, 89–113.

Tanimura, T., Isono, K., Takamura, T. and Shimada, I. (1982) Genetic dimorphism in taste sensitivity to trehalose in *Drosophila melanogaster*. *Journal of Comparative Physiology* 147, 433–437.

Taylor, R.A.J., Nault, L.R. and Styer, W.E. (1993) Experimental analysis of flight activity of three *Dalbulus* leafhoppers (Homoptera: Auchenarrhyncha) in relation to migration. *Annals of the Entomological Society of America* 86, 655–667.

Taylor, T.A. (1965) Observations on the bionomics of *Laspeyresia ptychora* Meyr. (Lepidoptera: Eucosmidae) infesting cowpea in Nigeria. *Bulletin of Entomological Research* 55, 761–773.

Taylor, T.A. (1969) On the population dynamics and flight activity of *Taeniothrips sjostedti* (Trybom) (Thysanoptera: Thripidae) on cowpea. *Bulletin of the Entomological Society of Nigeria* 2, 60–71.

Taylor, T.A. (1974) On the population dynamics of *Taeniothrips sjostedti* (Trybom) (Thysanoptera: Thripidae) on cowpea and an alternate host, *Centrosema pubescens* Benth., in Nigeria. *Revue de Zoologie Africaine* 88, 689–702.

Te Beest, D.O., Yang, X.B. and Cisar, C.R. (1992) The status of biological control of weeds with fungal pathogens. *Annual Review of Phytopathology* 30, 637–657.

Teng, P.S. (1992) Towards achieving an Asia–Pacific consensus on IPM. In: Ooi, P.A.C., Lim, G.S., Ho, T.H., Manalo, P.L. and Waage, J.K. (eds) *Integrated Pest Management in the Asia–Pacific Region*. CAB International, Wallingford, Oxon. UK, pp. 21–26.

Teng, P.S. and Heong, K.L. (1988) *Pesticide Management and Integrated Pest Management in Southeast Asia*. Consortium for International Crop Protection, College Park, Maryland, USA.

Teng, P.S., Klein-Gebbinck, H.W. and Pinnschmidt, H. (1991) An analysis of the blast pathosystem to guide modelling and forecasting. In: *Rice Blast Modelling and Forecasting, Selected Papers from the International Rice Research Conference*. International Rice Research Institute, Los Baños, Philippines, 99 pp.

Tepfer, M. (1993) Viral genes and transgenic plants. What are the potential environmental risks? Bio/Technology 11, 1125–1129.

Tette, J.P. and Jacobsen, B.J. (1992) Biologically intensive pest management in the tree fruit system. In: Zalom, F.G. and Fry, W.E. (eds) *Food, Crop Pests and the Environment: the Need and Potential for Biologically Intensive Integrated Pest Management*. APS Press, St Paul, Minnesota, USA, pp. 57–83.

Theodore, L., Ho, A. and Maroni, G. (1991) Recent evolutionary history of the metallothionein gene *mtn* in *Drosophila*. *Genetical Research* (Cambridge, UK) 58, 203–210.

Thomas, J.B., Bastiani, M.J., Bate, M. and Goodman, C.S. (1984) From grasshopper to *Drosophila*: a common plan for neuronal development. *Nature* 310, 203–207.

Thomas, M. and Waage, J.K. (1993) *Prospects for the Integration of Biological Control and Plant Resistance Breeding in IPM Systems for Resource-poor Farmers*. Study commissioned by International Institute of Biological Control and Technical Centre for Agriculture and Rural Cooperation.

Thompson, S.N. (1990) Nutritional considerations in propagation of entomophagous species. In: Baker, R.R. and Dunn, P.E. (eds) *New Directions in Biological Control: Alternatives for Suppressing Agricultural Pests and Diseases*. UCLA Symposia on Molecular and Cellular Biology New Series, Vol. 112, Alan R. Liss, pp. 389–404.

Thomson, N.J. (1987) Host plant resistance in cotton. *Journal of the Australian Institute of Agricultural Science* 53, 262–270.

Thottappilly, G., Monti, L., Mohan Raj, D.R. and Moore, A.W. (1992) *Biotechnology: Enhancing Research on Tropical Crops in Africa*. CTA/IITA copublication, IITA, Ibadan, Nigeria, 376 pp.

Tibayrenc, M., Neubauer, K., Barnabe, C., Guerrini, F., Skarecky, D. and Ayala, F.J. (1993) Genetic characterization of six parasitic protozoa: parity between random-primer DNA typing and multilocus enzyme electrophoresis. *Proceedings of the National Academy of Sciences USA*, 90, 1335–1339.

Tiedje, J.M., Colwell, R.K., Grossman, Y.L., Hodson, R.E., Lenski, R.E., Mack, R.M. and Regal, P.J. (1989) The planned introduction of genetically engineered organisms: ecological considerations and recommendations. *Ecology* 70, 298–315.

Tingey, S.V. and del Tufo, J.P. (1993) Genetic analysis with random amplified polymorphic DNA markers. *Plant Physiology* 101, 349–352.

Todd, J.L., Phelan, P.L. and Nault, R.L. (1990) Interaction between visual and olfactory stimuli during host-finding by the leafhopper *Dalbulus maidis* (Homoptera: Cicadellidae). *Journal of Chemical Ecology* 16, 2121–2133.

Todd, J.L., Madden, L.V. and Nault, L.R. (1991) Comparative growth and spatial distribution of *Dalbulus* leafhopper populations (Homoptera: Cicadellidae) in relation to maize phenology. *Environmental Entomology* 20, 556–564.

Toriyama, K. (1975) Recent progress of studies on horizontal resistance in rice breeding for blast resistance in Japan. In: *Horizontal Resistance to the Blast Disease of Rice*. CIAT ser. CE–9, Cali, Colombia, pp. 65–100.

Toung, Y.P.S., Hsieh, T.S. and Tu, C.P.D. (1990) *Drosophila* glutathione S-transferase 1–1 shares a region of sequence homology with the maize glutathione S-transferase III. *Proceedings of the National Academy of Sciences USA* 87, 31–35.

Treverrow, N. and Bedding, R.A. (1990) Control of the banana weevil borer, *Cosmopolites sordidus* (Germar) with entomopathogenic nematodes. In: *Proceedings and Abstracts of Vth International Colloquium on Invertebrate Pathology and Microbial Control*. Adelaide, Australia, p. 233.

Treverrow, N.L. and Bedding, R.A. (1993) Development of a system for the control of the banana weevil borer, *Cosmopolites sordidus* with entomopathogenic nematodes. In: Bedding, R.A., Akhurst, R.J. and Kaya, H.K. (eds) *Nematodes and the Biological Control of Insect Pests*. CSIRO Publications, Australia.

Triwidodo, H., Wiyaynti, R. and Hogg. D.B. (1992) The population dynamics and management program of white stem borer *Scirpophaga innotota* (Walker) (Lepidoptera: Pyralidae) in West Java, Indonesia. Presented at International Rice Research Conference, Los Baños, Phillipines, 15 pp.

Trowell, S.C., Lang, G.A. and Garsia, K.A. (1993) A *Heliothis* identification kit. In: Corey, S.A., Dall, D.J. and Milne, W.M. (eds) *Pest Control and Sustainable Agriculture*. CSIRO, Australia, pp. 176–179.

Trumble, J.T. (1985) Integrated pest management of *Liriomyza trifolii*: influence of avermectin, cyromazine, and methomyl on leafminer ecology in celery. *Agricultural Ecosystems and Environment* 12, 181–188.

Trumble, J.T. (1989) *Leafminer and Beet Armyworm Control on Celery. Final Report for 1988–89.* Project No. 9, University of California at Riverside, California, USA.

Trumble, J.T. (1990) Vegetable insect control with minimal use of insecticides. *Horticultural Science* 25, 159–164.

Trumble, J.T. (1991) *California Tomato Board Layman's Summary* 13–0.

Trumble, J.T. and Alvarado-Rodriguez, B. (1993) Development and economic evaluation of an IPM program for fresh market tomato production in Mexico. *Agricultural Ecosystems and Environment* 43, 267–284.

Trybom, F. (1908) Physapoda. In: Syostedt, Y. (ed.) *Wissenschaftliche Ergebnisse einer Schwedischen Zoologischen Expedition nach dem Kilimandjiaro, dem Meru, und den umgebenden Masaisteppen Deutsch-Ostafrica. 1905–1906*, Vol. 3. pp. 1–22.

Tryon, E.H. and Litsinger, J.A. (1988) *Feasibility of Using Locally Produced* Bacillus thuringiensis *to Control Tropical Insect Pests.* CICP/USAID Washington DC, USA, Project No. 936–4142, pp. 73–81.

Tsuchida, K., Geng, L.R., Kameoka, Y., Miyajima, N., Okano, K., Takada, N. and Maekawa, H. (1992) Fluorescence *in situ* hybridization as a tool for gene mapping in *Bombyx mori*. Abstract. In: *XIX International Congress of Entomology.* Beijing, China, p. 638.

Turelli, M. and Hoffmann, A.A. (1991) Rapid spread of an inherited incompatibility factor in Californian *Drosophila. Nature* 353, 440–442.

Turelli, M., Hoffmann, A.A. and McKechnie, S.W. (1992) Dynamics of cytoplasmic incompatibility and mtDNA variation in natural *Drosophila simulans* populations. *Genetics* 132, 713–723.

Ulluwishewa, R. (1992) Indigenous knowledge systems for sustainable development: the case of pest control by traditional paddy farmers in Sri Lanka. *Journal of Sustainable Agriculture* 3, 51–63.

UNCED (1992) Promoting sustainable agriculture and rural development. In: *Agenda 21* United Nations Conference on Environment and Development, Rio de Janeiro, Brazil, pp. 22–26.

UNDP (1992) *Farm Research Systems, India: Benefits of Diversity.* New York, USA, pp. 55–64.

US Department of Agriculture (1991) Part III. Proposed guidelines for research involving the planned introduction into the environment of organisms with deliberately modified hereditary traits: notice. *Federal Register* 56(22), 1 February, 4134–4151.

Useem, M., Setti, L. and Pincus, J. (1992) The science of Javanese management: organisational alignment in an Indonesian development programme. *Public Administration and Development* 12, 447–471.

Valent, B. and Chumley, F.G. (1991) Molecular genetic analysis of the rice blast fungus, *Magnaporthe grisea. Annual Review of Phytopathology* 29, 443–467.

Valero, M.V., Amador, L.R., Galindo, C., Figueroa, J., Bello, M.S., Murillo, L.A., Mora, A.L., Patarroyo, G., Rocha, C.L., Rojas, M., Aponte, J.J., Sarmiento, L.E., Lozada, D.M., Coronell, C.G., Ortega, N.M., Rosas, J.E., Alonso, P.L. and Patarroyo, M.E. (1993) Vaccination with SPf66, a chemically synthesised vaccine, against *Plasmodium falciparum* malaria in Colombia. *Lancet* 341, 705–710.

Van den Bosch (1978) *The Pesticide Conspiracy*. University of California Press, Berkeley, USA, 226 pp.

van der Kaay, H.J., Laarman, J.J., Curtis, C.F., Boorsma, E.G. and van Seventer, H.A. (1982) Susceptibility to *Plasmodium berghei* in a laboratory population of *Anopheles atroparvus* after the introduction of *Plasmodium* refractory genotypes. *Journal of Medical Entomology* 19, 536–541.

Van der Plank, J.E. (1968) *Disease Resistance in Plants*. Academic Press, New York, USA.

van Frankenhuyzen, K. (1993) The challenge of *Bacillus thuringiensis*. In: Entwistle, P.F., Cory, J.S., Bailey, M.J. and Higgs, S. (eds) Bacillus thuringiensis, *an Environmental Biopesticide: Theory and Practice*. J. Wiley & Sons, Chichester, UK, pp. 1–35.

Van Halteren, P. (1971) Insect pests of cowpea, *Vigna unguiculata* (L.) Walp. in the Accra plains. *Ghana Journal of Agricultural Science* 4, 121–123.

Van Rie, J., McGaughey, W.H., Johnson, D.E., Barnett, B.D and van Mellaert, H. (1990) Mechanism of insect resistance to the microbial insecticide *Bacillus thuringiensis*. *Science* 247, 72–74.

Verma, J.S. (1990) India agrochemical industry comes of age. *Farm Chemicals International* November, 20–21.

Vernick, K.D. and Collins, F.H. (1989) Association of a plasmodium-refractory phenotype with an esterase locus in *Anopheles gambiae*. *American Journal of Tropical Medicine and Hygiene* 40, 593–597.

Vickers, R.A., Rothschild, G.H.L. and Jones, E.L. (1985) Control of the oriental fruit moth, *Cydia molesta* (Busck) (Lepidoptera: Tortricidae), at a district level by mating disruption with synthetic sex pheromone. *Bulletin of Entomological Research* 75, 625–634.

VijayRaghavan, K., Palazzolo, M.J. and Rodrigues, V. (1991) The *Drosophila* nervous system as a model for analysing gene expression in complex organisms. *Current Science* 60, 562–569.

VijayRaghavan, K., Kaur, J., Paranjape, J. and Rodrigues, V. (1992) The *east* gene of *Drosophila* is expressed in the developing nervous system and is required for normal olfactory and gustatory responses of the adult. *Development Biology* 154, 23–36.

Villas Boas, G.L., Moscardi, F., Correa-Ferreira, B.S., Hoffmann-Campo, B.S., Corso, I.C. and Panizzi, A.R. (1985) *Indicacoes do manejo de pragas para percevejos*. Documentos No. 9, EMBRAPA–CNPSo, Londrina, Brasil, 15 pp.

Villas Boas, G.L., Gazzoni, D.L., Oliveira, E., Costa, N.P., Roessing, A.C., Franca-Neto, J.B. and Henning, A.A. (1990) Efeito de diferentes populacoes de percevejos sobre o rendimento e seus componentes, caracteristicas agronomicas e qualidade de sementes de soja. *Boletim de Pesquisa* (EMBRAPA–CNPSo, Londrina, Brasil) 1, 43.

Vogt, R. (1992) Functional and developmental specificity in pheromone and general odorant detection by Lepidoptera: binding proteins, degrading enzymes and membrane-bound receptors. Abstract. In: *XIX International Congress of Entomology*. Beijing, China, p. 98.

Waage, J.K. (1989) The population dynamics of pest–pesticide–natural enemy interactions. In: Jepson, P.C. (ed.) *Pesticides and Non-target Invertebrates.* Intercept, Wimborne, Dorset, UK, pp. 81–94.

Waage, J.K. (1991) Biodiversity as a resource for biological control. In: Hawksworth, D.L. (ed.) *The Biodiversity of Microorganisms and Invertebrates: Its Role in Sustainable Agriculture.* CAB International, Wallingford, Oxon, UK. pp. 149–163.

Waage, J.K. (1992) Biological control in the year 2000. In: Aziz, A., Kadir, S.A. and Barlow, H.S. (eds) *Pest Management and the Environment in 2000.* CAB International, Wallingford, UK, pp. 329–340.

Waage, J.K. (1993) *Making IPM Work: Developing Country Experience and Prospects. World Bank Agricultural Symposium January 1993.* World Bank Technical Paper, Washington DC, USA.

Waage, J.K., Carl, K.P., Mills, N.J. and Greathead, D.J. (1985) Rearing entomophagous insects. In: Singh, P. and Moore, R.F. (eds) *Handbook of Insect Rearing.* Elsevier, Netherlands, pp. 45–66.

Waibel, H. (1990) Pesticide subsidies and the diffusion of IPM in rice in Southeast Asia: the case of Thailand. *FAO Plant Protection Bulletin* 38(2), 105–111.

Walgate, R. (1990) *Miracle or Menace? Biotechnology and the Third World.* Panos Institute, London, UK.

Walker, T.S., Suckling, D.M., Shaw, P.W. and White, V. (1989) Evaluation of pheromone traps to reduce insecticide sprays in New Zealand apple orchards. In: *Applications of Pheromones to Pest Control. Proceedings of Joint CSIRO–DSIR Workshop, July 1988,* pp. 105–114.

Walker, V.K. (1989) Gene transfer in insects. *Advances in Cell Culture* 7, 87–124.

Wang, G., Mackill, D.J., Bonman, J.M., McCouch, S.R. and Nelson, R.J. (1994) RFLP mapping of genes conferring complete and partial resistance to blast resistance in a durably resistant rice cultivar. *Genetics* 136, 1421–1434.

Wang, G.L., Wing, R.A. and Paterson, A.H. (1993) PCR amplification from single seeds, facilitating DNA marker assisted breeding. *Nucleic Acids Research* 21, 25–27.

Wang, J.X. (1990) Use of the nematode *Steinernema carpocapsae,* to control the major apple pest *Carposina nipponensis* in China. In: *Proceedings and Abstracts of Vth International Colloquium on Invertebrate Pathology and Microbial Control, Adelaide, Australia,* p. 392.

Wang, J.X. (1993) Control of the peach fruit moth, *Carposina niponensis,* using entomopathogenic nematodes. In: Bedding, R.A., Akhurst, R.J. and Kaya, H.K. (eds) *Nematodes and the Biological Control of Insect Pests.* CSIRO Publications, Australia.

Wang, J.X. and Li, L.Y. (1987) Entomopathogenic nematodes research in China. *Review of Nematology* 10, 483–489.

Wang, J.Y., McCommas, S. and Syvanen, M. (1991) Molecular cloning of a glutathione S-transferase overproduced in an insecticide-resistant strain of the housefly (*Musca domestica*). *Molecular and General Genetics* 227, 260–266.

Ward, E.R., Uknes, S.J., Williams, S.C., Dincher, S.S., Wiederhold, D.L., Alexander, D.C., Ahl-Goy, P., Metraux, J.-P. and Ryals, J.A. (1991) Coordinate gene activity in response to agents that induce systemic acquired resistance. *Plant Cell* 3, 1085–1094.

Ward, E.R., Ryals, J.A. and Miflin, B.J. (1993) Chemical regulation of transgene expression in plants. *Plant Molecular Biology* 22, 361–366.

Warren, A.M. and Crampton, J.M. (1991) The *Aedes aegypti* genome: complexity and organization. *Genetical Research* 58, 225–232.

Warren, G.W., Carozzi, N.B., Desai, N. and Koziel, M.G. (1992) Field evaluation of transgenic tobacco containing a *Bacillus thuringiensis* insecticidal protein gene. *Journal of Economic Entomology* 85, 1651–1659.

Waters, L.C., Zelhof, A.C., Shaw, B.J. and Chang, L.Y. (1992) Possible involvement of the long terminal repeat of transposable element 17.6 in regulating expression of an insecticide resistance-associated P450 gene in *Drosophila. Proceedings of the National Academy of Sciences USA* 89, 4855–4859.

Watson, R.T., Albrittan, D.L., Anderson, S.O. and Lee-Bapty, L. (1992) *Methyl Bromide: Its Atmospheric Science, Technology and Economics*. UNEP, Nairobi, 41 pp.

Wattam, A.R. and Christensen, B.M. (1992) Further evidence that the genes controlling susceptibility of *Aedes aegypti* to filarial parasites function independently. *Journal of Parasitology* 78, 1092–1095.

Way, M. and Heong, K.L. (1994) The role of biodiversity in the dynamics and management of insect pests of tropical irrigated rice – a review. *Bulletin of Entomological Research* 84, 567–587.

Wayadande, A.C. and Nault, L.R. (1993) Leafhopper behaviour associated with maize chlorotic dwarf virus transmission to maize. *Phytopathology* 83, 522–526.

Webster, J.M. and Bronskil, J.F. (1968) Use of Gelgard M and an evaporation retardant to facilitate control of larch sawfly by a nematode–bacterium complex. *Journal of Economic Entomology* 61, 1370–1373.

Whalon, M.E., Miller, D.L., Hollingworth, R.M., Grafius, E.J. and Miller, J.R. (1993a) Laboratory selection of a resistant Colorado potato beetle (Coleoptera: Chrysomelidae) strain to the CryIIIA coleopteran specific delta endotoxin of *Bacillus thuringiensis. Journal of Economic Entomology* 86, 226–233.

Whalon, M.E., Rahardja, U. and Verakalasa, P. (1993b) Selection and management of *Bacillus thuringiensis* resistant Colorado potato beetle. In: Janson, R., Powelson, M., Ramon, K.V. and Zehnder, G. (eds) *Potato Pest Management: A Global Perspective*. APS Press.

Wheeler, H. and Diachun, S. (1983) Mechanisms of pathogenesis. In: Kommendahl, T. and Williams, P.H. (eds) *Challenging Problems in Plant Health*. American Phytopathological Society, St Paul, Minnesota, pp. 324–333.

White, R.F. (1979) Acetylsalicylic acid (aspirin) induces resistance to tobacco mosaic virus in tobacco. *Virology* 99, 410–412.

Whitten, M.J. (1979) The use of genetically selected strains for pest replacement or suppression. In: Hoy, M.A. and McKelvey, J.J. Jr (eds) *Genetics in Relation to Insect Management*. Rockefeller Foundation, New York, pp. 31–40.

Whitten, M.J. (1984) The theoretical basis of genetic control. In: Kerkut, G.A. and Gilbert, L.I. (eds) *Comprehensive Insect Physiology, Biochemistry, and Pharmacology*, Vol. 12. Pergamon Press, New York, USA.

Whitten, M.J. (1989) The relevance of molecular biology to pure and applied entomology. *Entomologia Experimentalis et Applicata* 53, 1–16.

Whitten, M.J. (1991) Australian insects in scientific research. In: Naumann, I.D. (ed.) *The Insects of Australia*, 2nd edn. Melbourne University Press, Carlton, Victoria, Australia, pp. 236–251.

Whitten, M.J. (1992) Pest management in 2000: what we might learn from the twentieth century. In: Aziz, S.A.K. and Barlow, H.S. (eds) *Pest Management and the Environment in 2000.* CAB International, Wallingford, Oxon., UK, pp. 9–46.

Whitten, M.J. and Hoy, M.A. (1994) Genetic improvement and other genetic considerations for improving the efficacy and success rate of biological control. In: Fisher, E. (ed.) *Principles and Application of Biological Control.* UC Press, Riverside, California, USA.

Whitten, M.J. and Oakeshott, J.G. (1990) Biocontrol of insects and weeds. In: Persley, G.J. (ed.) *Agricultural Biotechnology: Opportunities for International Development.* CAB International, Wallingford, Oxon., UK, pp. 123–142.

Wiebers, U. (1991) Agricultural technology and environmental safety: integrated pest management and pesticide regulation in developing Asia. Technical Department Asia Region, Agriculture Division, World Bank, Washington DC, USA, 78 pp. (mimeograph report).

Wien, H.C. and Rösingh, C. (1980) Ethylene evolution by thrips-infested cowpea provides a basis for thrips resistance screening with ethephon sprays. *Nature* (London, UK) 283, 192–194.

Williams, D.J. (1986) Mealybugs (Homoptera: Pseudococcidae) on cassava with special reference to those associated with wild and cultivated cassava in the Americas. In Herren, H.R., Hennessey, R.N. and Bitterli, R. (eds) *Biological Control of Host Plant Resistance to Control the Cassava Mealybug and Green Mite in Africa. Proceedings of the International Workshop IFAD, OAU/STRC, IITA, December 1982.* Ibadan, Nigeria, pp. 49–56.

Williams, J.G.K., Kubelik, A.R., Livak, K.J., Rafalski, J.A. and Tingey, S.V. (1990) DNA polymorphisms amplified by arbitrary primers are useful as genetic markers. *Nucleic Acids Research* 18(22), 6531–6535.

Williams, M.N.V., Pande, N., Nair, S., Mohan, M. and Bennett, J. (1991) Restriction fragment length polymorphism analysis of polymerase chain reaction products analyzed from mapped loci of rice (*Oryza sativa*) genomic DNA. *Theoretical Applied Genetics* 82, 489–498.

Williams, S., Friedrich, L., Dincher, S., Carozzi, N., Kessman, H., Ward, E. and Ryals, J. (1992) Chemical regulation of *Bacillus thuringiensis* delta-endotoxin expression in transgenic plants. *Bio/Technology* 10, 540–543.

Williamson, M. (1988) Potential effects of recombinant DNA organisms on ecosystems and their components. *Trends in Evolution and Ecology* 3, 32–35.

Williamson, M. (1992) Environmental risks from the release of genetically modified organisms (GMOS) – the need for molecular ecology. *Molecular Ecology* 1, 3–8.

Wilson, C., Pearson, R.K., Bellen, H.J., O'Kane, C.J., Grossniklaus, U. and Gehring, W.J. (1989) P-element mediated enhancer detection: an efficient method for isolating and characterizing developmentally regulated genes in *Drosophila. Genes and Development* 3, 1301–1313.

Wilson, L.T. and Waite, G.K. (1982) Feeding pattern of Australian *Heliothis* or cotton. *Environmental Entomology* 11, 297–300.

Wilson, S. (1991) IPM revival. *Agrichemical Age* April, 32–33.

Wilson, T.M.A. (1993) Strategies to protect crop plants against viruses: pathogen-derived resistance blossoms. *Proceedings of the National Academy of Sciences USA* 90, 3134–3141.

Winger, L., Smith, J.E., Nicholas, J., Carter, E.H., Tirawanchai, N. and Sinden, R.E. (1987) Ookinete antigens of *Plasmodium berghei*: the appearance of a 21 kd transmission blocking determinant on the developing ookinete. *Parasite Immunology* 10, 193–207.

Winks, R.G. (1989) Recent developments in fumigation technology, with emphasis on phosphines. In: *Proceedings from the International Conference on Fumigation and Controlled Atmosphere Storage of Grain, 14–18 February 1989, Singapore.*

Winks, R.G. (1992) The development of SIROFLO in Australia. In: *Proceedings from the International Conference on Controlled Atmospheres and Fumigation, 11–13 June 1992, Winnipeg, Canada.*

Wochok, Z. (1991) *Commercial Prospects for Genetic Manipulation in Crop Protection.* IMPACT AgBioIndustry, CAB International, Wallingford, Oxon., UK.

Wolfe, M. (1983) Genetic strategies and their value in disease control. In: Kommendahl, T. and Williams, P.H. (eds) *Challenging Problems in Plant Health.* American Phytopathological Society, St Paul, Minnesota, USA.

Wolfe, M.S., Barrett, J.A. and Jenkins, J.E.E. (1981) The use of cultivar mixtures for disease control. In: Jenkyn, F.J. and Plumb, R.T. (eds) *Strategies for the Control of Cereal Disease.* Blackwell Scientific Publications, Oxford, UK, pp. 73–80.

Wood, B.J. (1971) Development of integrated control programs for pests of tropical perennial crops in Malaysia. In: Huffaker, C.B. (ed.) *Biological Control.* Plenum Press, New York, USA, pp. 422–456.

Wood, H.A. and Granados, R.R. (1991) Genetically engineered baculoviruses as agents for pest control. *Annual Review of Microbiology* 45, 69–87.

Wood, H.A., Hughes, P.R., Van Beek, N. and Hamblin, M. (1990) An ecologically acceptable strategy for the use of genetically engineered baculovirus pesticide. In: Borkovec, A.B. and Masler, E.P. (eds) *Insect Neurochemistry and Neurophysiology.* Humana, Clitton, New Jersey, pp. 285–288.

Wood, M. (1992) Microbes blow those hornworms away. *Agricultural Research* June, 4–7.

Wood, R.J. and Newton, M.E. (1991) Sex ratio distortion caused by meiotic drive in mosquitoes. *American Naturalist* 137, 379–391.

Wood, R.J., Cook, L.M., Hamilton, A. and Whitelaw, A. (1977) Transporting the marker gene *re* (red eye) into a laboratory cage population of *Aedes aegypti* (Diptera: Culicidae), using meiotic drive at the *Md* locus. *Journal of Medical Entomology* 14, 461–464.

Woodburn, A.T. (1990) The current rice agrochemicals market. In: Grayson, B.T., Green, M.B. and Copping, L.G. (eds) *Proceedings of the Conference on Pest Management in Rice.* Elsevier Applied Science, London, UK, pp. 15–30.

World Health Organization (1991) *Report of the meeting 'Prospects for Malaria Control by Genetic Manipulation of Its Vectors', 27–31 January 1991, Tucson, Arizona, USA.* TDR/BCV/MAL-ENT/91.3, WHO, Geneva, Switzerland.

World Health Organization (1993) A Global Strategy for Malaria Control. WHO, Geneva, 30 pp.

Wu, C.T., Budding, M., Griffin, M.S. and Croop, J.M. (1991) Isolation and characterization of *Drosophila* multidrug resistance gene homologs. *Molecular and Cellular Biology* 11, 3940–3948.

Wyatt, G.R. (1991) Gene regulation in insect reproduction. *Invertebrate Reproduction and Development* 20, 1–35.

Xiao, J.H., Davidson, L., Matthes, H., Garnier, J.-M. and Chambon, P. (1991) Cloning expression and transcriptional properties of the human transcription factor TEF–1. *Cell* 65, 551–568.

Xu, T. and Rubin, G.R. (1993) Analysis of genetic mosaics in developing and adult *Drosophila* tissues. *Development* 117, 1223–1237.

Yang, H., Zhang, G., Zhang, S. and Hreng, J. (1993) Biological control of tree borers (Lepidoptera: Cossidae) in China with the nematode *Steinernema carpocapsae*. In: Bedding, R.A., Akhurst, R.A. and Kayal, H.K. (eds) *Nematodes and the Biological Control of Insect Pests*. CSIRO Publications, Melbourne, Australia.

Yang, X.Y., Yeo, S.L., Dick, T. and Chia, W. (1993) The role of *Drosophila* POU homeodomain gene in the specification of neural precursor cells in the developing embryonic central nervous system. *Genes and Development* 7, 504–516.

Yaseen, M. (1986) Exploration for natural enemies of *Phenacoccus manihoti* and *Mononychellus tanajoa*: the challenge, the achievements. In: Herren, H.R., Hennessey, R.N. and Bitterli, R. (eds) *Biological Control and Host Plant Resistance to Control the Cassava Mealybug and Green Mite in Africa. Proceedings of the International Workshop IFAD, OAU/STRC, IITA, December 1982, Ibadan, Nigeria*. pp. 81–102.

Yen, J.H. (1975) Transovarial transmission of *Rickettsia*-like microorganisms in mosquitoes. *Annals of the New York Academy of Sciences* 266, 152–161.

Yen, J.H. and Barr, A.R. (1973) The etiological agent of cytoplasmic incompatibility in *Culex pipiens*. *Journal of Invertebrate Pathology* 22, 242–250.

Yoshimura, S., Yoshimura, A., McCouch, S.R., Baraoidan, T.W., Nelson, R.J., Mew, T.W., and Iwata, N. (1995) Tagging and combining bacterial blight resistance genes in rice using RAPD asnd RFLP markers. *Molecular Breeding*, in press.

Young, N.D. and Tanksley, S.D. (1989) RFLP analysis of the size of chromosomal segments retained around the Tm–2 locus of tomato during backcross breeding. *Theoretical and Applied Genetics* 77, 95–101.

Young, S.Y. (1989) Problems associated with the production and use of viral pesticides. *Memorias do Instituto Oswaldo Cruz* 84(3), 67–73.

Yu, Z.H., Mackill, D.J. and Bonman, J.M. (1987) Inheritance of resistance to blast in some traditional and improved rice cultivars. *Phytopathology* 77, 323–326.

Yu, Z.H., Mackill, D.J., Bonman, J.M. and Tanksley, S.D. (1991) Tagging genes for blast resistance in rice via linkage to RFLP markers. *Theoretical and Applied Genetics* 81, 471–476.

Yu, Z., Podgwaite, J.D. and Wood, H.A. (1992) Genetic engineering of a *Lymantria dispar* nuclear polyhedrosis virus for expression of foreign genes. *Journal of General Virology* 73, 1509–1514.

Zacharopoulou, A., Frisardi, M., Savakis, C., Robinso, A.S., Tolias, P., Konsolaki, M., Komitopoulou, K. and Kafatos, F.C. (1992) The genome of the Mediterranean fruitfly *Ceratitis capitata*: localization of molecular markers by *in situ* hybridization to salivary gland polytene chromosomes. *Chromosoma* 101, 448–455.

Zadoks, J.C. (1989) *Development of Farming Systems. Evaluation of the Five-year Period 1980–1984.* Pudoc, Wageningen, The Netherlands, 90 pp.

Zadoks, J.C. (1993) Antipodes on crop protection in sustainable agriculture. In: Corey, S., Dall, D. and Milne, W. (eds) *Pest Control and Sustainable Agriculture.* CSIRO, Australia, pp. 3–12.

Zalokar, M. (1981) A method for injection and transplantation of nuclei and cells in *Drosophila* eggs. *Experientia* 37, 1354–1356.

Zalom, F.G. and Fry, W.E. (1992) Biologically intensive IPM for vegetables. In: Zalom, F.G. and Fry, W.E. (eds) *Food, Crop Pests and the Environment: the Need and Potential for Biologically Intensive Integrated Pest Management.* APS Press, St Paul, Minnesota, USA, pp. 107–165.

Zalom, F.G., Ford, R.E., Frisbie, R.E., Edwards, C.R. and Tette, J.P. (1992) Integrated pest management: addressing the economic and environmental issues of contemporary agriculture. In: Zalom, F.G. and Fry, W.E. (eds) *Food, Crop Pests and the Environment: the Need and Potential for Biologically Intensive Integrated Pest Management.* APS Press, St Paul, Minnesota, USA, pp. 1–13.

Zanotto, P.M.A. (1990) Polyhedrin gene of the *Anticarsia gemmatalis* nuclear polyhedrosis virus. MSc Thesis, University of Florida, Gainesville, USA.

Zehnder, G.W. and Gelernter, W.D. (1989) Activity of the M-ONE formulation of a new strain of *Bacillus thuringiensis* against Colorado potato beetle (Coleoptera: Chrysomelidae): relationship between susceptibility and insect life stage. *Journal of Economic Entomology* 82, 756–761.

Zhang, Q., Leach, J.E., Nelson, R.J., Wang, C. and Mew, T.W. (1995) Preliminary analysis of the population structure of *Xanthomonas oryzae* pv. *oryzae* in China. *Chinese Journal of Rice Science* 9(1), 7–14.

Zhao, J.J., Lazzarini, R.A. and Pick, L. (1993) The muse Hox1.3 gene is functionally equivalent to the *Drosophila sex-combs reduced* gene. *Genes and Development* 7, 343–354.

Zheng, L., Saunders, R.D.C., Fortini, D., Della-Torre, A., Coluzzi, M., Glover, D.M. and Kafatos, F.C. (1991) Low-resolution genome map of the malaria mosquito *Anopheles gambiae. Proceedings of the National Academy of Sciences USA* 88(1), 11187–11191.

Zinn, K., McAllister, L. and Goodman, C.S. (1988) Sequence analysis and neuronal expression of *fasciclin I* in grasshopper and *Drosophila. Cell* 53, 577–587.

Zraket, C.A., Barth, J.L., Heckel, D.G. and Abbott, A.G. (1990) Genetic linkage mapping with restriction fragment length polymorphism in the tobacco budworm, *Heliothis virescens.* In: Hagedorn, H.H. *et al.* (eds) *Molecular Insect Science.* Plenum Press, New York, USA, pp. 13–20.

Index

Note: Page numbers in *italic* refer to tables, figures and/or boxes